Key Methods in Geography

Key Methods in Geography

Edited by

Nicholas Clifford and Gill Valentine

SAGE Publications
London • Thousand Oaks • New Delhi

Contents

Data Collection in Physical Geography

Analysing and Representing Geographical Data

Contents

Notes on Contributors

Paul Aplin is Lecturer in Geographical Information Science (GIS) in the School of Geography at Nottingham University. He has interests in remote sensing, GIS and geostatistics and their application to physical geography. He is a Fellow of the Royal Geographical Society, a council member of the Remote Sensing and Photogrammetry Society (UK) and a member of the American Society for Photogrammetry and Remote Sensing. His current research projects include vegetation classification in southern Africa using multiscale remote sensing (funded by the Nuffield Foundation), and combining interferometry and per-field classification for temperate land-cover monitoring (funded by the National Space Development Agency of Japan).

Rob Bartram is a graduate of University College London (BA geography) and the University of Nottingham (PhD). He is a Lecturer at the University of Sheffield Department of Geography and has research and teaching interests in social and cultural geography. His recent research has looked at the aesthetics of postwar urban reconstruction, the landscape art of Stanley Spencer, the 'Brit Art' movement and theories of visual culture.

Michael Batty is Professor of Spatial Analysis and Planning in the Centre for Advanced Spatial Analysis (CASA), University College London. His research is in the development of computer-based technologies, specifically graphics-based and mathematical models for cities, and he has worked recently on applications of fractal geometry and cellular automata to urban structure. He is a Fellow of the British Academy and was formerly Director of the SUNY Buffalo site of NCGIA (1990–5). His most recent books are *Fractal Cities* (Academic Press, 1994) and *Spatial Analysis: GIS in a Modelling Environment* (Pearson GeoInformation, 1996). He is also the editor of the journal *Environment and Planning B*.

Iain S. Black is Lecturer in Geography at King's College London. He has written widely on the historical geography of money and banking in Britain between 1750 and 1950, with particular reference to the social, economic and architectural transformation of the City of

London. Recently, these interests have been extended to work on the role of imperial and colonial banking groups in the diffusion of British banking culture overseas between 1850 and 1950. He is currently working on a project funded by the British Academy examining the commercial architecture of Sir Edwin Lutyens in late imperial London.

Michael Bradford is Professor of Geography at the University of Manchester. He is currently Pro-Vice-Chancellor for Teaching and Learning. He was Head of the School of Geography from 1996 to 2000.

Myrna M. Breitbart is a Professor of Geography and Urban Studies and Dean of the School of Social Science at Hampshire College where she has taught since 1977. She is also a faculty affiliate to the Community Partnerships for Social Change Programme at the college. Her teaching and research interests as well as publications focus on the broad themes of participatory planning (with a special interest in youth), community economic development and struggles over urban public space. She is currently conducting research on the role of the arts in urban redevelopment with a special focus on Holyoke, Massachusetts. She has a strong commitment to community-based learning and participatory action research. She works closely with a number of women's development and housing organizations as well as urban youth and community art organizations.

Joanna Bullard is a Senior Lecturer in Geography at Loughborough University. She has undertaken fieldwork on glaciers, in tropical forests, along various coastlines and in sandy and rocky deserts around the world.

Adrian Chappell is a Lecturer in Geography at the University of Salford. He specializes in dryland geomorphology but his research into geostatistics spans many subdisciplines in geography. He obtained his doctorate from the University of London on the spatial variation of soil erosion in west Africa. He developed his research into geostatistics by diversifying its application to include the spatial and temporal variation of aeolian sediment transport in Australia and the UK; Australian continental dust; wandering gravel-bed rivers in the UK; and west African Sahel rainfall. His application of geostatistics is an attempt to improve the understanding of the spatial variation in geomorphic processes at several different scales, and to consider the implications for land-surface formation.

Nicholas J. Clifford completed his BA and PhD at Cambridge University and is now Professor of River Science at Nottingham University. His principal research interests are in fluvial geomorphology and the history of ideas and methods in geography. He is currently Director of the HARRP research group, which is dedicated to improving the scientific basis for sustainable river rehabilitation. He is a member of the editorial board of *Transactions, the Institute of British Geographers* and has produced diverse publications, such as *Turbulence: Perspectives on Flow and Sediment Transport* (with J. French and J. Hardisty; Wiley, 1993) and *Incredible Earth* (DK Books, 1996). He teaches courses in river form and processes, river restoration and the history and philosophy of geography.

Meghan Cope is an Associate Professor in the Department of Geography at the State University of New York–Buffalo. Her interests lie in the areas of gender, race and poverty and how these influence the everyday lives of urban Americans. She has done projects on employment, welfare, housing and education, all of which have been based on qualitative data and come from a feminist perspective. Her newest project is on young children's conceptualizations of urban space in inner-city Buffalo.

Marcus A. Doel is Professor of Human Geography at University of Wales Swansea. He is the author of *Poststructuralist Geographies* (Edinburgh University Press, 1999) and *Spaces of Consumption* (Sage, 2003). He has written widely on new theoretical directions in social and cultural geography, and especially the intersection of literary theory, continental philosophy and spatial science.

Danny Dorling is Professor of Quantitative Human Geography, School of Geography, University of Leeds. He was previously Reader and Lecturer in Geography, University of Bristol, and before that Joseph Rowntree Foundation Fellow, British Academy Fellow and then Senior Research Associate, University of Newcastle upon Tyne. His research interests currently concentrate on developing new methods of spatial micro-simulation with current funding from the Joseph Rowntree Foundation and British Telecommunications to develop a model of the population of Britain.

Richard Field is a Lecturer in Physical Geography in the School of Geography at Nottingham University. He gained his PhD from Imperial College London. His main interests are in global biodiversity patterns (particularly plants), island biogeography (with a special interest in the ecological dynamics of the Krakatau Islands, Indonesia) and biological invasions. He is a member of the British Ecological Society and the International Biogeography Society, a Fellow of the RGS and book reviews editor for *Journal of Biogeography and Global*

Ecology, Biogeography: A Journal of Macroecology and *Diversity and Distributions: A Journal of Biological Invasions and Biodiversity*. He also teaches courses in the interpretation of geographical data, and biogeography and environmental change.

Iain Hay is Professor and Head of the School of Geography, Population and Environmental Management at Flinders University, South Australia. His research work over the past two decades has focused on geographies of oppression (exploitation, marginalization, powerlessness, cultural imperialism, violence). Among his various journal editorial responsibilities, he is Asia-Pacific editor for *Ethics, Place and Environment*.

Mick Healey is Professor of Geography at the University of Gloucestershire. He is Director of the Geography Discipline Network, a consortium of higher education institutions whose aim is to research and develop teaching, learning and assessment practices in geography in higher education. His interests are in local economic development and pedagogic research. He is the Senior Geography Advisor for the National Subject Centre for Geography, Earth and Environmental Sciences. In 2000 he was awarded one of the first National Teaching Fellowships in England and Northern Ireland.

Robin A. Kearns completed his PhD at McMaster University, Canada, in 1987 and is Associate Professor in the School of Geography and Environmental Science, University of Auckland. His research focuses on the geography of health and health care, the cultural politics of place and research approaches in geography. His most recent book is *Culture/Place/Health* (Routledge, 2002), with Wilbert Gesler.

Stuart N. Lane is Professor of Physical Geography at the University of Leeds. Stuart completed his undergraduate degrees at the University of Cambridge, and his PhD at Cambridge and City University, London. He is a prolific author, with wide-ranging interests in geographical methodology and history. He is a fluvial geomorphologist, with major research projects in the process-form relationships in gravel-bedded rivers, and in the numerical simulation of river flows using complex three-dimensional computer codes.

Eric Laurier is ESRC Research Fellow at the University of Glasgow and has been involved as an ethnographer and analyst in a number of UK and European research projects. He has used an ethnomethodological approach to study work, mobility, city life, language, animal mind, technology, competence in interaction and community.

Robyn Longhurst is Senior Lecturer in Geography at the University of Waikato, New Zealand. Her areas of teaching and research include 'the body', feminist geography, the politics of knowledge production and qualitative methodologies. She is author of *Bodies: Exploring Fluid Boundaries* (2001) and a co-author of *Pleasure Zones: Bodies, Cities, Spaces* (2001).

John H. McKendrick is Senior Research Fellow in the Centre for Families and Relationships at the University of Edinburgh. He is a human geographer with interests in just about everything, but in particular the geographies of poverty, families and children. He struggled with SPSS as an undergraduate and postgraduate student and still uses SPSS to make sense of people's perceptions, experiences and attitudes of their life and the world in which they live.

Sara L. McLafferty is Professor of Geography at the University of Illinois at Urbana-Champaign. Her research explores the use of spatial analysis methods and GIS in analysing health and social issues in cities and women's access to social services and employment opportunities.

Miles Ogborn is Reader in Geography at Queen Mary, University of London. He has undertaken archival research on social policy in nineteenth-century Britain, the changing geographies of eighteenth-century London and the voyages of the English East India Company in the seventeenth century. He was awarded a Philip Leverhulme Prize in 2001.

Chris Perkins is Senior Lecturer in the School of Geography, University of Manchester, and is also Map Curator in John Rylands University Library Manchester. He is the author of five books and numerous academic papers relating to map production and use and is currently researching ethnographies of contemporary map use in the UK.

Ian Reid is Professor of Physical Geography at Loughborough University. His research interests span arid-zone river sedimentology, rift sedimentation in the Triassic North Sea and Pleistocene Dead Sea, and agricultural and forest hydrology. He has active research projects in Kenya, Israel and Honduras as well as the British Isles, and teaches courses on river systems and processes, physical habitat assessment in streams and water resource management.

Stephen Rice is Senior Lecturer in Physical Geography in the Department of Geography at Loughborough University. He was an undergraduate at Oxford University and gained his PhD from the University of British Columbia. His research interests fall within the general fields of fluvial geomorphology, fluvial sedimentology and river habitats, and he has active research projects examining the characterization of fluvial sediments using conventional sampling, photogrammetry and image analysis, and the role of tributary sediment inputs in punctuating benthic river ecosystems. He teaches courses in physical geography, earth-surface processes and landforms and quantitative methods.

Fiona M. Smith is a Lecturer in Human Geography at the University of Dundee. Her research currently focuses on gender and activism in post-communist societies and she is also interested in language and cross-cultural research. She has published widely on these issues and is co-author with Nina Laurie, Claire Dwyer and Sarah Holloway, of *Geographies of New Femininities* (Longman, 1999).

Catherine Souch gained her PhD from the University of British Columbia and is currently Chair and Associate Dean of Academic Affairs at Indiana University–Purdue University Indianapolis. She has research interests in palaeoenvironmental reconstruction using geomorphic evidence in lacustrine/wetland environment and in human impacts on the environment. Her current research projects include the wetland hydrology, sedimentology and geochemistry of the Indiana Dunes National Lakeshore, and lake sediment records of postglacial conditions. She teaches courses in research and field methods in geography, climatic change and physical systems of the environment.

Colin E. Thorn is Associate Professor of Geography in the Department of Geography, University of Illinois at Urbana-Champaign, where he teaches a course on contemporary social and environmental problems. His research interests are focused on periglacial geomorphology and conceptual issues in geomorphology. He is currently engaged in group research projects in Kärkevagge, Swedish Lapland, and the Jotunheimen Mountains of south-central Norway to examine the chemical weathering of coarse debris and the role of chemical weathering within pedogenesis. He is author of *Introduction to Theoretical Geomorphology* (Unwin Hyman, 1988).

Gill Valentine is Professor of Geography at the University of Sheffield where she teaches social and cultural geography and qualitative methods. Her research interests include geographies of children, young people and 'families', social identities and exclusion (especially in relation to lesbians and gay men and D/deaf people) and consumption. She is a recipient of the Philip Leverhulme Prize. She is (co)author/(co)editor of eight books, including *Key Concepts in*

Geography (Sage, 2003) and *Social Geographies* (Pearson, 2001), and co-editor of two journals: *Gender, Place and Culture* and *Social and Cultural Geography*.

Bettina van Hoven is a Researcher and Lecturer at the Department of Cultural Geography (University of Groningen, the Netherlands). Her key interests are gender, identity and power issues. Based on the research described in her chapter, her latest project addresses identity formation in confined places (i.e. prisons).

Paul White is a Professor of Geography at the University of Sheffield where he teaches social geography. His main research interests are in the settlement of international migrant groups in European cities. He is currently working on the settlement of the Japanese in London and Dusseldorf. He has published nine books, most recently *Paris* with D. Noin (Wiley, 1997).

List of Figures and Tables

Figures

Tables

Acknowledgements

We owe a huge debt of thanks to Robert Rojek at Sage for commissioning this book and for his editorial advice and support, and to all our contributors for their enthusiasm for the project. Sarah Holloway and Stephen Rice were involved in the original conception of this project and we are grateful for their input. Charlotte Kenten and Phil Soar provided invaluable help reformatting chapters. The quality of the final product is due to the efforts of David Kershaw the copy editor and Vanessa Harwood the production editor.

We are grateful to the following for kind permission to reproduce their material: Figures 13.1a, b, c images and data provided by Dr Henri Grissino-Mayer; Figure 18.1a black and white photograph of the University of Nottingham campus, © photography supplied by Infoterra; Figure 18.1b false colour composite satellite sensor (Landsat-7 Enhanced Thematic Mapper) image of Nottingham, data courtesy of Eros Data Centre, distributed by Infoterra; Figure 18.5a, b, c, d Landsat-7 Enhanced Thematic Mapper images, data courtesy of Eros Data Centre, distributed by Infoterra; Figure 22.2 data provided by J. Leys; Figure 23.1 digital data types: raster and vector map representations, adapted from the University of Melbourne's GIS Self-Learning Tool.

Every effort has been made to obtain permission to reproduce copyright material. If any proper acknowledgement has not been made we invite copyright holders to inform us of the oversight.

Gill Valentine would also like to acknowledge the support of the Philip Leverhulme Prize.

Getting Started in Geographical Research: how this book can help

Nicholas J. Clifford and Gill Valentine

Synopsis

Geography is a very diverse subject that includes studies of human behaviour and the physical environment. It is also a discipline that embraces a very diverse range of philosophical approaches to knowledge (from positivism to poststructuralism). As such geographers employ quantitative methods (statistics and mathematical modelling) and qualitative methods (a set of techniques that are used to explore subjective meanings, values and emotions such as interviewing, participant observation and visual imagery) or a combination of the two. These methods can be used in both extensive research designs (where the emphasis is on generalizing from patterns in large 'representative' datasets, which are assumed to represent the outcome of some underlying causal regularity or process) and intensive research designs (where the emphasis is on describing a single case study or small number of case studies with the maximum amount of detail, and where generality is achieved through interpretation). Yet, despite this diversity, all geographers, whatever their philosophical or methodological approach, must make common decisions and go through common processes when they are embarking on their research. This means doing preparatory work (a literature review, thinking about health and safety and research ethics); thinking through the practicalities of data collection (whether to do original fieldwork or rely on secondary sources; whether to use quantitative or qualitative methods or a combination of both); planning how to manage and analyse the data generated from these techniques; and thinking about how to present/write up the findings of the research. This chapter aims to guide you through these choices if you are doing research for a project or dissertation. In doing so, it explains the structure and content of this book and points you in the direction of which chapters to turn to for advice on different forms of research techniques and analysis.

The chapter is organized into the following sections:

- Introduction: the nature of geographical research
- Quantitative and qualitative approaches to geography
- Designing a geographical research project
- The philosophy of research and importance of research design
- Conclusion: how this book can help you get started.

INTRODUCTION: THE NATURE OF GEOGRAPHICAL RESEARCH

This book aims to help you prepare for, design and carry out geographical research, and to analyse and present your findings. Geographers have given attention to an enormous range of subject matter. Most aspects of the world, whether physically or environmentally determined, or politically, economically or culturally constructed, have been considered as suitable for geographical research. Moreover, the range of geographical inquiry continues to increase. Traditionally, geographers considered the contemporary human and physical worlds together with their historical configurations, thus extending geographies to the past as well as to the present. Now, in both physical and human geography, the range is even greater (see, for example, Walford and Haggett, 1995; Gregory, 2000; Thrift, 2002). Physical geographers have access to new techniques of absolute and relative environmental dating, and greater ability to gather, analyse and visualize large amounts of data. They can reconstruct palaeoenvironments and landform development, as well as model this into the future, over timescales ranging from years to geological epochs. In human geography, technological advances in areas such as GIS allow more flexible and more creative analysis of data, facilitating 'virtual geographies' which exist only in 'hyperspace'. For the less technically minded, the subject is probing areas traditionally within the domains of psychology and cultural anthropology: there are now, for example, imagined and mystic geographies, whose foundations, or connections with the 'real' world, are almost entirely interpretational, rather than empirical. All these new areas of geographical exploration bring challenges of interpretation as methods of research associated with them may be radically different (even fundamentally irreconcilable with one another), or so new that they have yet to be formalized into transferable schemes to inform other research programmes.

Until the 1980s, geography was cast largely as either a physical (environmental) science, a social science or some combination of the two. This implied a commonality of objective, if not entirely of method: there was a shared commitment to the goal of 'general' explanation. More recently, however, some would dispute the use of the term 'science' in any of its forms in certain areas of subject (for an excellent introduction to debates surrounding science – its meaning, construction and application – see Chalmers, 1990). Instead, the 'cultural' turn in human geography (which in part reflects the growing influence of feminist and poststructural approaches) has brought a new emphasis on meanings, representation, emotions and so on (see Chapters 10 and 28) that is more readily associated with the arts (these issues are discussed in more detail by Barnes (2001) and in various chapters in Holloway et al., 2003).

Given the breadth of geographical inquiry, it is not surprising that the subject is similarly broad with respect to the methods it employs and the philosophical and ethical stances it adopts. This book reflects the diversity of contemporary geography, both in the number, and the range, of chapters it contains. In this chapter, we want briefly to introduce you to different approaches to research methods and design and to offer some guidance on how you might develop your own research design for a geographical project using this book.

QUANTITATIVE AND QUALITATIVE APPROACHES TO GEOGRAPHY

The chapters in the book loosely deal with two forms of data collection/analysis: quantitative and qualitative techniques. Quantitative methods involve the use of mathematical modelling and statistical techniques to understand geographical phenomena. These techniques form the basis of most research in physical geography. They first began to be adopted by human geographers in the 1950s, but it was in the 1960s – a period dubbed the 'quantitative revolution' – that their application became both more widespread and more sophisticated in Anglo-American geography. It was at this time that, influenced by the 'scientific' approaches to human behaviour that were being adopted by social sciences such as economics and psychology, some human geographers began to be concerned with scientific rigour in their own research. In particular, they began to use quantitative methods to develop hypotheses and to explain, predict and model human spatial behaviour and decision-making (Johnston, 2003). (Collectively, the adoption of 'objective', quantified means of collecting data, hypothesis testing and generalizing explanations is known as positivism.) Much of this work was applied to planning and locational decision-making (Haggett, 1965; Abler et al., 1971).

In the 1970s, however, some geographers began to criticize positivist approaches to geography, particularly the application of 'objective' scientific methods that conceptualized people as rational actors (Cloke et al., 1991). Rather, geographers adopting a humanistic approach argued that human behaviour is, in fact, subjective, complex, messy, irrational and contradictory. As such, humanistic geographers began to draw on methods that would allow them to explore the meanings, emotions, intentions and values that make up our taken-for-granted lifeworlds (Ley, 1974; Seamon, 1979). These included methods such as in-depth interviews, participant observation and focus groups. At the same time Marxist geographers criticized the apolitical nature of positivist approaches, accusing those who adopted them of failing to recognize the way that scientific methods, and the spatial laws and models they produced, might reproduce capitalism (Harvey, 1973). More recently, feminist and poststructuralist approaches to geography have criticized the 'grand theories' of positivism and Marxism, and their failure to recognize people's multiple subjectivities. Instead, the emphasis is on refining qualitative methods to allow the voice of informants to be heard in ways which are non-exploitative or oppressive (WGSG, 1997; Moss, 2001) and to focus on the politics of knowledge production, particularly in terms of the positionality of the researcher and the way 'other' people and places are represented (see articles in a special issue of *Environment and Planning D*, 1992; Moss, 2001).

Humanistic inquiry did not just prompt interest in people's own account of their experiences, but also in how these experiences are represented in texts, literature, art, fiction and so on (Pocock, 1981). Again, such visual methodologies have also been informed and developed by the emergence of poststructuralist approaches to geography which have further stimulated human geographers' concerns with issues of representation.

Despite the evolving nature of geographical thought and practice, both quantitative and qualitative approaches remain important within the discipline of geography. While taken at face value they appear to be incompatible ways of 'doing' research, it is important not to see these two approaches as binary opposites. Subjective concerns often inform the development and use of quantitative methods. Likewise, it is also possible to work with qualitative material in quite scientific ways. Whatever methods are adopted, some degree of philosophical reflection is required to make sense of the research process (see Chapter 16). Equally, the two approaches are often combined in research designs in a process known as mixing methods (see below).

DESIGNING A GEOGRAPHICAL RESEARCH PROJECT

Faced with a bewildering array of possibilities, both in what to study and in how to approach this study, it may seem that geographical research is difficult to do well. However, the very range of geo-graphical inquiry is also a source of excitement and encouragement. The key is to harness this variety rather than to be overwhelmed by it. Essentially, geographical research requires perhaps more thought than any of the other human or physical academic disciplines. Whether this thought is exercised with the assistance of some formal scheme of how to structure the research programme, or whether it is exercised self-critically or reflexively in a much less formal sense, is less important than the awareness of the opportunities, limitations and context of the research question chosen, the appropriateness of the research methods selected, the range of techniques used to gather, sort and display information and, ultimately, the manner and intent with which the research findings are presented. For student projects, these questions are as much determined by practical considerations, such as the time available for the project or the funding to undertake the research. These limitations should be built into the project at an early stage so that the likely quality of the outcomes can be judged in advance. None of the constraints should be used after the research is completed to justify a partial answer or unnecessarily restricted project.

The 'scientific' view

Conventionally, geographical research programmes have been pre-sented as a sequence of steps or procedures (Haring and Lounsbury, 1983 – see below). These steps were based upon the premise that geography was an essentially scientific activity – that is, a subject identifying research questions, testing hypotheses regarding possible causal relationships and presenting the results with some sort of more general (normative) statement or context. The aim of separating tasks was to enable time (and money) to be budgeted effectively between each and to encourage a structuring of the thought processes under-pinning the research.

The steps identified in this form of 'scientific geographic research' (Haring and Lounsbury, 1983) are as follows:

- Formulation of the research problem – which means asking a question in a precise, testable manner, and which requires con-sideration of the place and timescale of the work.
- Definition of hypotheses – the generation of one or more assump-tions which are used as the basis of investigation, and which are subsequently tested by the research.
- Determination of the type of data to be collected – how much, in what manner they are to be sampled or the measurement to be done.

- Collection of data – either primary from the field or archive, or secondary from the analysis of published materials.
- Analysis and processing of the data – selecting appropriate quantitative and presentational techniques.
- Stating conclusions. Nowadays, this might also include the presentation of findings verbally or in publication.

Today, there is more recognition that these tasks are not truly independent and that an element of reflexivity might usefully be incorporated into this process. In some areas of the subject – particularly human geography – the entire notion of a formalized procedure or sequence would be considered unnecessary, and the notion of normative, problem-solving science would, at best, be considered applicable to a restricted range of subjects and methodology. Rather, as outlined above, many human geographers now reject or are sceptical of scientific approaches to human behaviour, preferring to adopt a more subjective approach to their research. Nevertheless, having said this, most qualitative research also involves many of the same steps outlined in the mechanical or scientific formulation above – albeit not conceptualized in quite the same way. For example, qualitative researchers also need to think about what research questions to ask, what data need to be collected and how this material should be analysed and presented. In other words, all research in geography – whatever its philosophical stance – involves thinking about the relationships among methods, techniques, analysis and interpretation. This important role is filled by research design.

The importance of research design

In its broadest sense, research design results from a series of decisions we make as researchers. These decisions flow from our knowledge of the academic literature (see Chapter 2), the research questions we want to ask, our conceptual framework and our knowledge of the advantages and disadvantages of different techniques (see Chapters 6–10 and 13–18). The research design should be an explicit part of the research: it should show that you have thought about how, what, where, when and why!

There are at least six key things you need to bear in mind to formulate a convincing research design.

1   THINK ABOUT WHAT RESEARCH QUESTIONS TO ASK

On the basis of your own thinking about the topic, the relevant theoretical and empirical literatures (see Chapter 2) and having consulted secondary material (see Chapter 5) – and, if possible, having discussed it with other students and your tutor – you need to move towards framing your specific research questions. For a human geographer, these might include questions about what discourses you can identify, what patterns of behaviour/activity you can determine, what events, beliefs and attitudes are shaping people's actions, who is affected by the issue under consideration and in what ways, and how, social relations are played out, etc. For a physical geographer, these might include questions concerning the rate of operation and location of a certain geomorphological process, the morphology of a selected set of landforms or the abundance and diversity of particular plant or animal species in a given area (many of the chapters in this volume provide examples of research problems).

It is important to have a strong focus to your research questions rather than adopting a scatter-gun approach asking a diverse range of unconnected questions. This also means bearing in mind the time and resource constraints on your research (see below). At the same time, as you develop a set of core aims it is also important to remain flexible and to remember that unanticipated themes can emerge during the course of fieldwork which redefine the relevance of different research questions. Likewise access or other practical problems can prevent some research aims being fulfilled and can lead to a shift in the focus of the work. As such, you should be aware that your research questions may evolve during the course of your project.

2 THINK ABOUT THE MOST APPROPRIATE METHOD(S) TO EMPLOY

There is no set recipe for this: different methods have particular strengths and collect different forms of empirical material. The most appropriate method(s) for your research will therefore depend on the questions you want to ask and the sort of information you want to generate. Chapters 6–11, 13–15 and 18 and 19 outline the advantages and disadvantages of core methods used by human and physical geographers. While many projects in human and physical geography involve going out into the field – for example to interview or observe people or to take samples or measurements – it is also possible to do your research without leaving your computer, living room or the library. For example, research can be based on visual imagery such as films and television programmes (see Chapters 10 and 28); secondary sources including contemporary and historical/archival material (see Chapters 5, 7 and 27); or GIS (Chapter 23). Some human geographers are also experimenting with conducting interviews and surveys by email or in chat-rooms (see a brief discussion of this in Chapter 6).

In the process of research design it is important not to view each of these methods as an either/or choice. Rather, it is possible (and often desirable) to mix methods. This process of drawing on different sources or perspectives is known as triangulation. The term comes from surveying where it describes using different bearings to give the correct position. Thus, researchers can use multiple methods or different sources of information to try to maximize an understanding of a research question. These might be both qualitative and quantitative (see, for example, Sporton, 1999). Different techniques should each contribute something unique to the project (perhaps addressing a different research question or collecting a new type of data) rather than merely being repetitive of each other.

3 THINK ABOUT WHAT DATA YOU WILL PRODUCE AND HOW TO MANAGE IT

An intrinsic element of your choice of method should not only involve reflecting on the technique itself but also how you intend to analyse and interpret the data you will produce. Chapters 5, 16, 19 and 21–24 all discuss how to analyse different forms of quantitative material, while Chapters 25–28 demonstrate how to bring a rigorous analysis to bear on interview transcripts/diary material, historical and archive sources and cultural material. For example, Chapters 19 and 22–24 discuss some of the issues you need to think about when deciding which statistical techniques to apply to quantitative data. Chapters 25 and 26 present alternative methods of coding interview transcripts/diary material: one manually and the other using computer software. While qualitative techniques emphasize quality, depth, richness and understanding, instead of the statistical representativeness and scientific rigour which are associated with quantitative techniques, this does not mean they can be used without any thought. Rather, they should be approached in as rigorous a way as quantitative techniques.

4 THINK ABOUT THE PRACTICALITIES OF DOING FIELDWORK

The nitty-gritty practicalities of who, what, when, where and for how long inevitably shape the choices we can make about our aims, methods, sample size and the amount of data we have the time to analyse and manage (see Chapter 15 on sampling and Chapter 6 on handling a large amount of qualitative data using computer software packages). Increasingly, the kind of work which is permissible is constrained by changing attitudes and legislative requirements with respect to safety and risk which ultimately define the range and scale of what you can achieve (see, for example, Chapter 4 on the health and safety limitations of fieldwork). It is important to bear in mind that the research written up by academics in journals and books is often conducted over several years and is commonly funded by substantial grants. Thus, the scale this sort of research is conducted on is very different from that at which student research projects must be pitched. It is not possible to replicate or develop fully in a three-month student dissertation or project all the objectives of a two-year piece of academic research that you may have uncovered in your literature review! Rather, it is often best to begin by identifying the limitations of your proposed study and recognizing what you will and will not be able to say at the end of it. Remember that doing qualitative or quantitative work in human geography, just like fieldwork in physical geography, requires a lot of concentration and mental energy. It is both stressful and tiring, so there is a limit to how much you can achieve in the field in any one day. Other practicalities, such as the availability of field equipment, tape recorders, cameras, transcribers or access to transport, can also define the parameters of your project.

Drawing up a time management chart or work schedule at the research design stage can be an effective way of working out how much you can achieve in your study, and later on can also serve as a useful indicator if you are slipping behind. While planning ahead (and in doing so, drawing on the experience of your tutor and other researchers) is crucial to developing an effective research design, it is also important to remember that you should always remain flexible.

5 THINK ABOUT THE ETHICAL ISSUES YOU NEED TO CONSIDER

An awareness of the ethical issues which are embedded in your proposed research questions and possible methodologies must underpin your final decisions about the research design. The most common ethical dilemmas in human geography focus around participation, consent, confidentiality/safeguarding personal information and giving something back (see Alderson, 1995; Valentine, 1999). In physical geography, ethical issues involve not only questions of consent (for example, to access field sites on private land) but also the potential impacts of the research techniques on the environment (for example, pollution, etc.). Thus, while ethical issues may seem routine or moral questions rather than anything which is intrinsic to the design of a research project, in practice, they actually underpin what we do. They can shape what questions we can ask, where we make observations, whom we talk to, and where, when and in what order. These choices in turn may have consequences for what sort of material we collect, how it can be analysed and used, and what we do with it when the project is at an end. As such ethics are not a politically correct add-on but should always be at the heart of any research design (see Chapter 3 for an overview of ethical issues, Chapter 11 for the specific ethics of participatory research and Chapter 12 about the specific ethical issues involved in working in different cultural contexts).

6 THINK ABOUT THE FORM IN WHICH YOUR RESEARCH IS TO BE PRESENTED

The scale and scope of your research design will partially be shaped by your motivations for doing the research and what you intend to use the findings for. If you are presenting your findings in a dissertation it will be very different from a piece of work that is to be presented as a report or in a verbal seminar. Chapter 29 outlines and illustrates these different forms of presentational style in detail, and Chapter 20 describes how to use maps in your work. Likewise, when you are designing, conducting and writing up your research, it is important to bear in mind the assessment criteria by which your findings will be judged. The final chapter in this volume (Chapter 30) explores some of the ways that your work might be assessed.

THE PHILOSOPHY OF RESEARCH AND IMPORTANCE OF RESEARCH DESIGN

The most basic, formal, distinction in research design is that between extensive and intensive approaches. The important aspects of these contrasting approaches have been explored in some detail for the social sciences by Sayer (1992). This book is an ideal and thought-provoking introduction to the ways in which research seeks to make sense of a complex world. Sayer reviews theories of causation and explanation in which 'events' (what we observe) are thought to reflect the operation of 'mechanisms' which, in turn, are determined by basic, underlying 'structures' in the world. The way in which explanations are obtained reflects differing degrees of 'concrete' and 'abstract' research – that is, how much our generalizations rely on observation and how much they rely on interpretation of the ways in which events, mechanisms and structures are related. In the physical sciences, more attention has been given to the broader topic of scientific explanation of which research design is a part, but recently within physical geography too, the implications of extensive and intensive research designs and the varying philosophies of research have begun to be explored (e.g. Richards, 1996, in the collection of papers edited by Rhoads and Thorn, 1996).

Both extensive and intensive designs are concerned with the relationship between individual observations drawn from measurement programmes or case studies, and the ability to generalize on the basis of these observations. The detailed distinctions are illustrated in Table 1.1, but the essential differences are as follows:

- In an *extensive research design*, the emphasis is on pattern and regularity in data, which are assumed to represent the outcome of some underlying (causal) regularity or process. Usually, large numbers of observations are taken from many case studies so as to ensure a 'representative' dataset, and this type of design is sometimes referred to as the 'large-n' type of study (see Chapter 15 on sampling).
- In an *intensive research design*, the emphasis is on describing a single or small number of case studies with the maximum amount

TABLE 1.1 The essential differences between extensive and intensive research designs

Notes	Intensive	Extensive
Research question	How? What? Why? In a certain case or example	How representative is a feature, pattern, or attribute of a population?
Type of explanation	Causes are elucidated through in-depth examination and interpretation	Representative generalizations are produced from repeated studies or large samples
Typical methods of research	Case study. Ethnography. Qualitative analyses	Questionnaires, large-scale surveys. Statistical analysis
Limitations	The relationships discovered will not be 'representative' or an average/generalization	Explanation is a generalization – it is difficult to relate to the individual observation. Generalization is specific to the group/population in question
Philosophy	Method and explanation rely on discovering the connection between events, mechanisms and causal properties	Explanation based upon formal relations of similarity and identification of taxonomic groups

Source: Based on Sayer (1992: Figure 13, p. 243)

of detail. This approach is therefore sometimes known as the 'small-n' type study. In anthropology, the term 'thick description' has been used (Geertz, 2000). In an intensive design, by thoroughly appreciating the operation of one physical or social system, or by immersion in one culture or social group, elements of a more fundamental, causal nature are sought. 'Explanation' is therefore concerned with disclosing the links among events, mechanisms and structures. General explanations are derived from identification of the structures underlying observation, and from the possible transferring of the linkages discovered from detailed 'instantiations'.

Importantly, both approaches may be undertaken in quantitative or qualitative fashions – there is no necessary distinction in the *techniques* used. The two approaches are, however, separated to some extent in their philosophical underpinnings and, more obviously, in the practical, logistical requirements they impose.

Philosophically, the extensive approach relies on the idea that the data pattern necessarily reflects an underlying cause, or process, which is obscured only by measurement error, or 'noise'. However, in the 'real' world, it is rare that one cause would lead directly, or simply, to another 'effect' – the chain of causation is more obscure, and 'noise' may be an essential part of the 'causation' , reflecting the presence of some other (unknown or uncontrolled) effect which merely mimics the apparent pattern. There is the related problem of being unable to explain individual occurrences on the basis of the 'average' behaviour of entire groups – the so-called ecological fallacy. In an intensive research design, there is a deeper appreciation of the 'layers' which separate observations from an underlying (causal) reality. As such (and at the risk of considerable oversimplification) extensive approaches have often been linked to positivist methodology and philosophy, and intensive approaches to realist methodologies and philosophies.

Practically, the different types of research design have clearly different requirements in both data type and amount and with respect to cost and time. The extensive design lends itself to situations where large amounts of data are already published or where large amounts of data can be generated from secondary sources. In many student projects, the need for many observations across comparative or contrasting field sites may be too daunting or logistically impossible if an attempt is made to mount an extensive research design based upon primary data sources. An exception to this is in laboratory-type studies, where a series of experiments may quickly build up a dataset representative of a wider range of conditions. The intensive design is perhaps more common, but care is needed to 'tease out' those aspects of the study which might disclose basic, causal processes.

CONCLUSION: HOW THIS BOOK CAN HELP YOU GET STARTED

Geographical research is complicated because geographical phenomena are many and varied, and because they may transcend multiple scales in space and time. Further, over the last few decades, geographers have adopted a diverse range of philosophic stances, methods and research designs in their efforts to understand and interpret the human and physical worlds. In order to make sense of this variety in a research context, a considerable amount of thought must go in to geographical research at all its stages. Although the prospect of embarking on your own research might seem daunting, this book has been put together to help make it easier for you. The first five chapters in this book all offer advice and guidance about how to prepare your research project. This chapter has explained the process of research design; Chapter 2 describes how to do a literature search to help define your topic and research questions; Chapter 3 raises some of the ethical issues you might need to consider in your research design; Chapter 4 explains the practical and logistical issues you need to plan for in terms of your own health and safety in the field; and Chapter 5 outlines some of the secondary data sources that might inform, or form the basis of your work. In addition, chapters 11 and 12 locate these issues within the specific situations of doing participatory research and working in different cultural contexts.

Other chapters explain how to use a range of quantitative and qualitative methods in human and physical geography, including questionnaire surveys (Chapter 6); finding historical data (Chapter 7); semi-structured interviews and focus groups (Chapter 8); participant observation (Chapter 9); visual imagery (Chapter 10); palaeo and historical data sources (Chapter 13); making observations and measurements in the field (Chapter 14); modelling (Chapter 17); and using GIS (Chapter 23). In addition, there is a specific chapter on sampling in physical and human geography (Chapter 15). You will obviously not want to use them all, but by reading across the chapters you will get a feel for the different ways that you might approach the same topic, and the advantages and disadvantages of different methods. When you have developed your own research design the chapter(s) appropriate to your chosen method(s) will give you practical advice about how to go about your research.

Further chapters then go on to explain how to analyse the sort of data collected by these diverse range of methods. Specifically, there are chapters on how to analyse qualitative interview and diary materials both manually and using computer software (Chapters 25 and 26); on how to analyse historical/archive sources (Chapter 27) and cultural criteria (Chapter 28); and on how to use statistics in a variety of different ways (Chapters 19, 21, 22, 24).

Finally, Chapters 20 and 29 explain how to present your findings in a variety of formats, how to use maps in your research and what sort of research criteria might be used to assess your work.

As with this chapter, each chapter in this book contains a synopsis at the beginning which briefly defines the content of the chapter and outlines the way the chapter is structured. At the end of each chapter is a summary of the key points raised in that chapter and an annotated list of recommended further readings. Several of the chapters also contain tips or useful exercises to develop your understanding of the topic in question (for example, see Chapters 2, 3 and 10).

Doing your own research can be one of the most rewarding aspects of a degree. It is your chance to explore something that really interests or motivates you and to contribute to geographical knowledge, so enjoy it – and good luck!

Summary

The key points raised in this chapter are as follows:

- Geographers are faced with a vast array of potential subject matter, techniques for data collection, visualization and analysis.
- Research design is crucial to link together data collection, methods and techniques, and to produce convincing, meaningful results.
- The basic choice in research design is between extensive and intensive designs. These have very different implications in terms of data collecton, analysis and interpretation, although both quantitative and qualitative methods may be used in either.
- Whichever form of research design is adopted, the ethical dimension of research must be considered.
- Practical issues of land access, time (or financial) constraints or field safety will, together with ethical issues, frequently determine the scope and kind of research adopted. While these are constraints, prior consideration of their likely effects will minimize the loss of intellectual integrity and merit in the project.

Further reading

- The different philosophies underlying human geography research are outlined in numerous volumes, including Cloke et al. (1991), Graham (1997) and Johnston (1997). Various chapters in Holloway et al. (2003) explain the tensions among geography as a physical science, social science or arts subject, and show how geographers' understandings of key concepts have evolved as approaches to geographical thought have developed.

- Sayer (1992) is a thorough and extensive treatment of 'methodology' and its connections to the way in which we make sense of the world through observations, experiments, surveys and experiences. It argues that our philosophy of the way in which the world is structured (how things come to be as they are seen) must inform our choice of research design and our choice of techniques for generalizing on the basis of the information which we collect from putting the research design into practice. The book depends on a particular ('realist') approach, but is an excellent starting point from which to relate philosophy to the practicalities of doing research and to the strengths and weaknesses of particular kinds of research methods.

- Chalmers (1990) is a student-centred volume which covers an enormous range of material dealing with more modern 'postpositivist' approaches to the philosophy of science. It examines the nature of scientific explanation (the relations among observations, experiments and generalization), the social and political dimensions of science and scientific research, and the way that scientific explanation and scientists have their own sociology of knowledge.

- The edited book by Limb and Dwyer (2001) provides a critical introduction to qualitative methodologies. Each method is illustrated with examples of the authors' own research experiences and practices.

- Moss (2001) is a collection of essays exploring feminist geography in practice. These essays share a particular concern with the ethics and politics of knowledge production, notably in terms of the positionality of the researcher and the way 'other' people and places are represented.

- Rhoads and Thorn (1996) is an edited volume of papers which were presented at the 27th Annual Binghampton Symposium in Geomorphology, held in 1996. It has some excellent and entertaining chapters dealing with changing ideas concerning science and its methods, the motivation of current geomorphological research and researchers, case studies of contrasting modelling approaches, and chapters dealing with differing approaches to explanation in geomorphology and the earth sciences.

- Gregory (2000) provides an up-to-date account of the way in which physical geography is structured, the research which physical geographers undertake and the methods they use. It seeks to present physical geography as a changing discipline, but one with strong connections to its past and with bright prospects for the future.

- The journal *Ethics, Place and Environment* contains articles dealing with all aspects of ethical concerns and practice within geography.

Note: Full details of the above can be found in the references list below.

References

Abler, R.F., Adams, J.S. and Gould, P.R. (1971) *Spatial Organization: The Geographer's View of the World*. Englewood Cliffs, NJ: Prentice Hall.

Alderson, P. (1995) *Listening to Children: Children, Ethics and Social Research*. Barnardo's: Ilford.

Barnes, T.J. (2001) 'Retheorizing economic geography: from the quantitative revolution to the "Cultural Turn"', *Annals of the Association of American Geographers*, 91: 546–65.

Chalmers, A.F. (1990) *Science and its Fabrication*. Buckingham: Open University Press.

Cloke, P., Philo, C. and Sadler, D. (eds) (1991) *Approaching Human Geography*. London: Paul Chapman.

Environment and Planning D: Society and Space (1992) Special issue on the politics of knowledge. Volume 10.

Geertz, C. (2000) *The Interpretation of cultures* (2nd edn). New York: Basic Books.

Graham, E. (1997) 'Philosophies underlying human geography research', in R. Flowerdew and D. Martin (eds) *Methods in Human Geography*. Harlow: Longman.

Gregory, K.J. (2000) *The Changing Nature of Physical Geography*. London: Arnold.

Haggett, P. (1965) *Locational Analysis in Human Geography*. London: Edward Arnold.

Haring, L.L. and Lounsbury, J.F. (1983) *Introduction to Scientific Geographical Research*. Dubuque, IA: W.C. Brown.

Harvey, D. (1973) *Social Justice and the City*. London: Edward Arnold.

Holloway, S.L., Rice, S. and Valentine, G. (eds) (2003) *Key Concepts in Geography*. London: Sage.

Johnston, R.J. (1997) *Geography and Geographers: Anglo-American Human Geography since 1945*. London: Arnold.

Johnston, R.J. (2003) 'Geography and the social science tradition', in S.L. Holloway et al. (eds) *Key Concepts in Geography*. London: Sage.

Ley, D. (1974) *The Black Inner City as Frontier Outpost: Image and Behaviour of a Philadelphia Neighbourhood. Monograph Series* 7. Washington, DC: Association of American Geographers.

Limb, M. and Dwyer, C. (eds) (2001) *Qualitative Methodologies for Geographers*. London: Arnold.

Moss, P. (ed.) (2001) *Feminist Geography in Practice*. Oxford: Blackwell.

Pocock, D.C.D. (ed.) (1981) *Humanistic Geography and Literature: Essays on the Experience of Place*. London: Croom Helm.

Rhoads, B.L. and Thorn, C.E. (eds) (1996) *The Scientific Nature of Geomorphology*. Chichester: Wiley.

Richards, K. (1996) 'Samples and cases: generalisation and explanation in geomorphology', in B.L. Rhoads and C.E. Thorn (eds) *The Scientific Nature of Geomorphology*. Chichester: Wiley, pp. 171–90.

Sayer, A. (1992) *Method in Social Science*. London: Routledge.

Seamon, D. (1979) *A Geography of a Lifeworld*. London: Croom Helm.

Sporton, D. (1999) 'Mixing methods of fertility research', *The Professional Geographer*, 51: 68–76.

Thrift, N. (2002) 'The future of geography', *Geoforum*, 33: 291–8.

Valentine, G. (1999) 'Being seen and heard? The ethical complexities of working with children and young people at home and at school', *Ethics, Place and Environment*, 2: 141–55.

Walford, R. and Haggett, P. (1995) 'Geography and geographical education. Some speculations for the twenty-first century', *Geography*, 80: 3–13.

WGSG (Women and Geography Study Group) (1997) *Feminist Geographies: Explorations in Diversity and Difference*. London: Longman.

2 How to Conduct a Literature Search

Mick Healey

Synopsis

Identifying the most relevant, up-to-date and reliable references is a critical stage in the preparation of essays, reports and dissertations, but it is a stage which is often undertaken unsystematically and in a hurry. This chapter is designed to help you improve the quality of your literature search.

The chapter is organized into the following sections:

- The purpose of searching the literature
- Making a start
- A framework for your search
- Managing your search
- Search tools
- Evaluating the literature.

THE PURPOSE OF SEARCHING THE LITERATURE

The purpose of this chapter is to support you in developing and using your literature search skills over a range of media, including paper, CD-ROMs and the web, not just books and journals. It is aimed primarily at undergraduate geography students needing to search the literature for research projects, dissertations and essays in human and physical geography. However, the search methods and principles are applicable to most subjects and, if you are a postgraduate geography student, you should also find it a useful refresher to get you started. Many sources are available worldwide, though details of accessing arrangements may vary. Country-specific sources are illustrated with selected examples of those available in the UK, North America and Australia.

Exercise 2.1   **Why read?**

Make a list of the reasons why you should read for a research project. Compare your list
with those in Box 2.1. Most also apply if you are preparing an essay.

Reading the literature is an important element of academic
research. It is a requirement with essays and projects as well as dissertations for you to relate your ideas to the wider literature on the
topic. Reading around the subject will also help you broaden and refine
your ideas, see examples of different writing styles and generally
improve your understanding of the discipline. When undertaking a
dissertation or thesis, reading will help you identify gaps, find case
studies in other areas which you may replicate and then compare with
your findings, and learn more about particular research methods and
their application in practice (Box 2.1). Effective reading may, of course,
take many different forms depending on your purpose – from skimming, through browsing, to in-depth textual analysis – and will rarely
involve just reading from the beginning to the end (Kneale, 1999).

Box 2.1   **Twenty reasons for reading for research**
1   Because it will give you ideas.
2   Because it will help you improve your writing style.
3   Because you need to understand what other researchers have done in your area.
4   To broaden your perspectives and set your work in context.
5   Because direct personal experience can never be enough.
6   Because your supervisor or manager expects you to.
7   So that you can drop names when you come to write up your research.
8   Because it's interesting.
9   To legitimate your arguments.
10   Because it may cause you to change your mind.
11   Because writers (and you will be one) need readers.
12   So that you can better understand the disciplinary traditions within which you are working.
13   So that you can become better at reading.
14   Because you don't like doing it.
15   So that you can effectively criticize what others have done.
16   It's a way of avoiding social contacts.
17   To learn about places you will never visit.
18   It keeps you off the streets.
19   To learn more about research methods and their application in practice.
20   In order to spot areas which have not been researched.
Source: Blaxter et al. (1996: 93)

MAKING A START

Your literature search strategy will vary with your purpose. Sometimes you may want to search for something specific, for example, a case study to illustrate an argument; in other situations a more general search may be required – for example, you might target identifying 15 major articles, which have been written on a particular topic, for an essay. Your search strategy may also vary with the level at which you are in the higher education system and your motivation. Identifying half a dozen up-to-date books on a topic may be appropriate at the beginning stage, while at graduate level you may need to explore whether any PhD theses have been written on the topic you are proposing to undertake for your thesis.

Exercise 2.2   Starting your search

You have been set a research project on a topic you know little about (e.g. organic farming). Before reading any further, write down the first three things you would do to find out what has already been written on the topic.

When I gave Exercise 2.2 to my students, the most common responses were to look in the subject section of the library catalogue, use a search engine on the Internet and ask a lecturer. These are all sensible strategies, though the usefulness of the Internet is, perhaps, exaggerated. However, apart from possibly asking the lecturer/professor who set the assignment for one or two references to get you going, usually the first things you should do are to identify and define the key terms in the assignment and construct a list of terms to use in your literature search. Only then is it appropriate to turn to the search tools, such as library catalogues, reference books, indexes, CD-ROMS, databases and websites, and seek help from a librarian.

Making a start is usually the most difficult stage of undertaking a research project or assignment. The issues involved in identifying your own research topic were discussed in Chapter 1. When you have a provisional idea about your topic and the research methods you may use, or when a research project or assignment is given to you, take a little while to plan your literature search. Defining the key terms in the topic or assignment is a good starting point. The dictionaries of human and physical geography are essential references for all geography students (Johnston et al., 2000; Thomas and Goudie, 2000). The indexes of appropriate textbooks will also help. These references will also help you identify search terms, as will a thesaurus, a good English dictionary and a high-quality encyclopedia. The Geobase subject classification is another source (see the section below on abstracts and reviews).

In identifying search terms, group them into three categories: broader, related and narrower. The first will be useful in searching for books, which may contain useful sections on your topic, while the second and third will be particularly helpful in identifying journal articles and websites and using indexes to books. Box 2.2 illustrates how to make a start with searching the literature for a research project on organic farming.

Box 2.2   Defining key terms and identifying search terms: an example

Topic: socioeconomic aspects of the geography of organic farming.

Definition: 'this system uses fewer purchased inputs compared with conventional farming, especially agri-chemicals and fertilizers, and consequently produces less food per hectare of farmland . . ., but is compensated by higher output prices' (Atkins and Bowler, 2001: 68–9).

Search terms:

Broader	Related	Narrower
Agricultural geography	Organic farming	Certified organic growers
Farm extensification	Organic agriculture	Organic organizations
Farm diversification	Organic production	Soil Association
Alternative farm systems	Organic growers	Organic food retailers
Food, geography of	Organic food	Organic food markets
Sustainable agriculture	Organic movement	Organic food shops

Note: 'Organic farming' is not listed in *The Dictionary of Human Geography*, although discussion of 'agricultural geography' and 'food, geography of' provides a useful context. Atkins and Bowler (2001) is the most recent text on the 'food and the environment' course reading list. The index identifies four mentions of 'organic farming'. These lead to a useful introduction to the topic and to several recent references, and are a source for some of the above search terms.

Tip

If you are collecting data as part of your research project remember to carry out a literature search on research methodologies and techniques as well as on your main topic. Many of the later chapters in this book will help you with this.

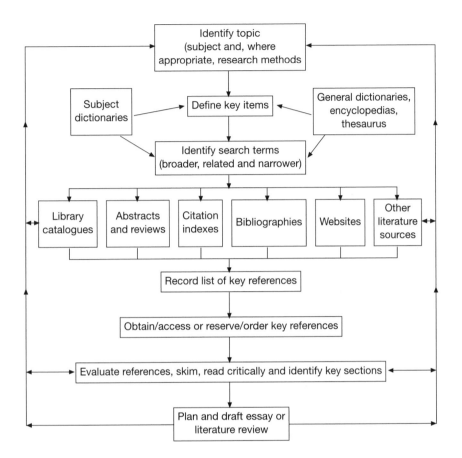

FIGURE 2.1 A framework for undertaking a literature search

A FRAMEWORK FOR YOUR SEARCH

A summary of how to search the literature is given in Figure 2.1 – which also provides a framework for the structure of this chapter. This figure might suggest that undertaking a literature search and writing an essay or a literature review is a linear process. The reality is much messier. There is frequent interaction between the different stages. As you begin to identify and scan the key references, your knowledge and understanding of the topic will increase, which will lead you to identify particular subtopics that you wish to investigate in more detail. A further search using new key terms may then be appropriate. Iteration is a key element of the search process. You should not give up after entering the first obvious key word in a search engine. Further thoughts are also likely to arise as you begin to draft your essay or literature review, which may call for additional searches. Saunders et al. (2000) capture some of this interactivity by conceptualizing the literature review process as a spiral.

> **Tip**
> Ensure that you use references appropriate for degree-level study. Your lecturer/ professor will not be impressed if you use textbooks, dictionaries and magazines aimed primarily at school/college-level study or, worse still, you cite *Microsoft Encarta*! Only a small proportion of websites are likely to be appropriate (see later). Many of the most appropriate references will be academic journal articles. Remember also to check the library catalogue to see that you have the latest edition of a textbook in the library (e.g. at the time of writing *The Dictionary of Human Geography* is in its fourth edition). Remember that earlier editions are usually the ones that your colleagues will leave on the open library shelves. A copy of the most recent edition of a popular textbook may be in the short loan collection.

> **Tip**
> A useful place to start building your list of references is the reading lists your lecturers/professors provide for their courses. In many institutions these are put on their intranet.

MANAGING YOUR SEARCH

The search process, as just indicated, is one that you will keep coming back to at various stages in your research. It is therefore sensible to keep a search diary, which includes the sources searched, the key words used and brief notes on the relevant references they reveal. This is, perhaps, best done using a wordprocessing package, which will enable you to list your key words, cut and paste the results from your online searches and keep track of which search engines and sources you have used.

> **Tip**
> Have a floppy disk available when you are searching. Many of the databases enable you to save your searches direct to disk at a PC. Some will also allow you to email them to yourself.

How long to spend on the search process depends on the purpose of your search. For example, are you seeking ten key references for an essay, or 50 or more references for a dissertation? Generally, the broader the topic and the more that has been written on it, the longer the search tends to take. This is because much of the effort is spent in trying to identify the key references from what may be a list of several hundred lesser important ones.

> **Tip**
> Aim to identify two to three times as many relevant references as you think you will need for your assignment/dissertation. Many may not exactly meet your needs when you obtain them and/or may not be accessible in the time you have available. If you are finding what appear to be too many relevant references, focus on the most recent ones and the references most frequently cited, and consider narrowing the search by, for example, focusing on a subtopic or restricting the geographical coverage. If you are finding too few relevant references, try some new search terms and consider broadening the topic or the geographical area. Also ask a librarian.

To avoid any possibility of plagiarism (that is, the unacknowledged use of the work of others), be sure to take down the full bibliographic details of the references you find, including, where relevant, the author(s)' name(s) and initials, year of publication (not print date), title of book, edition (if not the first), publisher and town/city of publisher (not printer), title of article/chapter, journal title, volume number and page numbers, and names of editors for edited books. Be sure to put all direct quotations in quotation marks and give the source, including the page number(s) (Mills, 1994). Pasting sections from a website directly into your work is also plagiarism if the source is not acknowledged.

> **Tip**
> Inconsistent referencing and missing or erroneous information are some of the most common comments made on student writing (see also Chapter 30). Most geogra-phers use the Harvard style of referencing, but departments vary on the format in which they like references to be cited. Guidance on how to provide correct refer-ences, including websites, is available from: www.unn.ac.uk/central/isd/cite/index.htm. It is best to acquire the habit of using one way of formatting references and to apply this *consistently* whenever you note a reference (down to the last comma, full stop and capital letter!).

> **Tip**
> When taking notes remember to put any sentences you copy (or paste from a website) in quotation marks and note down the page number(s) so that if you decide to use the author's direct words later you can acknowledge this properly.

SEARCH TOOLS

Various features are used to search CD-ROMs, databases and the web. As these vary it is important to check their 'help' facilities. Most allow you to use exact phrases. Simply place the phrase in double quotes (" ") (e.g. "organic farming"). Most search engines support the use of Boo-lean operators. The basic ones are represented by the words AND, OR and NOT (e.g. "organic farming" AND "UK"; "organic farming" OR "organic food"; "organic farming" NOT "North America"). The use of wild cards (*) may also be available. For example, 'farm*' will find records containing the word 'farm', including farm, farms, farmer, farmers and farming. It will also identify farmhouse, farmstead and farmyard.

Library catalogues

In most cases the first place to search for relevant books is the subject index of your library catalogue. You will usually need to use your broader list of search terms. Unfortunately the classification systems used in most libraries put geography books in several different sections of the library. But do not restrict yourself to books with 'geography' in the title. The integrative nature of the subject means that many books written, for example, for sociologists, economists, planners, earth scientists, hydrologists and ecologists, may be just as relevant. Once you have found the classification numbers of relevant books check the catalogue for other books with the same number and browse the relevant shelves. Looking at other books on the shelves near to the ones you are looking for often reveals other relevant references. Use your list of related and narrower search terms to explore the book indexes. Do not forget to also check where short loan and oversize books and pamphlets are shelved. Older books may be in store.

Tip
One of the quickest ways to generate a list of references is to find the latest book or article covering your topic and look at its reference list.

For a wider search try the combined catalogues of 21 of the largest UK university research libraries, which are available online through COPAC (copac.ac.uk/copac). The British Library Catalogue provides a national collection (blpc.bl.uk). The equivalent in the USA is the Library of Congress, which may be searched along with the catalogues of many other libraries in the USA and other countries via the Z39.50 Gateway (lcweb.loc.gov/z3950/gateway.html). To check book details try WorldCat (www.ref.uk.oclc.org:3000), which is a huge database of over 39 million references held in 17 000 libraries worldwide.

When you are away from the university it may be worth seeing whether your local university or public library has the book you require in stock. In the UK try the NISS Campus (www.niss.ac.uk/lis/obi/obi.html). The Library Catalogue of the Royal Geographical Society (www.rgs.org) was due online from autumn 2002. The library holds more than 150 000 books and receives over 800 periodical publications. You can use the library free if your department has an Educational Corporate Membership, otherwise there is a small fee charged per visit.

Abstracts and reviews

Evaluating the relevance of a book or journal article simply from its title is difficult. Abstracts give a clearer idea of their contents. The most useful set of abstracts for geographers is *Geobase*, which provides international coverage of the literature (particularly journal articles) on all aspects of geography, geology, ecology and international development. It is accessible online or through three CDs, dependent on the institution's subscription. The database provides coverage of over 1700 journals and contains close on one million records from 1980 onwards. The online version is updated monthly and the CDs quarterly. Book reviews which appear in the journals abstracted are also included. These are useful for evaluating the significance of books and finding out what has recently been published. *Environmental Abstracts* provides similar coverage for all environmental issues, and *Sociological Abstracts* is useful for some human geography topics. You should also check recent issues of review journals, particularly *Progress in Human Geography* and *Progress in Physical Geography*, for appropriate articles and updates on the literature in particular subfields.

> **Tip**
> The titles of some bibliographic sources change according to their format – hard copy, CD-ROM or online. Try (where available and accessible) to use the online version, which is usually the most up to date and the most flexible to use.

Citation indexes

One of the most useful tools for identifying journal articles is the ISI Web of Science. This consists of three citation indexes covering the social sciences, the arts and humanities, and science. As well as providing data on the number of times articles published in a wide range of journals are cited by authors of other articles, they also provide a valuable source for identifying journal articles, abstracts and reviews. They can thus be used for generating lists of articles with abstracts on particular topics, as well as identifying influential articles. You can also use the indexes to identify related records that share at least one cited reference with the retrieved item, and hence may be contributing to the discussion of related topics.

> **Tip**
> Some articles are cited frequently because they are heavily criticized, but they have nevertheless contributed to the debate. In the world of citation analyses the only real sin is largely to be ignored, which is the fate of most published papers. A few papers are ahead of their time and are not 'discovered' until several years after they have been published.

The three citation databases include articles published in over 8500 journals, plus book reviews and editorials. There are over 22 million records extending back to 1981, which are added to weekly. Virtually all higher education institutions in the UK and many elsewhere have taken out a subscription to the ISI Web of Science. In the UK access is through Mimas (www.mimas.ac.uk). Ask one of your librarians for your institution's ATHENS username and password. Access is also possible off-site if you first register with ATHENS on your institution's system.

Tip

Once you have identified key authors who are writing on your topic it is worth checking abstracts and citation indexes to see what else these authors have written, some of which may be on related topics.

Citation analyses are used to rank the impact that journals have on intellectual debate. They thus provide a crude guide as to which journals to browse through in the library and a possible basis for choosing between which of two otherwise apparently equally relevant articles to read first (Table 2.1). Lists of journal ranking may be obtained from ISI Journal Citation Reports (www.mimas.ac.uk/jcr), if your institution has a subscription. One of their limitations is that many of the

TABLE 2.1   Top geography journals, 2000

Rank	Journal title
1	*Transactions, Institute of British Geographers*
2	*Progress in Human Geography*
3	*Environment and Planning D*
4	*Annals of the Association of American Geographers*
5	*Economic Geography*
6	*Geoforum*
7 =	*Antipode*
7 =	*Political Geography*
9	*Area*
10	*Environment and Planning A*
11	*Journal of Geography in Higher Education*
12	*The Professional Geographer*
13	*Geografisker Annaler*

Note: Based on journals with an 'impact factor' (a measure of the frequency with which the 'average article' in a journal has been cited in a particular year) above 1.
Source: *Journal Citations Report, Social Sciences Edition* (2000)

key articles used by geographers are not published in mainline geography journals.

Bibliographies

A range of specialized bibliographies is available. The most useful are annotated. Some are in printed form. For example, the Countryside Agency Accession List includes details of new books, reports and pamphlets received at the Countryside Agency Library, and the Countryside Agency Selected Periodical Articles lists articles from recent journals under subject headings. An increasing number are available on the web at no charge and without registration (Table 2.2). Others may be available if your university has taken out a licence.

Tip

As you generate your list of references check whether your library holds the books, and if so whether they are on loan. If they are on loan put in a reservation request. In the case of journals check whether the library takes them. Also check whether your library has a subscription to the journals identified in your search for accessing the full text of articles online. If so, obtain passwords and check how to access them and whether you can do this off-site. If the library does not hold the book or journal, consider ordering the reference on inter-library loan (ILL). Journal articles can usually be ordered in a week to ten days. Recalling books, or ordering them through ILL, often takes longer.

Websites

An increasing amount of useful information is being placed on the web. However, identifying this from the huge amount of irrelevant and low-quality information is a time-consuming task. The general search engines, such as Google (www.google.co.uk) and AltaVista (uk.altavista.com), and Meta search engines, such as Dogpile (www.dogpile.com) and Ixquick (www.ixquick.com), which search other search engines' databases, are indispensable when searching for specific information, such as the URL address of an institution's website. However, topic searches can result in a vast number of sites of variable quality.

This is where Internet gateways or portals can be useful because they provide links to sites on particular subjects which have been evaluated for their quality. Most of the ones mentioned below can be accessed via Pinakes (www.hw.ac.uk/libWWW/irn/pinakes/pinakes.html). The JISC Resource Guide (www.jisc.ac.uk/subject/) also provides an entry point to many other sites. Among the more useful Internet gateways is BUBL (bubl.ac.uk/link/), which provides a catalogue of selected Internet sources for higher education classified by subject and Dewey classification number (BUBL LINK/5:15). It

TABLE 2.2 Examples of web-based geography bibliographies

Bibliography	Comments
Australian Heritage Bibliography www.ahc.gov.au/infores/HERA/ index.html	Full access is provided from this site but professional users will find the subscription service on CD-ROM (or accessing the Informit server at Melbourne) provides more flexibility in search formulation and output
Bibliography of Aeolian Research www.csrl.ars.usda.gov/wewc/ biblio/bar.htm	Coverage from 1930; updated monthly
Cold Regions Bibliography lcweb.loc.gov/rr/scitech/ coldregions/welcome.html	Organized by the Library of Congress, it contains over 208 000 bibliographic records – many with abstracts – and is increasing by approximately 6000 accessions annually
Development Studies – BLDS Bibliographic Database www.ids.ac.uk/blds/	Web-searchable version of the library catalogue and journal articles database of the British Library for Development Studies at the Institute of Development Studies, UK; contains records of over 116 000 documents
Disability and Geography Resource (DAGIN) www.swan.ac.uk/disability/dagin/ Authorad.html	Alphabetical list of references compiled for the Disability and Geography International Network in 1999
Economic and Business Geography Resources and Reading List faculty.washington.edu/krumme/ resources/resources_abc.html	List of hundreds of print and electronic resources arranged alphabetically by topic by Gunter Krumme at the University of Washington
Fieldwork in the Geography, Earth and Environmental Sciences Higher Education Curriculum www.glos.ac.uk/gdn/disabil/ fieldwk.htm	Annotated list of about 100 educational references compiled by the Geography Discipline Network in 2001
Gender in Geography www.emporia.edu/socsci/fembib/ index.htm	Alphabetical list contributed by members of the discussion list for Feminism in Geography
GIS Bibliography campus.esri.com/library/	The ESRI Virtual Campus Library provides a searchable database of over 16 500 references

TABLE 2.2 Continued

Bibliography	Comments
Population Index popindex.princeton.edu/index.html	Provides a searchable database containing over 40 000 abstracts of demographic literature published in the Population Index in the period 1986–2000
Transport-Transportation Research Information Services (TRIS Online) 199.79.179.82/sundev/search.cfm	Covers over 420 000 bibliographic records, with abstracts from the 1960s and 20 000 yearly additions; USA bias

includes many sites relevant to geographers. The Resource Discovery Network (RDN) (www.rdn.ac.uk) is another free Internet service dedicated to providing effective access to high-quality Internet resources for the learning, teaching and research community. They host a set of information gateways, including the Social Science Information Gateway (SOSIG) (www.sosig.ac.uk), which has a small geography section. A separate RDN Geography and Environment Gateway is under development. An older Geoinformation Gateway (www.geog.le.ac.uk/cti/info.html) is still available, but has not been updated since 1999. ELDIS provides a gateway to development information (nt1.ids.ac.uk/eldis/eldis.htm). However, the example search (Box 2.3) suggests that the information gateways may have some way to go yet before they are really useful, as they identified only a few sites and the ones identified varied between the two gateways examined. In the USA, Academic Info provides a gateway to college and research-level Internet resources for many subjects, including geography (www.academicinfo.net/geogmeta.html).

Box 2.3 Searching the literature: an example

Assignment: A 2000 word essay.

Topic: socioeconomic aspects of the geography of organic farming.

Library catalogues: 18 books on agricultural geography and sustainable agriculture were listed in the University of Gloucestershire *WebCat*; 12 specific ones on 'organic farming' (Dewey Classification: 333.76, 338.1, 630.68, 631.584). *COPAC*: key word search found the following number of books: organic – farming (356), agriculture (59), production (85), growers (18), food (91), movement (15).

Continued

Box 2.3   Continued

Abstracts and reviews: *Geobase*: key word search resulted in the following number of references: organic – farming (264), agriculture (64), food (17), production (91), growers (7), movement (7), but the majority were about scientific aspects. One useful article was found in a recent issue of *Progress in Human Geography*.

Citation indexes: *Social Science Citation Index*: key word search resulted in the following number of records: organic – farming (58), agriculture (31), production (10), growers (2), food (20), movement (3). *Science Citation Index* had 326 records relating to 'organic farming', but nearly all were about non-relevant aspects, such as the method of farming and the environmental impacts.

Bibliographies: a search for 'organic farming bibliographies' on *Google* found about 8510 records (3070, UK). Several useful references were also found from reference lists of books and articles identified by the above search tools and the economic geography reading list.

Websites: inputting 'organic farming' into two general search engines resulted in the following number of pages: *AltaVista* (46 501; 8370, UK); *Google* (about 102 000; about 22 000, UK). *BUBL* identified only four organic farming websites under the Dewey classification 631, while *SOSIG* identified six different websites on organic farming. Atkins and Bowler (2001) identify five organic food organizations in the UK. A *Google* search found specific websites for all of them.

Other sources: *Proquest Newspapers*: key word search found the following number of articles: organic – farming (252), agriculture (20), production (47), growers (21), food (472), movement (32). *Government Information Service* (a 'quick search') revealed 5645 matches, plus a separate section on 'organic farming' on the Department for Environment, Food and Rural Affairs pages.

Summary: though all the findings are presented above, decide at a fairly early stage, given the nature of the assignment (2000 words, 30% of marks for course) and the number of references identified, to focus on the UK. Decide to start with the most frequently cited and the most up-to-date references, websites and newspaper articles and those that appear to be the most comprehensive. Expect to find further references and undertake further more specific searches as you become more familiar with the topic. Make a short-list of references (two to three times as many as you think you are likely to need) to show your tutor to ask his or her advice on identifying key ones and any major omissions.

Many departments and institutions also provide links to websites relevant to areas in which they teach and research. The University of Colorado (Virtual Geography Department) (www.colorado.edu/ geography/virtdept/contents.html), the University of Utrecht (Geo-source) (www.library.uu.nl/geosource/cat10.html) and the Universities of Gottingen and Freiberg (Geo-Guide) (www.Geo-Guide.de) all provide extensive lists. A gateway site to Australian web servers is maintained by Charles Sturt University (www.csu.edu.au). There are separate geography and environment sections.

If you wish to develop your skills at finding and evaluating the quality of Internet resources, try the Internet Detective tutorial (www.sosig.ac.uk/desire/internet-detective.html). Two easy-to-use guides are the Internet Geographer (www.sosig.ac.uk/vts/geographer/ index.htm) and the Internet Earth Scientist (www.psigate.ac.uk/ vts/earth/index.htm). The Virtual Geography Department also provides some useful guidance (www.Colorado.EDU/geography/virtdept/ module/webwarmup/warmuptoc.html).

Other literature sources

For many topics newspapers can be a useful source of information, especially for up-to-date case studies. Proquest Newspapers provides access to UK broadsheets from 1996 or earlier. Several papers can be searched simultaneously. An institutional subscription is needed to access. Dialog@CARL is a North American subscription service which provides access to over 300 databases, including the full text of most major US newspapers. Many individual newspapers provide free online searchable databases. World Newspapers provides a searchable directory of online world newspapers, magazines and news sites in English (www.world-newspapers.com).

If you are undertaking your own thesis it is important to check whether anyone else has written a thesis on a similar topic by looking at the Index to Theses with Abstracts Accepted for Higher Degrees by the Universities of Great Britain and Ireland (www.theses.com). Dissertation Abstracts Online is a definitive subject, title and author guide to virtually every American dissertation accepted at an accredited institution since 1861. An institutional subscription is required to access it.

A wealth of information is available from central and local government via UK Government Online in the UK (www.ukonline.gov.uk) and usgovsearch in the USA (usgovsearch.northernlight.com/publibaccess). The Australian federal Department of the Environment maintains ERIN – Environmental Resources Information Network (www.erin.gov.au). A selection of UK official statistics is available via the Office for National Statistics (www.ons.gov.uk). Europa offers centralized access to European Union sites (europa.eu.int/index_en.htm).

EVALUATING THE LITERATURE

Do not be put off undertaking a systematic literature search, such as is illustrated in Box 2.3 because you feel you will not have time to read all the references you find. Indeed, you will not have time to read them all. However, the purpose of the literature search is to identify the most appropriate references for the task in hand (Table 2.3): 'learning how to determine the relevance and authority of a given resource for your research is one of the core skills of the research process' (Reference Services Divisions, Olin*Kroch*Uris Libraries, Cornell University Library, 2001: 3). Websites in particular need to be evaluated critically for their origin, purpose, authority and credibility.

If you follow the advice above you should have reduced the list of references several fold before you have even opened a book or journal or read a newspaper article or website, for example, by focusing on the most frequently cited and up-to-date references. The titles and abstracts will also help you to judge those references likely to be most relevant.

> **Tip**
> Avoid listing all the references you have found simply to try to impress your lecturer/ professor. You must use some relevant idea or material from each one to justify its inclusion. Ensure that you include a reference to each in the text, usually by following the Harvard system of referencing.

Exercise 2.3

Select four references from different sources on a topic that you are preparing, such as a website, a textbook, a journal article and a newspaper report, and use the criteria in Table 2.3 to evaluate their relevance, provenance and source reliability.

Exercise 2.4

Take a book or article relevant to your topic. You have five minutes to extract the key points it contains.

Researchers rarely read books from cover to cover and they read relatively few articles in their entirety. Like you, they do not have the

TABLE 2.3   Reducing your list of references to manageable proportions

Criterion	Possible (score 4 points)	More doubtful (score 2 points)	Probably forget it (score 0 points)
Relevance to my topic – judged by title and/or abstract (double the score for this criterion)	High	Moderate	Tangential
Up to date	Last 5 years	6–15 years old	Over 15 years old
Authority – the author or paper is cited in the references I have already read	Much	Recent paper not yet had time to be cited extensively	Older paper cited infrequently or not at all
Respectability and reliability of source publication	Published in major geographical publication or that of sister subject or something very close to my topic	Publication is not in geography or an allied field	Informal publication or unreliable Internet source
Nature of publication	Peer-reviewed academic journal or monograph	Textbook or conference proceedings	Popular magazine
Originality	Primary source of information – the authors generated this information using reliable and recognized methods	The authors take their information from clearly identified and reliable secondary sources	The authors assert facts and produce information, apparently, from thin air
Accessible	Instant by download or short walk to library	Obtainable with effort – reserve, inter-library loan	Unobtainable

Source: Modified from an idea by Martin Haigh (pers. comm., 29 January 2002)

time. They are practised at evaluating references in a few minutes by skimming the abstracts, executive summaries, publisher's blurbs, contents pages, indexes, introductions, conclusions and subheadings. This enables them to select the references that deserve more attention. Even then they will usually identify key sections by, for example, reading the first and last paragraphs of sections and the first and last sentences of paragraphs. This is not to suggest that all you need is a superficial knowledge of the literature; rather, that you should read selectively and critically to ensure that you obtain both a broad understanding of the topic and an in-depth knowledge of those parts of the literature that are particularly significant. If you are not familiar with the processes involved in critical and strategic reading, have a look at the relevant chapters in Blaxter et al. (1996) and Kneale (1999).

Exercise 2.5

In your next essay or research project try applying the framework for searching the literature outlined in this chapter. Please let me know how you get on by emailing me at mhealey@glos.ac.uk. I would be particularly interested in knowing what worked well, what blind alleys you took and what search tools you found most useful. Good hunting!

Summary

The aim of this chapter has been to identify effective and efficient ways systematically to search and evaluate the literature:

- The first stage is to define the key terms for your topic and to identify a range of search terms.
- You should then systematically search a range of sources, including library catalogues, abstracts and reviews, citation indexes, bibliographies and websites, being careful to keep a search diary.
- Having made a record of the references you have found you should evaluate each of them for such things as relevance, how up to date, authority, respectability, originality and accessibility.
- Although searching the literature needs to be systematic, it is also iterative and, as your knowledge and understanding of your topic and of the number and quality of the references you are identifying increase, you will inevitably need to make modifications to your search and repeat and refine many of the stages several times.
- For a quick guide, look at the exercises, boxes, tables, tips and the framework for undertaking a literature search shown in Figure 2.1.

Further reading

Few books focus only on searching the literature, though Hart (2001) is an exception; most are guides to study skills or how to research, which put the literature search process in a broader context.

- Baker (1999) presents a chapter on finding and searching information sources in education and social science. It is part of a guide to doing research.

- Blaxter et al. (1996) provide an excellent user-friendly guide on how to research; they include chapters on reading for research and writing up.

- Flowerdew and Martin (1997) provide a guide for human geographers doing research projects; they include chapters on finding previous work and secondary data sources.

- Hart (2001) provides a comprehensive guide for doing a literature search in the social sciences.

- Reference Services Divisions, Olin*Kroch*Uris Libraries, Cornell University Library (2001) provide a useful guide to assessing information sources critically.

Note: Full details of the above can be found in the references list below.

References

Atkins, P. and Bowler, I. (2001) *Food in Society: Economy, Culture and Geography*. London: Arnold.

Baker, S. (1999) 'Finding and searching information sources', in D. Bell (ed.) *Doing Your Research Project: A Guide for First-Time Researchers in Education and Social Science* (3rd edn). Buckingham: Open University Press, pp. 64–89.

Blaxter, L., Hughes, C. and Tight, M. (1996) *How to Research*. Buckingham: Open University Press.

Flowerdew, R. and Martin, D. (1997) *Methods in Human Geography: A Guide for Students Doing Research Projects*. Harlow: Longman.

Hart, C. (2001) *Doing a Literature Search: A Comprehensive Guide for the Social Sciences*. London: Sage.

Johnston, R.J., Gregory, D., Pratt, G. and Watts, M. (2000) *A Dictionary of Human Geography* (4th edn). Oxford: Blackwell.

Kneale, P.E. (1999) *Study Skills for Geography Students: A Practical Guide*. London: Arnold.

Mills, C. (1994) 'Acknowledging sources in written assignments', *Journal of Geography in Higher Education*, 18: 263–8.

Reference Services Divisions, Olin*Kroch*Uris Libraries, Cornell University Library (2001) *How to Critically Analyse Information Sources* (http://www.library.cornell.edu/okuref/research/skill26.htm. Accessed 5 July 2002).

Saunders, M.N.K., Lewis, P. and Thornhill, A. (2000) *Research Methods for Business Students* (2nd edn). Harlow: Financial Times/Prentice Hall.

Thomas, D. and Goudie, A. (2000) *A Dictionary of Physical Geography* (3rd edn). Oxford: Blackwell.

Acknowledgements

Gordon Clark, Martin Haigh, Ruth Healey, Martin Jenkins, Pauline Kneale, Ifan Shepherd and Chris Short all gave helpful comments on an earlier draft of this chapter. Useful suggestions for North American sources were given by Ken Foote, and Kevin McCracken and Bill Stinson advised on Australian sources.

| 3 | # Ethical Practice in Geographical Research |

Iain Hay

Synopsis

Ethical research in geography is characterized by practitioners who behave with integrity and who act in ways that are just, beneficent and respectful. Ethical geographers are sensitive to the diversity of moral communities within which they work and are ultimately responsible for the moral significance of their deeds. This chapter explains the importance of behaving ethically and provides some examples of ethical dilemmas.

The chapter is organized into the following sections:

- Introduction
- Why behave ethically?
- Principles of ethical behaviour and common ethical issues
- Truth or consequences? Teleological and deontological approaches to dealing with ethical dilemmas in your research
- Conclusion.

INTRODUCTION

To behave ethically in geographical research requires that you and I act in accordance with notions of right and wrong – that we conduct ourselves morally[1] (Mitchell and Draper, 1982). Ethical research is carried out by thoughtful, informed and reflexive geographers who act honourably because it is the 'right' thing to do, not because someone is making them do it (see, for example, Cloke et al., 2000; Dowling, 2000; Barker and Smith, 2001).

This chapter seeks to heighten your awareness of the reasons for, and principles underpinning, ethical research. Although I am conscious of the need to avoid ethical prescription, leaving you the opportunity to exercise and act on your own 'moral imagination' (Hay, 1998), I believe it is important to set out a range of specific ethical matters that colleagues and communities will commonly expect you to consider when preparing and conducting research. As a geographer, you must consider carefully the ethical significance of your actions in those contexts within which they have meaning and be prepared to take responsibility for your actions. In some instances (e.g. cross-cultural research), ethically reflexive practice includes acknowledging and working with (negotiating) different groups' ethical expectations in ways that yield satisfactory approaches for all parties involved (see Chapter 12).

The first part of the chapter introduces a few of the reasons why geographers need to behave ethically. I follow this with a discussion of some fundamental principles behind ethical behaviour and offer a few points of guidance that might assist you in your practice as an ethically reflexive geographer. As I note, however, no matter how well you might try to anticipate the ethical issues that might arise in your work, you are still likely to confront ethical dilemmas. In recognition of this, the chapter sets out a strategy that might help you resolve those dilemmas you face. I conclude with some suggestions on ways in which you might continue to develop as an ethical geographer. A number of real ethical cases are included in the chapter to illustrate points made and to provide some material for you to discuss.

In starting, there are two cautions I would like to offer. First, decisions about ethical practice are made in specific contexts. While all people and places deserve to be treated with integrity, justice and respect (see Smith, 2000a), ethical behaviour requires sensitivity to the expectations of people from diverse moral communities and acknowledgement of the webs of physical and social relationships within which the work is conducted (see Chapter 12). For these reasons, rules for ethical practice cannot be prescribed, particularly in work involving people (Hay, 1998). And as you will see from the real quandaries encountered by geographers that are interspersed through this chapter, the 'correct' resolution of most ethical dilemmas cannot be dictated. Secondly, simply because your peers, colleagues and institution (e.g. fellow students, professional geographers, university ethics committee) say that some behaviour or practice is ethical does not necessarily mean it is always so. That determination is ultimately up to you, your conscience and those people with whom you are working. In the end, you must take responsibility for the decisions you make.

WHY BEHAVE ETHICALLY?

Aside from any moral arguments that as human beings we should always act ethically, there are important practical arguments for geographical researchers to behave ethically. These fall into three main categories.

First, ethical behaviour protects the rights of individuals, communities and environments involved in, or affected by, our research. As social and physical scientists interested in helping to 'make the world a better place', we should avoid (or at least minimize) doing harm (Diener and Crandall, 1978; Mitchell and Draper, 1982; Peach, 1995). This does not always occur, a point made apparent by the Introduction to Macquarie University's (1997) (Australia) guidelines concerning research involving Aboriginal and Torres Strait Islander people:

> It has become evident in recent years that there has been a considerable misuse of academic research concerning Aboriginal and Torres Strait Islander people. As a consequence, many Aboriginal and Islander people have come to see research as a reason for their disempowerment. Instead of addressing Aboriginal and Torres Strait Islander problems, much research has focused upon matters of interest to science or white Australians.

Secondly, and perhaps a little more self-interestedly, ethical behaviour helps assure a favourable climate for the continued conduct of scientific inquiry. For instance, as Walsh (1992: 86) notes in a discussion of ethical issues in Pacific Island research, incautious practice and a lack of cultural awareness can lead to community denial of research privileges:

> Not infrequently, the blunderings of an inexperienced and culturally insensitive researcher creates problems for those researched and for [Pacific] Island governments, and some have been asked to leave. For these and similar reasons, all Island governments now require intending researchers to apply for research permits, may charge a non-refundable application fee, and it is not uncommon for a permit to be refused.

By behaving ethically, we maintain public trust. From that position of trust we may be able to continue research and to do so without causing suspicion or fear amongst those people who are our hosts (see Case Study 3.1) (Jorgensen, 1971; Mitchell and Draper, 1982; Walsh, 1992; Schuler et al., 1999).

Case Study 3.1   Waving, not drowning

While he was doing surveys of flood hazard perception in the USA, Bob Kates found that in rare cases his questions raised anxieties and fears in those people to whom he was speaking. Even the actions of the research team in measuring street elevations to calculate flood risk created rumours that led some people to believe that their homes were to be taken for highway expansion (Kates, 1994: 2).

For discussion
Is it Kates' responsibility to quash the rumours and suspicions his research engenders? Justify your answer.

Thirdly, growing public demands for accountability (AAAS, 1995) and the sentiment that institutions such as universities must protect themselves legally from the unethical or immoral actions of a student or employee mean there is greater emphasis on acting ethically than ever before.

Clearly, then, there are compelling moral and practical reasons for conducting research ethically. How are these brought to life?

PRINCIPLES OF ETHICAL BEHAVIOUR AND COMMON ETHICAL ISSUES

A growing number of organizations supporting research have established committees to scrutinize research proposals to ensure that work is conducted ethically. Because these committees regularly consider the possible ethical implications of research, their workings and the principles behind their operation might provide you with a useful starting point to ensure that your own research is ethical.

In general, the central guiding value for these committees is integrity, meaning a commitment to the honest conduct of research and to the communication of results. However, committees usually emphasize the desirability of three fundamental principles,[2] as set out in Box 3.1. These principles lead to a set of core questions (right-hand column) to guide your personal consideration of ethical matters. You will find it helpful always to consider these simple questions when reflecting on your work. They provide the beginnings of ethical practice. For instance, if you can sustain the argument that your work is just, is doing good and you are demonstrating respect for others, you are probably well on the way to conducting an ethical piece of work. If, by comparison, you believe that your work is just, yet you are not showing respect for others or you are doing harm, you may have an ethical dilemma on your hands. I set out a strategy for dealing with such situations later in this chapter.

Box 3.1 Principles of ethical behaviour

Justice: this gives emphasis to the distribution of benefits and burdens.	• Is this just?
Beneficence/non-maleficence: respectively, these mean 'doing good' and 'avoiding harm'. Our work should maximize benefits and minimize physical, emotional, economic and environmental harms and discomfort.	• Am I doing harm? Am I doing good?
Respect: individuals should be regarded as autonomous agents and anyone of diminished autonomy (e.g. intellectually disabled) should be protected. It is important to have consideration for the welfare, beliefs, rights, heritage and customs of people involved in research (National Commission, 1978; Canadian Tri-Council Working Group, 1997). Of course, respect should also extend to consideration of any discomfort, trauma or transformation affecting organisms or environments involved in the research.	• Am I showing respect?

The principles and questions set out in Box 3.1 are a good general framework but if you are just beginning to engage in research, some more specific advice on how to determine whether your research might be understood to be ethical will be helpful (see Case Study 3.2).

Typically, ethics committees give detailed consideration to five major issues when evaluating research proposals. These issues are set out in Box 3.2 and are accompanied by a set of 'prompts' for moral contemplation. As you can see, the 'prompts' offer no specific direction about the actions you should or should not take in any or all situations. To do so would be virtually impossible given the enormous variability of geographic research, the associated need for flexible research practices (see, for examples and discussion, Bauman, 1993; Schuler et al., 1999; Valentine, 1999; Valentine et al., 2001) and the recognition that society does not comprise an 'isolable, unitary, internally coherent whole' (Amit-Talai, in Matthews et al., 1998). It would also deny the need to negotiate specific issues with participants.[3]

When you are thinking about the issues in Box 3.2, consider them in terms of the value of integrity and the principles set out earlier –

Case Study 3.2   'They did that last week'

University student Ali bin Ahmed bin Saleh Al-Fulani carefully prepares a questionnaire survey for distribution to two groups of 16-year-old students in 'home groups' at two local high schools. The survey is central to the comparative work Ali is conducting as part of his thesis. In compliance with government regulations, Ali secures permission from the students' parents to conduct the survey. He also gets permission for his work from the university's Ethics Committee. The Ethics Committee requires him to include a covering letter to students which states that their participation in the study is voluntary and that no one is obliged to answer any of the questions asked. A few weeks before he intends to administer the questionnaire survey, Ali leaves near-final drafts of it with the students' teachers for comment. The draft copy of the questionnaire does not include the covering letter. It is Ali's intention to revise the questionnaire in the light of each teacher's comments and then return to the schools to administer the questionnaire during 'home group' meeting times. About a week after he leaves the survey forms with the teachers Ali calls them to find out if they have had an opportunity to comment on the questionnaire. The first teacher has just returned the questionnaire – with no amendments – by post. However, Ali finds that the second teacher had already made multiple copies of the forms and had administered the questionnaire to her student 'home group'. She asks Ali to come along to collect the completed forms. Ali scuttles off to the school immediately. He finds that the questionnaires had been completed fully by every student present in the home group. Only one student from the class of 30 had been absent so the response rate was 97% – an extraordinarily high rate. Ali feels he cannot ask the teacher to readminister the survey because she has already indicated several times that she is tired of his requests for assistance and access to the class.

For discussion

It would appear from the circumstances and from the very high response rate that students were not free to refuse to participate in the study. Is it ethical for Ali to use results that have been acquired without free and informed consent? Would your view change if the survey had dealt with some sensitive issue such as sexual assault or if the results had been acquired, without consent, through the use of physical force?

justice, beneficence and respect for others. For example, let us think about the prompt 'time for consideration of the study before consent is provided'. A human geographer might be expected to allow people more time to consider their involvement in a complex, long-term observational study than for a two-minute interview about their grocery shopping behaviour in a retail mall. In another example concerning harm minimization, a physical geographer might think about, and act on, the potential harm caused to nesting birds as a result of fieldwork involving weed clearance. Indeed, there could be no need to conduct the work if similar studies have been conducted previously.[4]

Box 3.2   Prompts for moral contemplation and action

Before and during your research, have you considered the following?

Consent[5]

- Amount of information provided to participants on matters such as purposes, methods, other participants, sponsors, demands, risks, time involved, discomforts, inconveniences and potential consequences.
- Accessibility and comprehensibility to prospective participants of information upon which consent decisions are made.
- Time for consideration of the study before consent is provided.
- Caution in research requiring deceit.[6]
- Caution in obtaining consent from people in dependent relationships.
- Recording of consent.
- Informed consent issues for others working on the project.

Confidentiality

- Disclosing identity of participants in the course of research and in the release of results.
- Consent and the collection of private or confidential information.
- Relationships between relevant privacy laws and any assurances made by researchers of confidentiality or privacy.
- Data storage during and after research (for example, field notes, completed surveys and tape-recorded interviews).

Harm

- Potential physical, psychological, cultural, social, financial, legal and environmentally harmful effects of the study or its results.
- Extent to which similar studies have been performed previously.
- Issues of harm for dependent populations.
- Relationship between the risks involved and the potential advantage of the work.
- Opportunities for participants to withdraw from the research after it has commenced.
- Competence of researchers and appropriateness of facilities.
- Representations of results.

Cultural awareness

- Personality, rights, wishes, beliefs and ethical views of the individual subjects and communities of which they are a part.

Continued

Box 3.2   Continued

Dissemination of results and feedback to participants

- Availability and comprehensibility of results to participants.
- Potential (mis)interpretations of the results.
- Potential (mis)uses of results.
- 'Ownership' of results.
- Sponsorship.
- Debriefing.
- Authorship.[7]

TRUTH OR CONSEQUENCES? TELEOLOGICAL AND DEONTOLOGICAL APPROACHES TO DEALING WITH ETHICAL DILEMMAS IN YOUR RESEARCH

No matter how well prepared you are, no matter how thoroughly you have prepared your research project, and no matter how properly you behave, it is likely that in your geographical research you will have to deal with a variety of ethical dilemmas (see, for example, Case Studies 3.3 and 3.4).

Case Study 3.3   You are being unco-operative

Catriona McDonald has recently completed an undergraduate research project that involved sending out a confidential questionnaire to members of a non-government welfare organization called ANZAC Helpers. Members of this organization are mainly World War Two veterans aged over 70 years, and a good deal of their volunteer work requires them to drive around familiar parts of a large metropolitan area. Catriona works part time for the organization and feels that her employment tenure is somewhat precarious. The research project has been completed and assessed formally by Ms McDonald's professor and Catriona is planning to present a modified report to ANZAC Helpers. In the work car-park one afternoon, Catriona meets Mr Montgomery Smythe, one of the organization's members and a man known to Catriona to be something of a troublemaker. After some small talk, Smythe asks what the study results are. Catriona outlines some of the findings, such as the percentage of the membership who are having trouble performing their voluntary duties for the organization due to old age and ill-health. Smythe then asks the student to tell him the names of those members who are having difficulties with their duties. It is possible that he could use the information to help encourage the implementation of strategies to help those members experiencing difficulties. It is also possible that he could campaign to have the same members redirected to less demanding volunteer roles that many of them are likely to find less fulfilling.

Continued

Case Study 3.3   *Continued*

For discussion
1   Should Ms McDonald give Mr Smythe the information he wants?
2   Given that some of the members of ANZAC Helpers might actually be putting their lives at risk by driving around the city, should Catriona disclose the names of those people she has discovered to be experiencing sight and hearing problems, for example, to the organization?

Case Study 3.4   The power of maps

Dr Tina Kong has recently commenced work as a post-doctoral fellow with a research organization applying GIS (geographical information systems) to illustrate and resolve significant social problems. This is a position in which full-time research is possible. It is also a position that can allow someone to begin to forge the beginnings of a noteworthy academic career for him or herself. However, much of that promise depends on producing good results and having them published in reputable journals. Tina decides to conduct work on environmental carcinogens (cancer-producing substances) in a major metropolitan area. She spends about two months of her two-year fellowship conducting some background research to assess the need for, and utility of, the work. After this early research, Tina resolves to use GIS to produce maps which will illustrate clearly those areas in which high levels of carcinogenic materials are likely to be found. At a meeting of interested parties to discuss the proposed research, one of the participants makes the observation that, if broadcast, the results of the study may cause considerable public alarm. For example, there may be widespread individual and institutional concern about public health and welfare; property values in areas with high levels of carcinogenic material may be adversely affected; past and present producers of carcinogenic pollutants may be exposed to liability suits; and local government authorities might react poorly to claims that there are toxic materials in their areas. Tina is cautioned by senior members of the research organization to proceed cautiously, if at all.

For discussion
Should Tina proceed with the project? Justify your answer. Should Tina 'cut her losses' and move into a less controversial area that might be supported by her colleagues?

How will you deal with such situations? To answer that fully, we will have to make a short excursion into the work of philosophers. But, first, I should point out that to resolve any dilemma you encounter it is likely that you will have to 'violate' one of the three ethical principles – justice, beneficence or respect for others – set out earlier in this chapter. That is why it is a dilemma! But you will probably feel more confident about the difficult decision you will have to make if it is well considered and informed by a basic appreciation of two key normative approaches to behaviour.

Box 3.3   Steps to resolving an ethical dilemma

How do you decide what to do if you are presented with an ethical dilemma? There are two major approaches that you might draw from. One focuses on the practical consequences of your actions (*teleological* or *consequentialist* approach) and might be summed up brutally in the phrase 'no harm, no foul'. In contrast, the *deontological* approach would lead you to ask whether an action is, in itself, right. For example, does an action uphold a promise or demonstrate loyalty? The essence of deonto-logical approaches is captured by Quinton's (1988: 216) phrase: 'Let justice be done though the heavens fall.' These two positions serve as useful starting points for a strategy for coping with ethical dilemmas.

1   *What are the options?*
List the full range of alternative courses of action available to you.

2   *Consider the consequences*
Think carefully about the range of positive and negative consequences associated with each of the different paths of action before you:

- Who/what will be helped by what you do?
- Who/what will be hurt?
- What kinds of benefits and harms are involved and what are their relative values? Some things (e.g. healthy bodies and beaches) are more valuable than others (e.g. new cars). Some harms (e.g. a violation of trust) are more significant than others (e.g. lying in a public meeting to protect a seal colony).
- What are the short-term and long-term implications?

Now, on the basis of your answers to these questions, which of your options produces the best combination of benefits maximization and harm minimization?

3   *Analyse the actions*
You now have to consider each of your options from a completely different perspective. Disregard the consequences, concentrating instead on the actions and looking for that option which seems problem-atic. How do the options measure up against such moral principles as honesty, fairness, equality and recognition of social and environmental vulnerability? In the case you are considering, is there a way to see one principle as more important than the others?

4   *Make* your *decision and act with commitment*
Now, bring together both parts of your analysis and make *your* informed decision. Act on your decision and assume responsibility for it. Be prepared to justify *your* choice of action. No one else is responsible for this action but you.

Continued

Box 3.3 Continued

5 *Evaluate the system*
Think about the circumstances which led to the dilemma with the intention of identifying and removing the conditions that allowed it to arise.

Source: Adapted from Stark-Adamec and Pettifor (1995); Center for Ethics and Business (2001)

Many philosophers suggest that two categories, teleological and deontological, exhaust the possible range of theories of right action (Davis, 1993). In summary, the former sees acts judged as ethical or otherwise on the basis of the consequences of those acts. The latter suggests, perhaps surprisingly, that what is 'right' is not necessarily good.

In the terms of teleology – also known as consequentialism – an action is morally right if it produces more good than evil (Frankena, 1973; Holden, 1979; Reynolds, 1979; Kimmel, 1988; White, 1988; Peach, 1995). It might therefore be appropriate to violate and make public the secret and sacred places of an indigenous community if doing so prevents the construction of a bridge through those sacred places, because disclosure yields a greater benefit than it 'costs'.

Deontological approaches reject this emphasis on consequences, suggesting that the balance of good over evil is insufficient to determine whether some behaviour is ethical. Instead, certain acts are seen as good in themselves and must be viewed as morally correct because, for example, they keep a promise, show gratitude or demonstrate loyalty to an unconditional command (Kimmel, 1988). It is possible, therefore, for something to be ethically correct even if it does not promote the greatest balance of good over evil. To illustrate the point: if we return to the example of the researcher made aware of the location of an indigenous group's secret, sacred places, a deontological view might require that researcher to maintain the trust with which he or she had been privileged, even if non-disclosure meant that construction of the bridge would destroy those sacred places.

Thus, we have two philosophical approaches that can point to potentially contradictory ethical ways of responding to a particular situation. This is not as debilitating as you might think. As Box 3.3 sets out, you can draw from these approaches to ensure that your response to an ethical dilemma is at least well considered, informed and defensible.

You can really give this scheme a workout by considering the (in)famous work of Laud Humphreys, set out as Case Study 3.5.

Case Study 3.5   The watch queen in the 'tea room'

In 1966–7, and as part of a study of homosexual behaviours in public spaces, Laud Humphreys acted as voyeuristic 'lookout' or 'watch queen' in public toilets ('tea rooms'). As a 'watch queen' he observed homosexual acts, sometimes warning men engaged in those acts of the presence of intruders. In the course of his observations Humphreys recorded the car licence plate numbers of the men who visited the 'tea room'. He subsequently learnt their names and addresses by presenting himself as a market researcher and requesting information from 'friendly policemen' (Humphreys, 1970: 38). One year later, Humphreys had changed his appearance, dress and car and got a job as a member of a public health survey team. In that capacity he interviewed the homosexual men he had observed, pretending that they had been selected randomly for the health study. This latter deception was necessary to avoid the problems associated with the fact that most of the sampled population were married and secretive about their homosexual activity (Humphreys, 1970: 41). After the study, Humphreys destroyed the names and addresses of the men he had interviewed in order to protect their anonymity. His study was subsequently published as a major work on human sexual behaviour[8] (Humphreys, 1970).

For discussion
1   It might be said that Humphreys' research in 'tea rooms' was a form of participant observation, a type of research which is often most successful when 'subjects' do not know they are being observed. Was it unethical for Humphreys to observe men engaged in homosexual acts in the 'tea room'? Does the fact that the behaviour was occurring in a public place make a difference to your argument? Why?
2   Was it ethical for Humphreys to seek and use name and address information – details that appear commonly in telephone books – from police officers who should not have released those details for non-official reasons? Would it have been acceptable if he had been able to acquire that same information without deceipt?
3   Upon completion of the research should Humphreys have advised those men who had been observed and interviewed that they had been used for the study? Why? Discuss the significance of your answer. Should only some research results be 'returned' to participants and not others? What criteria might one employ to make that determination? Why are those criteria (more) important (than others)?
4   Should Humphreys have destroyed the name and address information he used? How do we know he was not making the whole story up? How can someone else replicate or corroborate his findings without that information?
5   Humphreys' work offered a major social scientific insight into male homosexual behaviours. It might be argued that his book, *Tearoom Trade*, contributed to growing public understanding of one group in the broader community. Moreover, no apparent harm was done to those people whose behaviour was observed. Do you think, then, that the ends may have justified the means?

CONCLUSION

Being an ethical geographer is important. It helps to protect those people, places and organisms affected by our research and helps to ensure that we are able to continue to conduct socially and environmentally valuable work. The steps set out in this chapter to help you prepare for research and to deal with the ethical problems you may encounter should go some way to helping achieve these ends. However, your development as an ethically responsible geographer cannot stop with this chapter. It is important that you continue to heighten your awareness of ethical issues and develop your ability to act thoughtfully when confronted with dilemmas like those set out in this chapter. To that end, I shall conclude with some thoughts on ways in which you can continue to become a more ethical geographer.

Good luck!

Summary

What can you do to become a more ethical geographer?

- *Make sure your 'moral imagination' is active and engaged.* There will always be ethical issues in your research. Make ethics as normal a part of your research project discussions as how the stream gauge or questionnaire is working. Discuss ethical issues and possibilities with your colleagues. Learn to recognize ethical issues in context. Think about the potential moral significance of your own actions and those of other people. Remember that we live in a vast network of moral relationships. The meanings of particular behaviours and moral positions may sometimes be given or understood far from the places they might be expected (Smith, 1998) and interpreted from different ethical standpoints. Look for hidden value biases, moral logic and conflicting moral obligations. Make yourself aware of (local) ethical practices (Mehlinger 1986).
- *Develop your philosophical and analytical skills.* What is 'right' or 'good'? On what bases are those decisions made? Be prepared to think hard about difficult questions. For example, how can you evaluate prescriptive moral statements such as 'endangered species should (or should not) be protected' or 'research should (or should not) be conducted with the consent of all participants'?
- *Heighten your sense of moral obligation and personal responsibility.* 'Why should I be moral?' or 'Why should I think about ethics?' Embrace ethical thought and action as an element of your professional and social identity as a geographer. Come to terms with the idea that you need to act morally because it is the

'right' thing to do, not because someone is making you do it
(Mehlinger, 1986).

- *Expect – but do not passively accept – disagreement and
  ambiguity.* Ethical problems are almost inevitably associated with
  disagreements and ambiguities. However, do not let that
  expectation of ambiguity and disagreement provide justification for
  abandoning debate and critical thought. Learn to seek out the
  core of differences to see if disagreement might be reduced. Be
  committed enough to follow through on your own decisions.

Further reading

- A good place to start is the journal *Ethics, Place and Environment.* This publishes
  research, scholarship and debate on all aspects of geographical and environmental
  ethics. For example, in an article in the first issue of this journal, Hay (1998) argues for
  a 'flexible' approach to ethics and sets out ways in which moral imaginations –
  required for such an approach – might be stimulated and nurtured.

- Dowling (2000) provides a helpful introduction to some central ethical issues in
  geographical research (e.g. harm, consent) and makes a case for critical reflexivity (i.e.
  ongoing, self-conscious scrutiny of oneself as researcher and of the research
  process).

- Mitchell and Draper (1982) is a classic reference for geographers interested in issues
  of relevance and research ethics. Though dated, this volume is well worth a read.

- Proctor and Smith (1999) examine the place of geography in ethics and of ethics in
  geography through the themes of ethics and space; ethics and place; ethics and nature;
  and ethics and knowledge. It is useful if you have an interest in exploring broader
  ethical issues than those covered in this 'methods' chapter.

- Scheyvens and Leslie (2000) is a helpful article exploring ethical dimensions of power,
  gender and representation in overseas fieldwork. It is illustrated by several examples
  drawn from the authors' recent fieldwork practice.

- Smith (2000b) explores the interface among geography, ethics and morality. At its core
  is Smith's longstanding concern with the practice of morality.

- Workman et al. (1997) is a detailed and very prescriptive document intended to
  minimize harm done to environments and communities through fieldwork.

Note: Full details of the above can be found in the references list below.

NOTES

1 Although most of the principles discussed here apply to environmental research ethics,
   the chapter focuses most heavily on research involving humans. Readers especially

interested in environmental research ethics are strongly advised to consult ASTEC (1998).

2 These principles place a strong emphasis on individual autonomy. It is important to note that in some societies and situations (e.g. work with some indigenous groups and children), individual autonomy may be limited and influenced by related groups and other individuals who have authority over that individual. It is imperative, therefore, to consider the specific local contexts within which rights are understood (NHMRC, 2001).

3 Ethical dimensions of a research project should be negotiated between the researcher and the participant(s) to ensure that the work satisfies the moral and practical needs and expectations of those individuals and communities involved. However, the negotiations will be influenced by the different levels of power between the parties. In some situations a researcher will have more power than the informant (e.g. research work involving children) while in others (e.g. interview with the CEO of a large organization whose comments are critical to your study) informants may have more power. (For a discussion of some of these important issues and ways of dealing with them, see Matthews et al., 1998; Wilkinson, 1998; Valentine, 1999; Chouinard, 2000; Cloke et al., 2000; D'Cruz, 2000; Dowling, 2000; Dyck, 2000; Scheyvens and Leslie, 2000). In some contexts, there will also be certain community expectations of researcher conduct. For example, my university expects that observational studies of people in public places will cease if any single person in that place objects to the conduct of the study.

4 For detailed discussions of some of the ethical responsibilities of researchers involved in environmental fieldwork, see Workman et al. (1997) and ASTEC (1998: esp. pp. 21–4).

5 For an interesting discussion on the issue of consent, see Metzel (2000).

6 Wilton (2000) provides a fascinating reflection on geographical work involving deception.

7 Kearns et al. (1996) offer a concise and very helpful discussion of the ethics of authorship/ownership of knowledge.

8 In a postscript to his book, Humphreys (1970) convincingly addresses issues of misrepresentation, confidentiality, consequentiality and situation ethics associated with his research. His vocation as a religious minister might add, for some, strength to his arguments. For other discussion of this example, see Diener and Crandall (1978).

References

AAAS (American Association for the Advancement of Science) (1995) *Activities of the AAAS Scientific Freedom, Responsibility and Law Program* (available at http://www/aaas.org./spp/dspp/sfrl/fraud.htm).

ASTEC (Australian Science, Technology and Engineering Council) (1998) *Environmental Research Ethics*. Canberra: ASTEC.

Barker, J. and Smith, F. (2001) 'Power, positionality and practicality: carrying out fieldwork with children', *Ethics, Place and Environment*, 4: 142–7.

Bauman, Z. (1993) *Postmodern Ethics*. Oxford: Blackwell.

Canadian Tri-Council Working Group (1997) *Code of Ethical Conduct for Research Involving Humans* (available at http://www.hssfc.ca/Gen/Code ContentsEng.html).

Center for Ethics and Business (2001) *Resolving an Ethical Dilemma* (available at http://www.ethicsandbusiness.org/strategy.htm).

Chouinard, V. (2000) 'Getting ethical; for inclusive and engaged geographies of disability', *Ethics, Place and Environment*, 3: 70–80.

Cloke, P., Cooke, P., Cursons, J., Milbourne, P. and Widdowfield, R. (2000) 'Ethics, reflexivity and research: encounters with homeless people', *Ethics, Place and Environment*, 3: 133–54.

Davis, N.A. (1993) 'Contemporary deontology', in P. Singer (ed.) *A Companion to Ethics*. Oxford: Blackwell, pp. 205–18.

D'Cruz, H. (2000) 'Social work research as knowledge/power in practice', *Sociological Research Online*, 5 (available at http://www.socresonline.org.uk/5/1/dcruz.html).

Diener, E. and Crandall, R. (1978) *Ethics and Values in Social and Behavioural Research*. Chicago, IL: University of Chicago Press.

Dowling, R. (2000) 'Power, subjectivity and ethics in qualitative research', in I. Hay (ed.) *Qualitative Research Methods in Human Geography*. Melbourne: Oxford University Press, pp. 23–36.

Dyck, I. (2000) 'Putting ethical research into practice: issues of context', *Ethics, Place and Environment*, 3: 80–7.

Frankena, W.K. (1973) *Ethics* (2nd edn). Englewood Cliffs, NJ: Prentice Hall.

Hay, I. (1998) 'Making moral imaginations. Research ethics, pedagogy and professional human geography', *Ethics, Place and Environment*, 1: 55–76.

Holden, C. (1979) 'Ethics in social science research', *Science*, 26: 537–8.

Humphreys, L. (1970) *Tearoom Trade: A Study of Homosexual Encounters in Public Places*. London: Duckworth.

Jorgenson, A.J.G. (1971) 'On ethics and anthropology', *Current Anthropology*, 12: 326–34.

Kates, B. (1994) 'President's column', *Association of American Geographers' Newsletter*, 29: 1–2.

Kearns, R., Arnold, G., Laituri, M. and Le Heron, R. (1996) 'Exploring the politics of geographical authorship', *Area*, 28: 414–20.

Kimmel, A.J. (1988) *Ethics and Values in Applied Social Research*. London: Sage.

Macquarie University (1997) *Guidelines Concerning Aboriginal and Torres Strait Islander Research* (available at http://www.ro.mq.edu.au/HETHICS/ethicsa.html).

Matthews, H., Limb, M. and Taylor, M. (1998) 'The geography of children: some ethical and methodological considerations for project and dissertation work', *Journal of Geography in Higher Education*, 22: 311–24.

Mehlinger, H. (1986) 'The nature of moral education in the contemporary world', in M.J. Frazer and A. Kornhauser (eds) *Ethics and Responsibility in Science Education*. Oxford: ICSU Press, pp. 17–30.

Metzel, D. (2000) 'Research with the mentally incompetent: the dilemma of informed consent', *Ethics, Place and Environment*, 3: 87–90.

Mitchell, B. and Draper, D. (1982) *Relevance and Ethics in Geography*. London, Longman.

National Commission for the Protection of Human Subjects of Biomedical and Behavioural Research (1978) *Ethical Principles and Guidelines for the Protection of Human Subjects of Research (The Belmont Report)*. Department of Health, Education and Welfare Publication (OS) 78–0012. Washington, DC: US Government Printing Office.

NHMRC (Australian National Health and Medical Research Council) (2001) *National Statement on Ethical Conduct in Research Involving Humans –*

Preamble (available at http://www.health.gov.au/nhmrc/publications/humans/preamble.htm).

Peach, L. (1995) 'An introduction to ethical theory', in R.L. Penslar (ed.) *Research Ethics. Cases and Materials.* Bloomington, IN: Indiana University Press, pp. 13–26.

Proctor, J. and Smith, D.M. (eds) (1999) *Geography and Ethics: Journeys in a Moral Terrain,* London: Routledge.

Quinton, A. (1988) 'Deontology', in A. Bullock et al. (eds) *The Fontana Dictionary of Modern Thought* (2nd edn). Glasgow: Fontana, p. 216.

Reynolds, P.D. (1979) *Ethical Dilemmas and Social Science Research: An Analysis of Moral Issues Confronting Investigators in Research Using Human Participants.* San Francisco, CA: Jossey-Bass.

Scheyvens, R. and Leslie, H. (2000) 'Gender, ethics and empowerment. Dilemmas of development fieldwork', *Women's Studies International Forum,* 23: 119–30.

Schuler, S., Aberdeen, L. and Dyer, P. (1999) 'Sensitivity to cultural difference in tourism research: contingency in research design', *Tourism Management,* 20: 59–70.

Smith, D.M. (1998) 'How far should we care? On the spatial scope of beneficence', *Progress in Human Geography,* 22: 15–38.

Smith, D.M. (2000a) 'Moral progress in human geography: transcending the place of good fortune', *Progress in Human Geography,* 24: 1–18.

Smith, D.M. (2000b) *Moral Geographies. Ethics in a World of Difference.* Edinburgh: Edinburgh University Press.

Stark-Adamec, C. and Pettifor, J. (1995) *Ethical Decision Making for Practising Social Scientists. Putting Values into Practice.* Ottawa: Social Science Federation of Canada.

Valentine, G. (1999) 'Being seen and heard? The ethical complexities of working with children and young people at home and at school', *Ethics, Place and Environment,* 2: 141–55.

Valentine, G., Butler, R. and Skelton, T. (2001) 'The ethical and methodological complexities of doing research with "vulnerable" young people', *Ethics, Place and Environment,* 4: 119–25.

Walsh, A.C. (1992) 'Ethical matters in Pacific Island research', *New Zealand Geographer,* 48: 86.

White, T.I. (1988) *Right and Wrong. A Brief Guide to Understanding Ethics.* Englewood Cliffs, NJ: Prentice Hall.

Wilkinson, S. (1998) 'Focus groups in feminist research: power, interaction, and the co-construction of meaning', *Women's Studies International Forum,* 21: 111–25.

Wilton, R.D. (2000) ' "Sometimes it's OK to be a spy": ethics and politics in geographies of disability', *Ethics, Place and Environment,* 3: 91–7.

Workman, C., Gimingham, A. and Jermy, C. (eds) (1997) *Environmental Responsibilities for Expeditions: A Guide to Good Practice* (available at http://www.demon.co.uk/bes/ereguide.htm).

Acknowledgements

I would like to thank Clive Forster, Robin Kearns, Lisel O'Dwyer and Brian Stoffell for their helpful comments on earlier versions of this chapter.

Health and Safety in the Field

Joanna Bullard

Synopsis

Health and safety in the field concerns the practical steps that geographers can take to lessen the chances of an incident or accident causing harm to themselves and others during fieldwork. This chapter focuses on minimizing risks, reducing hazards and taking responsibility for health and safety during both group and independent fieldwork.

The chapter is organized into the following sections:

- Managing health and safety
- Know your field environment
- Know your own limitations
- Know your equipment
- What if something does go wrong?

INTRODUCTION

Fieldwork is often one of the most rewarding aspects of being a geography student. Many residential trips and one-day outings are remembered by students long after graduation. Fieldwork can, however, also be time-consuming, frustrating, difficult and potentially dangerous. This chapter offers some guidance about minimizing the risks and hazards involved in fieldwork. Following this guidance should reduce the dangers inherent in your chosen field activity and, consequently, should also make your time in the field more rewarding, more enjoyable and possibly a little easier.

As a geography student you may undertake fieldwork in a variety of contexts, including supervised and unsupervised, group and individual field activities, residential and non-residential field studies, local and overseas environments. These different contexts present a range of health and safety considerations and the roles and responsibilities of both students and staff may change accordingly.

This chapter discusses the assessment and management of risks associated with undertaking fieldwork. It covers evaluating and minimizing risk in a range of common field environments and emphasizes the importance of planning ahead. In particular, the chapter focuses on your responsibility, as an individual and as a member of a group undertaking fieldwork, to ensure that appropriate health and safety precautions are observed.

In addition to reading this chapter, note that every higher education institution has its own set of health and safety guidelines. You may have been required to sign a declaration stating that you have read and understood this document. One of your responsibilities as a student is to ensure that you follow any institutional or departmental guidelines when you are out in the field.

MANAGING HEALTH AND SAFETY

In most countries, national legislation provides the legal framework for health and safety management. For example, in the UK universities abide by the Health and Safety at Work Act 1974. In Australia, the Workplace Health and Safety Act 1995 has a similar role. These Acts place a duty upon employers to take steps to ensure, as far as possible, the health and safety of their employees and any other people affected by their activities including, in the case of universities, all students and members of the public. Additional or complementary legislation, either at the national or state scale, often requires that 'risk assessments' are undertaken to identify what should be done in order to carry out this duty. Assessments of risk are usually focused around 'risk' and 'hazard'. Put simply, hazards result from working in potentially dangerous environments and refer to environmental conditions, agents or substances that can cause harm. Risk is the chance that someone might be harmed by the hazard. During fieldwork hazards and risks can change rapidly – for example, as a result of changing weather conditions or political actions – and should be continually reassessed.

Many education institutions adopt a five-step approach to risk assessment:

1 *Identify the hazards*: during the fieldwork, what could cause harm – for example, slippery ground, high altitude, weather conditions, civil or political unrest?
2 *Identify who might be harmed and how*: this includes all the fieldworkers and members of the public.

3   *Evaluate and minimize the risks*: once suitable precautions have been taken (for example, wearing appropriate protective clothing), how likely is it that someone will be harmed?
4   *Record the findings*: write down the identified hazards and precautions to be taken.
5   *Review the assessment periodically*: situations change and so do the potential hazards and risks, so risk assessment is not an isolated task but an ongoing evaluation to be updated and revised as often as necessary.

A risk assessment should be completed for fieldwork conducted as part of your degree. The extent of your involvement in actually assessing the risks will vary according to the way in which the fieldwork is organized, but you have a responsibility to follow any precautions or safety measures laid down in the risk assessment.

Supervised and unsupervised group fieldwork

Group fieldwork will usually be undertaken as part of a course or module designed by a member of academic staff at your university. During supervised fieldwork one member of staff will take overall responsibility for field activities and will complete the risk assessment. The leader of the fieldwork will be responsible for ensuring that appropriate health and safety matters have been taken into consideration and are complied with by all accompanying members of staff and students. You are not, however, absolved of responsibility. Be aware that as an individual you have a responsibility for your own safety and also for the safety of others who may be affected by your actions including staff, other students and members of the public. You should follow any instructions given by the staff members. If you witness any accidents or foresee any unexpected problems these should be reported to the fieldwork leader as soon as possible. Failure to uphold your responsibilities may result in disciplinary measures and, in extreme cases, in a criminal prosecution. It is worth noting that your activities prior to fieldwork – for example, having a few drinks the evening before – can affect your performance and reliability in the field. Staff may refuse to let you participate in fieldwork if you are considered a danger to yourself or others.

On arrival at university you may have completed a confidential medical questionnaire. You may be asked to complete a more detailed questionnaire prior to undertaking any fieldwork, especially if it will be conducted overseas. These questionnaires should be used to give details of any diseases, conditions, disabilities or susceptibilities that you may have (e.g. diabetes, severe allergies, impaired hearing or vertigo). If you are involved in an incident or accident while on fieldwork this information may be vital in ensuring that you receive prompt and appropriate medical treatment. If you did not complete a questionnaire or equivalent, and do have medical needs that could compromise the health and safety of yourself or others, you should let staff know well before fieldwork commences so that they can take appropriate action where necessary.

During a field course you may undertake unsupervised group fieldwork. This commonly involves students working in groups of two or more. A member of staff may visit you during the field activities – for example, he or she may move from group to group along the course of a river or arrange to meet you at a particular location within a city. Alternatively you may be left entirely to your own devices and asked to book in or out of a field centre or to conduct work on campus or in your university city. In these situations you have a responsibility to the group of students with whom you are working. Keep together as a group, discuss any difficulties that you might have and pay attention to anyone who is feeling unwell or is having difficulty keeping up. You should follow any guidelines issued (for example, take a first-aid kit), have contact numbers for the university department or individual members of staff and know where you should be working and what time you are expected to check back in.

Individual, unsupervised fieldwork

You may be involved in individual, unsupervised fieldwork at any time during your university career, but this is most likely to be the case when you undertake fieldwork as part of a dissertation. Many universities now require every student to complete an individual risk assessment for his or her intended project. This can take many forms but will usually be organized around the five-step plan outlined above. Detailed discussion and examples of this procedure are provided by Higgitt and Bullard (1999).

The most important health and safety concerns associated with individual, unsupervised fieldwork arise from lone working, which should be avoided wherever possible. If you are conducting an individual project or dissertation a good strategy is to persuade a responsible friend or sibling to act as your field assistant. You might pair up with a friend from university and agree to help him or her survey a beach if he or she, in turn, will accompany you delivering questionnaires. If lone working is unavoidable or you are working with just one other person it is vital that you develop a strategy for dealing with emergencies should they arise. At the very least you should leave written details with a responsible adult of where you are going (for example, the map co-ordinates of your field site or the name and address of the person you intend to interview), of what you intend to do and of what time you expect to return. You should also write down what you expect the person to do in the event that you do not return on time. If your plans change you must let this person know. In risk assessment, lone working always counts as a hazardous activity. Where the site or conditions are themselves hazardous, lone working is liable to be dangerous.

Whether you are working in a group or as an individual there are some basic steps you can always take to minimize any health or safety risks. First, know the telephone numbers that can be used to contact the emergency services if necessary. In the UK, 999 or 112 will allow you to contact the fire, police and ambulance services, the coastguard and mine, mountain, cave and fell rescue services. Similar services can be contacted in the USA and Canada by dialing 911; in Australia by dialing 000; and in New Zealand, 111. If you are working overseas, find out local emergency numbers and the contact details of your embassy, high commission or consulate before you start working and keep them written down in a safe place (e.g. on a card inside your first-aid kit). Other emergency communication methods include the international Alpine Distress signal – six long whistle blasts or torch flashes or shouts for help in succession repeated at one-minute intervals – and the Morse code signal 'SOS' signalled by three short whistle blasts/torch flashes, three long blasts/flashes, three short blasts/flashes, pause, then repeat as necessary.

Secondly, have a basic kit that you take into the field with you every day. Never start any fieldwork in any environment without having with you a pen/pencil and paper and any personal medication. If you are working in a rural or remote environment you will also need a first-aid kit, mobile phone, whistle, compass, map, additional clothing (hat, gloves, socks, spare sweater), food and drink as a minimum. If you are working in a built-up or urban environment your field kit may include a first-aid kit, mobile phone, street map and personal alarm.

Thirdly, make sure you always know exactly where you are – this is relatively easy in a town centre but can be more problematic in an unfamiliar large city or a remote area with few landmarks. If the worst happens and you need help from the emergency services this will be a vital piece of information that you can provide.

When you carry out the risk assessment for your field project or set off on a field trip with other students, there are a number of other things you need to take into account to ensure the health and safety of everyone involved. These fall broadly into the categories of knowing your field environment, knowing your own limitations and knowing your equipment.

KNOW YOUR FIELD ENVIRONMENT

There are many different locations in which you might find yourself doing fieldwork, from a suburban street to a kibbutz, from a quarry to the top of a mountain. No matter where you intend to go there will always be some information that you can acquire before you start that will provide guidance as to how to act, dress and behave. Gathering this information together is important because, in many cases, you will need to complete a risk assessment for your field project *before* you reach the site. If the field site is nearby or very familiar to you then you can draw on your existing experiences when completing the risk assessment, but if the site is not somewhere you have been before or is overseas, you will need some help in anticipating the risks and hazards involved.

A key control on many field activities is the weather or climate of the area in which you are working. As part of your ongoing risk assessment you should consult a daily weather forecast throughout your fieldwork (see further reading). Specific hazards associated with weather and climate include hypothermia, frostbite, sunburn, dehydration, heat stroke and heat exhaustion. In most situations being aware of the predicted weather for the local area and having the correct clothing and equipment can reduce the risks associated with these hazards. Specialist information for mountain and coastal environments is available in many countries. When working at the coast make sure you know where the nearest coastguard station is and check local tide timetables.

If you are conducting people-orientated fieldwork at home or overseas you are advised to do some research on the local area and to make yourself familiar with any customs, political issues and religious beliefs that may affect the ways in which you conduct your research and how people react to your project. The aim is to minimize the risk of causing offence, which may lead to personal attack/injury or abuse. Take advice from local contacts about how to respect local customs and about any unrest in the area and dress appropriately for the environment and culture within which you are conducting your research. If possible avoid areas that are known to be 'unpleasant' and try not to enter unfamiliar neighbourhoods alone. A good street map is invaluable, plan your route and walk purposefully and with confidence.

When taking part in a supervised field trip, the trip leader should check local conditions and is likely to brief you about them. If you are told tide times or other specific pieces of information, such as neighbourhoods to avoid, make sure you write them down. Prior to the field trip you may be given a list of clothing and equipment to bring with you that is suitable for the expected conditions. Depending on the environment, this could include sturdy walking boots, waterproof jacket, gloves, sunglasses, sun hat, water bottle, long sleeves, etc. In some institutions you will not be allowed to take part in the field activities if you do not have appropriate clothing with you – this is not a minor issue, it is one with potentially profound health and safety implications.

Many government websites provide detailed information for nationals travelling overseas. This can include information on countries or regions to avoid, visa requirements, local customs, pre-travel inoculations and what to do should you be the victim of crime, fall ill or get into trouble of any sort. Make sure you are aware of any political, military or civil unrest in the country – local newspapers can be an invaluable source of information. If you are working alone or as a small independent group you should inform your nearest embassy, high commission or consulate of your presence in the country when you arrive to enable them to keep you informed of any hazardous situations. Nash (2000) provides a discussion of things to consider when planning independent overseas fieldwork and includes a list of essential information sources. Another useful reference both while planning and during an overseas field visit will be a guidebook to that country or area. Most popular guide books, such as those in the *Rough Guide* or *Lonely Planet* series, include sections on health risks, safety precautions and appropriate dress codes.

KNOW YOUR OWN LIMITATIONS

Many people underestimate the time needed to complete fieldwork tasks. The majority of incidents and accidents occur when people are tired. Make sure that the aims of your fieldwork are realistic and that you have allowed enough time to achieve them. Weather conditions can dramatically affect your efficiency in the field – it is hard to survey for six hours in the pouring rain but much easier (and more fun) if the sun is shining. A useful approach is to make a list of the minimum amount of work that you need do to achieve your basic goals and rank these tasks in order of priority. Then make a list of what would be useful or desirable if you have the time. Work through the list in order, achieve the minimum and then add to it if possible. Do not compromise your safety by trying to work longer hours in unsuitable conditions. If you are doing supervised fieldwork and are uncomfortable about undertaking certain field activities, tell the leader and ask whether or not there is an alternative. For example, if you are scared of heights and are very nervous about walking along a narrow path with a steep drop, ask if there is an alternative route or whether someone more experienced with the environment can walk with you.

Although you should try to work within your limitations, field-work is about facing challenges and doing things you might not normally do – for example, interviewing complete strangers or working in a challenging physical environment. There are many things that you can do before you head off into the field that will extend your skills and abilities and so reduce the risks involved. Some organizations offer courses and workshops and produce publications that may be useful (see further reading). If you intend to work in sparsely populated areas then you should consider taking a first-aid training course. These are run on a regular basis by such bodies as the St John Ambulance and St Andrew's Ambulance Association and may be available at your college or university as an evening class. Advice on expedition planning and organization is available from a number of sources including the Expedition Advisory Centre (UK) and the Explorers' Club (USA) (see further reading). More specialist courses, such as winter mountain skills and sea navigation, are run by specific organizations and societies.

KNOW YOUR EQUIPMENT

Whatever equipment you intend to use for your fieldwork, whether a tape recorder or a soil auger, make sure you know how to use it and how to carry it. Some items of equipment are very heavy and need two people to move them safely; even a simple surveying kit comprising staff, level and tripod can be awkward for one person to manage. Take advice from academic and support staff in your institution about the safest ways to collect samples from a river, to dig a deep soil pit or to conduct a house-to-house survey. If your equipment requires batteries and/or cassettes, take spare ones; if you are delivering questionnaires house to house make sure you have more copies than you think you need and a good street map. These types of precautions will prevent delays in your field programme and reduce the chance of you being tempted to work at unsuitable times – for example, delivering questionnaires after dark or working on a beach with a rapidly rising tide.

Mobile telephones have revolutionized the ways in which we communicate with each other. Although a mobile phone may be a useful way to contact help in a city, reception problems can render a telephone inoperable in isolated areas or in valleys. In addition, technological failure and (more commonly) lack of power can reduce your mobile phone to a useless piece of plastic or metal in any location. The same is true for personal alarms. Carrying a mobile phone or personal alarm is no substitute for a well thought-out system for informing other people of your whereabouts (your lone-working strategy).

If you are working in a rural or remote land area or at sea you may decide to take a hand-held Global Positioning System (GPS) with you. This type of device can provide accurate latitude/longitude co-ordinates and can be used for mapping and/or navigation. If you are using a GPS for navigation you must be aware of its limitations. Most hand-held GPSs are only accurate to a few metres latitude/longitude and are notoriously unreliable with regards to altitude readings. In addition they run on batteries, which may expire at an inconvenient moment and, like any technological device, can malfunction or break. If you do take a GPS with you into the field you should still ensure that you have a navigational compass and a map with you (if working overseas your compass should be calibrated for the country in which you are working) and that you know how to use it – for example, how to take a bearing and walk along it.

WHAT IF SOMETHING DOES GO WRONG?

If you take the precautions outlined in this chapter and carry out a full and careful assessment of risks before undertaking fieldwork, the chances of any problems should be minimized. However, risks can never be eliminated altogether. Vehicle accidents and breakdowns happen, people trip over and the weather changes. If you do find yourself in an unfortunate situation try to keep calm, assess your options and get help if necessary. Keeping calm is always easier said than done, but it is important that you try not to panic and not to induce panic in other people. Take some time to think about the situation, use your common sense and never place yourself or others in danger. If an incident or accident does happen that jeopardizes health and safety there are two key things that you, or members of your party, need to do. First, understand what has happened and, secondly, assess the severity of the situation.

When things start to go wrong try to understand what has happened as fully and quickly as possible. If you are lost on a hillside do you know where you took a wrong turning? If so, can you retrace your steps and get back on the right path? If your vehicle has broken down, do you know why (for example, a flat tyre)? Can you mend it safely? If someone is injured, do you know how the accident occurred? Was there a rockfall? Did they slide down a steep slope? This will help you to understand not only what their injuries might be but also whether or not the danger is still present. For example, an unstable cliff face may collapse as a series of multiple rockfalls. Look out for any dangers to yourself or to the casualty and never put yourself at risk. If necessary, can you treat the casualty using your first-aid kit?

If you cannot resolve your predicament then you need to assess the situation as accurately as possible before you seek help. If you are dealing with a casualty or casualties you cannot treat, try to determine what their injuries are. The emergency services will need to know how many casualties there are, whether they are conscious or unconscious and whether or not they are breathing and/or have a pulse. If the person has a known condition (such as asthma or a heart condition), and if you know how the incident occurred (for example, immersion in water or a traffic collision), inform the emergency services of this.

You also need to know where the incident has occurred. Take a note of your grid reference and landmarks if you are in a remote area, note any road numbers or junction details and have an idea of how far away from the nearest habitation you are. If you have to leave the scene to get help be sure to write this information down before you set off. Emergency services can trace your telephone call to any call box or motorway telephone but if you have to leave the site or are using a mobile phone to make the call you should be extra vigilant about establishing your exact location. If you are lost in hilly terrain on foot and cannot retrace your steps, stop and consider the safest route off the hill or mountain. Use your compass to set a bearing in that direction. Heading towards buildings, a water course or a road are often good options. If your vehicle has broken down in a remote area the best option is usually to stay with the vehicle until help arrives, especially in extreme weather conditions. Finally, if hazardous conditions such as poor visibility, unstable slopes, flooding, high tides or electricity, etc., have contributed to the accident or may affect the response from emergency services, inform them of this when you call.

When the crisis is over, try to recall and write down what went wrong and why, including times and dates, if possible. You should give this information to the safety officer in your department as it may be useful for establishing procedures to prevent similar future incidences.

Summary

* Fieldwork can be a risky business, but considering the variety of locations in which geographers undertake fieldwork and the amount of time they spend in the field, few incidents and accidents actually take place.
* Many of the precautions that are taken to protect health and safety in the field are common sense, and you would take the

same precautions while shopping at the weekend or going walking in the hills with friends – remember to apply this common sense when undertaking fieldwork in all situations (whether as part of a supervised group or as an unsupervised individual).

- Risk assessment techniques can be used to anticipate likely fieldwork hazards. Identify any precautions that could minimize the risks and ensure that these are fully implemented.

- Occasionally there may be special procedures associated with particular equipment or environments that you need to learn and apply – seek advice from tutors and/or specialist advisory services.

- You have a responsibility to initiate and/or follow appropriate health and safety guidelines and to take all reasonable precautions to ensure the health and safety of yourself and others in the field.

- By following the above advice, you may not only learn new skills and take on new responsibilities but, more importantly, you may also prevent any serious incidents or accidents occurring.

Further reading

There is not much written for students about health and safety in the field but two useful sources are as follows:

- Higgitt and Bullard (1999) detail why and how risk assessments are undertaken for undergraduate dissertations. Using two geography case studies, one human and one physical, their paper illustrates the types of hazards and risks that need to be considered.

- Nash (2000) is the first part of a guide to doing independent overseas fieldwork. It discusses where to go, what to do when you get there and also some of the health, safety and insurance issues that you should consider before setting off.

Note: Full details of the above can be found in the references list below.

Many government websites present comprehensive information about travelling overseas, including country-by-country guides to safety and security, local travel, entry requirements and health concerns. Some also include more general sections within the websites, including travellers' tips, how to get your mobile phone to work overseas and what to do if it all goes wrong. These sites are updated regularly, particularly with regard to political disturbances and natural disasters. The sites for UK, US and Australian citizens are listed below. Other nationals should consult the travel section of their home government website.

- Foreign and Commonwealth Office website (UK): http://www.fco.gov.uk/travel/
- Center for Disease Control and Prevention (health advice) (USA): http://www.cdc.gov/travel/
- US Department of State (travel advice): http://travel.state.gov/travel_warnings.html
- Australian Department for Foreign Affairs and Trade: http://www.dfat.gov.au/travel/

Up-to-date local weather forecasts are usually available in newspapers, by telephone (check local papers for the number) or from websites. For example:

- The Meteorological Office (UK): http://www.met-office.gov.uk/index.html
- National Weather Service (USA): http://www.nws.noaa.gov/
- Bureau of Meteorology (Australia): http://www.bom.gov.au/

If you are planning an expedition and need advice or want to develop particular skills before you leave, useful contacts include the following:

- The Expedition Advisory Centre (EAC) at the Royal Geographical Society (with the Institute of British Geographers) (http://www.rgs.org/category.php?Page=main expeditions) runs workshops and seminars on topics such as four-wheel drive training, people-orientated research techniques, risk assessment and crisis management, and produces publications offering advice on logistics and safety in a range of environments from tropical forests to deserts.
- The Explorers' Club (USA) (http://www.explorers.org/) can provide expedition planning assistance and has a lecture series which occasionally features sessions on field techniques.
- The International Union of Alpine Associations (UIAA) Mountain Medicine Centre provides advice on all aspects of high-altitude travel and safety. Information can be accessed via the British Mountaineering Council website: http://www.thebmc.co.uk/world/mm/mm0.htm

References

Higgitt, D. and Bullard, J.E. (1999) 'Assessing fieldwork risk for undergraduate projects', *Journal of Geography in Higher Education*, 23: 441–9.

Nash, D.J. (2000) 'Doing independent overseas fieldwork. 1. Practicalities and pitfalls', *Journal of Geography in Higher Education*, 24: 139–49.

5 Making Use of Secondary Data

Paul White

Synopsis

Secondary data consist of information that has already been collected for another purpose but which is available for others to use. It is an indispensable source of information for many student projects where resource limitations, such as time and money, often preclude data collection for an extensive area, a large population or comparison between places. The chapter first looks at the nature of secondary data: who collects it, what it consists of and why it is of interest to geographers. It then goes on to discuss how to access secondary data: where to obtain it (at scales ranging from the international to the local) and the use of Internet sources. Finally, it looks at the uses of secondary data: to provide a context for a study, for comparisons and as the prime evidence for analysis.

The chapter is organized into the following sections:

- Introduction: the nature of secondary data
- Accessing secondary data
- Utilizing secondary data
- Conclusion.

INTRODUCTION: THE NATURE OF SECONDARY DATA

We live in a world where things are counted. Arguably this has always been so. Some of the earliest writings, in various parts of the world, consisted of counts of people, goods, transactions and possessions. Conquering forces have always been keen to enumerate the attributes of what it is they have conquered. And wherever political power has needed financial funds, tax systems have produced data on wealth indicators of various sorts.

These exercises in data collection are multiplied many times over in the contemporary world, where it often seems that almost every aspect of human life is counted, and where aspects of these counts are stored in databases of some sort – births, examination performances, traffic flows, financial transactions, housing standards, consumption patterns and so on. Complex societies and economies count things more than do other populations with simpler structures and activities. The management, planning and provisioning of complex societies give rise to data needs, while in certain cases government agencies wish to gain insights into the lives of citizens for reasons of control. These exercises all produce secondary data which, once collected for one purpose, may be used for another. Such data can be either qualitative or quantitative (or a combination of the two) in nature, but the focus here is on quantitative data. (See Chapter 7 of this volume and also Kitchin and Tate, 2000, on qualitative secondary information.)

A number of general observations can be made about quantitative secondary data:

1 *The principal source of such data is normally government.* Non-governmental sources of secondary data certainly exist but tend to have only local spatial coverage or be very specific in their topics. Commercial data are often not made available to other researchers, for reasons of commercial confidentiality, or can only be acquired at a cost. Government-generated information covering multiple topics, however, is often collected with varied purposes in mind. Population censuses, one of the major government data sources of interest to geographers, normally provide a variety of data, reflecting a belief that it is a 'good thing' for government to have evidence available to support its policy-making, implementation and monitoring functions. A trend in many (but not all) parts of the world is the increasing availability of government data. Where at one time only limited parts of datasets were accessible to the general public (or the academic researcher), or where the costs of access were considerable (in terms of data purchase or the need to travel to inspect original data manuscripts), large amounts of data are now available to all via the Internet. Today the desk-bound researcher can call up the figures for the resident population of Mexican municipalities or the number of benefit claimants in the districts of an English city with a few clicks of a mouse.

2 In making use of secondary data, the *researcher needs to be aware that the information has been collected by someone else, for another purpose.* The questions asked might not have been exactly the ones that the researcher would have used, and the release of data to others may have involved the simplification or recategorization of information in ways that are potentially unhelpful.

3 Nevertheless, because of the massive financial investment that governments make in collecting data they are keen to maximize the data's utility. Census questions, data-recording systems and coding schedules are usually subject to intensive pretesting (Rose and O'Reilly, 1998). *Secondary data (particularly from government) tend to have been collected in ways that are much more robust than those available to the individual researcher.*

4 *Secondary data may already have been manipulated for particular, possibly political, purposes; hence, secondary data may not be entirely trustworthy.* Local census officers, for example, are sometimes believed to have inflated population sizes within their jurisdictions for reasons connected with local prestige or power. A strong case can be made that all government data inquiries reflect cultural perceptions which influence both what is asked and how the questions are answered (Hoggart et al., 2002).

5 *Available secondary data tend to be strongly spatially referenced.* Government pertains to spatially bounded units at a hierarchy of scales. Secondary data sources thus often emphasize the variable geography of society. However, the spatial 'containers' for some datasets may not be those the researcher would choose. Immigration data, to take one example, are generally only available at national levels so that a research project on the role of international migration in a single city would not find relevant data available. Some datasets (e.g. the results of extensive questionnaire surveys) are available at regional or other subnational levels, but it is generally only population census data that are available at the most detailed local level.

These general points indicate a series of advantages and disadvantages to the use of secondary data. Although the negative points must be borne in mind, it is possible to overstress them. For many geographical investigations, secondary data are indispensable, since the project could not proceed without them. This is particularly true for student projects where resources issues (of time, money and labour) are likely to preclude data collection for any study of an extensive area, a large population or involving comparisons between circumstances in two or more places or groups of people.

ACCESSING SECONDARY DATA

Unlike many of the issues dealt with in this book, questions of access to secondary data tend to be specific to the country in which a student

TABLE 5.1 Selected sources of official statistical data
A National statistical offices (in each case the web address starts http://www.)

Country	Web address
Argentina	indec.mecon.gov.ar
Australia	abs.gov.au
Brazil	ibge.gov.br
Canada	statcan.ca
China	stats.gov.cn/english/index.htm
Colombia	dane.gov.co
France	insee.fr
Germany	destatis.de
India	censusindia.net
Ireland	cso.ie
Italy	istat.it
Japan	stat.go.jp/english/index.htm
Mexico	inegi.gob.mx
Morocco	statistic.gov.ma
Netherlands	cbs.nl
New Zealand	stats.govt.nz
Poland	stat.gov.pl/english/
Portugal	ine.pt
South Africa	statssa.gov.za/default3.asp
Spain	ine.es
Sweden	scb.se
UK	statistics.gov.uk
USA	fedstats.gov

B International organizations

Organization	Web address
Europa (the European Union's statistics service)	http://europa.eu.int/comm/eurostat
Food and Agricultural Organization (UN)	http://apps.fao.org
United Nations Statistics Division	http://unstats.un.org/unsd
UNICEF	http://www.unicef.org/statis
World Health Organization (UN)	http://www3.who.int/whosis

is studying. Nevertheless, an attempt is made in what follows to raise a number of general issues concerning the availability of data.

For many student projects, at whatever scale, an obvious starting place is the website of the national statistical office of the country for which information is required. In many countries all available government data can be accessed via this office, even information gathered by disparate government departments and agencies. Table 5.1 provides a list of the web addresses of selected national statistical offices, as well as certain international agencies which either themselves gather or collate from other sources comparative information about major issues.

Amongst the international organizations, UNICEF is particularly useful for comparative development data on countries around the world. An additional source of useful statistical information, but which requires individual country entries to be compared, is the 'world factbook' produced by the American Central Intelligence Agency (http://www.cia.gov/cia/publications/factbook).

Students in England and Wales wishing to make use of detailed local data will find a mass of information available, although not always easy to access immediately. The results of the 1991 population census are all in computer databases, but a registration procedure is required, including agreements about the use to which the material will be put, before free access is granted (any student wishing to use the 1991 data should consult the following website for details: http://census.ac.uk/cdu/). Dale and Marsh (1993) provide a detailed discussion of what is available. The plans for the dissemination of the 2001 census of England and Wales envisage much wider open diffusion that will add census results to an already available set of information on neighbourhoods in Britain. Information on the availability of 2001 data is being developed on a dedicated census website (http://www.census.ac.uk/), but registration procedures will be needed for access to certain types of data.

The Office for National Statistics has collated a wealth of information on individual administrative wards from a range of government sources (http://www.statistics.gov.uk/neighbourhood.asp). The data available for all wards in the UK (including Northern Ireland) in mid-2002 are shown in Box 5.1. Clearly these reflect the administrative interests of government (for example in the amount of data available on claimants), as well as the ways in which statistical information is used to inform policy – particularly through the indices of deprivation used to identify areas to be targeted for particular initiatives. The use and calculation of such indices will be discussed later in this chapter.

Political devolution in the UK has always existed in terms of certain government data services, and students researching in Scotland and Northern Ireland have other statistical services available to them. For Scotland the site is http://www.gro-scotland.gov.uk (which includes an online 'data library'), while for Northern Ireland the site is http://www.nisra.gov.uk (containing impressive data maps as well as other sources).

The UK student might also consider the information available from UpMyStreet.com (http://www.upmystreet.com/), managers of a

**Box 5.1 Data available from the UK 'Neighbourhood Statistics'
(http://www.statistics.gov.uk/neighbourhood.asp)**

- Numbers of all forms of benefit claimants (e.g. family credit, income support, attendance allowance, disability, jobseekers, incapacity benefit, etc.).
- Numbers of VAT-registered enterprises, by sector and numbers of employees.
- Numbers in employment.
- Resident populations, with limited age structures.
- Numbers of births and deaths.
- Electorate.
- Indices of deprivation, absolute values on a number of dimensions, and rank within the 8414 wards in the UK.

major commercial database. This provides certain of the government data referred to earlier along with other local information such as school performance, supplemented by commercial advertising targeted to individual postcodes. Two other aspects of the database are of considerable interest. The first is information on house prices from recent sales within postal zones, extracted from Land Registry returns. The second is a classification of all neighbourhoods within one of 54 categories according to the ACORN system ('A Classification Of Residential Neighbourhoods'), accomplished on the basis of census data supplemented by the results of market research and lifestyle surveys about consumption patterns, likes and dislikes, and behaviour patterns. Such classifications fall under the broad heading of 'geo-demographics', discussed later in this chapter.

Census data, and other forms of spatially referenced information (including the ACORN classifications), can give rise to certain problems of interpretation for the unwary user. The most common error is that of the 'ecological fallacy'. This occurs where the aggregate results for an area are (mis)used by inferring that they apply to all individuals within that area. If in a neighbourhood the census shows that there is an above-average proportion of the elderly and an above-average proportion of houses without central heating, the ecological fallacy consists of inferring that the elderly live in unheated houses. It could be that the old people are wealthy and live in relative luxury surrounded by poorer younger families.

One way of avoiding the ecological fallacy is to use data on individuals rather than on aggregates. Instead of inferring individual characteristics from groups, the researcher can then examine actual people.

Inevitably the circumstances under which this is permitted are restrictive, but the 1991 and 2001 UK censuses produced an individual level sample (the 'Sample of Anonymized Records' or SARs) including 1% of all households in the UK and 2% of all individuals. Such samples are often known as 'microdata' and they are being increasingly made available in a number of countries. Canada has produced them in every census since 1971, while in the USA retrospective sample microdata have been generated from old census returns back to 1850 (see http://www.ipums.umn.edu/). The main US site allows students to download, by file transfer, information on sample populations in any state (see http://fisher.lib.virginia.edu/pums/). In any use of microdata, the user cannot expect a fine level of spatial detail. Because these are samples and contain potentially sensitive information on individuals, the spatial framework for their availability is coarse grained.

As indicated earlier, the ACORN system utilizes survey data from samples as well as 'population data' covering all cases. One further type of secondary data that students should be aware of is derived from sample questionnaire and other surveys. Some of these are carried out by government but many are carried out by private organizations (sometimes for government and sometimes for other purchasers of the data such as trade unions, lobby groups or public corporations). Certain surveys are carried out by academic researchers as part of major projects. Surveys often have special characteristics (certain of which also apply to the microdata series already mentioned).

First, they are samples and are thus subject to sampling errors. Secondly, surveys commonly inquire into opinions as well as facts. They thus bring in a whole range of attitudinal issues as well as the behavioural elements more commonly associated with the types of data so far discussed. Thirdly, surveys, especially when carried out by private polling organizations, sometimes deal with topics that are not of immediate interest to government, or which government regards as sensitive and to some extent untouchable. Finally, because sample sizes are often relatively small, surveys are not usually spatially referenced to detailed areas. Thus national or regional results may be published but rarely at a more local scale such as the neighbourhood – unless the survey was itself local in its execution (in which case it applies to only one place and should not be generalized).

For UK researchers, the UK Data Archive (housed at the University of Essex) provides online access (after registration) to a variety of survey data, including such major series as the British Social Attitudes Survey (Park et al., 2001), the Quarterly Labour Force Survey and the Scottish Household Survey (see http://www.data-archive.ac.uk/). In addition the Data Archive also acts as a depository for qualitative data such as interview transcripts. At an international scale, the University of Amsterdam maintains a useful website listing social science data-bases around the world (http://www.pscw.uva.nl/sociosite/databases.html).

The European Union operates a polling organization known as Eurobarometer which conducts surveys across the Union. These prin-cipally concern opinions on EU policies and integration, but other themes such as racism and xenophobia have also been investigated in recent years (Eurobarometer's home page is at http://europa.eu.int/comm/public_opinion). Commercial polling organizations (such as MORI, Gallup and NOP) conduct many interesting surveys each year, but detailed data are rarely made available without charge.

One distinctive way in which large-scale government surveys may be of particular use in a student project lies in the rigour with which such surveys are carried out. Questionnaires undertaken by students sometimes become problematic in the field because the wording of questions is not interpreted consistently by respondents, or because the questioner has failed to recognize all the possible answers in his or her coding schedule. As with the census, large-scale publicly funded surveys are rigorously pretested. The possibility is therefore raised in the next section of this chapter of students reusing (with suitable acknowledgements) questions that have formed part of such surveys. Two major 'question banks' that can be plundered in this way relate to British government surveys (http://qb.soc.surrey.ac.uk/docs/home47.htm) and to the Eurobarometer (http://www.nsd.uib.no/atle/eurob/eurosok.cfm).

It should be clear from this discussion, and from a few minutes spent investigating some of the web addresses given, that there is a huge variety of social scientific data available to a researcher without leaving the desk. Much of this information is free to access (albeit with registration procedures being needed for certain major British data sources). The issue to which this chapter now turns attention is the uses to which secondary data might be put.

UTILIZING SECONDARY DATA

Given the breadth of coverage and detailed nature of much available secondary data, it will be no surprise that the uses to which they can be put are extremely varied. Such data can be used to inform projects in which the primary focus of the research is to collect evidence by other means, but they can also be regarded as the primary form of information and then subjected to analysis of varying degrees of sophistication – from description to causal modelling. What follows here is organized under three subheadings, involving data as context, comparison and as the basis for analysis.

Data as context

The simplest level at which secondary data can be used is to provide a description of the characteristics of the place, space or group that is the focus of a research investigation being carried out by other means. Too many student projects discuss a particular situation in detail without ever contextualizing this within a wider world.

Secondary data can provide justifications for the choice of topic and location. Consider, for example, the following two projects:

- A study of reasons for the choice of a particular destination by retirement migrants.
- A study of the attitudes of young women to future family building.

In both cases a student could produce a self-contained piece of work in which all the evidence (for example, from questionnaires, in-depth interviews or focus group discussions) is collected by the student. However, much more rounded studies would result if they were con-textualized via (respectively):

- the use of census data to demonstrate the local importance of retirement migration, the origins of the migrants and the numbers of local elderly; and
- the use of population registration data and time series derived from them to demonstrate temporal changes in fertility and its charac-teristics in terms of age of mother, child parities, marital status and so on.

It should thus be clear that secondary data can be used to demonstrate that there is a research issue that merits being developed into a pro-ject, as well as providing a context for the work carried out. Secondary data are not here being subjected to any further analysis, but quoted as they are (albeit with some recognition of their utility and limitations). Such data may also be used to support a project that is inherently qualitative in nature but for which a justification in terms of import-ance is derived from statistical information.

The availability of such data for factual use should actually point to a recognition that, in any assigned work (and not just research projects), secondary data can support an argument. At the start of the twenty-first century it is no longer permissible for student essays to quote outdated sets of figures derived from ten-year-old textbooks in support of their contentions. Ideas derived from published texts should be tested against more recent data, wherever available, to examine their continuing plausibility.

Data as comparison

One of the most significant ways in which all forms of science (social science included) progress is through replication. This involves previous studies being repeated in different circumstances – for example, testing the same ideas but at a different place or with a repeat study in the original location some years after the initial investigation. Many student projects are implicitly based on a comparison of a local situation with a much wider aspect of the phenomenon. In all these endeavours there is a 'benchmark' against which deviation is being evaluated, measured and (possibly) interpreted or explained. This 'benchmark' is best defined (indeed, in many cases, it may only be defined) through the use of secondary data. The following projects are examples of this:

- A study of female employment characteristics in a local labour market (carried out through house-to-house questionnaires), aiming to consider whether or not they follow national patterns.
- A study of the spatial awareness of school children at a particular age (carried out through exercises undertaken in a school), aiming to consider whether there are differences in such awareness compared with a study carried out ten years earlier.

In the first case national data are needed. Problems may, of course, arise in that the available national data may not relate to exactly the same issues as the student is interested in but, given an awareness from the outset that such data will be used, the local survey can be adjusted to ask comparable questions.

The second case above is rather different because it involves treating the primary evidence produced by an earlier study as secondary evidence (the 'benchmark') for the new study. Questions used in the earlier study can be replicated in the new, and results can be compared, possibly involving statistical testing to evaluate the significance of observed differences.

This second example may involve wider possibilities of the replication of method. As indicated earlier, large-scale surveys have created considerable 'question banks' that are open for perusal and 'mining' for questions to add to student projects. The results obtained by the student can then be compared with the findings, almost certainly at a much larger scale, from the published surveys. To some students such comparisons might seem like 'cheating' – reusing materials that already exist rather than deriving new evidence. But it is partly to enable such comparisons that researchers in the past, including government statistical bodies, have made secondary data available.

Data as the basis for analysis

The remainder of this chapter is devoted to a third major use of secondary data in which such data provide the essential evidence without which the project could not go ahead. Consider the following three topics:

- Testing the idea that school league-table results are conditional upon the nature of the school's catchment area.
- Comparing the levels of residential segregation of different ethnic groups in a single city.
- Considering what social or economic factors correlate with mortality rates.

Each of these projects requires data that cannot be generated using the resources of an individual student researcher. In two of the projects (the first and third) there is an expectation that government data (school performance in one case, population mortality in the other) are to be matched with further data in an explanatory framework of some kind. Such uses of secondary data are extremely common, both in public sector research and in the academic community. Indeed, the three projects suggested above have been derived in part from existing studies (Herbert and Thomas, 1998; Peach, 1996; Townsend et al., 1987).

Data derived from secondary sources can often be entered into statistical manipulations in a straightforward manner. For example, population distribution by age groups can be compared between two areas using non-parametric statistics such as chi-square. However, there are also a number of specific techniques that have been developed to utilize secondary data in deeper ways to enhance their analytical value. We shall here briefly highlight four such important methods.

1 COMPOSITE INDICES

There is now a considerable history of geographers constructing overall 'scoring systems' for places – for example, in terms of standards of living (Knox, 1975). All the methods in use today can be traced back to certain common principles, one of which is the use of secondary data. Two of the best known of the resultant indices are the 'Townsend Index' for the measurement of deprivation (particularly in health research) and the Human Development Index for the international comparison of quality of life. Table 5.2 shows the variables considered in both cases.

The Townsend Index was developed for use in Great Britain and reflects certain cultural specificities of its origin (for example, its

TABLE 5.2   The characteristics of two composite indices

The Townsend Index	Human Development Index
Current unemployment	Life expectancy
Short-term wealth (car ownership)	Adult literacy
Long-term wealth (house ownership)	GNP per capita
Housing overcrowding	
Measured as standard deviations from the average on each variable, then summed	Measured as differences from the maximum on each variable, then summed

emphasis on house ownership as an index of long-term wealth, which would not be appropriate in societies such as Switzerland or Germany where long-term wealth is held in the form of financial investments). Data are collected on a set of areas (for example, wards within a city or local authority areas within a region). For each of the four variables in the index the mean and standard deviation values of the area datasets are calculated. The variables are defined as follows:

- The percentage of the working population that is currently unemployed.
- The percentage of households without a car.
- The percentage of households that does not own its own home.
- The percentage of households living in overcrowded conditions (sometimes taken as more than one person per room).

For each of these variables a 'high' value represents poor conditions. Each area is inspected to see where its actual value on each variable lies in relation to the means and standard deviations. An area that is much 'better' than the average on all variables will record uniformly low values. The composite index is derived from z scores:

$$z_a = (x_a - \mu) / \sigma$$

where x_a is the value of the given variable for area a; μ is the mean value for the given variable across all areas; and σ is the standard deviation for the given variable across all areas.

For each area the final Townsend Index score is obtained by summing the four z scores. Thus areas of low deprivation have high negative scores. Areas of high deprivation have high positive scores.

The Townsend Index represents a measure where the composite has been argued to encapsulate more than each of the individual variables. A number of variants to it have been made but with the basic structure still visible (Bradford et al., 1995). Townsend Index scores have been shown to be closely correlated with the distribution of various population subgroups in British cities, such as lone-parent families, the long-term sick and the unskilled (White, 2000). The provisos made earlier in this chapter about the ecological fallacy need to be borne in mind in interpreting such analyses.

As with the Townsend Index, so the Human Development Index (HDI) has been refined – in particular because of problems with international measures of gross domestic product. These are now generally given as 'purchasing power parities', precisely in order to facilitate international comparison. To some extent the HDI as currently used is an 'index of indices' with each of the three separate components resulting from analysis of raw data.

The methods involved in the creation of indices determine that the results should be regarded as relative to one another. The addition of an extra case necessitates the recalculation of the whole index. Similarly, it is not possible to compare the scores for the Townsend Index for neighbourhoods in different cities where these scores have been separately derived: one can say that two neighbourhoods are both affluent, but only in relation to the rest of the neighbourhoods in their respective cities – not in relation to each other.

2 MEASURES OF SPATIAL DISTRIBUTIONS

Social geographers, particularly those working on class and ethnicity, have developed a suite of methods for analysing the spatial distributions of population subgroups in comparison to the population as a whole. A series of single-number measures have been constructed to summarize such distributions and to show differences and similarities between them: these measures include the Indices of Dissimilarity and of Segregation. Once again, analysis leading to their calculation depends on secondary data, almost invariably from census results.

The basic premise is that, in a society characterized by no spatial separation between different subgroups, all subgroups will be similarly distributed. The measurement therefore consists in considering areas where this is not true. The two principal sets of output measures are as follows:

- *Indices of Dissimilarity*: these measure the differences between population subgroups. Values range from 0 to 100. At 0, two population distributions are identical. At 100, the two distributions are completely dissimilar.
- *Indices of Segregation*: these measure the extent to which a particular population subgroup is segregated within the total population. Again the values run from 0 to 100. At 0, there is no

segregation present. At 100, the subgroup is totally segregated and inhabits its own exclusive space.

As a reading of Peach (1996) will show, the number of indices of dissimilarity that can be calculated in any set of subgroups grows rapidly as the number of subgroups is increased. Thus for 6 subgroups (for example, the six socioeconomic groups commonly utilized in the UK and known as A, B, C1, C2, D, E) there are 15 Indices of Dissimilarity. For ten subgroups (as with the ethnicity groups in the 1991 census) there are 45. In contrast, there is only one Index of Segregation for each subgroup.

The calculation of dissimilarity and segregation indices (as with the composite indices referred to above) is informed by social theory, but only to the extent of justifying the method used. Whereas in the Townsend Index the individual variables are brought in because of their intuitive relationship with deprivation, in the analysis of spatial distributions there is an assumption that social distance and spatial distance are related. Other types of analyses of spatial distributions may not even start with these assumptions. This particularly applies to GIS-based analysis of spatially referenced secondary data. GIS analysis often operates from an inductive and empirical standpoint – looking for patterns and possible relationships rather than testing their theoretical existence (Martin, 1991).

3 CLASSIFICATIONS

Reference was made earlier to the ACORN classification system of UK postcode sectors. Such classifications are part of what is sometimes known as 'geodemographics'. A major aspect of company-based market research revolves around the generation and use of such tools. In terms of a student project the applicability is likely to be limited, but a number of simple suggestions might be made along the following lines:

* A student wants to interview residents of several different types of city districts about the use they make of the city centre.
* A student (perhaps working on a project in collaboration with a local council) wants to identify the zones within an extensive area (for example, a county) that might be targeted in a health campaign.

In both these cases some form of geodemographic classification might be a plausible way forward.

Classification methods are commonly based on multivariate statistical methodologies, such as the use of principal components or cluster analysis. Standard statistical packages such as SPSS can readily be utilized (see Chapter 24). The issue here relates to the secondary data on which classifications might be based. Any classification depends fundamentally on the nature of the variables that are put into the analysis. To take the case of the first project identified above, if the data that are put into the classification of city districts include a majority of variables dealing with housing characteristics, it is likely that any resultant classification will emphasize housing as an important discriminator. Classification methods are not 'objective' since they depend on the operator's choices of data. Reflection is therefore needed as to what might be the relevant lines of difference for the purposes of a particular project before the choices of data for classification are made.

4 CORRELATION AND REGRESSION

The fourth and final major set of analytical techniques a student might want to perform on secondary data involve the consideration of relationships between datasets and the possibility of producing statistical explanations of causality. Such explanations may not necessarily be backed by social theory: it is often easier to show that there is an empirical relationship between two datasets or variables than to provide convincing theoretical arguments as to why these relationships come about. Nevertheless, there are many areas of geographical research that depend heavily on the analysis of such relationships: the field of electoral geography, for example, would be very much reduced if research of this kind was excluded.

Many projects seek ultimately to suggest reasons for the appearance of a particular pattern (spatial or otherwise) in a single variable. To the example quoted at the start of this section (variations in school league table performance) can be added many others – such as spatial variations in the electoral success of a particular party or (at an international scale) in medal performance in the Olympic Games. In each of these the generalized form of the project is to calibrate a regression model in the format:

| Variations in phenomenon to be explained | *result from* | the influences (singly or collectively) of a set of other measurable variables |

In statistical terms this is the familiar relationship whereby a series of 'independent variables' condition the level of a related 'dependent variable'.

Certain specific issues arise in the use of secondary data for such analyses. First, there may be no other way of examining the hypothesized relationships other than through the use of secondary data. No researcher can expect to collect school performance measures as part of a primary data-gathering exercise, nor is there any need to do so. Similarly, there are issues of scale in most of the problems just outlined: the datasets needed to operationalize the project generally cover a lot of spatial ground.

A second, and often problematic, issue, however, is the temporal coincidence (or lack of it) of particular datasets. Consider the use of population census data in certain of the projects just outlined. In the UK (and in many other countries) the census is a decennial exercise. Elections take place at more frequent intervals, while league tables of school or hospital performance are produced annually. In a study in which a student aims to explain voting for the Scottish National Party by reference to the social and economic characteristics of constituencies there is an immediate dilemma. If the bulk of the constituency data are drawn from the 1991 census, should the study be limited to the 1992 general election (the closest one to the census date), or can the study use more recent elections even though the temporal coincidence of the dependent variable (voting patterns) and independent (census-based) variables is becoming ever more remote? The purist argument would privilege the former solution while the pragmatic view would be to mix datasets from different years on the basis that there is no plausible alternative. Most census users in government, commerce and academia accept that the data they use are often somewhat elderly. Some attempts are made to update them by various estimation and correction techniques, but only for major variables (such as the population sizes of administrative areas).

A third issue concerning the use of secondary data in analysis may be the lack of spatial coincidence of datasets. School catchments, for example, are drawn for admission purposes and do not necessarily coincide with postcode sectors or administrative wards. Certain spatial units in one dataset may need amalgamation to fit the spatial framework of another before correlation and regression analysis can take place. This is often known as the 'modifiable areal unit problem' (Morphet, 1993). The relevant procedures can best be accomplished via the use of a GIS, although 'traditional' methods of 'eyeballing' areas and allocating them manually to coincide with others can still be effective, if time-consuming.

Fourthly, even with the abundant nature of secondary data today, there are often still gaps in availability (Dorling and Simpson, 1999). The researcher may hypothesize that household income is an important explanatory factor for the problem at hand, but data on this are not available in most countries at anything less than regional level. Income questions are regarded as too sensitive (and too liable to result in false answers) to be posed in most government surveys, although they are used (generally in the form of predefined bands) in commercial questionnaires. Other issues that are seen as sensitive (such as sexual orientation or certain disease histories) are similarly rarely raised. In some cases researchers can identify 'surrogate' variables to be used as replacements for those that are unobtainable. Thus, for example, in the Townsend Index discussed earlier, car ownership and house ownership are both used as indicators of wealth in circumstances where data on actual household income and capital, were they available, would be better incorporated into the technique instead.

Finally, the interpreter of the results of correlation and regression analysis on secondary data has to be alert to the dangers of inferring cause from correlation.

CONCLUSION

There is room for a consideration of secondary data in almost any student project on contemporary aspects of human geography. The uses of such data vary from the contextual at one end of the scale to the fundamental basis for analysis at the other. Consideration of what is already known about a topic should play a role alongside the literature review that identifies what is already thought about it. Early examination of available data can help in the formulation of a research project, help to identify gaps in understanding, provide a justification for the choice of areas, groups or case studies for research, and can demonstrate the importance of the field in which a project is set. Such uses of secondary data do not necessarily mean that the resulting project need follow a positivist-empirical methodology. The data may be seen as providing the 'extensive' basis and context for a more 'intensive' investigation that takes a very different approach.

Public opinion, political discourses, policy discussions and commercial actions are all today increasingly based on a reading (or sometimes a misreading) of secondary data. Information is gathered for multitudinous purposes and used by many different sections of society. In contributing to the well-being of society, geographers need to be able to investigate issues at scales for which only secondary data provide plausible sources of evidence; they need to be able to present carefully supported conclusions to the policy community; and they need to be able to advance aspects of their own discipline through the understanding of variability at a variety of scales. For these endeavours secondary data analysis is an essential tool for progress.

Summary

* Secondary data are primarily, but not exclusively, collected and made available by governments.

- Researchers have no control over what or how information is collected.
- Many secondary datasets contain information that is strongly spatially referenced.
- Increasing amounts of secondary data are becoming publicly available, particularly via the Internet.
- Secondary data can be used to provide a context for a wide range of geographical studies, they can be used in comparisons and they can provide the basis for analysis.
- Problems over the inference of cause from outcome must be borne in mind, as in the use of many other geographical methods.

Further reading

- Hakim (1982) is a wide-ranging classic text dealing with a number of different ways in which secondary data can be used, including the reanalysis of statistical data, the use of existing questionnaire results and the use and abuse of official data.

- Peach (1981) provides a good illustration of the debates about causality that arise from the discerning of patterns or outcomes in secondary data and the need to theorize process as a supplementary stage in secondary analysis.

- Townsend et al. (1987). Townsend's name is associated with the deprivation index he suggested. This study demonstrates the richness of the social geographical argument that can be developed from the analysis of official data, in this case in the sphere of health in the UK.

- Walford (2002) provides a recent compendium of secondary source locations, as well as further consideration of the issues involved in secondary analysis.

Note: Full details of the above can be found in the references list below.

References

Bradford, M.G. et al. (1995) 'Constructing an urban deprivation index: a way of meeting the need for flexibility', *Environment and Planning A*, 27: 519–34.

Dale, A. and Marsh, C. (1993) *The 1991 Census User's Guide*. London: HMSO.

Dorling, D.F.L. and Simpson, S. (1999) 'Introduction to statistics in society', in D.F.L. Dorling and S. Simpson (eds) *Statistics in Society: The Arithmetic of Politics*. London: Arnold, pp. 1–5.

Hakim, C. (1982) *Secondary Analysis in Social Research*. London: Allen & Unwin.

Herbert, D.T. and Thomas, C.J. (1998) 'School performance league tables and social geography', *Applied Geography*, 18: 199–210.

Hoggart, K., Lees, L. and Davies, A. (2002) *Researching Human Geography*. London: Arnold.

Kitchin, R. and Tate, N.J. (2000) *Conducting Research into Human Geography: Theory, Methodology and Practice*. Harlow: Prentice Hall.

Knox, P.L. (1975) *Social Well-Being: A Spatial Perspective*. Oxford: Oxford University Press.

Martin, D. (1991) *Geographic Information Systems and their Socioeconomic Applications*. London: Routledge.

Morphet, C. (1993) 'The mapping of small-area census data – a consideration of the effects of enumeration district boundaries', *Environment and Planning A*, 25: 1267–77.

Park, C. et al. (2001) *British Social Attitudes: Public Policy, Social Ties: The 18th Report*. London: Sage.

Peach, C. (1981) 'Conflicting interpretations of segregation', in P. Jackson and S. Smith (eds) *Social Interaction and Ethnic Segregation*. London: Academic Press, pp. 19–33.

Peach, C. (1996) 'Does Britain have ghettos?' *Transactions, Institute of British Geographers*, 21: 216–35.

Rose, D. and O'Reilly, K. (1998) *ESRC Review of Government Social Classifications*. London: Office for National Statistics/Swindon: Economic and Social Research Council.

Townsend, P., Phillimore, P. and Beattie, A. (1987) *Health and Deprivation: Inequality and the North*. London: Croom Helm.

Walford, N. (2002) *Geographical Data: Characteristics and Sources*. Chichester: Wiley.

White, P. (2000) 'Who lives in deprived areas in British cities?' *Geocarrefour (Revue de Géographie de Lyon)*, 75: 107–16.

6 Conducting Questionnaire Surveys

Sara L. McLafferty

Synopsis

Questionnaire survey research is a research method for gathering information about the characteristics, behaviours and/or attitudes of a population by administering a standardized set of questions, or questionnaire, to a sample of individuals. In geography, questionnaire surveys have been used to explore people's perceptions, attitudes, experiences, behaviours and spatial interactions in diverse place contexts. This chapter explains the basics of why and how to carry out survey research.

The chapter is organized into the following sections:

- Introduction
- Questionnaire design
- Strategies for conducting questionnaire surveys
- Sampling
- Conclusion.

INTRODUCTION

Survey research has been an important tool in geography for several decades. The goal of survey research is to acquire information about the characteristics, behaviours and attitudes of a population by administering a standardized questionnaire, or survey, to a sample of individuals. Surveys have been used to address a wide range of geographical issues, including perceptions of risk from natural hazards; social networks and coping behaviours among people with HIV/AIDS; environmental attitudes; travel patterns and behaviours; mental maps; power relations in industrial firms; gender differences in household responsibilities; and access to employment. In geography, questionnaire surveys were first used in the field of behavioural geography to examine people's environmental perceptions, travel behaviour and consumer choices (Rushton, 1969; Gould and White, 1974). Survey research methods quickly spread to other branches of human geography, and today they are an essential component of the human geographer's toolkit.

Questionnaire survey research is just one method for collecting information about people or institutions. When does it make sense to conduct a questionnaire, rather than relying on secondary data (see Chapter 5) or information collected by observational methods (see Chapter 9)? Survey research is particularly useful for eliciting people's attitudes and opinions about social, political and environmental issues such as neighbourhood quality of life, or environmental problems and risks. This style of research is also valuable for finding out about complex behaviours and social interactions. Finally, survey research is a tool for gathering information about people's lives that is not available from published sources – e.g. data on diet, health and employment characteristics. In developing countries where government data sources are often out of date and of poor quality, questionnaire surveys are a primary means of collecting data on people and their characteristics.

Before embarking on survey research, it is critically important to have a clear understanding of the research problem of interest. What are the objectives of the research? What key questions or issues are to be addressed? What people or institutions make up the target population? What are the geographical area and time period of interest? These issues underpin how the survey is designed and administered. Surveys can be expensive and time-consuming to conduct, so the quality and type of information gathered are all important.

Although each survey deals with a unique topic, in a unique population, the process of conducting survey research involves a common set of issues. The first step is *survey design*. Researchers must develop questions and create a survey instrument that both achieves the goals of the research and is clear and easy to understand for respondents. Secondly, we need to decide how the survey will be administered. Postal (or mail-back) questionnaires and telephone interviews are just a few of the many *strategies for conducting surveys*. Thirdly, survey research involves *sampling* – identifying a sample of people to receive and respond to the questionnaire. This chapter provides a brief introduction to each of these issues, drawing upon examples from geographic research.

QUESTIONNAIRE DESIGN

Questionnaires are at the heart of survey research. Each questionnaire is tailor made to fit a research project, including a series of questions that address the topic of interest. Decades of survey research have shown that the design and wording of questions can have significant effects on the answers obtained. There are well established procedures for developing a 'good' questionnaire that includes clear and effective questions (Fowler, 2002).

Good questions are ones that provide useful information about what the researcher is trying to measure. Although this may appear to be simple, straightforward advice, it is often challenging to implement. Questions can range from factual questions that ask people to provide information to opinion questions that assess attitudes and preferences. Writing good questions requires not only thinking about what information we are trying to obtain but also anticipating how the study population will interpret particular questions. Let's examine the following question: 'Are you concerned about environmental degradation in your neighbourhood?' This item raises more questions than it answers. What does a person mean when he or she says that he or she is concerned? What does environmental degradation mean? Do people understand it? How does each respondent define his or her neighbourhood? Questions should be clear and easy to understand for survey respondents, and they should provide useful, consistent information for research purposes.

One of the most important rules in preparing survey questions is: keep it simple. Avoid complex phrases and long words that might confuse respondents. Do not ask two questions in one. The question: 'Did you choose your home because it is close to work and inexpensive?' creates confusion because there is no obvious response if only one characteristic is important. Jargon and specialized technical terms cause problems in survey questions. Terms like 'accessibility' or 'power' or 'GIS' are well known among geographers, but ambiguous and confusing for most respondents. Don't assume that respondents are familiar with geographic concepts! Define terms as clearly as possible and avoid vague, all-encompassing concepts. Asking people about their involvement in community activities, without specifying what kinds of activities and what level of involvement, is unlikely to produce useful responses. It is better to ask a series of questions about specific types of involvement rather than a single vague question. Finally, one should avoid negative words in questions. Words like 'no' and 'not' tend to confuse respondents (Babbie, 2001) (see Box 6.1).

Responses to survey questions are as important as the questions themselves. *Open-ended* questions allow participants to craft their own responses, whereas *fixed-response* questions offer a limited set of responses. Open-ended questions have several advantages. Respondents are not constrained in answering questions. They can express in their own words the fullest possible range of attitudes, preferences and emotions. Respondents' 'true' viewpoints may be better represented.

Box 6.1   Guidelines for designing survey questions

Basic principles:

- Keep it simple.
- Define terms clearly.
- Use the simplest possible wording.

Things to avoid:

- Long, complex questions.
- Two or more questions in one.
- Jargon.
- Biased or emotionally charged terms.
- Negative words like 'not' or 'none'.

Open-ended questions provide qualitative information that can be analysed with qualitative methodologies, as discussed elsewhere in this volume. Increasingly, geographers are using open-ended responses in questionnaire surveys as part of the broader shift towards qualitative methodologies. Relying on a mix of open-ended and fixed-response questions, Gilbert (1998) analysed survival strategies among working, poor women and their use of place-based social networks. The fixed-response questions provided data on the demographic and household characteristics of the women and their social interaction patterns, while the open-ended questions offered detailed insights about women's coping strategies and life circumstances.

Fixed-response questions are commonly used in survey research, and the principles for designing such questions have been in place for decades. There are several advantages to fixed responses. First, the fixed alternatives act as a guide for respondents, making it easier for them to answer questions. Secondly, the responses are easier to analyse and interpret because they fall into a limited set of categories (Fink and Kosecoff, 1998). The downside is that such responses lack the detail, richness and personal viewpoints that can be gained from open-ended questions.

A simple type of fixed-response question is the factual question that asks about, say, age, income, time budgets or activity patterns. Responses may be numerical or involve checklists, categories or yes/no answers. The key in framing these types of questions is to anticipate all possible responses. As in all phases of survey design, it is important to think about the kind of information needed for research as well as characteristics of the study population that might influence their responses. A 'don't know' or 'other' option is generally included to allow for the fullest range of responses. For numerical information (age, income), one must decide between creating categorical responses (i.e. age < 15, 15–25) or recording the actual numerical value. Creating categories involves a loss of information – a shift from interval to ordinal data – but the categorical information may be easier to analyse. Also, for sensitive topics such as age, respondents are more likely to answer if the choice involves a broad category rather than a specific number.

Finding out about attitudes and opinions involves more complex kinds of fixed-response formats. In general, respondents are asked to provide a rating on an ordinal scale that represents a wide range of possible responses. The Likert scale presents a range of responses anchored by two extreme, opposing positions (Robinson, 1998). For example, residents may be asked to rate the quality of the schools in their neighbourhood from 'excellent' to 'satisfactory' to 'poor'. The two extreme positions, 'excellent' and 'poor', serve as anchor points for the scale, and any number of alternative responses can be included in between (Box 6.2). It is best to use an odd number of responses – 3, 5 and 7 are common – so that the middle value represents a neutral opinion. Respondents often want the option of giving a neutral answer when they do not have strong feelings one way or the other. Odd-numbered scales give such an option, whereas even-numbered scales force the response to one side. Another approach is to present the scale as a continuous line connecting the two anchors. Respondents are asked to draw a tick mark on the line at the location representing their opinion, and the distance along the line shows the strength of opinion. This gives maximum flexibility, but respondents are often confused about the process and it is difficult to compare results among respondents. Consequently most researchers work with fixed Likert scales.

Box 6.2   Examples of Likert-type responses

Please rate the quality of schools in your neighbourhood:

Excellent	_____			Poor	(continuous)
Excellent		Satisfactory		Poor	(three-point scale)
Excellent	Good	Satisfactory	Fair	Poor	(five-point scale)

Attitudinal scales can be difficult to evaluate because there is no 'objective' standard for knowing whether or not a response is accurate. However, researchers can take several steps to improve validity. In general, it is better to offer respondents more possible answers than fewer – i.e. a five-point scale provides more information than a 3-point scale. But as the number of categories increases, respondents lose their ability to discriminate among categories and the responses lose meaning. An intermediate number of categories (five or seven are commonly used) works best. Because responses often vary depending on how a question is worded, another good practice is to use multiple questions, with different wording and formats, to measure the same concept. By comparing responses across questions one can check if people give consistent responses. If so, the responses can be averaged or combined statistically to represent the underlying concept or attitude. This strategy was used in a study of health status differences among residents of three contrasting urban neighbourhoods in Glasgow (Sooman and MacIntyre, 1995). To measure health status, the authors asked a series of questions about symptoms and perceptions for a wide range of mental and physical health conditions, and responses were combined statistically to explore variations across neighbourhoods and social class groups.

Questionnaires should also include a clear set of instructions to guide individual responses. For self-administered surveys – those that do not involve an interviewer – the questionnaire has to be self-explanatory. The instructions for respondents must be written in simple, direct language and be as clear and explicit as possible so that the questionnaire can be filled out without assistance.

Fixed-response questions work best in self-administered questionnaires, and the design and layout of the questionnaire are critically important. For questionnaires involving interviewers, the key is to have a clear and consistent set of instructions for interviewers to follow. There are well tested guidelines for designing and formatting interviewer-administered questionnaires (Fowler, 2002).

The final and critically important step in questionnaire construction is pretesting (pilot-testing). In this phase, we test the questionnaire on a small group of people to check the questions, responses, layout and instructions. Are the questions understandable? Does the questionnaire allow all possible responses? Are the instructions clear and easy to follow? Is the questionnaire too long? Do any questions make respondents uncomfortable? Pretesting often reveals flaws in the questionnaire that were not obvious to researchers. The questionnaire is then modified, and it may be pretested again before going to the full sample. Several pretests may be needed to achieve a well designed questionnaire. For interview-based surveys, pretesting has other benefits. It builds interviewing skills and helps interviewers develop confidence and rapport with respondents. In sum, pretesting is an essential step in ensuring a successful questionnaire survey.

STRATEGIES FOR CONDUCTING QUESTIONNAIRE SURVEYS

There are many strategies for conducting questionnaire surveys. Among the traditional methods are telephone surveys, face-to-face interviews and postal surveys. Advances in computer technology have stimulated the growth of Internet and email-based survey research. Survey strategies differ along many dimensions – from practical issues like cost and time, to issues affecting the quality and quantity of information that can be collected. Some survey strategies require the use of interviewers whereas others utilize self-administered questionnaires.

Face-to-face interviews

Face-to-face interviews are one of the most flexible survey strategies. They can accommodate virtually any type of question and question-naire. The interviewer can ask questions in complex sequences, administer long questionnaires, clarify vague responses and, with open-ended questions, probe to reveal hidden meanings. The personal contact between interviewer and respondent often results in more meaningful answers and generates a higher rate of response. Interviews require careful planning. Interviewers need training and prepara-tion to ensure that the process is consistent across interviewers. Thus, face-to-face interviews are generally the most expensive and time-consuming survey strategies. Another drawback is the potential for interviewer-induced bias. The unequal relationship between inter-viewer and respondent, embedded in issues of gender, 'race', ethnicity and power, can influence responses (Kobayashi, 1994).

Telephone interviews

Telephone interviews are widely used in market research and are becoming a more common strategy in social science research. They combine the personal touch of interviews with the more efficient and lower-cost format of the telephone. In many places, firms can be hired to conduct telephone surveys, saving researchers the time and expense of training interviewers and setting up phone banks. Phone surveys, however, are generally limited to short questionnaires with fixed-response questions. Such surveys miss people who do not have tele-phones or who are frequently away from home. Finally, although the interviewer and interviewee are only connected remotely, issues of power and bias can creep into phone surveys.

Postal surveys

Postal (or mail) surveys are self-administered questionnaires distributed in a post-out, post-back format. A stamped, addressed envelope is included for returning the completed survey, and reminder notes may be sent later to encourage people to respond. For interviewees, there is no time pressure to respond; forms can be completed at a convenient time. The main weakness of postal surveys is the low response rate. Typically, less than 30% of questionnaires will be completed and returned, and those who respond may not be representative of the target survey population. People with low levels of education or busy lives are less likely to respond. The unevenness of responses often violates random or stratified sampling plans and makes it difficult to estimate sample sizes. Finally, low response means that more surveys must be sent out, increasing the cost of the survey effort.

Drop and pick-up questionnaires

A related strategy is the drop and pick-up questionnaire. This involves leaving self-administered questionnaires at people's homes and picking the surveys up at a later date. The person dropping off the surveys can give simple instructions and a brief description of the survey effort. The personal contact in dropping off the survey gives response rates close to those for face-to-face interviews, but with much less time and interviewer training. Thus, the method combines the strengths of interview and self-administered strategies. This comes at a cost – the costs are substantially higher than are those for postal or telephone surveys, though still less than those for personal interviews.

Internet surveys

A new approach is the Internet survey, which is similar to a postal survey but conducted via email or the Internet. The questionnaire can have the same format as a standard postal questionnaire or it may be an 'intelligent', computer-assisted questionnaire that checks and directs people's responses (Couper et al., 1998). An important advantage for geography research is that the questionnaire can include detailed colour graphics, such as maps, photographs, video clips and animations. Internet surveys are in their infancy and have not been widely evaluated. Their strengths and limitations depend on how the survey is designed and administered (O'Lear, 1996). Distributing the questionnaire via the Internet or email raises a host of sampling issues. Who are the respondents? Do they represent the target population? Clearly people without access to email and the Internet will be left out of the sample. Beyond that, how will a list of email addresses of potential respondents be generated? What types of people respond and don't respond to Internet surveys? Although many questions remain, Internet surveys represent a significant innovation whose use is expanding rapidly and will continue to expand in the years to come.

Each survey strategy has distinct advantages and disadvantages and the 'best' choice varies from one research project to another. Choosing a survey strategy involves weighing practical considerations, such as time and cost constraints, with research considerations, such as response rate, types of questions and the need (or lack of it) for interviewer skills. Frequently, the research context limits one's choice. Surveys in developing countries often rely on personal interviews (Awanyo, 2001); map perception surveys often use computer-assisted questionnaires and the Internet. Regardless, researchers should note the limitations of the chosen method and attempt to minimize their effects.

SAMPLING

Sampling is a key issue in survey research because who responds to a survey can have a significant impact on the results. The sample is the subset of people to whom the questionnaire will be administered. Typically the sample is selected to represent some larger population of interest – the group of people or institutions that are the subject of the research. Populations can be very broad – e.g. 'all people in the UK' – or they can be quite specific, for example, 'married women with children who work outside the home and live in Chicago'. Populations are bounded in time and space, representing a group of people or institutions in a particular geographical area over a particular time period. Effective sampling requires that this population of interest be clearly defined.

The first step in sampling is to identify the sampling frame – those individuals who have a chance to be included in the sample (Fowler, 2002). The sampling frame may include the entire population or a subset of the population. Sometimes the design of the survey limits the sampling frame. For instance, in a telephone survey drawn from a telephone directory, the sampling frame only includes households that have telephones and whose telephone numbers are listed in the directory. Similarly, an Internet survey excludes people who do not have access to the Internet or who do not use it. The resulting sample will be biased if those excluded from the sampling frame differ significantly from those included.

Sampling also involves decisions about how to choose the sample and sample size. Commonly used sampling procedures include random sampling, in which individuals are selected at random, and systematic sampling, which involves choosing individuals at regular intervals – i.e. every tenth name in a telephone directory (Robinson, 1998). The former ensures that each individual has the same chance of being selected, whereas the latter provides even coverage of the population within the sampling frame. Sometimes the population consists of subgroups that are of particular interest – for example, different neighborhoods in a city or ethnic groups in a population. If these subgroups differ in size, random sampling will result in the smaller subgroups being under-represented in the sample.

Stratified sampling procedures ensure that the sample adequately represents various subgroups. In stratified sampling, we first divide the population into subgroups and then choose samples randomly or systematically from each subgroup. Surveys that explore differences among groups or geographical areas often rely on stratified sampling. A recent study by Fan (2002) utilized stratified sampling to examine differences in labour market experiences among three groups – temporary migrants, permanent migrants and non-migrants – in Guangzhou City in China. The sample consisted of more than 1500 respondents, stratified to represent not only the three migrant groups but also various occupational groups and districts within the city (Fan, 2002). Respondents were chosen randomly within each occupational and geographical group, with adjustments to ensure that the three migrant groups were appropriately represented.

Another important issue is how large a sample should be. Large sample sizes give more precise estimates of population characteristics and they provide more information for addressing the research problem. However, large samples also mean more questionnaires and more time and effort spent in interviewing and analysis. The cost of survey research increases proportionately with sample size. In choosing a sample size, analysts must trade off the benefits of added information and better estimates with the costs of administering and analysing the surveys.

One way to decide on a sample size is to focus on subgroups rather than the population as a whole. The sample must be large enough to provide reasonably accurate estimates for each of the subgroups that are being compared and analysed. Large sample sizes shrink quickly when divided into subgroups. For example, in analysing travel patterns by gender, urban/rural residence and three categories of ethnic origin, there will be 12 $(2 \times 2 \times 3)$ subgroups. An overall sample size of 100 yields just eight responses on average for each subgroup, which is too small to produce reliable subgroup estimates. To avoid this problem, the researcher should first identify the various subgroups and choose an adequate sample size for each. This is easy to do in a stratified sampling design. With other sampling procedures the issue is trickier. Small subgroups will be missed unless the overall sample size is large. Researchers should carefully assess the various subgroups of interest in choosing a sample size.

Sample size decisions also involve thinking about how much precision or confidence is needed in various estimates. Precision always increases with sample size, but the improvements in precision decrease at larger sample sizes. The benefits of larger samples begin to level off at sample sizes of 150–200 (Fowler, 2002).

It is also important to think about how the survey data will be analysed. Typically researchers use statistical procedures such as chi-square and analysis of variance (ANOVA) in analysing survey responses. These procedures require sample sizes of approximately 25 or more. Larger sample sizes are needed for multivariate statistical procedures such as multiple regression analysis and logistic regression. If separate statistical analyses will be performed for different subgroups in the sample, each subgroup must have a sample size that is sufficient for statistical analysis.

Finally, sample size decisions also involve very real budgetary and time constraints and these may well be beyond the researcher's control. In sum, there is no single answer to the sample size decision. The decision involves anticipating what the data will be used for and how they will be analysed, and balancing those considerations with the realities of money and time.

Sampling decisions are important because they can introduce various sources of bias into a research project. Sampling bias arises when the sample size is not large enough accurately to represent the study population or subgroups within it. More importantly, the sampling frame may be biased, as occurs in telephone or Internet surveys. Many survey procedures under-represent disadvantaged populations, including poor and homeless people and ethnic and racial minorities. Special efforts are needed to ensure that these groups are not excluded from the research project. Finally, non-response bias occurs when those who refuse to respond differ significantly from those who do respond. Non-response often correlates with age, social class, education and political beliefs, resulting in a sample that is not representative of the study population. Although non-response bias cannot be eliminated, its effects can be minimized with a good sampling design. Because survey results are often highly dependent on the characteristics of the sample, bias is a crucial issue in sampling and survey design.

CONCLUSION

Conducting questionnaire surveys involves a series of steps, including designing and pretesting the questionnaire; choosing a survey strategy; identifying a sample of potential respondents; and administering the survey. These complex decisions are closely interconnected. The design of the questionnaire affects whether or not face-to-face interviews are needed. For many projects, financial constraints dictate the use of postal or telephone surveys and relatively small sample sizes. Thus, in any survey project there is a continual give and take among various decisions, framed by the goals and constraints of the research endeavour.

Questionnaire surveys have well-known limitations, as discussed at various points in this chapter. For geographical research, poorly worded questions, ambiguous responses and non-response bias are all issues that raise major concerns. Some geographers contend that survey information is of limited value, especially when compared to the rich and detailed information that can be gleaned from depth interviews and participant observation (Winchester, 1999). A more balanced view recognizes the strengths of questionnaire surveys – their ability to gather information from large samples, about large and diverse populations; their ability to incorporate both open and fixed questions and their use of trained interviewers to elicit information; and, finally, in the Internet era, their ability to reach widely dispersed populations with innovative, computer-assisted, graphically based questionnaires. Despite their limitations, surveys remain the most efficient and effective tool for collecting population-based information.

Questionnaire surveys have a long history in geographic research, a history that continues to evolve as the discipline of geography changes. During the 1970s, survey methods facilitated the shift away from statistical analysis of secondary data towards behavioural and environmental perception research. During the 1980s and 1990s, survey methods became less popular as a result of a 'qualitative turn' in human geography. Today, as geographers search for a common ground between quantitative and qualitative methods, questionnaire surveys are playing an important role in innovative 'mixed' methodologies (Sporton, 1999). As these developments unfold, questionnaire surveys will continue to provide a rich array of information about people's lives and well-being in their diverse geographical contexts.

Summary

* Questionnaire surveys are useful for gathering information about people's characteristics, perceptions, attitudes and behaviours.
* Before embarking on survey research, clearly identify the goals and objectives of the research project. Decide what information you are trying to gather via the questionnaire survey.

- Survey research involves three key steps: designing the questionnaire, choosing a survey strategy and choosing the survey respondents (sampling).
- The questionnaire should be designed to acquire useful information about the research problem of interest. Questionnaires can include both open-ended and fixed-response questions. In either case, the questions should be clearly and simply worded and should avoid jargon.
- The types of survey strategies include face-to-face interviews, telephone surveys, postal surveys, drop and pick-up surveys and Internet surveys. Each has distinct advantages and disadvantages, and the choice among them depends on the type of questionnaire, desired response rate, and time and budgetary constraints.
- Sampling involves identifying the group of people to whom the questionnaire will be administered. The sample should be selected to represent well the target population and to minimize non-response bias. The sample size should be large enough to represent various subgroups in the population and to allow effective statistical analysis of results.

Further reading

A great deal has been written about questionnaire surveys in the social sciences. The following are just a few of the sources I have found useful:

- Babbie (1990) is a comprehensive, well written book that covers both the theory and practice of survey research and its role in the social sciences.

- Fink and Kosecoff (1998) provide a very clear primer on how to conduct questionnaire surveys, focusing on 'how to' and practical advice.

- Fowler (2002) is an excellent, detailed, up-to-date discussion of survey research methods with an extensive bibliography. This book emphasizes methodological topics such as non-response bias, validity, questionnaire evaluation and ethical issues.

- Survey Research Laboratory, University of Illinois at Chicago – *Sites Related to Survey Research* (http://www.srl.uic.edu/srllink/srllink.htm). This is a comprehensive Internet site for survey research, including links to journals, organizations, sample questionnaires, software packages for analysing survey data, and sampling-related software and websites.

Note: Full details of the above can be found in the references list below.

References

Awanyo, L. (2001) 'Labor, ecology and a failed agenda of market incentives: the political ecology of agrarian reforms in Ghana', *Annals of the Association of American Geographers*, 91: 92–121.

Babbie, E. (1990) *Survey Research Methods* (2nd edn). Belmont, CA: Wadsworth.

Babbie, E. (2001) *The Practice of Social Research* (9th edn). Belmont, CA: Wadsworth.

Couper, M., Barkin, R., Bethlemhem, J., Clark, C., Martin, J., Nicholls, W. and O'Reilly, J. (eds) (1998) *Computer-Assisted Survey Information Collection*. New York: Wiley.

Fan, C. (2002) 'The elite, the natives, and the outsiders: migration and labor market segmentation in urban China', *Annals of the Association of American Geographers*, 92: 103–24.

Fink, A. and Kosecoff, J. (1998) *How to Conduct Surveys: A Step-by-Step Guide* (2nd edn). Thousand Oaks, CA: Sage.

Fowler, F. (2002) *Survey Research Methods*. (3rd edn). Thousand Oaks, CA: Sage.

Gilbert, M. (1998) ' "Race", space and power: the survival strategies of working poor women', *Annals of the Association of American Geographers*, 88: 595–621.

Gould, P. and White, R. (1974) *Mental Maps*. Baltimore, MD: Penguin Books.

Kobayashi, A. (1994) 'Colouring the field: gender, "race" and the politics of fieldwork', *The Professional Geographer*, 46: 73–9.

O'Lear, S. (1996) 'Using electronic mail surveys for geographic research: lessons from a survey of Russian environmentalists', *The Professional Geographer*, 48: 209–17.

Robinson, G. (1998) *Methods and Techniques in Human Geography*. New York: Wiley.

Rushton, G. (1969) 'Analysis of spatial behavior by revealed space preference', *Annals of the Association of American Geographers*, 59: 391–406.

Sooman, A. and MacIntyre, S. (1995) 'Health and perceptions of the local environment in socially contrasting neighbourhoods in Glasgow', *Health and Place*, 1: 15–16.

Sporton, D. (1999) 'Mixing methods in fertility research', *The Professional Geographer*, 51: 68–75.

Winchester, H.P.M. (1999) 'Interviews and questionnaires as mixed methods in population geography: the case of lone fathers in Newcastle, Australia', *The Professional Geographer*, 51: 60–7.

7 Finding Historical Data

Miles Ogborn

Synopsis

Historical data in geography are materials that can be used to provide interpretations and analyses of the geographies of past periods. Finding such data is a matter of their survival and of discovering where the data are now kept and how the data can be accessed. Historical material is not just something found in libraries and archives but also includes letters, personal diaries, photographs, works of art, etc.

This chapter is organized into the following sections:

- Introduction
- Sources for questions and questions for sources
- Survival and archives
- Finding historical sources
- Access and data collection.

INTRODUCTION

There is a huge variety of historical data that can be used within human geography. As well as all sorts of written and numerical material which can tell us about social and economic history – diaries perhaps, or the census – historical geographers have also used methods as diverse as oral history (interviewing people about what they and others did in the past – see Rose, 1990), dendrohistory (determining the types of the wood used in building to understand regional economies, societies and cultures – see Biger and Liphschitz, 1995) and the cultural interpretation of visual images or maps (see, for example, Heffernan, 1991, for an analysis of French representations of desert landscapes in the nineteenth century, and Harley, 2001). This range suggests that many of the issues raised by the use of historical data are the same as for work on contemporary human geographies (see Chapters 10 and 28). However, there is a particular set of questions which always need to be considered in relation to historical work: what sources of data have been kept, where and how they are kept, who can get access to them and what you should do when you get access to them? While this chapter cannot give you a guide as to how to find exactly the sources you are looking for, it can give you an indication of how to begin to look and what sorts of issues you will need to consider. To do so it will deal first with the relationship between *research questions* and *historical sources*; secondly, with the question of the *survival* of material from the past; thirdly, with the practicalities of *finding historical sources* in libraries and archives; and, finally, with the issues of *access* to archives and the *collection* of historical data. The place to start is deciding what you are looking for.

SOURCES FOR QUESTIONS AND QUESTIONS FOR SOURCES

In any historical work there are always limitations on the sources you can use. As one historical geographer has put it, 'the dead don't answer questionnaires' (Baker, 1997: 231), so you are restricted by the 'survival' of sources: whether that is those people who have survived to tell you about the past or, more usually, which maps, documents, pictures, sound recordings, physical objects or landscape features have survived for you to interpret. This means that anyone considering historical work needs to think both about the research questions they want to answer and about whether the sources they will need to use to provide those answers exist and are accessible. Indeed, it is sensible to think about your research questions in terms of the available sources as well as thinking about the availability of sources for particular research questions which may have been devised through more theoretical work or through reviewing the existing literature.

You can come at this problem from either end. It is possible to derive a set of research questions about the past from just a few interests and ideas. For example, following feminist debates and contemporary concerns it would be interesting to ask about women's use of public space in the nineteenth-century city (see, for example, Domosh, 1998; Walkowitz, 1998; Rappaport, 2001). You might want to know what sorts of things women did in public space (and what sorts of public space they did those things in): was it leisure, shopping, paid work, education, political activism or charity work? You might need to ask whether this differed by social class or age. You might also want to know what people thought of the women who did those things, and what the women who did them thought of them too. All these possibilities can be worked through without considering the sources of information you might use. However, beginning to do so means making some choices and ruling lines of inquiry in or out on the basis of the sources that can be found to answer the questions. For example, you would have to decide on which city or cities to look at. This is a theoretical question: should it be an industrial city (Manchester, Lyons, Chicago), a capital city (London, Paris, Washington), an imperial or colonial city (Calcutta, Melbourne, Buenos Aires) or some other sort of city? It is also a practical question: can you read the necessary languages and get to where the sources are to be found? You would also have to consider which sources would tell you about the activities you want to focus on. What are the sources for the history of women shopping in nineteenth-century Calcutta? What are the sources for women's charitable work in nineteenth-century Melbourne (see Gleeson, 1995)? In this way you can generate a sense of what historical data you are looking for and can begin to think about where you might look for it. In doing so you will realize that different sources will answer different sorts of questions, and that some questions are more easily answered than others. For example, due to differences in work, leisure and education in the nineteenth century it is much more likely that you would find reflections by middle- and upper-class women about their lives and activities than for working-class women, and much more likely that you would find European women writing about life in the colonial city than African or Asian women (Blunt and Rose, 1994; Bondi and Domosh, 1998). This does not mean that the material does not exist at all, that it is not important to look for it (you never know what you might find) or that it is not possible to expand the definition of what counts as a source when asking new sorts of questions. It does mean that you have to keep formulating research questions and considering how you can find data that will help you answer them.

This means thinking about the other end of the problem too. What are the sources that are accessible to you, and what sorts of questions will they provide answers to? In the absence of any really clearly defined research questions – but with some interests in particular periods, places, events or activities – you can begin by looking at a source or set of sources and thinking about what sorts of questions they could answer. This does not have to involve anything more elaborate than a trip to the nearest library, but it is a way of tying the material available to you into wider research questions and broader literatures. So, for example, if you picked up Elizabeth Gaskell's novel *Mary Barton* (1848) about life in early nineteenth-century Manchester (Gaskell, 1970), or Edwin Chadwick's (1842) *Report on the Sanitary Condition of the Labouring Population of Great Britain* (Chadwick, 1965), or a set of letters written from or to your great, great grandmother who lived in Liverpool in the 1880s, you could begin to ask questions about women's lives in nineteenth-century British cities. You could also begin to think about the research questions to which these sources might begin to provide answers, what other sources you would need and what wider theoretical and substantive literatures you would need to cover in order to make sense of them.

In fact, it soon becomes apparent that you need to do both at the same time: working out what sort of sources will answer the questions you want to ask and deciding what sorts of questions are appropriate for the sources you have available. This means that there is a need to think about ways of finding sources that allow both general and specific searches to be carried out, and ways of quickly evaluating the sources that you have found prior to your full analysis or interpretation of them (see Chapter 27).

SURVIVAL AND ARCHIVES

It is obvious that not everything that happens leaves a record (in writing, sound or image), and equally obvious that not every record that is made survives and is stored away for later use. If any place where such records are kept so that they can be used as sources of information is thought of as an 'archive', we must include our own personal archives – of letters, photographs, perhaps even diaries – as well as the official archives of companies, organizations and public bodies – and the libraries and galleries – local and national libraries and art galleries, university libraries, sound libraries, map libraries and picture libraries – where books, recordings and images are kept for later enjoyment and use. What is created and what survives in these archives is a social and political process which can tell us much about the conditions under which information of different sorts is produced, used and evaluated (Ogborn, forthcoming a). As was pointed out above, a middle-class woman in the nineteenth century was more likely to have kept a diary than a working-class woman, and that diary was more likely to have survived in a family or public archive, or even in a later published form, than that of her working-class counterpart. One woman had more time, space and power than the other to construct this sort of version of her life for herself, and possibly others. Yet it does not always work this way. For example, there is no record in the Indian Mughal empire's archival record of many of the earliest encounters with English traders in India in the seventeenth century – we only have the letters and journals of English merchants and diplomats. This is not because they were more powerful than the Mughal emperors; far from it. They do not figure in the Mughal record because they were deemed to be an insignificant force in India compared with the other Asian rulers and traders with whom the

Mughal leaders had to deal. Their activities in India were not worth recording. Although these cases work out differently, in each one the creation and survival of a record are at least in part a matter of power. For a fragile manuscript, magnetic tape or photograph to survive, someone has to think it is worth keeping and have the ability to keep it secure and legible (Ogborn, forthcoming b).

Yet it is also more than just a matter of power. Archives, understood in this broad sense, are the sites of memory. They are the places – whether a cardboard box in the attic or an imposing public building – where people can begin to construct accounts of the past. This means that they are also full of emotion because they are the places where people's lives are remembered, and where we have a responsibility to think carefully about how we reconstruct those lives in the present and for the future. What is held within these archives and how we can use that information are shaped by the commitments of many people to maintaining a record of the past. This can take many forms (Samuel, 1994), but each one contains within the selection of material that is kept, stored and catalogued – the word 'archived' serving to cover all this – a commitment to remembering the past, a valuing of certain sorts of relationships and representations, and a sense of how that material might be used. Trying to understand each archive and each source within it in terms of both power relations and emotional investment in the past can help us to understand the historical data we are using better.

In the remainder of the chapter I want to deal with some very broad categories of archive and archived historical sources, how to find them and what questions to ask about using them.

FINDING HISTORICAL SOURCES

Libraries

As was pointed out above, the search for historical data can certainly begin with the libraries you have access to – whether these are local libraries, university libraries or specialist libraries. I have made a distinction here between libraries and archival collections to deal with the difference between printed sources which can be available in many places and those, usually manuscript (meaning handwritten), sources which are by their nature only available in one place. This does not mean that libraries do not also hold archival collections; many of them do. What it does mean is that they are certainly places where printed sources of various kinds (including printed editions of manuscript sources) are available to you.

BOOKS

It may be an obvious point but printed books from the periods and places, and on the subjects, that you are interested in are a crucial source. There are, for example, many different ways in which geographers have begun to make use of fictional literary representations, particularly novels, to explore the geographies of the past (Sharp, 2000). For example, both Mandy Morris (1996) and Richard Phillips (1995) have worked on ideas of gender, childhood and nature in two quite different children's books: Frances Hodgson Burnett's *The Secret Garden* (1911) and R.M. Ballantyne's *The Fur Trappers* (1856). In a quite different context, David Schmid (1995) and Philip Howell (1998) have explored the representations of the city in detective fiction. There are also many examples from literary studies in all languages of interpretations which are attentive to questions of space, place and landscape. Beyond the fiction section, it is worth considering whether the books which you might otherwise pass over as 'out of date' could become the sources for a historical study. The resurgence of interest in the history of geography and of geographical thought has been based upon using old geography books and periodicals to try to understand and explain the sorts of geographical ideas – and the representations of people, places and environments – that were part of understandings of geography in different periods and places (for example, Livingstone, 1992; Matless, 1998; Mayhew, 2000; Barnes and Hannah, 2001; Driver, 2001; Withers, 2001). Indeed, the same sorts of ideas and methods can be applied to any set of books or periodicals from the past on any subject you are interested in. One example is Chris Philo's (1987) examination of the changing ideas of madness and its treatment in the now discontinued *Asylum Journal*, a periodical for those involved in what he calls the 'mad business' of nineteenth-century England and Wales. Another example is Mona Domosh's (2001) use of urban exposés of New York in the 1860s to interpret the sorts of ideas about women and public space outlined earlier in this chapter (see also Howell, 2001). Finding these books is simply a matter of using the library catalogue and scanning along the shelves in the sections you are interested in. However, you also need to be aware that any one library collection may not have all the books which you will require: perhaps all those by one author, or those which are referred to by the authors you are interested in as influences on them or that put arguments they want to challenge. Finding these texts involves some detective work, and one useful resource is the public catalogue of the British Library (available at http://www.blpc.bl.uk) since you can use this to search for books by author and subject whether or not you have access to that particular London library. You should also be aware that there are many specialist libraries for all sorts of subjects which may be able to provide access to collections of books that will answer your research questions. Some libraries specialize in terms of their subject matter – for example, in Britain, the Wellcome Library for the History of Medicine, the Cornish Studies Library or the Manx National Heritage Library and, in the USA, the Library for Caribbean Research in New York or the library of the Black Film Center in Bloomington, Indiana. Other libraries are defined by the sort of material they hold – for example, map libraries like the collections at the British Library, the National Library of Wales, the Royal Geographical Society with the Institute of British Geographers and the Newberry Library in Chicago, or newspaper libraries such as the British Library Newspaper Library at Colindale in north London (http://www.bl.uk/collections/newspapers.html). These libraries can be found in the same way as archives (see below).

PRINTED SOURCES

Many libraries, especially university libraries, also hold printed sources of various sorts. We might certainly include printed maps under this heading. Another good example, for Britain, are the nineteenth-century parliamentary papers. These are the record of the inquiries and reports made by contemporary politicians, civil servants and reformers on a whole range of subjects of concern: poverty, prisons, prostitution, the conditions of work in factories, public health and so on. Through the collection of opinion and statistics they tried to reach conclusions about what could or should be done. As a result, they provide a very rich source of information on the subjects they were dealing with and on contemporary ideas about those issues and how they could be addressed (see, for example, Driver, 1993, on the poor law; Ogborn, 1995, on prisons; and Kearns, 1984, on public health). Your library may have them in their original series or in the facsimile editions produced by the Irish University Press. It should also be noted that other countries have comparable forms of official publication on all sorts of issues of concern to them (see the *Checklist of United States Public Documents, 1789–1909* (US Government, 1911) and, for an example that uses US government inquiries into early twentieth-century immigration, see King, 2000).

Parliamentary papers were printed in the nineteenth century to make them available to as many people as possible, and other institutions have also used printing to do the same for manuscript sources. For example, there are printed versions of seventeenth- to nineteenth-century diaries, memoirs and journals (for example, those of Samuel Pepys (1970–83), Giacomo Casanova (1997) and Fanny Burney (1972–84)). There are also series like the Hakluyt Society's publications of travellers' accounts or the Chetham Society which, since 1843, has published material relating to the counties of Lancashire and Cheshire (see http://chethams.org.uk). The former series, which has produced over 350 volumes since the mid-nineteenth century, provides annotated (and sometimes translated) transcripts of ships' logs, journals and letters for a huge range of voyagers to and from a wide range of places, giving easy access to material which would otherwise be only available to a few in forms that are difficult to read and understand (for an example of work based on these and other printed versions of manuscript sources, see Ogborn, 2002).

Archival collections

There is no cast-iron distinction that can be made between libraries and archives. What I want to stress here is that archival collections can be thought of as holding material that is unique to them. Because of that, and also due to the increasing importance of visual sources in historical geography, I include art galleries in this section as well. This is not simply because of the particular set of things that archival collections and galleries have, but also because of the nature of much of the material they hold which – in forms like handwriting or oil painting – are one-offs which have to change their form to be reproduced as printed sources or photographs. Having said that, archival collections can be full of all sorts of material, both rare and commonplace, and another aspect of them is that their collections have often been put together in relation to a particular individual, family, institution or theme. People have their own archives, as do charities, businesses and public bodies (even university geography departments – see Withers, 2002). The largest archives are those of governments or states (and those of Europe and North America the largest of those) which have been established as gatherers and collectors of material for some considerable time, and they are often organized according to the government departments that collected and archived the material.

Finding whether there is archival material on a historical subject you are interested in means thinking carefully about who – in terms of individuals or organizations – would have produced information about it at the time and where that might now be stored (if it has survived at all), whether it is still with the person or organization concerned or deposited in a public archive. As with finding artistic works it means searching for those who produced the sources and for thematic collections within which they are now held (for examples of work in historical geography that use visual images, see Stephen Daniels' (1999) study of the landscape garden designs of a particular individual, Humphry Repton, and James Ryan's (1997) study of photography in the British Empire, which uses lots of thematic collections of photographs including the Royal Geographical Society with the Institute of British Geographers picture library: http://www.rgs.org). In both cases this means careful attention to individuals and to the institutional contexts within which they operated in order to track down appropriate archives.

There is, therefore, a huge number and variety of archives. ARCHON (the Historical Manuscripts Commission's principal information gateway for users of manuscript sources: http://www.hmc.gov.uk/archon) lists nearly two thousand archives and libraries of all sorts in the UK and the Republic of Ireland from the Abbot Hall Art Gallery and Museum in Kendal to the Zoological Society of London. There are many thousands of others all over the world. Finding out which ones hold sources that might be useful for your project is made easier by reference guides in both book form and online. A few of these are listed below:

- Janet Foster and Julia Sheppard's (2000) *British Archives* is a single-volume guide with entries on over a thousand archives giving contact details, information on opening times, access, finding aids and facilities, as well as brief outlines of major holdings. The entries are organized alphabetically by town, with an index by archive name, an Index to Collections which is predominantly made up of personal and organizational names, and a Guide to Key Subjects which offers broader categories. There are other similar guides for other countries, including the National Historical Publications and Records Commission (1988) *Directory of Archives and Manuscript Repositories in the United States*, which is organized by state, gives full contact details along with brief descriptions of the main holdings and has a full subject index.
- *The National Register of Archives* (http://www.hmc.gov.uk/nra): the NRA's indexes – available through this website – contain references to the papers of about 150 000 corporate bodies, persons and families which are held in archives and libraries across the UK. The indexes can be searched by corporate name, personal name, family name or place-name, or they can be browsed alphabetically. This means that you need to know the names of the people and organizations you want to trace since there is no subject key-word search to allow searching by topic. References in NRA are linked directly to ARCHON.
- *ARCHON* (http://www.hmc.gov.uk/archon): this site gives contact details (including addresses, contact names, telephone and fax numbers, email and website addresses) for nearly two thousand archives and museums, as well as cross-referencing them to their NRA listings. Similar access for Australia is made available through the *Directory of Archives in Australia* (http://www.asap.unimelb.ed.au/asa/directory) with details of nearly five hundred repositories. There are also sites such as *Libdex* (http://www.

libdex.com) and *Libweb* (http://sunsite.berkley.edu/Libweb) which provide access to many thousands of library home pages and online catalogues worldwide.

- *Access to Archives* (A2A) (http://www.a2a.pro.gov.uk) is a database of catalogues of UK archives. It allows you to search the titles that appear in those catalogues. This means that you can do subject key-word searches. However, searches for the names of places, people and organizations are likely to be more productive as they are more specific and are likely to appear in the titles of archival records. You should also be aware that this site contains the catalogues of a relatively small (but growing) proportion of all the archives in the UK. Again, a similar service is provided for Australian archives by the National Library of Australia's *Register of Australian Archives and Manuscripts* (http://www/nla.gov.au/raam).

In many cases, where the project is based to any extent upon material gathered by government departments, the sources will be held by the appropriate national archives. These have very extensive holdings of material and, in many cases (including that of Britain), their catalogues can be searched online: see, for example, the British Public Record Office (in Kew in London) at http://www.pro.gov.uk; the French Archives Nationales (in Paris) at http://www.archives nationales.culture.gouv.fr; the US National Archives and Records Administration (in Washington, DC) at http://www.nara.gov; and the National Archives of Canada (in Ottawa) at http://www.archives.ca.

It is also useful to be aware that some archives are dedicated to specific forms of material. In Britain, much oral history material and other audio sources are held at the National Sound Archive at the British Museum (http://www.bl.uk/collections/sound-archive/nsa. html). Moving images can be found at the National Film and Television Archive at the British Film Institute (http://www.bfi.org.uk), and the Mass Observation archive at Sussex University (http://www.sussex.ac.uk/library/massobs) holds a collection, gathered since 1937, of material based on observations of everyday life.

Finally, historical source material is increasingly becoming available in electronic and online formats. For example, the British Calendar of State Papers, Colonial for North America and the West Indies for 1574–1739 and a vast Transatlantic Slave Trade database are both available on CD-ROM. Online resources include the Great Britain Historical GIS project (http://www.geog.port.ac.uk/gbhgis/index.htm) which is making available both geographically located social, economic and demographic statistics (mainly from the period 1851–1939) and the GIS through which they can be mapped. For an earlier period, Robert Shoemaker and Tim Hitchcock's Old Bailey Proceedings

Online project (http://www.shef.ac.uk/hri/bailey.htm) will make available the full proceedings of London's central criminal court for the period 1670–1834 and the University of Edinburgh's Charting the Nation: Maps of Scotland and Associated Archives, 1550–1740 project (http://www.chartingthenation.lib.ed.ac.uk) has over 2000 cartographic images and descriptions on its website. There is also information on a wide range of data deposited by historical researchers with the History Data Service at Essex University and searchable via their website at http://hds.essex.ac.uk.

ACCESS AND DATA COLLECTION

In most cases the uniqueness of the material, the level of development of online services and the expense of photocopying or photographing texts and images mean that you will have to go to the data rather than having them come to you. This means thinking about some practicalities. First, is the material available to you? Different archives have different rules on who is allowed to access the material. Private archives of families, companies or charities are not obliged to allow you access. You will have to find out who is in charge and explain your project and the reasons why you want to see the material they have. They may restrict certain items or restrict the use you can make of them – for example, not using people's names or not publishing work based on their collection without permission. Public archives often have rules which restrict material for time periods from 30 to 100 years for reasons of confidentiality and sensitivity, but they are also committed to making their collections accessible to public use. Secondly, when is the archive open? Opening times vary, and you need to make sure that the archive is open when you get there and not going to close shortly after! Finally, is it worth the trip? Often items in lists and catalogues sound more interesting and useful than they are. You need to find out as much as possible about them before you commit your time and resources to a research visit. You can write to, telephone or email the archive and ask about the material you want to see. You will want to ask for a description of what is there in qualitative and quantitative terms. The archivists know their collections better than anyone so they can be a lot of help to you. If you have to travel a long way and stay away from home for an extended period, archival research can be a costly business. Make sure you have thought about the resource implications of your research before you devise a project on 'race' and urban change in the 1940s which can only be undertaken in the municipal archives of the City of Los Angeles.

Once you have located the material and decided that an archive needs to be visited, there are a few hints about working methods which might come in handy. First, talk to the archivists early on and ask them about the material you are interested in. They may also be able to point you towards other useful material that you have not located in advance. Secondly, find out how that particular archive works – what are the opening hours, how do the catalogues work, how do you order material to read, how much can you have on your desk at any one time? You need to ensure that you have a steady flow of material coming to you. If you find yourself waiting around with nothing to read, use the time productively by searching the catalogues or the book shelves for other relevant material, or by talking to other researchers. Thirdly, work out how long the archival research is going to take you. You should, perhaps even before you get to the archive, have a full list of the material you want to look at. After you have looked at roughly a tenth of that material, you should assess how long the job will take if you keep working at that rate. On the basis of those back-of-the-envelope calculations and the time (and resources) you have available for the research, make a judgement about whether you need to begin sampling the material differently or prioritizing what to read in a different way. It is very unlikely that you will be able to speed up significantly unless you change how you are doing the research, since the limits are usually set by the material you are using. You need to come away from the archive with what is most useful to you in the time available. Fourthly, keep a clear record of what you have read and what you still have to cover. You must also always ensure that both the notes you take on the material and your lists of what to read use the reference system which the archive uses, whereby each item will have an individual reference number (the one you use to order it in the reading room). If you have to recheck a quotation or some figures then you have to know, at any point in the future, where they came from so you can go back to it. And when you use those data in your work you will also have to tell others where the data came from using this reference number. Finally, always take a pencil, a pencil sharpener, an eraser and some paper. Most archives only let you use pencils and, whether you are equipped with a laptop or not, you will certainly need one for jotting down research notes and so on. There is nothing worse than spending the first precious hours of archival research trying to find a pencil in a place you don't know.

Summary

- Work out what sort of sources will answer the questions you want to ask, and decide what sorts of questions are appropriate for the sources you have available.

- Remember that the nature of the material that survives in 'archives' is a matter of both power relations and an emotional investment in the past.
- Historical sources can come in many different forms – literary, visual, printed official publications, manuscripts – and there are many different libraries, galleries, archives and museums that contain this material.
- Find out about access to the archives and plan your collection strategy carefully.

Further reading

- Discussions of archival research methods in historical geography can be found in Baker (1997) and Ogborn (forthcoming a).

- Useful surveys of the sorts of work that has been done and is being done in historical geography can be found in Dodgshon and Butlin (1990), Graham and Nash (2000) and in articles published in the *Journal of Historical Geography*.

- Details of British archives and their collections can be browsed through in Foster and Sheppard (2000) (this is its fourth edition). Previous editions (written by Janet Foster) are still useful if in danger of having some out-of-date details.

Note: Full details of the above can be found in the references list below.

References

Baker, A.R.H. (1997) ' "The dead don't answer questionnaires": researching and writing historical geography', *Journal of Geography in Higher Education*, 21: 231–43.

Barnes, T.J. and Hannah, M. (2001) 'The place of numbers: histories, geographies, and theories of quantification', *Environment and Planning D: Society and Space*, 19: 379–83.

Biger, G. and Liphschitz, N. (1995) 'Foreign tree species as construction timber in nineteenth-century Palestine', *Journal of Historical Geography*, 21: 262–77.

Blunt, A. and Rose, G. (1994) *Writing Women and Space: Colonial and Postcolonial Geographies*. New York: Guilford Press.

Bondi, L. and Domosh, M. (1998) 'On the contours of public space: a tale of three women', *Antipode*, 30: 270– 89.

Burney, F. (1972–84) *The Journals and Letters of Fanny Burney (Madame D'Arblay)*. Oxford: Clarendon Press (12 volumes).

Casanova, G. (1997) *History of My Life*. Baltimore, MD: Johns Hopkins University Press (6 volumes, translated by W.R. Trask).

Chadwick, E. (1965, originally published in 1842) *Report on the Sanitary Condition of the Labouring Population of Great Britain, 1842* (edited by M.W. Flinn). Edinburgh: Edinburgh University Press.

Daniels, S. (1999) *Humphry Repton: Landscape Gardening and the Geography of Georgian England.* New Haven, CT: Yale University Press.

Dodgshon, R.A. and Butlin, R.A. (1990) *An Historical Geography of England and Wales* (2nd edn). London: Academic Press.

Domosh, M. (1998) 'Those "gorgeous incongruities": polite politics and public space on the streets of nineteenth-century New York City', *Annals of the Association of American Geographers*, 88: 209–26.

Domosh, M. (2001) 'The "women of New York": a fashionable moral geography', *Environment and Planning D: Society and Space*, 19: 573–92.

Driver, F. (1993) *Power and Pauperism: The Workhouse System, 1834–1884.* Cambridge: Cambridge University Press.

Driver, F. (2001) *Geography Militant: Cultures of Exploration and Empire.* Oxford: Blackwell.

Foster, J. and Sheppard, J. (2000) *British Archives: A Guide to Archival Resources in the United Kingdom.* Basingstoke: Macmillan.

Gaskell, E. (1970, originally published in 1848) *Mary Barton: A Tale of Manchester Life.* Harmondsworth: Penguin Books.

Gleeson, B.J. (1995) 'A public space for women: the case of charity in colonial Melbourne', *Area*, 27: 193–207.

Graham, B. and Nash, C. (2000) *Modern Historical Geographies.* London: Prentice Hall.

Harley, J.B. (2001) *The New Nature of Maps: Essays in the History of Cartography.* Baltimore, MD: Johns Hopkins University Press.

Heffernan, M.J. (1991) 'The desert in French orientalist painting during the nineteenth century', *Landscape Research*, 16: 37–42.

Howell, P.M.R. (1998) 'Crime and the city solution: crime fiction, urban knowledge, and radical geography', *Antipode*, 30: 357–78.

Howell, P.M.R. (2001) 'Sex and the city of bachelors: sporting guidebooks and urban knowledge in nineteenth-century Britain and America', *Ecumene*, 8: 20–50.

Kearns, G. (1984) 'Cholera and public health reform: the significance of the geographical patterns', *Bulletin of the Society for the Social History of Medicine*, 35: 30–2.

King, D. (2000) *Making Americans: Immigration, Race, and the Origins of the Diverse Democracy.* Cambridge, MA: Harvard University Press.

Livingstone, D.N. (1992) *The Geographical Tradition.* Oxford: Blackwell.

Matless, D. (1998) *Landscape and Englishness.* London: Reaktion.

Mayhew, R.J. (2000) *Enlightenment Geography: The Political Languages of British Geography, 1650–1850.* Basingstoke: Macmillan.

Morris, M.S. (1996) ' "Tha'lt be like a blush-rose when tha' grows up, my little lass": English cultural and gendered identity in *The Secret Garden*', *Environment and Planning D: Society and Space*, 14: 59–78.

National Historical Publications and Records Commission (1988) *Directory of Archives and Manuscript Repositories in the United States* (2nd edn). Phoenix, AZ: Oryx Press.

Ogborn, M. (1995) 'Discipline, government and law: separate confinement in the prisons of England and Wales, 1830–1877', *Transactions, Institute of British Geographers*, 20: 295–311.

Ogborn, M. (2002) 'Writing travels: power, knowledge and ritual on the English East India Company's early voyages', *Transactions, Institute of British Geographers*, 27: 155–71.

Ogborn, M. (forthcoming a) 'Knowledge is power: using archival research to interpret state formation', in A. Blunt et al. (eds) *Cultural Geography in Practice*. London: Arnold.

Ogborn, M. (forthcoming b) 'Archives', in S. Harrison, S. Pile and N. Thrift (eds) *Patterned Ground: Ecologies of Nature and Culture*. London: Routledge.

Pepys, S. (1970–83) *The Diary of Samuel Pepys*. London: Bell (11 volumes, edited by R. Latham and W. Matthews).

Phillips, R.S. (1995) 'Spaces of adventure and the cultural politics of masculinity: R.M. Ballantyne and *The Young Fur Traders*', *Environment and Planning D: Society and Space*, 13: 591–608.

Philo, C.P. (1987)'Fit localities for an asylum: the historical geography of the nineteenth-century "mad business" in England as viewed through the pages of the *Asylum Journal*', *Journal of Historical Geography*, 13: 398–415.

Rappaport, E.D. (2001) *Shopping for Pleasure: Women in the Making of London's West End*. Princeton, NJ: Princeton University Press.

Rose, G. (1990) 'Imagining Poplar in the 1920s: contested concepts of community', *Journal of Historical Geography*, 16: 415–37.

Ryan, J.R. (1997) *Picturing Empire: Photography and the Visualisation of the British Empire*. London: Reaktion.

Samuel, R. (1994) *Theatres of Memory. Volume I. Past and Present in Contemporary Culture*. London: Verso.

Schmid, D. (1995) 'Imagining safe urban space: the contribution of detective fiction to radical geography', *Antipode*, 27: 242–69.

Sharp, J.P. (2000) 'Towards a critical analysis of fictive geographies', *Area*, 32: 327–34.

US Government (1911) *Checklist of United States Public Documents, 1789–1909*. Washington, DC: US Government Printing Office.

Walkowitz, J.R. (1998) 'Going public: shopping, street harassment, and streetwalking in late Victorian London', *Representations*, 62: 1–30.

Withers, C.W.J. (2001) *Geography, Science and National Identity: Scotland Since 1520*. Cambridge: Cambridge University Press.

Withers, C.W.J. (2002) 'Constructing "the geographical archive" ', *Area*, 34: 303–11.

8 Semi-structured Interviews and Focus Groups

Robyn Longhurst

Synopsis

A semi-structured interview is a verbal interchange where one person, the interviewer, attempts to elicit information from another person by asking questions. Although the interviewer prepares a list of predetermined questions, semi-structured interviews unfold in a conversational manner offering participants the chance to explore issues they feel are important. A focus group is a group of people, usually between 6 and 12, who meet in an informal setting to talk about a particular topic that has been set by the researcher. The facilitator keeps the group on the topic but is otherwise non-directive, allowing the group to explore the subject from as many angles as they please. This chapter explains how to go about conducting both interviews and focus groups.

The chapter is organized into the following sections:

* Introduction
* What are semi-structured interviews and focus groups?
* Formulating questions
* Selecting and recruiting participants
* Where to meet
* Recording and transcribing discussions
* Ethical issues
* Conclusion.

INTRODUCTION

Talking with people is an excellent way of gathering information. Sometimes in our everyday lives, however, we tend to talk too quickly, not listen carefully enough and interrupt others. Semi-structured interviews (sometimes referred to as informal, conversational or 'soft' interviews) and focus groups (sometimes referred to as focus group interviews) are about talking with people but in ways that are self-conscious, orderly and partially structured. Krueger and Casey (2000: xi) explain that focus group interviewing (and we could add here, semi-structured interviewing) is about talking but it is also

> ... about listening. It is about paying attention. It is about being open to hear what people have to say. It is about being nonjudgmental. It is about creating a comfortable environment for people to share. It is about being careful and systematic with the things people tell you.

Over the last few decades there has emerged in geography interesting debates (especially amongst feminist geographers) about the utility and validity of qualitative methods, including semi-structured interviews and focus groups (Eyles, 1988; Pile, 1991; Schoenberger, 1991; McDowell, 1992a; Nast, 1994). Many geographers have moved towards what Sayer and Morgan (1985) call 'intensive methods' to examine the power relations and social processes constituted in geographical patterns.

Geographers employ a range of intensive or qualitative methods. Some included in this book are participant observation, keeping a research diary and visual methodologies. Semi-structured interviews, however, are probably one of the most commonly used qualitative methods (Kitchin and Tate, 2000: 213). Focus groups are not as commonly used but they have become increasing popular over the last decade (see *Area*, 1996: Vol. 28, which contains an introduction and five articles on focus groups).

Geographers have used focus groups to collect data on a diverse range of subjects. As early as 1988 Burgess, Limb and Harrison used focus groups (which they called 'small groups') to explore people's environmental values (Burgess et al., 1988a; 1988b). A decade later Miller et al. (1998) conducted focus groups (as well as surveys and ethnographic research) on shopping in northern London to explore links between shopping and identity. Myers and Macnaghten (1998) conducted focus groups to investigate 'rhetorics of environmental sustainability'. Wolch et al. (2000) ran a series of focus groups in Los Angeles with an aim of exploring the role played by cultural difference in shaping attitudes towards animals in the city.

Geographers have also used semi-structured interviews to collect data on an equally diverse range of subjects. Winchester (1999: 61) conducted interviews (and questionnaires) to gather a range of information about the characteristics of 'lone fathers' and the causes of marital breakdown and post-marital conflict in Newcastle, Australia. Valentine (1999) interviewed couples, some together, some apart, in order to understand further gender relations in households. Johnston

(2001) conducted interviews (and focus groups) with participants (or subjects – see McDowell, 1992b: footnote 4, on the contested nature of the terms 'participant' and 'subject') and organizers at a gay pride parade in Auckland, New Zealand. Johnston was interested in the relationship between people taking part in the parade (hosts) and people watching the parade (guests). Punch (2000) conducted interviews (and participant observation) with children and their families in Churquiales, a rural community in the south of Bolivia. She wanted to 'document the ways in which children devise ways to contest adult's [sic] power and control in their lives' (Punch, 2000: 48).

In this chapter I define briefly what I mean by semi-structured interviews and focus groups. These two methods share some characteristics in common; in other ways they are dissimilar. I also discuss how to plan and conduct semi-structured interviews and focus groups. This discussion includes formulating a schedule of questions, selecting and recruiting participants, choosing a location, transcribing data and thinking through some of the ethical issues and power relations involved in conducting semi-structured interviews and focus groups. Throughout the chapter empirical examples are used in an attempt to illustrate key arguments.

WHAT ARE SEMI-STRUCTURED INTERVIEWS AND FOCUS GROUPS?

Interviews, explains Dunn (2000: 51), are verbal interchanges where one person, the interviewer, attempts to elicit information from another person. Basically there are three types of interviews: structured, unstructured and semi-structured, which can be placed along a continuum. Dunn (2000: 52) explains:

> Structured interviews follow a predetermined and standardised list of questions. The questions are always asked in the same order. At the other end of the continuum are unstructured forms of interviewing such as oral histories. The conversation in these interviews is actually directed by the informant rather than by the set questions. In the middle of this continuum are semi-structured interviews. This form of interviewing has some degree of predetermined order but still ensures flexibility in the way issues are addressed by the informant.

Semi-structured interviews and focus groups are similar in that they are conversational and informal in tone. Both allow for an open response in the participants' own words rather than a 'yes or no' type answer.

A focus group is a group of people, usually between 6 and 12, who meet in an informal setting to talk about a particular topic that has been set by the researcher (for other definitions see Krueger, 1988;

Morgan, 1988; Merton and Kendall, 1990; Stewart and Shamdasani, 1990; Greenbaum, 1993; Johnston et al., 2000: 272). The method has its roots in market research. The facilitator or moderator of focus groups keeps the group on the topic but is otherwise non-directive, allowing the group to explore the subject from as many angles as they please. Often researchers attempt to construct as homogeneous a group as possible (but not always – see Goss and Leinback, 1996). The idea is to attempt to simulate a group of friends or people who have things in common and feel relaxed talking to each other. When Honey-field (1997; also see Campbell et al., 1999) conducted research on representations of place and masculinity in television advertising for beer he carried out two focus groups: one with five women, one with seven men. In both groups the participants had either met before, were friends or lived together as 'flatmates'.

Focus groups tend to last between one and two hours. A key characteristic is the interaction between members of the group (Morgan, 1988: 12; Cameron, 2000). This makes them different from semi-structured interviews which rely on the interaction between interviewer and interviewee. Focus groups are also different from interviews in that it is possible to gather the opinions of a large number of people for comparatively little time and expense.

Focus groups are often recommended to researchers wishing to orientate themselves to a new field (Morgan, 1988; Greenbaum, 1993). For example, in 1992 I began some research on pregnant women's experiences of public spaces in Hamilton, New Zealand. There was no existing research on this topic so I wanted to establish some of the parameters of the project before using other methods. I did not know what words pregnant women in Hamilton used to refer to their pregnant bodies – tummies? stomachs? breasts? boobs? – therefore, it would have been difficult to conduct interviews. Focus groups provided an excellent opportunity to gather preliminary information about the topic (see Longhurst, 1996, for an account of these focus groups).

Both semi-structured interviews and focus groups can be used as 'stand-alone methods', as a supplement to other methods or as a means for triangulation in multi-methods research. Researchers often draw on a range of methods and theories. Valentine (1997: 112) explains:

> Often researchers draw on many different perspectives or sources in the course of their work. This is known as triangulation. The term comes from surveying, where it describes using different bearings to give the correct position. In the same way researchers can use multiple methods or different sources to try and maximize their understanding of a research question.

To sum up thus far, semi-structured interviews and focus groups can be used for a range of research, are reasonably informal or conversational in nature and are flexible in that they can be used in conjunction with a variety of other methods and theories. It is also evident that semi-structured interviews and focus groups are more than just 'chats'. The researcher needs to formulate questions, select and recruit participants, choose a location and transcribe data while at the same time remaining cognizant of the ethical issues and power relations involved in qualitative research. In the section that follows I address these topics.

FORMULATING QUESTIONS

Dunn (2000: 53) explains: 'It is not possible to formulate a strict guide to good practice for every interview [and focus group] context.' Every interview and focus group requires its own preparation, thought and practice. It is a social interaction and there are no hard and fast rules one can follow (Valentine, 1997). Nevertheless there are certain procedures that researchers are well advised to heed.

To begin, researchers need to brief themselves fully on the topic. Having done this it is important to work out a list of themes or questions to ask participants. People who are very confident at interviewing or running focus groups often equip themselves with just a list of themes. Personally, I like to be prepared with actual questions in case the conversation dries up. Questions may be designed to elicit information that is 'factual', descriptive, thoughtful or emotional. A combination of different types of questions can be effective depending on the research topic. Researchers often start with a question that participants are likely to feel comfortable answering. More difficult, sensitive or thought-provoking questions are best left to the second half of the interview or focus group when participants are feeling more comfortable. In Box 8.1 is a list of questions I drew up in order to examine large/fat/overweight people's experiences of place. This schedule could be used for semi-structured interviews or focus groups. Follow-up questions are in parentheses.

I would not necessarily ask these questions in the order listed. Allowing the discussion to unfold in a conversational manner offers participants the chance to explore issues they feel are important. At the end of the interview or focus group, however, I would check my schedule to make sure that all the questions had been covered at some stage during the interview or focus group.

It is important to remember that it can take time for participants to 'warm up' to semi-structured interviews and focus groups. If possible, therefore, it is worth offering drinks and food as a way of relaxing

---

Box 8.1   Semi-structured interview and focus group schedule

- Can you remember a time in your life when you were **not** large/fat/ overweight? (Tell me about that. How did people respond to you then?)
- Are there places that you avoid on account of being large? (Why? How do you feel if you do visit these places?)
- Are there places where you feel comfortable or a sense of belonging on account of your size? (Tell me about these places and how you feel in them.)
- In New Zealand there is a strong tradition of spending time at the beach. Do you go to the beach? (Explain. What is it like for you at the beach?)
- Describe your experience of clothes shopping. (Where do you shop? Are shop assistants helpful? Are the changing rooms comfortable? Do you ever feel that other shoppers judge you on account of your size?)
- When you shop for groceries or eat out in a public space, how do you feel? (Why?)
- Are there any issues concerning your size that arise at work? (What are these issues?)
- Do you feel cramped in some spaces? (For example, movie theatre seats, small cars, planes?)
- Do you exercise? (If so, what do you do and where do you do it?)
- Have you made any modifications to your home to suit your size? (For example, altered doorways, selected particular furniture, arranged furniture in specific ways, modified bathroom/toileting facilities. Explain.)
- Do you imagine that your life would be different if you were smaller? (Explain.)
- Are there any issues that you would like to raise that you feel are important but that you haven't had a chance to explore in this interview/focus group?

---

people. It is also useful at the beginning of a focus group to engage participants in some kind of activity that focuses their attention on the discussion topic. For example, participants might be asked to draw a picture, respond to a photograph or imagine a particular situation. This technique tends to be used more by market researchers but it can also prove effective for social scientists. Kitzinger (1994) presented focus group members with a pack of cards bearing statements about who might be 'at risk' from AIDS. She asked the group to sort the cards into different piles indicating the degree of 'risk' attached to each 'type of person'. Kitzinger (1994: 107) explains that '[s]uch exercises involve people in working together with minimal input from the facilitator and encourage participants to concentrate on one another (rather than on the group facilitator) during the subsequent discussion'.

SELECTING AND RECRUITING PARTICIPANTS

Selecting participants for semi-structured interviews and focus groups is vitally important. Usually people are chosen on the basis of their experience related to the research topic (Cameron, 2000). Swenson et al. (1992: 462, cited in Cameron, 2000: 89) call this 'purposeful sampling'. When using quantitative methods the aim is often to choose a random or representative sample, to be 'objective' and to be able to replicate the data. This is not the case when using qualitative methods. Valentine (1997: 111, emphasis in original) explains that, unlike with most questionnaires, 'the aim of an interview [and a focus group] is *not* to be representative (a common but mistaken criticism of this technique) but to understand how individual people experience and make sense of their own lives'.

For example, if you were studying 'racial violence' you might anticipate interviewing and/or running focus groups with people from different ethnic groups, especially those thought to be involved in the violence. However, you might also want to examine the ways in which people's ethnic or racial identities intersect with other identities such as gender, sexuality, 'migrant status' and age in order to explore more fully the processes shaping racial violence. It is not only participants' identities that need to be considered, however, when conducting research. Valentine (1997: 113) makes the important point that 'When you are thinking about who you want to interview it is important to reflect on who you are and how your own identity will shape the interactions that you have with others.' She explains this is what academics describe as being *reflexive* or recognizing your own *positionality* (see England, 1994: 82).

There are many strategies for recruiting participants for semi-structured interviews and focus groups. Some strategies work for both methods while others are more appropriate for one or the other. If you are recruiting participants for interviews it is 'common practice to carry out a simple questionnaire survey to gather basic factual information and to include a request at the end of the questionnaire asking respondents who are willing to take part in a follow-up interview to give their address and telephone number' (Valentine, 1997: 114). It is also possible to advertise for participants in local newspapers or on radio stations requesting interested parties to contact you.

Another method for recruiting participants for interviews is 'cold calling' – that is, calling on people (usually strangers) to ask if they would be prepared to be interviewed. When I was studying the ways in which managers in the central business districts of Auckland and Edinburgh present themselves at work (their dress, comportment and grooming) I called into retail stores and businesses, introduced myself and asked to speak with the manager. I then explained the research and requested an interview. This can be a nerve-racking process because interviewers often get a high refusal rate. In my research on managers, however, approximately 70% of those approached agreed to take part, resulting in 26 interviews (see Longhurst, 2001).

As mentioned already, focus groups are often made up of people who share something in common or know each other. Group membership lists, therefore, can be a useful tool for recruiting. People who already know each other through sports clubs, social clubs, community activities, church groups or work can make an ideal focus group. When I conducted focus groups on men's experiences of domestic bathrooms (a private space rarely discussed by geographers) I succeeded in enlisting (with the help of friends) four groups of men. The first group belonged to the same rugby club, the second were colleagues in a government department, the third were 'job-seekers' and the fourth were family/friends.

Another route useful for securing participants for focus groups is what Krueger (1988: 94) refers to as 'recruiting on location' or 'on-site recruiting'. I used this strategy to recruit first-time pregnant women to talk about their experiences of public places. Pregnant women were approached at antenatal classes, midwives' clinics and doctors' surgeries. These women 'opened doors' to me speaking with other pregnant women. Social scientists refer to this as 'snowballing': 'This term describes using one contact to help you recruit another contact, who in turn can put you in touch with someone else' (Valentine, 1997: 116).

WHERE TO MEET

Not only is it necessary to decide how to select and recruit participants but also to decide where to conduct the interview or focus group meeting. It comes as no surprise to most geographers that where an interview or focus group is held can make a difference (Denzin, 1970). Ideally, the setting should be relatively neutral. I once made the mistake of helping to facilitate a focus group about the quality of service offered by a local council at the council offices. The discussion did not flow freely and it soon became apparent that the participants felt hesitant (understandably) about criticizing the council while in one of their rooms. However, it is worth noting that 'In most cases if you are talking to business people or officials from institutions and organizations you will have no choice but to interview them in their own office' (Valentine, 1997: 117; but also see McDowell, 1997, on interviewing investment bankers in the City of London. Being in the environs you are studying can also prove useful).

It is not always possible to conduct interviews and focus groups in 'the perfect setting' but if at all possible aim to find a place that is neutral, informal (but not noisy) and easily accessible. For example, if you are conducting a reasonably small focus group it is possible to sit comfortably around a dining-room table (see Fine and Macpherson, 1992, for an account of a focus group that took place 'over dinner'). Needless to say, if it is a larger focus group a larger space will be required, perhaps a room at a school, church or club. The main consideration for both semi-structured interviews and focus groups is that interviewees feel comfortable in the space. It is important that the interviewer also feels comfortable (see also Chapter 4). Valentine (1997: 117) warns: 'For your own safety never arrange interviews with people you do not feel comfortable with or agree to meet strangers in places where you feel vulnerable.'

RECORDING AND TRANSCRIBING DISCUSSIONS

When conducting semi-structured interviews or focus groups it is possible to take notes or to audio/video record the discussion. I usually audio(tape) the proceedings. This allows me to focus fully on the interaction instead of feeling pressure to get the participants' words recorded in my notebook (see Valentine, 1997). Directly after the interview I document the general tone of the conversation, the key themes that emerged and anything that particularly impressed or surprised me in the conversation. Taking these notes, in a sense, is a form of data analysis (for information on qualitative data analysis, see Miles and Huberman, 1994; Kitchin and Tate, 2000).

It is advantageous to transcribe interviews and focus groups as soon as possible after conducting them (for how to code a transcript, see Chapter 25). Hearing the taped conversation when it is still fresh in your mind makes transcription much easier. Focus groups, especially large groups, can be difficult to transcribe because each speaker, including the facilitator, needs to be identified. In Box 8.2 is an example of a transcript from a focus group of men who met to discuss their experiences of domestic bathrooms and toilets. Note the 'dynamism and energy as people respond to the contributions of others' (Cameron, 2000: 84). In this focus group excerpt one of the participants puts a question to other group members. Wayne's question about bidets (small, low basins for washing the genitals and anal areas) spearheads a discussion on cultural difference. Note the various transcription codes: the starts of overlap in talk are marked by a double oblique //; pauses are marked with a dot in parenthesis (.); non-verbal actions, gestures and facial expressions are noted in square brackets; and loud exclamations are in **bold** typeface (for more detailed transcription codes, see Dunn, 2000: 74).

Box 8.2 Transcription of a focus group

Wayne: The other question you didn't ask is: has anyone got a bidet? And if they have, then the other question is: does anybody know how bidets are supposed to be used? (.)

Brent: I've never understood how it works.

Robert: No, nor have I.

Christopher: Well, I've got an idea about how it works but it just doesn't seem very efficient.

Robert: Crocodile Dundee just uses a water fountain [laugher].

Brent: Exactly.

Facilitator: That's an interesting question. What do you think about using, I mean, how do you use a bidet?

Robert: It's supposed to wash your bloody bottom [sic] out isn't it?

Christopher: Mm, that's the whole object. And then you use a towel?

Robert: I don't know, I think so, instead of wiping //

Wayne: // In Asia for instance all you get is a bloody hole in the ground and a dipper and you dip the water out and go 'woof' and over and it washes it clean.

Robert: **Oh my god! That's terrible!**

Christopher: It's another culture.

Wayne: And that is probably why Asians are so good at squatting on their heels 'cause they are used to it. You look at a kid, a tiny tot, and they can squat on their heels all day long.

Source: Audio-tape excerpt from a focus group conducted by David Vincent in 1999 (see Longhurst, 2001)

As this transcript illustrates, sometimes data can be 'sensitive'. It is not surprising, therefore, that there are numerous ethical issues to consider when conducting semi-structured interviews and focus groups (see Chapter 3).

ETHICAL ISSUES

Two important ethical issues are confidentiality and anonymity. Participants need to be assured that all the data collected will remain secure under lock or on a computer database accessible by password only; that information supplied will remain confidential and participants will remain anonymous (unless they desire otherwise); and that participants have the right to withdraw from the research at any time without explanation. It is also sound research practice to offer to provide participants with a summary of the research results at the completion of the project and to follow through on this commitment. This summary might take the form of a hard copy or an electronic copy posted on a website (for example, see Ruth Bankey's site http:// www.geo.ed.ac.uk/~rba/ on her research on agoraphobia).

Focus groups pose a further complication in relation to confidentiality because not only is the researcher privy to information but also members of the group. Therefore, participants need to be asked to treat discussions as confidential. Cameron (2000: 90) explains:

> As this [confidentiality] cannot be guaranteed, it is appropriate to remind people to disclose only those things they would feel comfortable about being repeated outside the group. Of course, you should always weigh up whether a topic is too controversial or sensitive for discussion in a focus group and is better handled through another technique, like individual in-depth interviews.

Another ethical issue is that participants in the course of an interview or focus group may express sexist, racist or other offensive views. In an earlier quotation, Krueger and Casey (2000: xi) claim that researchers ought to listen, pay attention and be non-judgemental. Sometimes, however, being non-judgemental might simply reproduce and even legitimize interviewees' discrimination through complicity (see Valentine, 1997). Researchers need to think carefully about how to deal with such situations because there are no easy solutions.

Researchers also need to think carefully about how to interview or run focus groups in different cultural contexts (see Chapter 12). For example, 'First World' researchers investigating 'Third World' 'subjects' need to be highly sensitive to local codes of conduct (Valentine, 1997). In short, there is a web of ethical issues and power relations that need to be teased out when conducting semi-structured interviews and focus groups. Feminist geographers in particular have made a useful contribution in this area (for example, see McDowell, 1992b; Dyck, 1993; England, 1994; Gibson-Graham, 1994; Katz, 1992; 1994; Kobayashi, 1994).

CONCLUSION

In this chapter I have outlined two qualitative methods – semi-structured interviews and focus groups – and how they can be employed in geographical research. Both methods involve talking with people in a semi-structured manner. However, whereas semi-structured interviews rely on the interaction between interviewee and interviewer, focus groups rely on interactions amongst interviewees. Both methods make a significant contribution to geographic research, especially now that discussions about meaning, identity, subjectivity, politics, knowledge, power and representation are high on many geographers' agendas. Critically examining the construction of knowledge and discourse in geography (see Rose, 1993) has led to an interest in developing alternative methodological strategies coupled with greater reflexivity about the process of research. Semi-structured interviews and focus groups are useful for investigating complex behaviours, opinions and emotions and for collecting a diversity of experiences. These methods do not offer researchers a route to 'the truth' but they do offer a route to partial insights into what people do and think.

Summary

- Semi-structured interviews and focus groups are about talking with people but in ways that are self-conscious, orderly and partially structured.
- These methods are useful for investigating complex behaviours, opinions and emotions and for collecting a diversity of experiences.
- Every interview and focus group requires its own preparation, thought and practice.
- There are a range of methods that can be used for recruiting participants, including advertising for participants, accessing membership lists, on-site recruiting and 'cold calling'.
- Interviews/focus groups ought to be conducted in a place where participants and the interviewer feel comfortable.
- When conducting semi-structured interviews or focus groups take notes and/or audio/video record the discussion.
- There is a web of ethical issues and power relations that need to be teased out when using these methods.
- Semi-structured interviews and focus groups make a significant contribution to geographic research, especially now that discussions about meaning, identity, subjectivity, politics,

knowledge, power and representation are high on many
geographers' agendas.

Further reading

There are numerous excellent books, book chapters and articles on semi-structured
interviews (and interviewing more generally) and focus groups written by geographers and
other social scientists. I have listed below some of the more recently published titles:

- Krueger and Casey (2000). This popular book, first published in 1988, has been
  reprinted three times. The third edition is easy to read, well illustrated and offers
  numerous examples of how to use focus groups. It is one of the most comprehensive
  guides on focus groups available.

- Cameron (2000) offers a geographer's perspective on focus groups, explaining the
  various ways they have been used, how to plan and conduct them and how to analyse
  and present results.

- *Area* (1996: Vol. 28) contains an introduction by Goss and five articles on focus groups
  (by Burgess; Zeigler, Brunn and Johnston; Holbrook and Jackson; Longhurst; and Goss
  and Leinback). The collection illustrates effectively the range of research carried out by
  geographers using focus groups.

- Valentine's (1997) chapter on 'conversational interviews' is highly readable and
  provides advice on whom to talk to, how to recruit participants and where to hold
  interviews. Valentine raises interesting questions about the ethics and politics of
  interviewing and alerts readers to some of the potential pitfalls that can occur in
  research.

- Dunn (2000) discusses structured, semi-structured and unstructured interviewing in
  geography, critically assessing the relative strengths and weaknesses of each method.
  His chapter provides advice on interview design, practice, transcription, data analysis
  and presentation. Like Valentine, Dunn has a useful guide at the end of the chapter to
  further reading.

Note: Full details of the above can be found in the references list below.

References

Area (1996) 28(2). 'Introduction to focus groups' by J.D. Goss and five papers
    on using focus groups in human geography by Burgess; Zeigler, Brunn and
    Johnston; Holbrook and Jackson; Longhurst; and Goss and Leinback.
Bankey, R. (2002) *Summary of Current and Previous Research* (available at
    http://www.geo.ed.ac.uk/ ~ rba. Accessed 20 February 2002).
Burgess, J. (1996) 'Focusing on fear: the use of focus groups in a project for the
    Community Forest Unit, Countryside Commission', *Area*, 28: 130–5.

Burgess, J., Limb, M. and Harrison, C.M. (1988a) 'Exploring environmental values through the medium of small groups. 1. Theory and practice', *Environment and Planning A*, 20: 309–26.

Burgess, J., Limb, M. and Harrison C.M. (1988b) 'Exploring environmental values through the medium of small groups. 2. Illustrations of a group at work', *Environment and Planning A*, 20: 457–76.

Cameron, J. (2000) 'Focusing on the focus group', in I. Hay (ed.) *Qualitative Research Methods in Human Geography*. Melbourne: Oxford University Press, pp. 83–102.

Campbell, H., Law, R. and Honeyfield, J. (1999) ' "What it means to be a man": hegemonic masculinity and the reinvention of beer', in R. Law et al. (eds) *Masculinities in Aotearoa/New Zealand*. Palmerston North: Dunmore Press, pp. 166–86.

Denzin, N. (1970) *The Research Act: A Theoretical Introduction to Social Research*. Chicago, IL: Aldine.

Dunn, K. (2000) 'Interviewing', in I. Hay (ed.) *Qualitative Research Methods in Human Geography*. Melbourne: Oxford University Press, pp. 50–82.

Dyck, I. (1993) 'Ethnography: a feminist method?' *The Canadian Geographer*, 37: 52–7.

England, K. (1994) 'Getting personal: reflexivity, positionality and feminist research', *The Professional Geographer*, 46: 80–9.

Eyles, J. (1988) 'Interpreting the geographical world: qualitative approaches in geographical research', in J. Eyles and D.M. Smith (eds) *Qualitative Methods in Human Geography*. Cambridge: Polity Press, pp. 1–16.

Fine, M. and Macpherson, P. (1992) 'Over dinner: feminism and adolescent female bodies', in M. Fine (ed.) *Disruptive Voices: The Possibilities of Feminist Research*. East Lansing, MI: University of Michigan Press, pp. 175–203.

Gibson-Graham J.K. (1994) ' "Stuffed if I know!": reflections on post-modern feminist social research', *Gender, Place and Culture*, 1: 205–24.

Goss, J.D. and Leinback, T.R. (1996) 'Focus groups as alternative research practice: experience with transmigrants in Indonesia', *Area*, 28: 115–23.

Greenbaum, T. (1993) *The Handbook for Focus Group Research*. Lexington, MA: Lexington Books.

Honeyfield, J. (1997) 'Red blooded blood brothers: representations of place and hard man masculinity in television advertisements for beer.' Masters thesis, University of Waikato, New Zealand.

Johnston, L. (2001) '(Other) bodies and tourism studies', *Annals of Tourism Research*, 28: 180–201.

Johnston, R.J., Gregory, D., Pratt, G. and Watts, M. (eds) (2000) *The Dictionary of Human Geography* (4th edn). Malden, MA: Blackwell.

Katz, C. (1992) 'All the world is staged: intellectuals and the projects of ethnography', *Environment and Planning D: Society and Space*, 10: 495–510.

Katz, C. (1994) 'Playing the field: questions of fieldwork in geography', *The Professional Geographer*, 46: 67–72.

Kitchen, R. and Tate, N.J. (2000) *Conducting Research into Human Geography*. Edinburgh Gate: Pearson.

Kitzinger, J. (1994) 'The methodology of focus groups: the importance of interaction between research participants', *Sociology of Health and Illness*, 16: 103–21.

Kobayashi, A. (1994) 'Coloring the field: gender, "race", and the politics of fieldwork', *The Professional Geographer*, 46: 73–9.

Krueger, R.A. (1988) *Focus Groups: A Practical Guide for Applied Research*. Thousand Oaks, CA: Sage.

Krueger, R.A. and Casey, M.A. (2000) *Focus Groups. A Practical Guide for Applied Research* (3rd edn). Thousand Oaks, CA: Sage.

Longhurst, R. (1996) 'Refocusing groups: pregnant women's geographical experiences of Hamilton, New Zealand/Aotearoa', *Area*, 28: 143–9.

Longhurst, R. (2001) *Bodies: Exploring Fluid Boundaries*. London: Routledge.

McDowell, L. (1992a) 'Valid games?' *The Professional Geographer*, 44: 219–22.

McDowell, L. (1992b) 'Doing gender: feminism, feminists and research methods in human geography', *Transactions, Institute of British Geographers*, 17: 399–416.

McDowell, L. (1997) *Capital Culture. Gender at Work in the City*. Oxford: Blackwell.

Merton, R.K. and Kendall, P.L. (1990) *The Focused Interview: A Manual of Problems and Procedures* (2nd edn). New York: Free Press.

Miles, M.B. and Huberman, A.M. (1994) *Qualitative Data Analysis: An Expanded Sourcebook*. Thousand Oaks, CA: Sage.

Miller, D., Jackson, P., Thrift, N., Holbrook, B. and Rowlands, N. (1998) *Shopping, Place and Identity*. London: Routledge.

Morgan, D.L. (1987) *Focus Groups as Qualitative Research*. London: Sage.

Morgan, D.L. (1988) 'Focus groups as qualitative research', *Qualitative Research Methods*, 16.

Myers, G. and Macnaghten, P. (1998) 'Rhetorics of environmental sustainability: commonplaces and places', *Environment and Planning A*, 30: 335–53.

Nast, H. (1994) 'Opening remarks on "women in the field" ', *The Professional Geographer*, 46: 54–5.

Pile, S. (1991) 'Practising interpretative geography', *Transactions of the Institute of British Geographers*, 16: 458–69.

Punch, S. (2000) 'Children's strategies for creating playspaces', in S.L. Holloway and G. Valentine (eds) *Children's Geographies. Playing, Living, Learning*. London and New York: Routledge, pp. 48–62.

Rose, G. (1993) *Feminism and Geography: The Limits of Geographical Knowledge*. Cambridge: Polity Press.

Sayer, A. and Morgan, K. (1985) 'A modern industry in a reclining region: links between method, theory and policy', in D. Massey and R. Meegan (eds) *Politics and Method*. London: Methuen, pp. 147–68.

Schoenberger, E. (1991) 'The corporate interview as a research method in economic geography', *The Professional Geographer*, 43: 180–9.

Stewart, D.W. and Shamdasani, P.N. (1990) *Focus Groups: Theory and Practice*. Newbury Park, CA: Sage.

Swenson, J.D., Griswold, W.F. and Kleiber, P.B. (1992) 'Focus groups: method of inquiry/intervention', *Small Group Research*, 24: 459–74.

Valentine, G. (1997) '"Tell me about" . . .: using interviews as a research methodology', in R. Flowerdew and D. Martin (eds) *Methods in Human Geography: A Guide for Students Doing a Research Project*. Edinburgh Gate: Addison Wesley Longman, pp. 110–26.

Valentine, G. (1999) 'Doing household research: interviewing couples together and apart', *Area*, 31: 67–74.

Winchester, H.P.M. (1999) 'Interviews and questionnaires as mixed methods in population geography: the case of lone fathers in Newcastle, Australia', *The Professional Geographer*, 51: 60–7.

Wolch, J., Brownlow, A. and Lassiter, U. (2000) 'Constructing the animal worlds of inner-city Los Angeles', in C. Philo and C. Wilbert (eds) *Animal Spaces, Beastly Places*. London and New York: Routledge, pp. 71–97.

9 Participant Observation

Eric Laurier

Synopsis

Participant observation involves spending time being, living or working with people or communities in order to understand them. In other words, it is, as the name implies, a method based on participating and observing in which fieldnotes or video notes are used as a method of data collection. The basis of this approach is to get, and stay, as close to the spatial phenomenon being studied as possible and it is thereby quite distinct from methodologies that emphasize distance and objectivity.

The chapter is organized into the following sections:

* What is participant observation?
* Commentators and players
* Doing participant observation/becoming the phenomenon
* Adequate commentaries on culture and society
* Results: respecifying the generalities of social science
* Final words of advice.

WHAT IS PARTICIPANT OBSERVATION?

> I have no great quickness of apprehension or wit . . . my power to follow a long and purely abstract train of thought is very limited . . . [but] I am superior to the common run of men in noticing things which easily escape attention, and in observing them carefully. (Charles Darwin, Preface to *The Expression of the Emotions in Man and Animals*)

Participant observation is perhaps the easiest method in the world to use since it is ubiquitous and we can all already do it. From the moment we are born we are in various ways observing the world around us and trying to participate in it. Children acquiring language for the first time listen and watch what, when and how their parents are doing what they are doing. They observe greetings and have greetings directed at them, and attempt to participate by, at first, looking and, later, waving and making sounds that approximate, and eventually are, hellos and goodbyes. It is, of course, not just children who use this method to acquire skills: surgical students spend a great deal of time observing surgery and are gradually entered into the practical demands of actually doing surgery as fully participating surgeons. International migrants finding themselves in foreign countries have a massive task of observing the multitude of activities and exactly how they are done in order to fit in as participants. Amongst other background knowledge they have to acquire the locals' ways of getting everyday things done, such as greetings, ordering coffee, queuing for buses, making small talk, paying their taxes and so on.

So far so good: participant observation is easy – it does not require the mastering of arcane skills or technical lexicons. And yet there is a catch (well, there had to be one, right?). You may have guessed already from the mention of surgeons and migrants that participant observation as a way of engaging with life-worlds will unfortunately present you with particular challenges. Also for better and for worse, because it is not an *external* method administered on research subjects, such as a questionnaire or a lab test, participant observation has *no preset formal steps* to doing it. Or rather the stages that anyone doing participant observation must go through are the stages *which arise out of the phenomenon and settings* you are investigating.

If you do not know how to do, or be something, then learning how to do or be that thing will be as hard for you as for anyone attempting to participate in it. Think of the effort and time required to do informative participant observation studies of air-traffic control (Harper and Hughes, 1993), mathematical problem-solving or playing jazz music (and these expert cultures have been studied through participant observation.) And yet participant observation can be turned to such seemingly 'simple' spatial phenomena as shopping in the supermarket or going clubbing (Malbon, 1999) or walking in the city (Watson, 1993). In these latter examples you may already be able to do them and the demands on you will then be to provide a commentary that describes them in revealing and interesting ways. Key to your success in doing participant observation is, as Charles Darwin says of his own powers, that you notice things that otherwise escape attention and that you observe *carefully*.

COMMENTATORS AND PLAYERS

A common mistake, made as often by well qualified social researchers as by those new to participant observation, is to take observation as the dominant part of participant observation. This is to some extent a legacy of scientistic ideas about 'objective observers' who watch their research subjects in a detached, emotionless manner and are thereby able to provide objective descriptions of what was occurring. Even though most researchers no longer pursue this kind of objective observation many still underestimate the importance of participation, proposing that it is sufficient to watch what is going on and then write down their observations.

To give you a sense of why it may not be enough simply to observe, let us move on to what your participant observation should produce: *commentary*. Being able to comment on the culture, society and geography of various spaces and places is indeed the major requirement of doing geography. In that sense all geographers are commentators and many of them exceptionally good ones. If we think about sports commentators for a moment, as against social and cultural commentators, we can see that they are seldom the ones playing the game: they are sitting to one side observing it. Some sports commentators provide exasperatingly bad and irrelevant commentary because they have never played the game they are commenting on. One ingredient of a decent sports commentator is that he or she should be, or have been, a player to offer any kind of insight into the game. Knowing *how* to play the game in no way guarantees insightful remarks since many of the outstanding players and competitors have very little to say. Shifting back to social and cultural research, it is the case that far too many of its researchers are only commentators and have never played. The point that is being reiterated here is that the best participant observation is generally done by those who have been involved in and tried to do and/or be a part of the things they are observing.

DOING PARTICIPANT OBSERVATION/BECOMING THE PHENOMENON

Despite my having suggested that there is no template for doing participant observation there are features in its course, which while not prespecifying what is to be done, will give you a sense of whether you are making any progress or not. If you have never been involved in the event, activity, phenomenon, group or whatever you are investigating then, on your arrival, you will find yourself cast into some category. You will be called, if not a 'greenhorn', 'beginner', then something worse such as an 'incompetent', 'tourist' or 'outsider'. This is not simply a pejorative term for you but also assigning you to a type which is related to what you are expected to be able to do, the perspective you will have on events and what you will need to be taught. This is also not such a bad place to start since people's expectations of you will not be too high and you will be expected to be observing so as to learn how to become one of the group/company/band/players. Recording your observations at this point is vital since, if all goes well, they should have changed by the time you are finished (see Box 9.1). They are also, at this point, the perspective of 'any person' who may well be whom you will wish to write your report for at the end of your fieldwork. Without keeping a record of your own struggles to get 'the knowledge' you are likely to forget the lay member's perspective once you no longer have it. Consequently you will no longer appreciate what it is that may seem odd, irrational or otherwise mysterious to those, like yourself, now in the know.

Just how long it takes to become competent in what you choose to study, and indeed whether it is possible to reach that state, will vary according to what you choose to study using this method. Should you choose to study a supermarket as a shopper or stacking the shelves as

---

Box 9.1   Recording observations

In a café which I was studying as part of a community project, the counter staff talked really loudly and greeted customers coming through the door. I noticed this the first time I went in and noted it down. It seemed odd that the staff should allow their conversations to be overheard by everyone. I also noticed that they greeted a lot of customers by their name, calling out: 'Hi Betty!', 'Just take a seat Mr Stewart and I'll be right with you.' As my fieldwork in the neighbourhood continued over its six-month period I gained the status of a 'regular' (one kind of participant status) in the café where I no longer noticed how loudly they talked. I had become used to the place. I had a favourite table. I had a favourite toasted sandwich. Not only that, I had become one of the people whom the staff greeted by name and would say with a smile: 'A toasty today?' When writing up the research I returned to ponder the loudness of their talk, the things they said, the fact that they remembered names and favourite foods, and began to see its part in creating the ambience of a 'place where everybody knows your name'. Greeting people on their arrival also made known to anyone there who were the 'locals' and who were passing trade. The loud talk tuned you into local gossip about who was marrying whom, who had been expelled from school, where the roadworks were, what people ate and drank there. 'Loud talk' was, then, an everyday method used by the staff and enjoyed by the regulars as giving a feeling of community and offering some local knowledge (for a much longer description, see Laurier et al., 2001).

a member of the staff then it is not so demanding to become competent in these activities. However, other communities of practice, such as mathematicians, rural villages or jazz piano players, may take considerable time and effort before you will be recognized as an accepted member of their groups. For whatever activity you decide to participate in there will be different ordinary and expert ways in which you are instructed in how to do it, from the more formal (i.e. lessons, workshops, courses, rulebooks, etc.) to the informal (tips, jokes, brief chats). For some rural villages you will always be an 'incomer' even if you reside in them until your dying day. A successful participant observation does not turn entirely on becoming *excellent* in the activity (becoming an Olympic athlete or leading cardiac surgeon) or passing as, say, a homeless Italian-American or even as a member of the opposite sex. Yet by the end you should possess a degree of the particular know-how, appropriate conduct and common knowledge of the place and/or people you have chosen to study.

Here is a brief extract where David Sudnow (1978: 82) offers a sense of what he was doing in what he called 'going for the jazz'. He had already acquired a basic competency in piano but was trying to *become* a jazz piano player, so he was, among other methods, starting to look very carefully at what the experienced players did with their body at the piano:

> At live performances I had watched the very rapid improvisation players whose records had served as my models, but their body idioms in no way seemed connected in details to the nature of their melodies, and my occasional attempts to emulate the former had no appreciable bearing on my success with the latter. This, for example, had a little shoulder tic, but mimicking that (which I found myself doing after a night of watching him) did not make his sorts of melodies happen. Another sat tightly hunched over the piano, playing furiously fast, but assuming that posture seemed to have no intrinsic relationship to getting my jazz to happen as his did.

Sudnow spent a *decade* learning, first, how to play a piano; secondly, how to play jazz that was recognized by other jazz musicians as jazz; and, finally, how to instruct students in jazz piano. The above quotation gives only a hint of the experiences involved in Sudnow's odyssey. However long you devote to your study, and it's more likely to be days or weeks rather than years, what is important is to keep notes, audio-visual records and ideally a journal along the way. Basically, record as much as you possibly can, and even more importantly, try to write straightforward and detailed descriptions of the phenomena you are interested in. Record what you see and can make sense of and what puzzles or upsets you (with a camera or camcorder if possible). Sudnow video taped himself playing the piano so that he could then repeatedly watch what his hands were doing on the piano keyboard in relation to the production of the music. These recordings will form the base materials for your commentary at the end of your time doing fieldwork. It is not impossible to write a delightful report on your participant observation without having kept a record of it at the time but it is considerably more difficult and the details will inevitably slip away. In my own research practice I try to take photographs and shoot video wherever and whenever possible. Video in particular provides you with some 'stuff' to work with once you are writing up your research. It is a 'retrievable dataset' and reviewable to find unanticipated details that you could not formulate in words at the time, nor may even have noticed, since you were too busy being engaged in the situation. Photos and video clips are also very helpful for presenting your results to lay and expert audiences. With video, audio and still photographs you can share your data and I would strongly encourage you to do data review sessions in a group to see what different people are able to find in the visual material you are presenting them with. Note down what you all see in common and try to consider how you do so. Materials that may initially seem quite uninteresting should, by the end of their close viewing and description, provide you with surprises. Drawing reflective conclusions from the materials you have gathered remains reliant on your insight.

Fieldnotes

The notes in Box 9.2 were taken in the passenger seat of a company car after a conversation with a regional manager about her boss and some of her co-workers.

Fieldnotes or, rather, certainly *my* fieldnotes are often badly written, dull, cryptic and not the kind of thing I would want to show anyone. However, I have included a 'cheat' sample of a not too dire section to give you at least a flavour of how roughly notes tend to be written at the time. What is important to bear in mind is that notes should be taken because you will not be able to recall sufficient details of what happens or what people say during a lengthy engagement with them. As it happens, getting your notebook out is also quite useful in showing your researcher status in the field or setting so that you will be seen to be at work doing your research and (may) be taken seriously as a result. You should be aware that there are sensitive settings where you will want to keep your status as a researcher more low key. In such an event you may end up scribbling notes in the toilet or in the bus afterwards, or some other hidden place, and there are plenty of amusing stories from experienced geographers about doing so. The notes reproduced in Box 9.2 were written up while the regional manager was out of the car dropping in to visit one of her clients.

Box 9.2 Fieldnotes

Transcript

. . . She feels equally that she is not competent in that 'talk', not as 'real' saleswoman – so it's a question of identity, there's also a lingering suspicion of the salesman and the real salesman's attitudes. Somehow they aren't trustworthy – once all other comparative morals have disappeared into doing sales. They're described as 'smooth', 'able to talk themselves out of anything'. M describes herself as 'all over the place' and I guess by contrast, rough around the edges. These very misfootings in her performance reinforce her authenticity however? She's also fairly hierarchical – at lunch suggesting she'd never tell tales to her employers that were personal or about losing face – being incompetent. Since her boss's story about going out on her first date with her current boyfriend – with a hangover and then having to nip away half-way through for a camouflaged barf. For M there are further reasons why she's doing so much just now and why she's so exhausted. It's the learning cycle of the job – at first blissfully ignorant of all that was involved and fluking her way through and then as she learns about the job realising there's a whole lot more she should be doing – symbolised by the territorial strategy. And I guess further . . .

As importantly, note-taking can help you concentrate during situations which may otherwise allow your mind to wander. The uses of note-taking to help you pay attention will, I am sure, be familiar from your use of them to do so during university lectures. To repeat myself the most significant use of your fieldnotes will be *after the event*, in helping you recall the details of the situation which you wish to describe, analyse or reflect on. As you can see from the example of my notes they generally need to be tidied up, expanded upon and details frequently need to be filled in which will allow your jottings to be decrypted.

Video notes

The video footage shown in Box 9.3 was taken later during the same research study as the notebook extract given in Box 9.2. From repeated watchings, the notes shown below the video clips were eventually written up.

Digital video (DV) records of events during fieldwork also require writing up afterwards so, even though they save you making notes at the time, you cannot escape making some notes eventually. The clips you are interest in can be imported from DV tapes to computers that have the appropriate connectivity (usually firewire). Once video is imported then, using various software packages (i.e. Quicktime, Media-player, Realproducer, Adobe Premiere, Final Cut Pro), stills can be extracted to be arranged in sequence as in Box 9.3. If you are working towards multimedia documents of your research, the sequences can be kept as moving 'real time' audio-visual data and inserted into html files. There are plenty of technical guides to this available online and, given the rapid changes in video technology, I will not go into further technical detail on dealing with digital video. Broadly it is worth pointing out that it has been getting easier and easier to use and geographers have barely scratched the surface of what might be done with it for describing social practices or presenting their results.

As with using a notebook, using a camcorder makes your status as a researcher highly visible, only more so. When you switch the camera on people frequently feel obliged to make faces or talk to camera as if they were starring in a docu-soap. For instance, Marge talks to the camera to some extent in the extract shown in Box 9.3 when she waggles the vodka flavourings towards me and the camcorder. Serious documentary makers emphasize getting the groups or individuals you wish to video familiar with the presence of the camera to the point of ignoring it and urging them to do what they would normally do were the camera not there. In fact many anthropologists hand over the camcorder and allow their communities to video themselves in an attempt to hand over control of the video-making to those being represented. In my own practice I tend to hold on to the camera since I have a rough idea of what I am after and am anxious to avoid giving busy people any more work to do than they already have. In the above example I had been videoing Marge for more than a week. In the clips

Box 9.3   Video notes

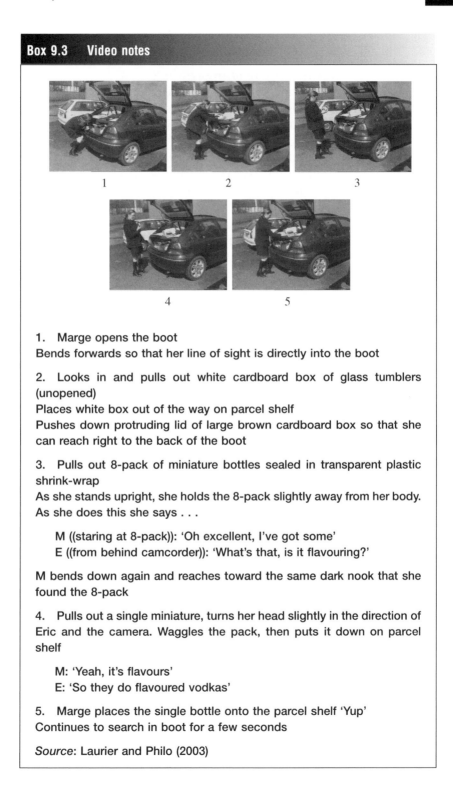

1. Marge opens the boot
Bends forwards so that her line of sight is directly into the boot

2. Looks in and pulls out white cardboard box of glass tumblers (unopened)
Places white box out of the way on parcel shelf
Pushes down protruding lid of large brown cardboard box so that she can reach right to the back of the boot

3. Pulls out 8-pack of miniature bottles sealed in transparent plastic shrink-wrap
As she stands upright, she holds the 8-pack slightly away from her body. As she does this she says . . .

> M ((staring at 8-pack)): 'Oh excellent, I've got some'
> E ((from behind camcorder)): 'What's that, is it flavouring?'

M bends down again and reaches toward the same dark nook that she found the 8-pack

4. Pulls out a single miniature, turns her head slightly in the direction of Eric and the camera. Waggles the pack, then puts it down on parcel shelf

> M: 'Yeah, it's flavours'
> E: 'So they do flavoured vodkas'

5. Marge places the single bottle onto the parcel shelf 'Yup'
Continues to search in boot for a few seconds

Source: Laurier and Philo (2003)

and their accompanying description I was trying capture something of her work as it happens 'naturalistically' (in contrast to experimentally or by questionnaire or interview).

Video clips or transcribed sequences of stills, as I have noted already, assist in sharing your original data. In my notes on video clips I try to put down what is happening without jumping to any particular conclusions. As you can see from the example in Box 9.3, they are perhaps starkly descriptive, yet they are what social life looks like 'in the flesh', so to speak. They are also a first stage towards your analysis, much like tabulating the descriptive statistics from a numerical study. Unlike statistics they do not give you 'results' at this stage but are reliant at this point on your further description and analysis of what is occurring. In the example with Marge, Chris Philo and myself were investigating how a sales region was organized through the mundane and time-consuming work with cardboard boxes and a car boot being packed, unpacked, opened and repacked at different clients' venues.

ADEQUATE COMMENTARIES ON CULTURE AND SOCIETY

First, an oft-quoted remark from Harvey Sacks (1992: 83) on reading his students' reports after he had sent them to do participant observation work on people exchanging glances:

> Let me make a couple of remarks about the problem of 'feigning ignorance'. I found in these papers that people [i.e. the students in his class] will occasionally say things like, 'I didn't really know what was going on, but I made the inference that he was looking at her because she's an attractive girl.' So one claims to not really know. And here's a first thought I have. I can fully well understand how you come to say that. It's part of the way in which what's called your education here gets in the way of your doing what you in fact know how to do. And you begin to call things 'concepts' and acts 'inferences', when nothing of the sort is involved. And that nothing of the sort is involved, is perfectly clear in that if it were the case that you didn't know what was going on – if you were the usual made up observer, the man from Mars – then the question of what you would see would be a far more obscure matter than that she was an attractive girl, perhaps. How would you go about seeing in the first place that one was looking at the other, seeing what they were looking at, and locating those features which are perhaps relevant?

The warning Sacks is making to his students is not to exercise a kind of professional scepticism ('feigning ignorance') which subverts the intelligibility of the things and actions they are able to observe. In this case a guy checking out an attractive girl. Sacks is warning of the dangers of acting like a Martian who has landed in a city and is without the myriad methods and experiences we have for making sense of our local environment. Moreover Sacks is saying that what we see in any setting is tied to the fact that we are participants and that there are classifications that we *are* able to make almost instantly and definitely pre-theoretically as part of our natural attitude to the world. In doing participant observation of places we are already competent inhabitants of and can take their appearance for granted, the solution to doing adequate descriptions of them is not to import strange labels for the things we see or hear or otherwise sense almost instantly. It is the categories that the locals do and would use to describe their observations that we are interested in.

If you are a 'local' already you have huge advantages in providing adequate descriptions of how and why things get done in the way they get done. Yet you are also at the disadvantage of no longer noticing how such things get done because they are so familiar as to be *seen but unnoticed* and you may never have attempted to make them into any kind of formal description (Garfinkel and Sacks, 1986). The exercise Sacks set his students is one that you might also 'try at home without supervision' (see Exercise 9.1).

Exercise 9.1

In a public place such as a busy street, university library or park, set yourself down with a notebook, camera or camcorder. Watch the people there and look for exchanges of glances between people who are not otherwise interacting with one another. Write notes at the time and afterwards if you have video to re-view and check your observations. The notes are to be on what kind of persons exchange glances with what kinds of second (or third persons). Try to consider what kinds of actions take place during exchanges: recognition, snubs, reprimands, warnings, challenges, thanks. Can exchanged glances be hostile? Friendly? Flirty? Defensive? And if so, how so?

After gathering your observations they can be further described and analysed in order for you to consider the social categories you have used. For instance, that a 'well dressed woman' exchanged 'friendly' glances with a 'mother and baby', or that a 'teenage girl avoided the glance of another teenage boy'. What you will start to show in your analysis is something of what you know and use already to make sense of your everyday life. Moreover, if you carry out this exercise you should start to get a grasp on how it is that we use glances and how you are able to see what someone else sees. That is, you can see what it is they are looking at, not so much from working out the exact focus of their look but by seeing what it is in the scene that they would glance at or who they would exchange glances with (i.e. a beggar lying in a doorway, a librarian dropping a box of journals, a kid on a skateboard). This exercise is done with a minimum level of participation by you as a researcher and a minimum level of disruption to the place you are investigating. While easy to do, the test is to get a really good description done that makes available how glancing works as a social and cultural activity. Your description will be adequate when you show some parts of how glancing works in the particular observations you have in hand. This exercise in doing a participant observation is illustrative of a broad type of participant observations, which are those of our common or everyday life-world.

Things will be slightly different if you have pursued participant observations of new and uncommon sets of skills such as mathematical proving, walking a police officer's beat (Sacks, 1972), learning mountaineering, nursing the dying, living among the homeless on the streets (Rose et al., 1965), dropping in at a drop-in centre for the mentally ill (Parr, 2000) or using a car as a workplace (Laurier and Philo, 2003). From these more practically ambitious projects, as I suggested earlier, your adequacy as a commentator turns on you having learnt things which lay members and indeed geographers cannot be expected to know. This certainly makes delivering 'news' easier since unlike 'exchanging glances' or 'answering the telephone' or 'buying a newspaper', not everyone knows how these more obscure, expert, secret or exotic activities are done. Not everyone could tell whether what you were saying about these activities indicated that you really know what you were talking about or whether it was sense or nonsense. For that reason you should consider testing the veracity of your descriptions on the people they were purported to represent.

Competence in the particular field you have selected will be one way in which your comments will attain a reasonable degree of adequacy. You yourself and other competent members like you will be able to see your comments as closely tied to the activities they are describing. Ideally you will be able to show some things that are in many ways known already but have simply never been closely described and analysed before. Doing adequate descriptions is already a challenge but in this second case is certainly no easy matter. It will be additionally hard because you will be trying to write or speak to two different audiences: towards the more abstract concerns of human geographers and other researchers and, just as importantly, writing for and as a member of the group you are describing.

RESULTS: RESPECIFYING THE GENERALITIES OF SOCIAL SCIENCE

Human geography, like most other social sciences, has a host of big topics such as power, class, race, sexuality and gender. If like Charles Darwin your 'power to follow a long and purely abstract train of thought is very limited' then these classic topics may not best be pursued as purely theoretical matters. As big-picture issues they sit uncomfortably with the more modest and small-picture concerns of actually doing a participant observation. Are you really going to resolve disputes that have dogged the social sciences for a century from your study of a drop-in centre? Perhaps not, but you may be able to respecify what appear to be big abstract problems into worldly, ordinary practical problems.

And what might your results look like? Your results should be ones that you could *not* have guessed from the big topics. Nor should you have been able to imagine them before doing your study, or else why bother going to have a look if you can work it out without ever consulting anything in the world? Darwin's methods were to pay close attention to animals in all kinds of places, including his pets at home, and observe them in extended detail in ways and to ends that had not been pursued before. Despite their quite ordinary non-technical provenance, Darwin's patient observations of animal life overturned our view of ourselves, our origins and our relation to nature and God. Now *that* is a result. But what has been made of his work should not be confused with the actual lowly and lengthy observations he documented that ended up being drawn into these larger conflicts. His 'results' were mostly descriptions, alongside sketches and still photographs, interspersed with reiterations of longstanding problems in biology and zoology and occasional bursts of inspiration and insight.

The strengths of participant observation are, hopefully, quite clear by now in that it is easy to do and it provides a more direct access to phenomena than some of the more complex methodologies of social science. It allows you to build detailed descriptions from the ground up that should be recognizable to the groups whose lives you have entered into. Its limits are that it does not have a handle that you can turn to make results pop out. Nor can it be shoe-horned in as a replacement for statistical methods since it will provide only very weak answers to the kinds of questions that could be hypothesis tested.

The kind of evidence that arises out of a one-off description allows your study to bring into view certain types of phenomena that are too complex for methodologies that seek and detect general features. Good data from a participant observation can be and usually are a particular instance of some practice or event or feature that elicits your interest. Sometimes the instance is an exception to the rule that teaches you the rule such as when someone does a foot fault at Wimbledon. At that moment you discover a little bit more about the game of tennis that you never knew before. It is not a well-known nor commonly broken rule but it is there in the game. Sometimes an instance is simply one that you find recurring all the time, such as in the video notes example from earlier. Marge packed and repacked her car boot hundreds of times a month. Looking at one instance of it revealed methods that she used each and every time she did her packing. If you consider that in each and every place the locals have at hand just what they need, then and there, to produce locally the spatial phenomena and interactional events you are observing, then it becomes apparent that wherever you start there is material for your analysis.

There is no need for you to climb a ladder to get a view nobody else has. As Wittgenstein (1980) remarks:

> I might say: if the place I want to get to could only be reached by a ladder, I would give up trying to get there. For the place I really have to get to is a place I must already be at now. Anything I might reach by climbing a ladder does not interest me.

FINAL WORDS OF ADVICE

It is with a mild hint of self-irony that my finishing words are: avoid reading books which claim to describe 'how to do' participant observation. If you must, read just the one and then throw it away afterwards. A preparatory way to learn how this kind of fieldwork is done, if you cannot get hold of someone else who has done participant observation already, is to read an actual study which has been based in participant observation (see the further reading at the end of the chapter). None of these is a substitute for doing a study yourself. Much like learning to play the piano or work out a mathematical proof or describe what a strawberry tastes like, you have to take a bite. Being told how playing the piano or maths or tasting is done, in a book, does not and cannot provide what you need to know. Participant observation is not difficult, nor obscure, though the topics, places, people, subjects and more to which you apply it may be. Since it acquires the shape and scale of its phenomena, in your first studies choose things you reckon you can handle.

Summary

* Participant observation is a simple skill of doing and watching that we all do as part of our everyday lives without realizing it.
* It is important to participate as well as just observe.
* This approach can be applied to new places and practices but can also be used to make visible familiar places and practices.
* As a participant observer you need to keep written fieldnotes and/or video notes of your research.

- The data collected need to be analysed like other empirical material and can be used to understand and make sense of more abstract problems.

Further reading

The following books and articles provide good examples of participant observation in practice:

- Crang (1994) is an example of participant observation used to examine how a waiter's work gets done and how looking closely at this work teaches us about surveillance and display in workplaces.

- Goffman (1961) is a classic from a very readable author, by turns thought-provoking, amusing and disturbing. It is not just geographers who have found this book of value; it is a key text on many courses in sociology, social psychology, psychiatric nursing and doctoring.

- Harper et al. (2000) is based on two of the authors spending time working alongside the employees of new telephone-banking facilities and traditional banks and building societies. This study shows how participant observation can be carried out in a business environment.

- Liberman (1985). Participant observation is perhaps best known as a way of studying 'exotic' cultures where the everyday is rendered strange. This study offers us insights into the practical reason that shapes normal life among aborigines.

- Livingston (1987) is a very accessible and practical introduction to what is a sometimes puzzling and certainly distinctive approach in the social sciences which heavily utilizes participant observation.

- Sudnow (1978). Mentioned earlier in the main text, this book was a best-seller when released and is read by both academics and musicians. In addition, if you can play the piano then it can be used as a tutorial in the basics of jazz. For those unfamiliar with playing music, some sections will be hard to grasp since it relies on a basic knowledge of notation, chords, etc.

- Wieder (1974). Based on the author's residence as a researcher in a halfway house for ex-convicts, this study illuminates how the 'convict's code' is used as a device for making sense of and producing events at the halfway house. It provides a good basis for seeing how a particular place and its inhabitants organize their everyday lives.

Note: Full details of the above can be found in the references list below.

References

Crang, P. (1994) 'It's showtime: on the workplace geographies of display in a restaurant in South East England', *Environment and Planning D: Society and Space*, 12: 675–704.

Garfinkel, H. and Sacks, H. (1986) 'On formal structures of practical actions', in
    H. Garfinkel (ed.) *Ethnomethodological Studies of Work*. London: Routledge &
    Kegan Paul, pp. 160–93.
Goffman, E. (1961) *Asylums*. Harmondsworth: Penguin Books.
Harper, R. and Hughes, J.A. (1993) ' "What a f-ing system! Send 'em all to the
    same place and then expect us to stop 'em hitting": making technology work
    in air traffic control', in G. Button (ed.) *Working Order: Studies of Work,
    Interaction and Technology*. London: Routledge, pp. 127–44.
Harper, R., Randall, D. and Rouncefield, M. (2000) *Organisational Change and
    Retail Finance: An Ethnographic Perspective*. London: Routledge.
Laurier, E. and Philo, C. (2003) 'The region in the boot: mobilising lone subjects
    and multiple objects', *Environment and Planning D: Society and Space*.
Laurier, E., Whyte, A. and Buckner, K. (2001) 'An ethnography of a cafe:
    informality, table arrangements and background noise', *Journal of Mundane
    Behaviour*, 2 (http://mundanebehavior.org/issues/v2n2/laurier.htm).
Liberman, K. (1985) *Understanding Interaction in Central Australia: An Ethno-
    methodological Study of Australian Aboriginal People*. London: Routledge and
    Kegan Paul.
Livingston, E. (1987) *Making Sense of Ethnomethodology*. London: Routledge
    and Kegan Paul.
Lynch, M. (1985) *Art and Artifact in Laboratory Science: A Study of Shop Work
    and Shop Talk in a Research Laboratory*. London: Routledge.
Malbon, B. (1999) *Clubbing: Dancing, Ecstasy and Vitality*. London: Routledge.
Parr, H. (2000) 'Interpreting the hidden social geographies of mental health: a
    selective discussion of inclusion and exclusion in semi-institutional places',
    Health and Place, 6: 225–38.
Rose, E., Gorman, A., Leuthold, F., Singer, I.J., Barnett, G., Bittner, E. and
    O'Leary, J.C. (1965) *The Unattached Society: An Account of the Life on
    Larimer Street among Homeless Men*. Denver, CO: Institute of Behavioral
    Science, University of Colorado.
Sacks, H. (1972) 'Notes on police assessment of moral character', in D. Sudnow
    (ed.) *Studies in Social Interaction*. Glencoe, IL: Free Press, pp. 280–93.
Sacks, H. (1992) *Lectures on Conversation. Vol. 1*. Oxford: Blackwell.
Sudnow, D. (1978) *Ways of the Hand: The Organization of Improvised Conduct*.
    London: MIT Press.
Watson, D.R. (1993) 'Fear and loathing on West 42nd Street: a response to
    William Kornblum's account of Times Square', in J.R.E. Lee and D.R. Watson
    (eds) *Final Report: 'Plan Urbain' Interaction in Public Space (Appendix II)*.
    Manchester: Manchester University, Department of Sociology.
Wieder, D.L. (1974) *Language and Social Reality: The Case of Telling the Convict
    Code*. The Hague: Mouton.
Wittggenstein, L. (1980) *Culture and Value*. Trans. P. Winch. Blackwell: Oxford.

10 Geography and the Interpretation of Visual Imagery

Rob Bartram

Synopsis

'Visual imagery' is a broad term that embraces cinematic film, photography, promotional materials, computer games, etc. Geographers are interested in the way key concepts like space, place and landscape are used and created in visual imagery. The visual methodologies they adopt are based on philosophical and theoretical ideas about the interpretation of visual imagery that have been developed in other disciplines. Visual methodologies involve thinking not only about the choice and production of images and image aesthetics but also about audiences and the sensory affects of visual imagery.

The chapter is organized into the following sections:

- Introduction
- Visual imagery: some key philosophical and theoretical ideas
- Visual imagery and cultural meaning
- Interpreting visual imagery
- Additional thoughts – image affects.

INTRODUCTION

Geography is a highly visual subject. Geographers place great emphasis on the importance of visual imagery to substantiate claims about landscape, place and process – in lectures, in research papers and in coursework. At a populist level, through journals such as the *National Geographic*, geography is as much about the quality of visual representation as textual or cartographic detail. Indeed, my early interest in geography stemmed precisely from an interest in the photography that accompanied such publications. Yet while visual imagery has formed

a dominant way of expressing geographical knowledge, our interpretation of visual imagery has sometimes lacked a critical awareness:
all too often, visual imagery has been used as a straightforward
reflection of reality, with no sense of how, when and by whom the
image has been produced. I think this can be partly explained by the
complexity of philosophical and theoretical debates that are involved
with the interpretation of visual imagery. I am sure this has acted as a
deterrent to geographers who have wanted to know more. But I am
also fairly certain that the lack of critical engagement with visual
imagery can be attributed to an unfortunate misconception about
geography – that only cultural geographers should be concerned with
the interpretation of visual imagery. The main purpose of this chapter,
then, is to put forward some rudimentary ideas about how geographers
can go about interpreting visual imagery, irrespective of their interests
in the discipline. I will begin this chapter by unravelling some of the
key philosophical and theoretical ideas relating to the interpretation
of visual imagery. I will then explore the relationship between visual
imagery and cultural meaning and provide some practical advice
about choosing and using appropriate visual materials. For the interpretative method itself, I will highlight three areas of concern: the
production of the image, image aesthetics and interpreting audiences.
Finally, I will offer some additional thoughts on the sensory affects of
visual imagery.

It should be noted from the outset that the ideas expressed in this
chapter are not put forward as a rigid framework for the interpretation
of visual imagery. Rather, they provide some useful pointers as to how
you might work with visual materials, and how you might allow ideas
and issues to emerge through the interpretation process itself. Some of
you might be concerned about some of the implicit ambiguities of this
'non-prescriptive' methodology. Don't be. One of the most stimulating things about interpreting visual imagery is that it is never predictable. You never quite know what you will end up seeing.

VISUAL IMAGERY: SOME KEY PHILOSOPHICAL AND THEORETICAL IDEAS

Throughout this chapter I use the phrase 'visual imagery'. It is a broad
term of reference that embraces diverse visual forms, from cinematic
film and photography to promotional materials, art and computer
games. In other relevant literature, you will sometimes find reference
to 'visual culture' or, simply, 'the visual'. Although I prefer to use the
term 'visual imagery' to describe a myriad of visual forms, this does
not suppose that all types of visual imagery can be lumped together
and approached and interpreted in the same way. Indeed, you need to

be sensitive to the codes, conventions and genres associated with each particular visual image.

Geographers have shown a keen interest in visual imagery in the last 15 years with some notable work being carried out on cinematic geographies (Aitken and Zonn, 1994; Clarke, 1997), landscape art (Cosgrove and Daniels, 1988; Daniels, 1993), landscape photography (Kinsman, 1995) and place advertising (Burgess and Wood, 1988). Geography has clearly become attuned to the dominance of visual culture, 'where knowledge as well as many forms of entertainment are visually constructed, and where what we see is as important, if not more important, than what we hear or read' (Rose, 2001: 1). Geographers have been preoccupied with understanding how power relations work in the production and reception of visual imagery. For example, Gillian Rose (1995) has argued that Gainsborough's portrait of *Mr and Mrs Andrews*, painted in 1748, has come to symbolize the quintessential English landscape, despite its partial, 'southern' perspective.

The attempt to interpret visual imagery has been characterized by the forging of interdependent links between the aesthetics of the image (for example, symbolic content, colour, atmosphere, perspective) and wider social meaning. So, in the case of Gainsborough's portrait of *Mr and Mrs Andrews*, it has been argued that the portrait is not just a likeness of the two landowners, or indeed their land, but a portrait that we can interpret as symbolic of English national identity.

Geography's interest in visual imagery can be related to the broader contours of twentieth-century philosophy that have challenged the dominance of visual culture and its implicit separation of viewer and viewed, subject and object, individual and world. To put this simply, there is a range of moral and intellectual dilemmas posed by how we view the world, and how this relates to individual and collective empowerment. What makes this debate all the more intriguing is that for some commentators such as Jean Baudrillard, the contemporary world is becoming 'hypervisible' – modern technological developments in media and telecommunications have made much of the world present to us in real time (see Zurbrugg, 1997). With such an array of visual images, how do we know what is real or authentic? More importantly, how are these visual images manipulated to exercise power? Meanwhile, Paul Virilio has expressed concern over the speed of telecommunication transmission and how this might induce a form of 'social inertia' – i.e. an inability to be socially and politically active in a purposeful manner (Virilio, 2000).

We don't need to get too bogged down in this debate, but it is important to bear in mind that interpreting visual imagery carries with it the weight of extensive philosophical debate. I will leave it up to you to decide how far you want to pursue your interest in the philosophical debate, although there are several references at the end of this chapter that should help if you want to take your interests further. I will take just two ideas from this preliminary discussion and expand on them in further detail: what is the relationship between visual imagery and cultural meaning? How do we go about interpreting the meaning of visual imagery?

VISUAL IMAGERY AND CULTURAL MEANING

It should already be apparent that there is a strong link between visual imagery and cultural meaning. To be more precise, visual imagery *always* produces cultural meaning, whether it involves passing a photograph around family and friends in your sitting room or watching a blockbuster film at the cinema. We can understand this relationship in terms of 'sign' and 'signification' – the visual form that has been 'encoded' with meaning, and its 'decoding', or interpretation. Cultural signs have systems of reference points – referents – that allow us to interpret their complex and interconnected meaning despite their often minimalist representation. Put simply, it allows certain signs to be symbolic of additional or associated ideas and images. Think of how corporate logos such as those used by Shell or Nike, have become reduced to very basic signs. The Nike 'swoosh' says it all; it symbolizes all that we need to know as consumers when we purchase Nike goods. Modern art and even advertising work on very similar principles – we can reduce complex meaning to a single colour, texture or form. For example, Damien Hirst's art has become well known for its simplicity. In the past, his work has included the suspension of cows, sheep and pigs in formaldehyde preservative liquid. There is no intrinsic meaning *in* these basic compositions, but their symbolism prompts us to ask all sorts of questions about western cultural views on animal welfare, and even about life and death.

Cultural signs are also seldom static. They connect, and are continually connected, to other signs, and we would be foolish to think that we can capture them, pin them down to a 'true', essential meaning. As you will come to appreciate if you interpret visual imagery, it is easier to suggest *how* cultural meaning is manipulated than discover its essential truth. Furthermore, we should not assume that cultural signs have an original condition or form that we should privilege with greater meaning: reproductions of well-known visual images can be powerful in their own right. For example, John Constable's *The Hay Wain*, painted in 1820–1, has been reproduced on tea towels, calendars, biscuit tins and other consumer goods. The reproductions of *The Hay Wain* are just as important in terms of cultural meaning as the original painting.

INTERPRETING VISUAL IMAGERY

Students interpreting visual imagery frequently ask me to provide them with a specific method they can use. Although we can describe the interpretation of visual imagery as form of 'textual analysis' (see Chapter 28), there is no singular or formal visual methodology that can be referred to. Besides, it is far more important to think about the pragmatics of interpreting visual imagery than to think of interpretation as a strictly applied science. So on this basis, here are some practical things to think about.

Choosing and using an appropriate visual image to interpret

If you are fortunate enough to be given the freedom to choose a visual image to work with, my advice is to select an image that you have either produced yourself (typically, from your own photography) or an image that particularly inspires or interests you. This will ensure that you will overcome a hurdle often identified by geography students – that without necessary training in media or arts-based courses, they lack the necessary skills and knowledge to interpret visual imagery. If you are already familiar with the visual image, this can sometimes generate sufficient confidence for a detailed interpretation to unfold. While students increasingly plum for cinema film and occasionally landscape art and photography to conduct their interpretation, you could also think about documentary film, modern art, computer games, Internet websites or 'non-commercial' photography and video as potential sources of interesting visual imagery.

You should then think about doing some background research. Basic information can be tracked down by using a search engine on the Internet. You will find that most contemporary films, artists and photographers have their own websites, although there are some generic websites such as the International Movie Database (IDMb). Look out for critical reviews, perhaps through newspaper websites or through e-journals. Of course, there is the university library – be prepared to explore the journals that do not form part of your regular geography reading list such as *Frieze, Modern Painters* and *Sight and Sound*. Finally, you will find exhibition and gallery catalogues a fruitful source of information.

You need to be careful with your use of visual imagery. For example, if you intend to use visual imagery for a seminar presentation, you need to spend time thinking about how your choice of visual image will relate to the oral part of the presentation. If you are interpreting art or photography, it is better to choose just a few, indicative images and explain them carefully and in depth, than try to present too many and provide only a superficial interpretation. Equally, if you intend to interpret a film, do not attempt to show clips of longer than, say, ten minutes. Quite often, students feel compelled to show an extended clip in an attempt to explain the whole film. A brief synopsis of the film at the beginning of your presentation will suffice. If you have time and editing facilities to hand, you can of course put together the most important scenes from the film on one video.

Interpreting the visual image

Once you have chosen an image, or have been presented with an image to work with, there are few pragmatic issues to bear in mind before moving on to the more detailed aspects of interpretation. You need to be familiar with the visual image before you interpret it. This involves spending time watching and viewing, making a note of your initial impressions and taking time away from the image for further reflection. As much as anything, this sharpens the eye and makes you attentive to minute detail. It is, after all, the minute detail that provides the most interesting and intriguing ideas about visual imagery (see Exercise 10.1).

Exercise 10.1   **Interpreting historical photographs**

This activity is designed to hone your interpretative skills and involves the use of historical photographs that depict a local scene that is familiar to you. Historical photographs can be sourced from local libraries, the university library or from local publications. Take just one photograph of a street, park, building or agricultural land, and try to locate the scene on a detailed map. Once the scene has been located, try to replicate the same shot by taking your own photograph. When you have got the two photographs side by side, begin to compare and contrast them. Don't just concentrate on the obvious differences between the two images but look at the more subtle changes as well. For example, look at the differences in terms of architectural design, building materials, landscaping and boundaries. This exercise will accustom you to the depth of interpretation that can be achieved and the most appropriate method of recording this information. If the exercise is performed well, you should be able to express ideas about how and when the photograph was taken, and how this might influence the representation of the scene chosen.

We can now turn to the detailed part of the interpretation. This will involve three areas of concern:

1   production of the image;
2   image aesthetics;
3   interpreting audiences.

It might well be that you focus on only one of the areas of concern. If this is the case, you should still bear in mind that the three areas are interlinked.

Production of the image

Taking the production of the image as the first area of concern, we can begin to ask some interpretative questions:

- Who produced the image?
- What do we know about the producer of this image in terms of background education and training, and social identities (age, ethnicity, gender, sexuality)? Do these influences inspire the production of the image?
- Who commissioned the image, when was it made and when was it exhibited/shown/broadcast?

Image aesthetics

Next we can consider image aesthetics:

- Describe the image (easier than it sounds!): try to convey a sense of how the image works in terms of colour, composition, atmosphere, angle of view, perspective, speed and tempo, sound, narrative.
- Try to identify the symbolic elements of the visual image – how does the image relate to other cultural images and ideas?
- How does the visual image relate to specific cultural genres? Is it indicative of a particular moment in cultural history? Does it represent a departure from the conventions of its genre?
- How does the exhibition/display/broadcast of the visual imagery affect its critical reception?

Interpreting audiences

While the first two areas of concern are relatively straightforward, the interpretation of audience engagement with visual imagery is far more complicated and can require extensive, painstaking research. Indeed, geographers, with a few notable exceptions (for example, Burgess et al., 1991), have preferred to limit their interest to the production of the image and image aesthetics. A compromise can be reached. We can *infer* how, why and where audiences engage with visual imagery. We can also make inferences about the intended audiences by asking questions about the availability of the image. For a more detailed interpretation of audience engagement with visual imagery, we would have to use focus groups, in-depth interviews or questionnaires involving participating audiences (see Chapters 8 and 6, respectively) (Exercise 10.2).

Exercise 10.2 Interpreting David Lean's film adaptation of *Oliver Twist* (1948)

Let me explain how I initiate the interpretation of visual imagery by using cinematic film. I use an exercise that involves a basic interpretation of *Oliver Twist*, a well-known film. The exercise is designed to inspire confidence in the interpretative procedures already discussed in this chapter, and to provide a sense of the depth of interpretation that can be achieved. At the beginning of the exercise, I point out that it takes time to develop a sophisticated level of interpretation, to know what to look for and what to do when you hit the inevitable 'dead-end'. After explaining in detail the key areas of concern, I then show my students a series of short clips from David Lean's 1948 film adaptation of *Oliver Twist*. The film should be fairly familiar to the students because it is shown quite frequently on terrestrial TV and, if it is not familiar to them, they should at least be familiar with the plot and main characters from reading the novel. I make sure the students get to see a variety of landscapes, scenes and settings used in the film, from the opening sequence that shows the bleak, wide panoramas of the moorland that Oliver's mother has to struggle across to get to the workhouse, through to the closing scene that reveals the labyrinth of dark and dingy alleyways that make up the East End of London. I show the scenes sequentially so that I can provide a brief overview of the story at the same time. I then get the students to work in twos or threes to answer the following questions about the film's aesthetics:

• How does David Lean create different moods and atmospheres in this film? (The kinds of prompts I use relate to dialogue, soundtrack, camera angle, lighting, film continuity and visual symbols.)
• How does David Lean make use of landscape to create these moods and atmospheres? (The prompts I use to address this question relate to scale of the scene or setting and how this might be enhanced through cinematic effects by the use of camera lenses, lighting, etc.)

This film is particularly useful to inspire confidence in interpreting visual imagery, not least because symbolic landscapes can be clearly identified. It can also be used to have a wider discussion about the importance of relating the production of visual imagery to the appropriate moment in cultural history: some students are initially tempted to make the links between this film and the Victorian period in which it is set. However, they soon realize that the film is more appropriately discussed in the context of British postwar culture. The film clearly relates to a sense of postwar optimism and pride. For example, David Lean's dramatic use of St Paul's Cathedral as a backdrop can be related to the symbolic status of the cathedral during the war. St Paul's Cathedral became emblematic of the national war effort, and specifically the 'Blitz spirit' of London's East End communities.

This exercise requires little background preparation and even with a basic under-standing of British postwar culture, some of the important links between the image aesthetics and the context of production can be recognized.

ADDITIONAL THOUGHTS – IMAGE AFFECTS

So far I have emphasized how we can interpret visual imagery by using the concepts of sign and signification. There are problems with using these concepts, not least because they rely on the assumption that the viewer is detached from the image and able to *extract* meaning from it at will. As Jean Baudrillard and other commentators have recognized, when we try to extract meaning from visual imagery, sometimes the visual image becomes just *too* good to be true because it yields meaning all too easily. This is what can happen when we turn to a visual image already knowing the meanings that we want to find. Interpreting visual imagery then becomes a ritual of 'discovering' what we already know.

There are alternative ways of interpreting visual imagery and we can turn to these if we reconceptualize vision itself. If we accept that viewing the world is also about being in the world and not detached from it, visualizing imagery becomes meaningful in its own right. Let me explain this carefully. To view the image is to experience a visual sensation. We can 'enter' into a film, a photograph, a painting and 'feel' movement, tempo, touch, smell, taste, noise and atmosphere. The image is a sensory affect, whether it causes us to laugh or cry, experience fear or engenders compassion. It doesn't matter that these images are invariably not real, they still stimulate the senses. How many times have you watched a hospital-based soap opera on TV and had to turn away at the sight of a gory operation? The operation is realistic but it is not real, and yet we experience real feelings when we watch it.

Visualizing the image is just one sensory affect. We can also acknowledge the importance of bodily senses in the making of the image itself. For example, artists such as Stanley Spencer wrote extensively about the significance of painting to his personal well-being. The act of painting, and the skills and techniques that it entailed, was Stanley Spencer's way of sensing and making sense of the world. It allowed him to be immersed in community life, took him to remote and idyllic spots in the countryside and allowed him to contemplate the rhythm and routines of his own life.

So we should not be dismissive of the sensory affects of visual imagery. On the contrary, we should recognize that they determine how we think, feel and act, and that we produce distinctive geographies accordingly. Anthropologists, psychologists and therapists have long understood the sensory affects of visual imagery. There is no reason why geography cannot embrace them too (see Exercise 10.3).

Exercise 10.3　　**The moving image**

This activity requires you to think about visual 'affects'. Choose a piece of art, photography or a film clip that you find emotionally powerful. For example, I find Edvard Munch's *The Scream* a particularly harrowing image – you can almost *feel* yourself being sucked into the vortex of pain in this image.

When you have found an appropriate image, write about the thoughts and feelings that it inspires in the form of expressive prose, preferably over different time-frames (this will get you to think about visual 'affects' being transformative and emergent over time). Write specifically about the impact of colour, texture, 'framing', scale and some of the more abstract feelings that you might have about movement, tempo, touch, smell, taste, noise and atmosphere. If you give enough time to this part of the exercise, you find that you have a rich, descriptive account.

In small discussion groups, try to share some of your ideas and perhaps reflect on why your chosen image engendered the particular feelings it did. As a concluding part of the discussion, you might want to discuss the problem of 'translating' your thoughts and feelings to the written word, and the more general problem of writing expressively.

The exercise lends itself to the interpretation of modern and mostly abstract art, although appropriate images could be taken from contemporary cinema or even the Internet.

Summary

* Geographers need to engage with the interpretation of visual imagery in a critical fashion.
* The interpretation of visual imagery is intricately linked to philosophical and theoretical debates about the production and experience of culture.
* Visual imagery can be explored through an understanding of the relationship between the 'sign' and 'signification'.
* We can interpret visual imagery by asking detailed, pragmatic questions about the production of images, image aesthetics and audiences.
* Visual images affect. They inspire us to think, feel and act, and we produce distinctive geographies accordingly.

Further reading

There is a growing number of books and articles written on the interpretation of visual imagery. I have highlighted some of the most important contributions written by geographers below:

* Rose (1996) is by far the most accessible explanation of visual methodologies. The article contains a table of key questions you could refer to when interpreting visual imagery. However, these questions should be used as prompts and cues for interpretation rather than 'applied' systematically.

* Rose (2001) takes the explanation of visual methodologies further by making necessary connections to theories of the visual.

The remaining books and articles relate to specific forms of visual imagery. In all cases, they exemplify the interpretative method outlined in this chapter. On photography, see

Kinsman (1995) and Crang (1997); on film, see Short (1991), Aitken (1994), Aitken and Zonn (1994) and Clarke (1997); on art, see Cosgrove and Daniels (1988), Daniels (1993) and Crouch and Toogood (1999).

Note: Full details of the above can be found in the references list below.

References

Aitken, S.C. (1994) 'I'd rather watch the movie than read the book', *Journal of Geography in Higher Education*, 18: 291–307.

Aitken, S.C. and Zonn, L.E. (1994) *Place, Power, Situation and Spectacle: A Geography of Film*. London: Rowman & Littlefield.

Burgess, J., Harrision, C. and Maiteny, P. (1991) 'Contested meanings: the consumption of news about nature conservation', *Media, Culture and Society*, 13: 499–519.

Burgess, J. and Wood, P. (1988) 'Decoding Docklands: place advertising and the "decision-making" strategies of the small firm', in J. Eyles and D.M. Smith (eds) *Qualitative Methods in Human Geography*. Cambridge: Polity Press, pp. 94–117.

Clarke, D. (ed.) (1997) *The Cinematic City*. London: Routledge.

Cosgrove, D. and Daniels, S. (eds) (1988) *The Iconography of Landscape: Essays on the Symbolic Representation, Design and Use of Past Environments*. Cambridge: Cambridge University Press.

Crang, M. (1997) 'Picturing practices: research through the tourist gaze', *Progress in Human Geography*, 21: 359–73.

Crouch, D. and Toogood, M. (1999) 'Everyday abstraction: geographical knowledge in the art of Peter Lanyon', *Ecumene*, 6: 72–89.

Daniels, S. (1993) *Fields of Vision: Landscape Imagery and National Identity in England and the United States*. Cambridge: Polity Press.

Kinsman, P. (1995) 'Landscape, race and national identity: the photography of Ingrid Pollard', *Area*, 27: 300–10.

Rose, G. (1995) 'Place and identity: a sense of place', in D. Massey and P. Jess (eds) *A Place in the World? Places, Cultures and Globalization*, Milton Keynes: Open University Press, pp. 87–132.

Rose, G. (1996) 'Teaching visualised geographies: towards a methodology for the interpretation of materials', *Journal of Geography in Higher Education*, 20: 218–94.

Rose, G. (2001) *Visual Methodologies: An Introduction to the Interpretation of Visual Materials*. London: Sage.

Short, J.R. (1991) *Imagined Country: Society, Culture and Environment*. London: Routledge.

Virilio, P. (2000) *The Information Bomb*. London: Verso.

Zurbrugg, N. (1997) *Jean Baudrillard: Art and Artefact*. London: Sage.

Participatory Research Methods

Myrna M. Breitbart

Synopsis

Participatory research seeks to democratize research design by studying an issue or phenomenon with the full engagement of those affected by it. It involves working collaboratively to develop a research agenda, to collect data, to engage in critical analysis and to design actions to improve people's lives or effect social change.

The chapter is organized into the following sections:

- What is participatory research and why do it?
- Participatory action research principles
- Formulation of the research questions
- Research design and data collection methods
- Some areas of challenge and concern
- With all the constraints, why do it?

WHAT IS PARTICIPATORY RESEARCH AND WHY DO IT?

In 1990, Kenneth Reardon, a professor of urban planning, discussed the prospect of doing research in East St Louis with Ms Ceola Davis, a 30-year-old community outreach worker. She greeted him with the following: 'The last thing we need is another university person coming to East St Louis to tell us what any sixth grader here already knows' (Reardon, 2002: 17). At the second meeting, according to Reardon, Ms Davis indicated on what terms she and others *might* be willing to work in partnership with university faculty and students. In what came to be known as the 'Ceola Accords', she laid down the following terms:

- Local residents, not folks from the university, would decide what issues would be addressed through the partnership.
- Local residents would be actively involved in all stages of the planning process.

- The university would have to agree to a six-month probationary period and, if passed, a minimum five-year commitment.
- The resident group wanted help incorporating as a non-profit (Reardon, 2002: 18).

The suspicion with which Ms Davis viewed outsiders coming to do research in her community is not unusual. Communities are often treated as laboratories, provided no role in the research process and benefit little from the results of studies conducted within their borders. In the case of East St Louis, by the time Reardon arrived, there were already 60 reports sitting on the shelf that had not resulted in any improvements to the neighbourhood. The partnership that has since developed between the city and the university has nevertheless become one of the best examples of participatory research (see Box 11.1).

Participatory research or, as it is more commonly known, partici-patory action research (PAR), involves the study of a particular issue or phenomenon with the full engagement of those affected by it. Its most distinguishing features are a commitment to the democratiza-tion and demystification of research, and the utilization of results to improve the lives of community collaborators. Most definitions com-bine data collection, critical inquiry and action. Where PAR departs most from other forms of social research that also focus on effecting change is in the *means* used to achieve this end. In PAR, the *means* by which data are co-generated and interpretations debated are a key part of the change process.

PAR was developed in part as a response to exploitative research and as a component of a much larger radical social agenda. There is a vast collection of writing on its history, theory and practice (see, for example, Fals Borda and Rahman, 1991; Park et al., 1993; Greenwood and Levin, 1998; Reason and Bradbury, 2001). Many participatory researchers situate their practice within a broader tradition of libera-tionist movements and early critiques of international development work that questioned the purposes, ethics and outcomes of social research done on *behalf* of other people. Fals-Borda and Paulo Freire, in particular, stress the role of participatory research in recovering knowledge 'from below' (Freire, 1970; Fals-Borda, 1982; Fals-Borda and Rahman, 1991). Feminists have also been active in promoting partici-patory research (see Maguire, 1987; Gluck and Patai, 1991; McDowell, 1992; Stacey, 1998). They believe research 'has an obligation to create social spaces in which people can make meaningful contributions to their own well-being and not serve as objects of investigation' (Benmayer, 1991: 160). PAR has also sought to address the root causes of social injustice and was a practice embedded in the Civil Rights and Black Nationalist movements (Bell, 2001: 49).

Box 11.1   The East St Louis Action Research Project

The East St Louis Action Research Project (ESLARP) began in 1987 when a state representative in Illinois challenged the president of the University in Champaign–Urbana (UICU) to serve the research and education needs of this low-income neighbourhood. What began as a series of research projects within the University's Architecture, Landscape Architecture, and Urban and Regional Planning programmes eventually evolved into an ongoing collaboration between many disciplinary departments at UICU and several community-based organizations in East St Louis. Ken Reardon, a former professor at UICU, now at Cornell University, is most responsible for establishing the ESLARP project, and has written about both its accomplishments and its challenges (Reardon, 1997; 2002).

When the collaboration began, community leaders set the ground rules. All research was to be driven by resident-defined needs, and the skills of analysis were to be transferred to community participants by academics. The project began with the collection of data that were used to produce a map of the untapped resources and problems in the neighbourhood. Based on early discussions of this preliminary information, decisions were made about what additional data to collect and how. Utilizing photographs, landownership records, more GIS-produced maps of physical conditions and census data, the city of East St Louis was compared with its suburbs. Patterns of uneven development, documented in a very real and accessible format, became the basis for further research and actions to produce change. School performance was evaluated, street clean-ups were begun and direct actions were organized to move the city to address its problems. Additional research and planning were undertaken to create recreational space for children, affordable housing and to redirect proposed light rail-lines through the neighbourhood. Community partners have also made significant demands on the university to the point of requiring them to create a satellite neighborhood college for adult learners. East St Louis residents choose the curriculum, which includes courses on developing affordable housing, political economy and the ABCs of Organizing. Still under way, this participatory action research project combines technical and capacity building with attention to addressing real problems.

William Bunge's 'Geographical Expeditions' in Detroit at the end of the 1960s and early 1970s provide early examples of the application of participatory data collection within the field of geography. Bunge sought to elicit the perspectives of people living in poverty through creative data collection and analysis, creating 'folk geographers' of neighbourhood residents (Bunge, 1977). Bunge's work embraced some key concepts of PAR:

- The importance of field research as an educational tool.
- The power of knowledge when used to design social actions to make a difference in people's lives.
- The acceptance of professionals as both teachers and students.

One notable product, the 'Geography of the Children of Detroit', creatively maps the landscape from the perspective of children (Field Notes, 1971). While children were not active participants in the research, adult residents were involved in all aspects of data collection.

In the years since Bunge's ground-breaking work, a number of geographers and environmental educators have employed participatory research techniques. Colin Ward pioneered participatory research with youth and the use of the built environment as a resource for critical learning through his work for the Town and Country Planning Association (Ward and Fyson, 1973; Ward, 1978; Breitbart, 1992). Roger Hart took up this work in the 1970s and has continued to develop highly effective models for engaging children from multiple cultural backgrounds in genuine PAR projects around the world (Hart, 1997).

PARTICIPATORY ACTION RESEARCH PRINCIPLES

PAR is not a specific methodology with exact procedures, nor is it about data-collection alone. In a participatory data collection process, however, great value is placed on the knowledge that 'ordinary' people possess (Park, 1993: 3). For this reason, participatory research often relies on less formal data-collection methods and seeks to foster a community's capacity to problem solve and design actions without having to rely solely on outside experts.

The most basic and distinguishing principle of participatory research is sustained dialogue between external and community researchers. It is through dialogue that formal academic knowledge is meant to work 'in a dialectical tension' with popular insider knowledge to produce a more complete understanding of a situation or environment. This form of dialogue involves not only the exchange of information and ideas but also of feelings and values (Park, 1993: 12). PAR is also about sharing power and involves a commitment to see that the outcomes benefit the community in measurable ways by building on assets. In practice, this means that community members should be hired and trained, whenever possible, as research partners. They should have the power to define research objectives and offer opinions about data collection methods, utilizing their lived experience as a beginning basis for the investigation.

Participatory research is often initiated *outside* the university, with academics assuming the role of consultant or collaborator (supplying information and technical skills). More important than technical are personal skills (e.g. the ability to facilitate conversation without dictating choice and an awareness of one's own biases and their impact on research). The research is clearly *not* participatory if you *do everything* – frame questions, conduct the research and write up and disseminate the results. This is true even if you spend a lot of time with community members asking their opinions or having them review your findings.

It is important to remember, too, that participatory inquiry can be used as a tool to build knowledge of community assets and strengths as well as to identify or address problems. The Holyoke Community Arts Inventory that I am currently involved in is part of a broader study of new urban revitalization strategies that use the arts to promote economic and community development. As a participatory research project, we are collaborating with Holyoke artists in an exploration of their own cultural assets. This involves a complicated search for the diverse meanings that residents give to the creative pursuits in their lives. The data collected, as well as the process of generating and making it public, are both meant to contribute to the social change and planning process.

In general, the literature on participatory research provides considerably more detail about the ideology and politics of the approach than about the research process (Reason, 1994: 329). Yet it is the application of the above principles within stages of inquiry that may otherwise be quite similar to traditional social research that best distinguishes its practice. A great challenge in writing this chapter has been the recognition that every participatory research project differs markedly in the methods it employs.

FORMULATION OF THE RESEARCH QUESTIONS

In participatory research, the process of formulating a research topic originates with the people who are affected most by it. In this early stage, community members may form partnerships with faculty and students. It is not uncommon for community-based organizations to want to interview prospective collaborators. These initial meetings prior to data collection can be an important time when knowledge is shared and trust is built.

Data collection in PAR begins with the lived experiences of the collaborators and the acquisition by outside researchers of basic knowledge about the neighbourhood or geographic entity under study. It is also common for co-researchers to try to create a shared base of knowledge. Many techniques can be employed to do this. For the first year of a three-year participatory research project on the labour-managed work co-operatives of Mondragón, Spain, Davydd Greenwood and his team read and discussed existing literature on co-operatives together. They also shared their own individual expertise on particular subjects related to their research (Greenwood, 1991: 95).

Deciding what data to collect and how to collect them depends on whether or not a specific question or topic has been targeted. If the research question is undecided, certain methods may provide an informational basis upon which to determine this question. Some common tools include brainstorming, focus groups, neighbourhood walks and photography story boards. The Holyoke Community Arts Inventory began with a focus group attended by city residents who discussed how their involvement in arts-related activities currently impacts their lives. They also shared future visions of the city were the arts and culture to play a larger role. Ideas generated during the discussion helped to define a research project, which focused on producing a community arts map and learning more about indigenous talent so that this information could be taken into account in current urban revitalization plans.

In working with adults or youth, brainstorming workshops, social mapping and model construction provide other effective ways for drawing out participants' existing knowledge. Whenever the young people who work with YouthPower begin a new project (see Box 11.2), their initial brainstorming sessions are designed to generate opinions about which aspects of their environment they feel are most in need of improvement. One project that resulted in the Youth Vision Map for the Future of Holyoke began with a youth-led walk around the city. Participants recorded on film those aspects of the environment they felt had potential and those in need of immediate attention. These images were then placed on a base map for later assessment and analysis. Similar data-gathering techniques and resource inventories were employed in the East St Louis Action Research Project (ESLARP), where residents used disposable cameras to identify important issues and sites in the neighbourhood with 'untapped' resources. In both projects, a preliminary gathering of information and opinions was used to help participants prioritize the issues they would like to study further.

The *YouthPower Guide* (Urban Places Project, 2000), a PAR project involving Holyoke youth and university faculty and students, provides a step-by-step guide for engaging youth in community planning. With respect to data collection, it includes over 20 activities that can be completed in a one- to two-hour session and that are all youth led. Some of the basic rules it sets out for brainstorming apply to work with adults as well. These include:

Box 11.2   Youth-driven community research and planning in Holyoke, Massachusetts

This participatory action research project is an ongoing effort on the part of youth and collaborating adults in the city of Holyoke, Massachusetts, to acquire more information about the city's strengths and weaknesses and to plan for its future. As one of the oldest planned industrial cities in the USA, Holyoke now has many assets but also suffers the effects of a global economy and decades of deindustrialization. It is also a city divided socially between an older white immigrant and newer Puerto Rican population.

While many organizations are involved in ongoing planning, one after-school programme, El Arco Iris (the Rainbow), has been especially important in the promotion of a participative process. El Arco Iris offers young people aged 10–19 years an opportunity to take art classes and develop skills as youth leaders. TMAD/YMAD (Teens Making a Difference and Youth Making a Difference), along with YouthPower, are adjunct programmes begun by co-directors José Colon and Imre Kepes to initiate community-based research and to design urban improvements. Since they began, young people have conducted extensive research on the city of Holyoke, identified a range of problems and opportunities, and brainstormed possible solutions. My work with this programme began over ten years ago when I initiated a banner design project that combined neighbourhood exploration and critical assessment with the design of visual and public representations of young people's visions for the future of the city (Breitbart, 1995; 1998). I have continued to work with Hampshire College students and the staff of our community-based learning programme to support subsequent youth-initiated planning efforts.

What is most inspiring about this work is the extent to which young people have provided a role model for adult decision-makers in the city around participatory research and alternative planning practices. They have provided the spark for revitalization initiatives that might otherwise have been overlooked. Since 1995, YouthPower has contributed to the development of a Master Plan for Holyoke and directly enhanced the quality of life through the reclamation of vacant lots for parks and community gardens, as well as the production of murals and banners. YouthPower has also become the most active citizen group in the city to help plan a large-scale urban initiative to build a walkway and arts corridor along the historic canals, as one way to tie downtown to its neighbourhoods. It has also been a driving force behind a variety of collaborative research and planning projects with other youth organizations. This has led to the creation of periodic Youth Summits and a Youth Commission to advise the city. YouthPower has collaborated with faculty

Continued

Box 11.2   Continued

and students from the University of Massachusetts in the production of
a book for other youth, *The YouthPower Guide: How to Make Your
Community Better* (Urban Places Project, 2000). Written in an engaging
style, the book provides many techniques for doing participatory, community-based research that results in actions to improve the environment and quality of life.

- beginning with ice-breakers (strategies to get people talking to each other);
- making sure the goal is to generate ideas not make decisions;
- encouraging everyone to participate;
- writing down ideas without debating them; and
- not stopping the process because of occasional silences, as these are often the precursor of more good ideas.

To this list, I would add that any events with children also need to include refreshments and lots of breaks!

Some of the 'dos' and 'don'ts' of brainstorming are illustrated in a meeting convened to design a walkway along Holyoke's historic canals that involved architects, adult residents and youth. Everyone was given colour-coded stickers used to divide themselves into brainstorming teams. Participants were also given red and black stickers to mark their most favourite and next favourite design ideas. Following introductions, the session began with one of the architects showing slides and talking about what factors generate a 'sense of place' in other cities that have redesigned waterfront areas, such as San Antonio, Texas. Youth peer leaders then led a slide presentation of the prior data they had collected about key problem areas in Holyoke. Groups divided up for further discussion. When called back together to share ideas, interesting patterns emerged. Adults, it appears, had come up with many ideas that bore a striking similarity to those shared in the slide presentation by the architect – coffee carts with umbrellas, a microbrewery, wild metal sculptures floating in the canals. Obviously, the presentation of earlier images, though well intentioned, served to limit rather than expand the adults' imaginations, thus defeating the purpose of brainstorming.

RESEARCH DESIGN AND DATA COLLECTION METHODS

Once some preliminary sharing of knowledge takes place, participatory research requires decisions about what further information to

collect and how to collect it. This can lead to discussions about how to train individuals who may be unfamiliar with the chosen research techniques. The choice of research methods in all stages of data collection is driven, however, by more than the likelihood that they will produce desired information. Data collection methodologies are also chosen to ignite a process of personal and social change, liberating old knowledge and nourishing the critical and creative capacities of participants.

Participatory research requires consideration of how the methods chosen further these goals, demystify the research process, encourage maximum involvement and contribute to the increased equalization of power among participants. Full participation in these decisions requires discussion about everything from the wording of questions to their sequencing. It also necessitates an acceptance of democratic process and a full understanding of the decision-making structure and rules of discourse, no matter how young or old the participants (Hart, 1997). Some important things to consider in these deliberations are the range of skills that participants have to offer and the diversity of the group, including the mix of age, gender and cultural or racial background. Issues of safety and access to equipment or transportation may also come into play. In almost all cases, more than one investigative approach is chosen so as to draw upon the widest range of skills (Green, 1997: 61).

Both quantitative (see Chapter 6) and qualitative data-collection methods (see Chapters 8 and 9) are used in participatory research. Common qualitative data collection methods that go beyond brainstorming or photographic documentation and are used with children and adults include interviewing, oral history, drawing, social mapping and local surveys or environmental inventories. Participatory research with adults may also include the use of story-telling and focus groups.

Since many of these data collection techniques are described elsewhere in this book, I will focus here on their specific use in participatory research. Let's begin with *interviews* (see Chapter 8). In the Roofless Women's Action Research Mobilization, a collaborative research project involving the University of Massachusetts–Boston, homeless women, the city and other community-based women's organizations, interviews and focus groups were used to collect data on the causes of women's homelessness (Williams, 1997: 14–17). Based on the experiences of those in the research team who had been homeless, it was decided that they would do individual anonymous interviews with currently homeless women because their prior experiences would elicit trust. Other members of the research team collaborated in the conduct of focus groups with formerly homeless women. Group discussions allowed the women to reflect on the experience of having

been homeless. Some thought here was obviously given to a purposeful division of labour. It is also common in PAR for academics and
community researchers, adults and youth to do paired interviews.
These not only draw on the diverse experiences of those working
together but also allow for differing interpretations of the results
later on.

Oral histories and *group story-telling* are common data collection
vehicles in PAR that enable partners to share perspectives and generate empirical data. In discussing her collaboration as an academic on
a participatory research project with adult women in the El Barrio
Popular Education Program in New York City, Benmayer (1991:
159–60) describes how she and the other women on the research team
discussed and wrote about what they were learning collectively. The
organic process of introducing questions to the group, examining them
and then reformulating the questions and beginning again contrasts
with the more typical linear progression of much social research. It
reflects the fact that PAR can be a 'continuous educational process'
that does not necessarily begin and end with one project (Park, 1993:
15). Ongoing evaluation and critique can generate new questions,
issues and strategies that build upon a deepening understanding of an
issue or topic.

There are many examples in PAR of the use of qualitative surveys.
In the Immigration Law Enforcement Monitoring Program, researchers with the American Friends Service Committee and a local
university collaborated with community-based organizations to collect data on the geography of human rights violations along the Mexico–USA border. Mexican immigrants had found that their anecdotal
information was insufficient to prompt a response and that a graduate
student-designed computer database was too difficult to analyse. The
solution was to design their own documentation form based on their
experiences – one that allowed the people surveyed to categorize their
abuse experiences by type, location, immigrant status, etc. The group
then hired a computer programmer to put this form online at four sites
along the border accessed by immigrant workers and their families.
The result was a highly detailed human rights survey that could be
used to analyse and address the abuses (Williams, 1997: 8–13).

There are many ways to generate spatial representations, from
crude drawings to annotated photographic essays, and various forms of
mapping. *Social mapping* is a particularly useful low-technology data
collection tool that has been widely employed by geographers engaged
in participatory research as a repository for the raw data resulting from
environmental inventories. Social maps encourage critical reflection
and policy analysis and help to set research agendas. They can also
suggest plans for action. All social mapping begins with the acquisition of a good base map of the area you wish to survey. Planning

FIGURE 11.1 Collaborative social mapping in Holyoke

departments are usually accommodating in sharing large-format maps. In Holyoke (Figure 11.1), youth and adults, divided into small groups, enhance the information on base maps by designing their own icons or using self-stick dots of varying colours to designate aspects of the environment they consider important – for example, red dots for areas you like; green dots for locations you hang out in and enjoy; blue dots for places you avoid, etc.

Analysis of the geographical information on social maps compiled by multiple research partners often involves discussion of the relationship between the 'real' and perceived characteristics of an environment. In the early stages of participatory research, these contrasting spatial representations can be useful in identifying the problems or assets of an area from different perspectives based on participants' gender, age, social class or ethnicity. This information can provoke debate and help decide on foci for further research. In doing participatory research with residents of a neighbourhood in the Bronx, New York, Roger Hart used templates representing such things as dangerous places, places used by teenagers alone, etc., to map the differing perceptions of the neighbourhood by diverse groups of residents. The resulting composite map used tapes of the conversations and analysis of spatial patterns to make recommendations for new types of recreational spaces (Hart et al., 1991).

It is helpful for researchers to walk or drive around the study area with cameras prior to social mapping. Annotated photographs can then be used in place of dots or icons to represent and categorize environmental data. In a collaboration between Hampshire College interns and youth from Nuestras Raices, a community gardening and development organization, co-researchers combed the city with papers, pads and cameras to collect data on sites they considered ecologically sound and those they felt were potentially dangerous. Expanding on a set of symbols already devised for the international Green Map project (www.greenmap.com), they produced a map of their findings that included in their definition of 'green' sites, Puerto Rican restaurants and locally owned businesses. Health information compiled through neighbourhood interviews was also mapped, suggesting geographic patterns of illness (such as asthma) that necessitated further examination. The Holyoke Youth Green Map was also used to begin to acquire vacant lots for future community gardens.

The data collection process in PAR has its own educative value, in part because it necessitates the pooling of information and perspectives among diverse research partners. The commonalities and differences in perspective among researchers that visual displays such as social maps present can generate fruitful discussions that imply avenues for further research or action. However, the meaning of information collected through social mapping is not always readily apparent, nor is it always possible to decide upon a focus for research based solely on a visual or graphic representation. Additional discussion among participants is necessary to uncover varied interpretations and help define priorities.

The Youth Vision Map for the Future of Holyoke resulted from an amalgamation of ideas drawn from many brainstorming sessions. It contains bright visual symbols, placed at various sites, of what modifications young people would like to introduce into the urban landscape. Examples include murals, a Store for Safety (discretely to provide youth with safe-sex products), a Teen Café (to hang out in, access cheap food and listen to music) and a Youth Van (staffed by adults and youth to bring youth information about health issues and youth programmes). In this case, because the data suggested several possibilities for action, a summit for all the youth organizations involved in the data collection provided an opportunity for each to discuss what *its* priorities were. The results were shared and each group went home with its own project to work on.

GIS (geographic information systems) is another form of mapping that can be used by geographers as a tool for data analysis, especially with adults. However, the time involved in training and the access required to expensive technology may limit its use. One possibility is for a PAR team with access to a geography department to ask for technical assistance with GIS to analyse their field-generated data. The ESLARP provides many GIS-generated maps for the use of community organizations on its website. It also furnishes limited instruction on how to employ GIS to analyse environmental, economic and social data.

In general, while PAR draws heavily upon a variety of social research methods, it minimizes approaches that are beyond the material or technical resources of the individuals involved. With quantitative data, this may mean the use of descriptive statistics that help to tell a story, allow comparisons to be made or identify causal relationships without requiring a significant level of quantitative expertise. In the ESLARP, census and other statistical data are put online by students and faculty, along with the software that allows community residents to pose questions, plug in the data and produce a result in a number of different formats, including charts, diagrams and maps. Another strategy is to divide the labour so that, while some researchers are collecting information, others are taking those data and putting them in a graphic or statistical form that allows them to be analysed.

Students who want to undertake participatory research should have some working knowledge of basic social research techniques, as described in this book. A community-based learning course, where the reading and analysis of geographic theory and case studies are combined with work on a hands-on field project, is one effective way to gather experience using such techniques. Impressive examples of courses linked to the East St Louis Action Research Project can be found at www.eslarp.uiuc.edu. If data collection methods are also unfamiliar to community researchers, they too should be trained. This transfer of skills of geographic analysis was the core of Bunge's work in Detroit, and is currently under way in Holyoke, as young people are taught the techniques of social mapping, and in the ESLARP, where adult residents interested in acquiring social research skills are offered free courses.

While PAR does not require everyone to be equally involved in all phases of research, there is a strong commitment to encouraging active participation. Utilizing a variety of research methods and a division of labour that consciously seeks to make use of the particular strengths of each collaborator is one way to assure widespread participation in the collection of information and its exchange.

SOME AREAS OF CHALLENGE AND CONCERN

Participatory data collection presents a number of challenges. One is co-ordination of the different skills and levels of participation that

FIGURE 11.2 Youth present their vision map for the future of Holyoke

each partner brings to the process. These differences may be due to 'cultures of silence' inculcated in those whose ideas have never been seriously valued or who lack experience expressing their views. Academics used to a culture of expression may also tend to silence others by jumping in to fill the gaps. Good facilitation skills are therefore essential.

One of the strengths of the ongoing youth projects in Holyoke is the commitment of its facilitators to providing young people with the tools of oral presentation (e.g. making outlines of main notes, using visual aids such as slides), and then allowing the young people to 'run the show' (Figure 11.2). This often involves the exercise of some serious restraint on the part of adults if a young person occasionally misses a point, speaks too quickly or otherwise flounders.

The fact that participatory research seeks everyone's full involvement can present further challenges, given differing levels of personal commitment. Community partners are often over-extended or lack money and resources. Students and faculty also face competing demands on their time and work within a time frame that differs markedly from the 'real world'. The fact that most colleges operate on what I would call the 'Brigadoon' principle, disappearing at crucial times of the year only to reappear and try to pick up where they left off after seasonal breaks, can make PAR projects very difficult to sustain.

Participatory research is time-consuming and predicated on trusting relationships and a commitment to the project's duration. While it may be possible to incorporate this into a semester-long class, it is more commonly used for long-term projects.

Since a project will be evaluated by all participants, it must also simultaneously try to meet divergent goals. These can include the building of community and the development of a critical under-standing of the assets and problems of a place. The writing you may be asked to do as part of a participatory research project may be quite different from that which would be done for a course paper or a professional geography journal. Tensions can develop between 'prac-tically orientated' community partners and 'theoretically orientated' students or faculty (Perkins and Wandersman, 1997).

Just as Ceola Davis made demands on Ken Reardon and his students, community collaborators can also set down the criteria for investigation that academic partners must agree to. These demands can constrain the parameters of a project or redirect its focus. Addi-tionally, since one project is not likely to address all the needs that stakeholders articulate, participatory research almost always begets more projects, soliciting the continued involvement of outside research partners. This issue of sustainability of the partnership presents its own challenges.

WITH ALL THE CONSTRAINTS, WHY DO IT?

For those of us who do participatory research, it is a humbling experience. After years of working with local neighbourhoods on collaborative projects, Randy Stoeker (1997) still confronts what he refers to as the 'haunting question' of how he can conduct this type of research so that it is both 'empowering and liberatory'. He never-theless urges students and faculty not to be so concerned about 'doing the right thing' that they become 'paralysed.' Community collab-orators of the likes of Ceola Davis or the youth at El Arco Iris will always tell you when you have erred, and will always give you the benefit of the doubt if you are honest, respectful and follow through. Furthermore, there are no 'pure' forms of PAR, only degrees of participative practice.

The primary reasons why I and other proponents continue to advocate participatory research as a method rest with our commit-ment to its political aims. We also believe that the data it produces are more likely to be useful, accurate and lead to actions that address people's real needs and desires. When adults and youth in Holyoke came together to consider design elements to incorporate into a proposed walkway along the industrial canals, adults in the room came up with many fanciful ideas for metal sculptures, umbrella food carts and so forth. Youth, in contrast, voiced a range of practical concerns that had emerged from their research prior to the brainstorming session. This included the need for bike racks so they could access the area from neighbourhoods far away, rubbish cans to collect the litter, a police presence so their parents would allow them to come and gardens to enhance the beauty of the space. While the architects may have categorized these ideas as 'details to be thought about later', to the young people they represented key design features crucial to a successful plan.

Nadinne Cruz, Director of the Haas Center for Public Service at Stanford University, posed a thought-provoking question at a recent service learning conference: 'what if our institutions' exercise of social responsibility is to assess how adequate the knowledge of our faculty and students is for addressing critical issues in the world?' In spite of the complexities, experience with participatory research suggests that we are more likely to address this challenge if we collaborate in methodology design, data collection, analysis and action with those individuals and groups most likely to gain from the investigative process.

Summary

- In PAR the collaborative means by which data are co-generated, interpreted and used to design actions play a key role in social transformation.
- In PAR university researchers and their collaborators share knowledge, power and a decision-making role.
- Participatory research begins with the lived experiences of the collaborators and the acquisition by outside researchers of basic knowledge about the geographic entity under study.
- In addition to producing information, PAR data collection methods should draw on the experiences and nourish the critical and creative capacities of all participants.
- Utilizing a variety of data collection methods and a division of labour that consciously builds upon the strengths of each member of the team is one way to assure widespread participation.

Further reading

The following books provide a range of examples of different forms of PAR:

- Hart (1997) provides an introduction to children's participation in community-based environmental planning and identifies certain principles and practices that have worked in diverse cultural settings around the world. It includes detailed case studies and information about concrete methods of participative practice, such as mapping, interviews and surveys. The book is also visually rich with photographs and diagrams.

- Park et al. (1993) begins with a Foreword by Paulo Freire, one of the world's experts on participative research as a vehicle for personal and political transformation. Cases referenced here are drawn from the North American experience and address the relationships among power and knowledge, research methods and social action. An appendix identifies key organizations that promote participatory research.

- Reason and Bradbury's (2001) comprehensive collection of articles on action approaches to social science is directed at an academic audience. It is divided into four sections that address theories and methods of participatory research as well as the application of these approaches and the skills necessary for implementation. It also explores the role of universities in action research.

Note: Full details of the above can be found in the references list below.

References

Bell, E. (2001) 'Infusing race into the US discourse on action research', in P. Reason and H. Bradbury (eds) *Handbook of Action Research: Participative Inquiry and Practice*. London: Sage, pp. 48–58.

Benmayer, R. (1991) 'Testimony, action research, and empowerment: Puerto Rican women and popular education', in S. Gluck and D. Patai (eds) *Women's Words: The Feminist Practice of Oral History*. New York: Routledge, pp. 159–74.

Breitbart, M. (1992) ' "Calling up the community": exploring the subversive terrain of urban environmental education', in J. Miller and P. Glazer (eds) *Words That Ring Like Trumpets*. Amherst, MA: Hampshire College, pp. 78–94.

Breitbart, M. (1995) 'Banners for the street: reclaiming space and designing change with urban youth', *Journal of Education and Planning Research*, 15: 101–14.

Breitbart, M. (1998) ' "Dana's mystical tunnel": young people's designs for survival and change in the city', in T. Skelton and G. Valentine (eds) *Cool Places: Geographies of Youth Cultures*. London: Routledge, pp. 305–27.

Bunge, W. (1977) 'The first years of the Detroit Geographical Expedition: personal report', in R. Peet (ed.) *Radical Geography*. London: Methuen, pp. 31–9.

Fals-Borda, O. (1982) 'Participation research and rural social change', *Journal of Rural Cooperation*, 10: 25–40.

Fals-Borda, O. and Rahman, M. (1991) *Action and Knowledge: Breaking the Monopoly with Participatory Research*. New York: Apex.

Field Notes (1971) *The Geography of the Children of Detroit. Discussion Paper 3*. East Lansing, MI: Detroit Geographical Expedition and Institute.

Freire, P. (1970) *Pedagogy of the Oppressed*. New York: Seabury.

Gluck, S. and Patai, D. (eds) (1991) *Women's Words: The Feminist Practice of Oral History*. New York: Routledge.

Green, L.W. (1997) 'Background on participatory research', in D. Murphy, M. Scammel and R. Sclove (eds) *Doing Community-based Research: A Reader.* Amherst, MA: LOKA Institute, pp. 53–66.

Greenwood, D. (1991) 'Collective reflective practice through participatory action research: a case study from the Fagor cooperatives of Mondragón', in D. Schön (ed.) *The Reflective Turn: Case Studies in and on Educational Practice.* New York: Teachers College Press, pp. 84–107.

Greenwood, D. and Levin, M. (1998) *Introduction to Action Research: Social Research for Social Change.* Thousand Oaks, CA: Sage.

Hart, R. (1997) *Children's Participation.* London: Earthscan.

Hart, R., Iltus, S. and Mora, R. (1991) 'Safe play for west farms'. Unpublished paper.

Maquire, P. (1987) *Doing Participatory Research: A Feminist Approach.* Amherst, MA: Center for International Education, University of Massachusetts.

McDowell, L. (1992) 'Doing gender: feminism, feminists and research methods in human geography', *Transactions, Institute of British Geographers*, 17: 399–416.

Park, P. (1993) ' What is participatory research? A theoretical and methodological perspective', in P. Park et al. (eds) *Voices for Change: Participatory Research in the US and Canada.* Westport, CT: Greenwood Press, pp. 1–20.

Park, P., Brydon-Miller, M., Hall, B. and Jackson, T. (eds) (1993) *Voices for Change: Participatory Research in the US and Canada.* Westport, CT: Greenwood Press.

Perkins, D. and Wandersman, A. (1997) 'You'll have to work to overcome our suspicions', in D. Murphy et al. (eds) *Doing Community-based Research: A Reader.* Amherst, MA: LOKA Institute, pp. 93–102.

Reardon, K. (1997) 'Institutionalizing community service learning at a major research university: the case of the East St Louis Action Research Project', *Michigan Journal of Community Service Learning*, 4: 130–6.

Reardon, K. (2002) 'Making waves along the Mississippi: the East St Louis Action', *New Village: Building Sustainable Cultures*, 3: 16–23.

Reason, P. (1994) 'Three approaches to participatory inquiry', in N. Denzin and Y. Lincoln (eds) *Handbook of Qualitative Research.* Thousand Oaks, CA: Sage, pp. 324–39.

Reason, P. and Bradbury, H. (eds) (2001) *Handbook of Action Research: Participative Inquiry and Practice.* London: Sage.

Stacey, J. (1998) 'Can there be a feminist ethnography?' *Women's Studies*, 11: 21–7.

Stoeker, R. (1997) *Are Academics Irrelevant? Roles for Scholars in Participatory Research* (available at http://uac.rdp.utoledo.edu/comm-org/papers98/pr.htm. Accessed 30 June 1999).

Urban Places Project (eds) (2000) *The YouthPower Guide: How to Make your Community Better.* Amherst, MA: UMass Extension.

Ward, C. (1978) *The Child in the City.* New York: Pantheon.

Ward, C. and Fyson, A. (1973) *Streetwork.* London: Routledge & Kegan Paul.

Williams, L. (1997) *Grassroots Participatory Research: A Working Report from a Gathering of Practitioners.* Knoxville, TN: Community Partnership Center.

12 Working in Different Cultures

Fiona M. Smith

Synopsis

Cross-cultural research is the term used to describe researching 'other' cultures using other languages. This may even work in distant places but can also include working with 'other' communities closer to home. It requires a sensitivity to cultural similarities and differences, unequal power relations, fieldwork ethics, the practicalities and politics of language use, the position of the researcher and care in writing up the research. All these issues are explored in this chapter. The chapter is organized into the following sections:

- Fieldwork in 'different' cultures: understanding difference and sameness
- Working with different languages
- Working with difference, power relations and positionality
- (Re)presenting 'other' cultures
- Conclusion.

FIELDWORK IN 'DIFFERENT' CULTURES: UNDERSTANDING DIFFERENCE AND SAMENESS

The first encounters many geography undergraduates have with fieldwork in different cultural contexts are on organized field courses, by undertaking fieldwork for dissertations or by participating in an expedition abroad (Nash, 2000a; 2000b; see also Chapter 4). At the University of Dundee, we take students to southeast Spain. It is somewhere 'different' from Scotland for students to experience, providing 'an intense learning curve of different environments and cultures', as one student recently put it. Such a 'classic' approach to fieldwork shapes many undergraduate careers in the subject. However, encounters with 'different' cultures need not always involve travel to distant places. Nairn et al. (2000) outline how New Zealand students participated in a field trip which involved meeting migrants to New Zealand from different cultures and thereby seeing New Zealand from a 'different' perspective.

To the popular imagination travel to 'other' places and cultures almost defines 'geography' as a subject, particularly as it is presented in publications such as the *National Geographic* or in television programmes about grand adventures. These contribute to notions about the romance of fieldwork (Stoddart, 1986; Nairn, 1996; Madge and Bee, 1999). Fieldwork, particularly if abroad or in a difficult setting, is often seen as 'character building' (Sparke, 1996: 212), a rite of passage to becoming a 'proper' geographer (Rose, 1992). As attractive as these images are, however, as geographers we have a responsibility to think critically about how we undertake fieldwork in different cultures, whether these are distant from home or just down the road. As a useful starting point, Nash (2000a: 146) argues that 'all geographers undertaking fieldwork overseas [or in more local 'different' cultures] need to be sensitive to local attitudes and customs' in a manner that 'respects the cultural as well as the physical environments you encounter'. Nash suggests a code of ethics applicable to such fieldwork (Box 12.1), which highlights the need to think about the 'difference' or 'similarity' of other cultures, to consider uneven and unequal power relations, and to move away from 'ethnocentric' approaches to fieldwork.

In the heading for this section the word 'different' is deliberately placed in inverted commas to raise the question of how we think of cultures as 'other' or 'different'. Cultures and their relations to each other are understood in a variety of ways (Hall, 1995; Skelton and Allen, 1999). Other cultures are sometimes regarded as unchanging, even 'traditional' or 'primitive', with change caused by outside forces (usually from the West) (Twyman et al., 1999). Such approaches emphasize the 'exotic appeal' of other places or their difference and danger. This was found in the representations of other cultures produced by Europeans during colonialism and imperial rule which emphasized their strangeness and exotic qualities, the apparent danger posed by 'savage' peoples and the 'need' for the 'civilizing' and modernizing influence of European colonial cultures. This mix of approaches is often summarized under the term 'orientalism', from the work of Edward Said (1978; see also Driver, 1999; Phillips, 1999). Contemporary accounts where westerners assume their own experiences are the model to which other cultures aspire or see the other culture as a backdrop for their own enrichment can also be 'orientalist'. Think about how western travellers sometimes focus on the excitement and adventure provided by other countries with little understanding of the cultures of the countries themselves. Such approaches produce 'self-centred' or 'ethnocentric' geographies, where

Box 12.1   A code of ethics for tourists which is equally applicable to fieldwork in different cultures

- Travel in a spirit of humility and with a genuine desire to learn more about the people of your host country. Be sensitively aware of the feelings of other people, thus preventing what might be offensive behaviour on your part. This applies very much to photography.
- Cultivate the habit of listening and observing, rather than merely hearing and seeing.
- Realize that often the people in the country you visit have time concepts and thought patterns different from your own. This does not make them inferior, only different.
- Instead of looking for the 'beach paradise', discover the enrichment of seeing a different way of life, through other eyes.
- Acquaint yourself with local customs. What is courteous in one country may be quite the reverse in another – people will be happy to help you.
- Instead of the western practice of 'knowing all the answers', cultivate the habit of asking questions.
- Remember that you are only one of thousands of visitors to this country and do not expect special privileges. If you really want your experience to be a 'home away from home', it is foolish to waste money on travelling.
- When you are shopping, remember that the 'bargain' you obtained was possible only because of the low wages paid to the maker.
- Do not make promises to people in your host county unless you can carry them through.
- Spend time reflecting on your daily experience in an attempt to deepen your understanding. It has been said that 'what enriches you may rob and violate others'.

Source: Nash (2000a), after Weaver (1998)

one's own culture is set as the measure for all others (Cloke, 1999: 43).

Other approaches regard cultures as more dynamic and inter-connected. Some argue globalization erases differences between cultures in the technologically and culturally interconnected 'global village' or highlight the emergence of one globalized consumer culture built around brands such as McDonald's, Coca-Cola or Nike. These approaches see 'other' cultures as 'just like us'. However, this rather overlooks the diversity of the experiences of cultural, economic and technological globalization in which geographers are fundamentally interested (Kiely, 2000). One event may be watched on television by people across the world, but its significance for them will vary, as will their ability to influence responses to the event. Taking diversity into account provides a third set of ideas about culture where, instead of thinking of other cultures as either 'strange' and 'unchanging', or as 'just the same as us', we seek to understand other cultures in and of themselves while also understanding how local places and cultures are connected to national and global processes in uneven and unequal ways (Massey and Jess, 1995). For example, cities are important centres in the new economic geographies of globalization, but they are unevenly connected to these processes – New York is a command centre of global finance, Mexico City is less central. Urban residents in these two cities vary in their ability to connect to these processes, depending on their position in the labour market, access to education, class, gender, ethnicity and so on. Researchers should consider how research participants may be situated in these uneven social relations so that the research does not exacerbate or perpetuate inequalities or stereotypes. And the researcher should be aware of how his or her research is often made possible by his or her own relatively privileged position in these wider processes (Sidaway, 1992; Laurie et al., 1999; Skelton, 2001).

A further approach to culture considers how the encounters, relocations and flows of globalization may stretch out cultural relations across space to produce 'hybrid' cultural forms reflecting multiple forms of belonging in and across different places particularly associated with experiences of transnational communities, migrants and diaspora populations. An example of a hybrid cultural form in the British context is bhangra music, 'a fusion of Punjabi folk music with hip-hop, soul and house' developed out of British-Asian cultures (Dwyer, 1999: 292). Another would be the creation of hyphenated identities such as 'Black-British', 'Afro-American' or 'Japanese-Canadian' (Hall, 1995). These suggest 'different' cultures are also found in close proximity to each other, perhaps most obviously in the diverse ethnic geographies of major cities. However, Mark McGuinness (2000) suggests there is a danger in focusing on what appear 'obviously different' cultures and that, in the North American or European context, for example, a diversity of 'white' cultures should also be analysed rather than seeing them as the 'norm' and 'other' cultures as those that are novel or different.

Rather than cultures being slow-to-change, fixed sets of beliefs, values and behaviours, with permanent connection to places, Stuart Hall (1995: 187) argues culture is 'a meeting point where different influences, traditions and forces intersect', formed by 'the juxtaposition and co-presence of different cultural forces and discourses and their effects' and consisting of 'changing cultural practices and meanings'. This suggests our 'own' cultures are as caught up in change as other cultures, and that connections between cultures and places are often highly dynamic. However, it is worth noting the many instances where people emphasize fixed, homogeneous and bounded cultures in the face of such fluidity, often involving claims to landscapes, territories or places of which nationalist movements are useful examples (Storey, 2001).

Using more fluid conceptualizations of culture means we have to question what ideas about similarity and difference we bring to our cross-cultural studies. Nevertheless, it is important not to overplay the changing and dynamic nature of cultures. Instead we need to pay attention to how cultures are embedded in, and part of, ongoing global inequalities – for example, between developed and less developed countries. It is also important to explore how different people's experiences of culture 'will be affected by the multiple aspects of their identity – "race", gender, age, sexuality, class, caste position, religion, geography and so forth – and [are] likely to alter in various circumstances' (Skelton and Allen, 1999: 4). The remainder of the chapter looks at specific ways in which geographers grapple with these challenges in cross-cultural research, starting with work in different languages.

WORKING WITH DIFFERENT LANGUAGES

Many people around the globe speak English, either as their first language or as a second or third language because of its importance for business, administration, education and travel. Thus students who speak English are in a fortunate position but reliance on this can make native English speakers lazy in learning other languages and there are many contexts where fieldwork in another country will be hampered if you do not make some effort to communicate in the relevant language. While this can seem daunting, it is possible to make a big difference to fieldwork with relatively simple strategies which pay attention to language issues.

For example, students investigating the local labour market of the tourist resort of Benidorm on the University of Dundee field course produced a bi-lingual questionnaire before going to Spain, getting the English translated into Spanish by a Spanish teacher they knew. The questionnaire explained in both languages who they were and what they were doing and then gave dual-language versions of each question. Using the bi-lingual questionnaire they were able to find out about the employment experiences not only of expatriate English speakers and of Spaniards with high-level English skills but also of people with limited English who found accessing better paid jobs in a resort dominated by British tourists very difficult. The questionnaire also stimulated further conversations, with friends or relatives of those who were completing it who spoke more English joining in to translate parts of the discussion between the students and the person completing the questionnaire.

So making the effort to communicate in another language, even if it is not perfect, can be very productive, and is often valued by research participants as a sign of real interest in their culture and community. These students understood as a result how an ability to speak English alongside Spanish allowed some people to gain better paid, more highly skilled jobs, while others were marginalized economically because of this language issue. Other students observed how the use of different European languages in shops, services and newspapers, and the extent to which migrants learnt Spanish reflected the emergence of 'transnational communities' of British, Dutch or Norwegian migrants who were seeking their own place in the sun (O'Reilly, 2000). Some students interviewed migrant workers from Morocco, Algeria and Colombia who hold marginal jobs as casual labour on farms and construction sites. By putting together their language skills to hold conversations in French and English, with some Spanish thrown in, the students gained some insight into the lives of these marginalized workers, some of whom, because they could speak Spanish, were in slightly better positions. Finally, some students studied the importance of the regional language in road signs, place-names and school books which reflects the upsurge of regional cultural identities and politics in the last 20–25 years. None of these examples involved students being fluent in Spanish (although this clearly would help) but they all show it is possible to work in fairly simple ways with an awareness of language and the politics around language use.

Of course, it is possible to spend time learning the relevant language or to study somewhere you already speak the language. This is useful if fieldwork is taking place over a longer period, since even everyday life can be difficult without at least some basic language skills. However, while knowing a language fluently helps in understanding and being understood, even then translating concepts and ideas between languages requires care. I studied German for my degree, along with geography, and have undertaken much of my research in Germany. One project necessitated in-depth interviews with people in an eastern German city involved in local action groups concerned with urban redevelopment (Smith, 1999). In English I would probably describe these as 'community groups', but in German the groups called themselves 'citizen initiatives' or saw themselves as part of the 'civic movement', which has a different set of implications. Since I was interested in how the activists viewed their local political action, it was important to work with their terms and not to impose mine. So I deliberately analysed the interviews in German, making clear notes on how the research participants used these particular ideas, and only translated interview extracts when I produced the final version of the papers to be published in English, including footnotes explaining the significance of different terms. This approach, rather than analysing my version of the participants' words, arguably produced a richer insight into the local political geographies of urban development and allowed me to explore different cultural understandings of citizen activism.

An alternative to learning the language is to use translators and interpreters. Twyman et al. (1999) discuss the use of translators in their study of society–environment links in the management of range-lands in Botswana. The British fieldworkers found the range of local languages too great and depended on high-level language skills supplied by their translators. Drawing on Sturge (1997: 21) and Tambiah (1990: 3), Twyman et al. (1999: 320) argue that 'translation is a practice of intercultural communication . . . in which we understand other cultures as far as possible in their own terms but in our language'. This neat formulation suggests translation involves 'mapping ideas and practices' between cultures (Twyman et al., 1999: 320). Rather than only recording answers to questions once they were translated into English, they taped and transcribed the whole process of communication back and forwards between the various research participants, including the translator. Analysing this revealed how the person interpreting often had to summarize roughly the meaning of what the interviewee said in order to let the interview proceed, but after the event the interviewer and the translator could discuss in more detail the ideas articulated. This revealed that it was actually very difficult to 'map' the ideas and meanings of the research participants on to English language terms. In fact many interviewees were already talking in what was not their first language and used a variety of different language terms to communicate their attitudes and practices in land and livestock management. In writing up their research, Twyman et al. (1999) addressed the processes of translation, interpretation and the mapping of meaning between and across cultures, noting particularly where mapping these ideas was problematic or provided new insights.

These examples suggest it is sometimes possible to work in relatively simple ways with different languages, even in contexts where we are not fluent in the language, and through this to gain an insight into the cultures studied. However, if we want detailed understandings of the processes involved and of the meanings people attach to these processes, we can pay attention to the possibilities and problems of translation and interpretation between languages. Furthermore, we can be aware of the politics of language use –

especially where abilities to communicate in one language or another confer status or access to particular benefits or privileges.

WORKING WITH DIFFERENCE, POWER RELATIONS AND POSITIONALITY

Moving on to more general questions of difference, unequal power relations and the position of the researcher, I now consider a range of practical responses by geographers which illustrate the challenges and possibilities of cross-cultural research.

During the early 1990s, while researching urban change in eastern Germany, I became aware of the unease many eastern Germans felt about comparing their experiences to those of western Germans, or western culture more generally. Claims about the 'victory' of the West after the collapse of communism and the practical need to adapt to western German administrative and legal frameworks served to devalue their cultural and political experiences, seeing them as no longer relevant to the future. Instead of accepting the dominant politics of postcommunist 'transition' which posited western norms as the only possible goal of change (Pickles and Smith, 1998; Burawoy and Verdery, 1999), I tried to be open to the diverse experiences on which people drew to develop new local politics. One day I attended a meeting with a community group which was developing a local industrial heritage museum. Group members discussed the problems of industrial decline and unemployment their city faced. Then one person asked why I was interested in their city and asked what Glasgow, my home city, was like in comparison. Since several group members had been academics in the past I assumed they would be interested in the similarities and differences between the two cities. However, what followed was not a discussion of these similarities. Instead several people rebuffed my attempts at finding commonalities, claiming the situation in Glasgow was in no way like that in their city.

Initially, I was embarrassed and felt I had lapsed into the problematic stance of what they saw as a 'typical westerner', comparing everything outside western society (usually unfavourably) to the 'normal' West. On further reflection it seemed the group members were partially correct, since the pace of change in their city was greater than in Glasgow. However, what was more important was that I had set up a framework where I claimed to be interested in their experiences in and of themselves, without any claims to 'know best'. I had deliberately tried not to essentialize differences between 'East' and 'West', or between capitalism and communism, instead seeking connections as well as differences. Into this framework I had then introduced a note suggesting what was new, difficult and often very painful for them, with many experiencing redundancy, was not so unusual, special, or particular. My comparisons were seen as belittling the severity of people's experiences, denying their individual significance. As I developed my analysis, it seemed to me that this moment was not just one where I made a 'mistake' in the fieldwork. The fact some group members found my comparisons unacceptable highlighted precisely the politics of naming processes, cultures and experiences as 'similar' or 'different' which are tied in strongly to the negotiation of postcommunist transition and German unification and which could not be escaped as I undertook the research.

Our research can never escape from the power relations shaping the situations in which we research. We must address these carefully and take account of them in the choices we make in our research practices as well as in the interpretations we develop. One strategy for addressing such inequalities is to work through the complex positionality of the researcher, subjecting the research process itself to scrutiny and not assuming the researcher is a disembodied presence, removed from the research process (see also Chapter 8). This does not mean adopting a self-centred view of research where 'other cultures and other people' become merely the 'exotic backdrops of authorial self-discovery' (Lancaster, 1996: 131, cited in Twyman et al., 1999: 315). Nor does it mean assuming the researcher can know exactly how he or she is viewed by the research participants, or to account for the significance of every element of his or her own identity in the research (Rose, 1997). Rather as researchers 'we must recognise and take account of our own position, as well as that of our research participants, and write this in to our research practice' (McDowell, 1992: 409) in ways that are sensitive to the difference our presence makes in the research, and how the process of research itself can shape social relations.

Many researchers provide detailed accounts of positionality (*The Professional Geographer*, 1994; Sparke, 1996; Madge et al., 1997; Laurie et al., 1999; Limb and Dwyer, 2001). Tracey Skelton (2001: 89) provides a definition:

> By positionality I mean things like our 'race' and gender . . . but also our class experiences, our levels of education, our sexuality, our age, our ableness, whether we are a parent or not. All of these have a bearing upon who we are, how our identities are formed and how we do our research. We are not neutral, scientific observers, untouched by the emotional and political contexts of places where we do our research.

This means being aware of how aspects of our own identities are significant or might change as we 'travel' (spatially or culturally) to different contexts. For Skelton, not having children as a young researcher marked her out as different from the women she interviewed about gender relations in Monserrat. Rather than avoiding their questions about this, she used this as a point of discussion, recognizing that some of the women felt luckier or more mature than her: 'I found this a healthy way of letting the power I had in the interview context – I was the one asking questions – dissipate and shift into complex positions within the interviews' (Skelton, 2001: 91).

At times, thinking about positionality leads researchers to question whether it is appropriate to undertake studies where they are 'outsiders', especially where their outsider status might perpetuate the ways less powerful groups and cultures have been represented by those in more powerful positions, such as westerners representing the experiences of people in developing countries (Sidaway, 1992; Madge, 1994; Radcliffe, 1994), people who are white representing those from ethnic minorities (Dwyer, 1999) or men writing about women's experiences (Sparke, 1996). Some conclude that they should not undertake particular research. Kim England (1994) withdrew from researching the lesbian community in Toronto, Canada, because she felt she was too much of an 'outsider' and the context of homophobia in Canada meant that her trying to 'speak for' or 'give voice to' the lesbian community led to the danger of her colonizing the experiences of this group of women. Instead she felt it more appropriate to leave research of the community to other lesbian women. However, Robina Mohammad's (2001) discussion of research on Pakistani women in Britain suggests that the need to consider positionality does not disappear where we appear to be 'insiders', since we are also partly 'outsiders' by the very fact that we are engaged in research, and other aspects of our own identities (such as dress, accent or education) can be markers of our difference as well as our similarities. Rather than her apparent 'insider' status allowing access to 'the truth', Mohammad suggests that within her study participants presented 'multiple truths'. The challenge was to understand 'which truth' was being told and whose interests were being served by particular representations. Similarly, participants in Claire Dwyer's study of young Muslim women in Britain argued that 'they might have responded differently to an Asian or Muslim interviewer – but they might not necessarily have given "better truths" ' (in Laurie et al., 1999: 49).

To some extent, then, we are always involved in working with 'different' cultures and must negotiate the power relations of similarity and difference in our research whether these cultures are 'remote' or close at hand. As Heidi Nast (1994: 57) argues, 'we can never *not* work with "others" who are separate and different from ourselves; difference is an essential aspect of all social interactions that requires that we are always everywhere in between or negotiating the worlds of me and not-me'. The challenge is to 'write this into our research practice' (McDowell, 1992: 409).

(RE)PRESENTING OTHER CULTURES

Finally, it is important to consider how to represent the people and places studied in the field report or dissertation. Writing to thank those who have helped and, where appropriate, sending a copy of the report is a good start. If you worked in another language or your academic findings would be inappropriate (people might be interested in what you discovered, but not necessarily in your latest theoretical insight), a revised feedback report might be more appropriate, as might be giving a presentation, writing an article for the local newspaper or being interviewed on local radio. You might decide to work more collaboratively with people from local universities or ask participants to comment on your analysis, seeing whether they think you have understood correctly what is going on. It does not automatically follow that they are right and you are wrong, but you can work with that difference of interpretation as a way of decentring your position as the apparently all-knowing researcher, compared to the research participants. Such consideration applies to visual representations as well as to written texts, as the following example illustrates.

American anthropologist, Kathleen Kuehnast (2000), had been researching the economic burdens faced by women in regions affected by farm privatization in Kyrgyzstan, a central Asian post-Soviet state. For the cover of her report she chose a photograph of 'a Kyrgyz elderly woman and her daughter-in-law, holding a baby, each dressed in the warm clothing of semi-nomadic herders' (Kuehnast, 2000: 105). She was surprised when some women in government jobs, whom she knew, were offended by the photograph, feeling it presented their country in a poor light. After initially dismissing this as the unwillingness of higher-status women to address problems facing nomadic women in their own country, Kuehnast wondered why else the photograph was problematic. Perhaps it failed to represent women's achievements in education and employment, buying in too strongly to the notion of women as victims of communist repression, and going against 'well-ingrained Soviet ideologies that portrayed the female worker as strong, competent, and vital' (p. 107). Alternatively, many of the women interviewed by Kuehnast saw in the western media images of glamorous women, which flooded the country after independence, an ideal of 'western', 'modern' or 'American' womanhood (with the 'right' clothes, makeup and leisured lifestyle). In this context, the portrayal of Kyrgyz women in traditional clothing and as working women may have seemed to illustrate too sharply the apparent 'failings' of Kyrgyz women in adopting suitably 'westernized' or 'modernized' gender identities.

Representation is fundamentally problematic. Even where the researcher has deliberately chosen to get close to the research subject, often through the use of qualitative research methods (Limb and Dwyer, 2001), the 'analysis, writing up and dissemination of information often forces us to detach ourselves, switch back to "Western mode" to produce texts and develop "distance" to use information' (Madge, 1994: 95). As Kuehnast found, even analysis and representation which are meant to be helpful are not immune to being problematic for the research participants. Each researcher therefore must make his or her own choices, often resulting in pragmatic responses to his or her situation. Clare Madge (1994: 96), in her research on the Gambia, decided that some of the information she gathered talking to people who became her friends could not be included in her analysis, since to use that information would 'betray the trust of my friends'. Audrey Kobayashi (2001) opted to stop working on 'other' cultures outside Canada and to work with Japanese Canadians within Canada in collaborative activist research. Tracey Skelton at times felt it was impossible to write about her research on Montserrat without reproducing the unequal colonial relations she was trying to combat, so for some time she decided not to write on these topics. However, in the end she decided that 'as part of the politics of reflective and politically conscious feminist and/or cross-cultural research, we have to continue our research projects, we must publish and disseminate our research. If we do not, others without political anxieties and sensitivities about their fieldwork processes take the space' (Skelton, 2001: 95).

CONCLUSION

There are many challenges in working in different languages and different cultures and even experienced geographers do not always get it right. However, the key issue is to pay attention to the issues raised in this chapter as we plan and undertake fieldwork, and analyse and represent what we find, keeping in mind 'why we are doing it and what the research we do means to other people' (Skelton, 2001: 96). When done well, research in different languages and different cultures can be incredibly enriching and challenge us to think about difference and diversity in productive and sensitive ways.

Summary

- Cross-cultural research is challenging, enriching and rewarding.
- Concepts of cultural difference should inform such research, taking into account the fluidity of cultures, unequal social relations, the need to avoid ethnocentrism and an openness to hybrid cultural forms.
- Simple strategies may address language differences. Attention to language use and the articulation of meanings provides insight into the 'other' culture. Translation requires careful consideration of how meanings 'map' between cultures.
- Addressing the power relations surrounding research and writing the positionality of the researcher into our research accounts both help avoid ethnocentrism. 'Outsiders' and 'insiders' should consider carefully their relation to the research. This may provide a variety of accounts without necessarily producing the single 'correct' answer to interpreting a particular situation.
- Choices about representation in written, verbal or visual formats should avoid reinforcing unequal power relations or stereotypes and be informed by the ethics and politics of the research.

Further reading

This guide to further reading identifies references for the key themes of the practicalities of fieldwork abroad; debates about cultural change; language; and negotiating power relations:

- Two articles by Nash (2000a, 2000b) discuss practical issues in undertaking independent fieldwork abroad. The first addresses establishing contacts, legal requirements for visas, collecting and exporting samples, health and safety issues, and training. The second considers budgeting and fundraising.

- Hall (1995) and Skelton and Allen (1999) explore current debates about contemporary cultural change.

- Smith (1996) and Twyman et al. (1999) consider translation between different languages, discussing how translation itself can become part of the focus for analysis and issues around working with a translator.

- Sidaway (1992) and Madge et al. (1997) provide detailed examples of negotiating power relations, positionality and representation in cross-cultural research.

Note: Full details of the above can be found in the references list below.

References

Burawoy, M. and Verdery, K. (eds) (1999) *Uncertain Transition: Ethnographies of Change in the Postsocialist World*. Lanham, MD: Rowman & Littlefield.

Cloke, P. (1999) 'Self-Other,' in P. Cloke et al. (eds) *Introducing Human Geographies*. London: Arnold, pp. 43–53.

Driver, F. (1999) 'Imaginative geographies', in P. Cloke et al. (eds) *Introducing Human Geographies*. London: Arnold, pp. 209–16.

Dwyer, C. (1999) 'Migrations and diasporas', in P. Cloke et al. (eds) *Introducing Human Geographies*. London: Arnold, pp. 287–95.

England, K. (1994) 'Getting personal: reflexivity, positionality and feminist research', *The Professional Geographer*, 46: 80–9.

Hall, S. (1995) 'New cultures for old', in D. Massey and P. Jess (eds) *A Place in the World? Places, Cultures and Globalization*. Milton Keynes: Open University Press, pp. 175–213.

Kiely, R. (2000) 'Globalization: from domination to resistance', *Third World Quarterly* 21: 1059–70.

Kobayashi, A. (2001) 'Negotiating the personal and the political in critical qualitative research', in M. Limb and C. Dwyer (eds) *Qualitative Methodologies for Geographers*. London: Arnold, pp. 55–70.

Kuehnast, K. (2000) 'Ethnographic encounters in post-Soviet Kyrgyzstan: dilemmas of gender, poverty and the Cold War', in H. de Soto and N. Dudwick (eds) *Fieldwork Dilemmas: Anthropologists in Postsocialist States*. Madison, WI: University of Wisconsin Press, pp. 100–18.

Lancaster, R.N. (1996) 'The use and abuse of reflexivity', *American Ethnologist*, 23: 130–2.

Laurie, N., Dwyer, C., Holloway, S.L. and Smith, F.M. (1999) *Geographies of New Femininities*. Harlow: Longman.

Limb, M. and Dwyer, C. (eds) (2001) *Qualitative Methodologies for Geographers*. London: Arnold.

Madge, C. (1994) 'The ethics of research in the "Third World" ', in E. Robson and K. Willis (eds) *DARG Monograph No. 8: Postgraduate Fieldwork in Developing Areas*. London: Developing Areas Research Group of the Institute of British Geographers, pp. 91–102.

Madge, C. and Bee, A. (1999) 'Women, science and identity: interviews with female physical geographers', *Area*, 31: 335–48.

Madge, C., Raghuram, P., Skelton, T., Willis, K. and Williams, J. (1997) 'Methods and methodologies in feminist geographies: politics, practice and power', in Women and Geography Study Group, *Feminist Geographies: Explorations in Diversity and Difference*. Harlow: Longman, pp. 86–111.

Massey, D. and Jess, P. (1995) 'Places and cultures in an uneven world', in D. Massey and P. Jess (eds) *A Place in the World? Places, Cultures and Globalization*. Milton Keynes: Open University Press, pp. 215–40.

McDowell, L. (1992) 'Doing gender: feminism, feminists and research methods in human geography', *Transactions, Institute of British Geographers*, 16: 400–19.

McGuinness, M. (2000) 'Geography matters? Whiteness and contemporary geography', *Area*, 32: 225–30.

Mohammad, R. (2001) ' "Insiders" and/or "outsiders": positionality, theory and praxis', in M. Limb and C. Dwyer (eds) *Qualitative Methodologies for Geographers*. London: Arnold, pp. 101–17.

Nairn, K. (1996) 'Parties of geography fieldtrips: embodied fieldwork', *New Zealand Women's Studies Journal*, 12: 88–97.

Nairn, K., Higgitt, D. and Vanneste, D. (2000) 'International perspectives on fieldcourses', *Journal of Geography in Higher Education*, 24: 246–54.

Nash, D.J. (2000a) 'Doing independent overseas fieldwork. 1. Practicalities and pitfalls', *Journal of Geography in Higher Education*, 24: 139–49.

Nash, D.J. (2000b) 'Doing independent overseas fieldwork. 2. Getting funded', *Journal of Geography in Higher Education*, 24: 425–33.

Nast, H. (1994) 'Opening remarks on "Women in the Field" ', *The Professional Geographer*, 46: 54–66.

O'Reilly, K. (2000) *The British on the Costa del Sol: Transnational Identities and Local Communities*. London: Routledge.

Phillips, R. (1999) 'Colonialism and postcolonialism', in P. Cloke et al. (eds) *Introducing Human Geographies*. London: Arnold, pp. 277–86.

Pickles, J. and Smith, A. (eds) (1998) *Theorizing Transition: The Political Economy of Post-Communist Transformations*. London: Routledge.

Radcliffe, S. (1994) '(Representing) post-colonial women: authority, difference and feminisms', *Area*, 6(1): 25–32.

Rose, G. (1992) 'Geography as a science of observation: the landscape, the gaze and masculinity', in F. Driver and G. Rose (eds) *Nature and Science: Essays in the History of Geographical Knowledge*. London: IBG Historical Geography Research Group, pp. 8–18.

Rose, G. (1997) 'Situating knowledges: positionality, reflexivity and other tactics', *Progress in Human Geography*, 21: 305–20.

Said, E. (1978) *Orientalism*. Harmondsworth: Penguin Books.

Sidaway, J. (1992) 'In other worlds: on the politics of research by "first world" geographers in the "third world" ', *Area*, 24: 403–8.

Skelton, T. (2001) 'Cross-cultural research: issues of power, positionality and "race" ', in M. Limb and C. Dwyer (eds) *Qualitative Methodologies for Geographers*. London: Arnold, pp. 87–100.

Skelton, T. and Allen, T. (eds) (1999) *Culture and Global Change*. London: Routledge.

Smith, F.M. (1996) 'Problematizing language: limitations and possibilities in "foreign language" research', *Area*, 28: 160–6.

Smith, F.M. (1999) 'The neighbourhood as site for contesting German reunification', in J. Sharp et al. (eds) *Entanglements of Power: Geographies of Domination and Resistance*. London: Routledge, pp. 122–47.

Sparke, M. (1996) 'Displacing the field in fieldwork: masculinity, metaphor and space', in N. Duncan (ed.) *BodySpace*. London: Routledge, pp. 212–33.

Stoddart, D.R. (1986) *On Geography and its History*. Oxford: Blackwell.

Storey, D. (2001) *Territory: The Claiming of Space*. Harlow: Prentice Hall.

Sturge, K. (1997) 'Translation strategies in ethnography', *Translator*, 3: 21–38.

Tambiah, S.J. (1990) *Magic, Science, Religion, and the Scope of Rationality*. Cambridge: Cambridge University Press.

Twyman, C., Morrisson, J. and Sporton, D. (1999) 'The final fifth: autobiography, reflexivity and interpretation in cross-cultural research', *Area*, 31: 313–26.

Weaver, D.B. (1998) *Ecotourism in the Less Developed World*. Wallingford: Cab International.

13 Getting Information about the Past: Palaeo and Historical Data Sources

Catherine Souch

Synopsis

Climatic change may be natural, anthropogentic (forced by human activity) or a combination of the two. Changes in climate also take place over a wide variety of time and space scales. The instrumental record (where climatic variables are measured directly) extends over only a tiny fraction of earth history, and also provides only a partial spatial coverage. As a result, reconstructions of climatic and environmental conditions frequently rely on proxy, or surrogate, data. Proxy records are historical (documentary) archives or natural (physical, chemical or biological) systems that are dependent on climate and which incorporate in their structure some measure of this dependency so it is preserved through time. The challenge in interpreting these data is to separate the effects of climate (the signal) from all the other non-climatic influences (the noise).

The chapter is organized into the following sections:

- Introduction
- Proxy data sources
- Analysis of proxy records
- Issues to consider with proxy data and reconstructions
- Conclusion.

INTRODUCTION

Variability is an intrinsic property of the earth's natural environment. Atmospheric conditions, for example, vary over time periods which range from less than seconds to billions of years (the age of the earth). While some of these changes are periodic, linked to daily, annual and millennial astronomical cycles, others are more episodic, related to processes and feedbacks internal to the earth–atmosphere system. Today, significant attention focuses on natural climate change and variability, not only to gain insight into past environments but also to provide a baseline against which to document and evaluate the ever-increasing impacts of human activities (Houghton et al., 2001).

The instrumental record is generally accepted to be too short to provide a complete picture of climate variability and change. Direct observations of the environment, the atmosphere, hydrosphere, lithosphere and biosphere have been conducted for only a tiny fraction of the earth's ~4.5 billion year history (Bradley, 1999). Routine (instrument-based) observations of climate began in western Europe during the late seventeenth and early eighteenth centuries, but many countries only established hydrometeorological networks in the late nineteenth century, after the Vienna Meteorological Congress in 1873. Today, surface data remain sparse for both polar regions, and regular measurements are available for only two-thirds of the surface of the Southern Hemisphere. Thus instrumental data provide an inadequate perspective (both temporally and spatially) on natural and human-induced environmental change and variability and the development of conditions today.

If we are to understand how large natural climate change can be; how rapidly climate change can occur; the effects of climate change on other environmental processes at regional and global scales; and/or the extent of the effects of human activities, the record of climate information has to be extended back before the era of direct measurements. To do this we must rely on proxies – measurable properties of biological, chemical or physical systems – that provide quantitative information about past temperature, rainfall, ice volumes, etc. Comprehensive reviews of proxy data sources and their interpretation are provided by Bradley and Jones (1995), Lamb (1995), Bradley (1999), Jones et al. (2001) and Ruddiman (2001).

As in any field, the tools we use shape our understanding. Proxy data do not provide a direct record of climate; rather, they have acted as a filter transforming climatic conditions at a point/period in time into a physical, chemical or biological signature. Each line of evidence differs in many important regards – for example, the climate information it can yield, its spatial (geographic) coverage and representativeness, the period for which information can be reconstructed (temporal coverage) and its resolution (the ability to resolve events accurately in time). These are key issues to consider when evaluating proxy data and the reconstructions based upon them. Each is discussed further below.

Exactly which proxy data are used in a study depends on the research question being asked, the location of the study region and an array of logistical issues and constraints (access to equipment, laboratory facilities and technical expertise). Often, independent proxies are found in close proximity. Thus they can be, and should be, used in combination to provide complementary insight into past environmental conditions.

PROXY DATA SOURCES

The two general classes of climate proxies most commonly used to yield quantitative reconstructions of past climate are biotic proxies, which are based on the composition of plant and animal groups and/or measures of their growth rates, and geological proxies, which quantify changes (physical or chemical) in earth's materials that have accumulated through time, most commonly sediments in oceans or lakes, or ice in polar or alpine glaciers. These proxies, organized by the natural archives in which they are stored, are described in Table 13.1. Attention to where each proxy is found, and the length and detail of the records they can yield (their spatial and temporal coverage and resolution), is considered further below.

Where biologic material is preserved through time – for example, through annual growth rings/layers in living systems such as trees and corals, or in lakes, bogs, estuaries or ocean sediments – records of past biotic changes, which may be related to climate, can be interpreted. Given fossils of plants tend to be more abundant than those of animals in continental geologic records, vegetation plays a critical role in the reconstruction of ancient climates (millions to billions of years). Warmer climates tens of millions of years ago are inferred from the presence of palm-like trees (temperature-sensitive species) in northern latitudes, for example (Ruddiman, 2001). For the younger continental record, the relative abundance of species, based on plant remains in the form of either macrofossils (cones, seeds, leaves) or pollen (which is very resistant to decay because of its protective coating), yields sensitive records of terrestrial climate. In the oceans, the relative abundance and distribution of four common groups of shell-forming animal and plant plankton (formanifera, coccoliths, diatoms and radiolaria) are used for climate reconstructions.

Geological proxies rely largely on interpreting changes in the volume and/or physical/chemical properties of earth materials (sediment or snow/ice) accumulating through time. Virtually all continental sedimentary deposits (aeolian, glacial, fluvial, lacustrine) convey some sort of palaeoclimatic signal. However, it is often difficult to identify the unique combination of climatic events leading to the formation of such deposits. Much more detailed and precise

TABLE 13.1 Common sources of proxy data for palaeoclimatic interpretations and their key characteristics. High-resolution proxy data sources provide information at daily, seasonal and annual timescales, often for specific locations and/or small regions. Low-resolution data, in contrast, provide integrated climate data for timescales of centuries to millennia

Archive	Proxy records	Climatic elements	Temporal resolution/coverage	Geographic coverage
High-resolution, local-regional scale patterns				
Historical	Written records of weather or related phenomena, or phenological information	Temperature, precipitation, snowfall, frost, floods	Detailed information (day, hour, season). Oldest records ~5000 y. Most records <10^2 y and discontinuous	Global. Longest records East Asia, Middle East and Europe. Local information
Tree rings	Ring widths, cell structure/density, isotopic composition	Temperature, rainfall, drought, runoff, cycles such as El Niño and Pacific Decadal Oscillation	Continuous records. Ring width resolution 1 year; may contain seasonal information (cell structure 1–5 weeks). Oldest records ~8000 y southwest USA; 10^2–10^3 y common	Continental areas excluding desert and tundra regions. Mainly mid-latitudes. Information on local/regional climates
Lake sediments	Varves (annual layers): Thickness/composition Relative abundance of pollen and macrofossils Diatoms, ostracods and other aquatic biota	Streamflow/snow melt, rainfall, Temperature	Continuous records. Resolution 10–10^2 y. Most records <10^4 y. Oldest non-varved records ~10^5 y	Varves – need strong seasonality and absence of post-depositional mixing. Deep anoxic environments. Other lake records, most continents
Corals	Growth rates, isotopes, trace elements	Sea-surface temperatures, adjacent continental rainfall and runoff, ocean circulation, tropical winds	Annual resolution. Coverage 10 y	Tropical/subtropical oceans

TABLE 13.1 *continued*

Archive	Proxy records	Climatic elements	Temporal resolution/ coverage	Geographic coverage
Ice cores	Ice fabric, stable isotopes (H and O), gas content of air bubbles, trace element and micro-particle concentrations	Snowfall, temperatures, humidity, wind speed and atmospheric circulation, atmospheric composition	Continuous records. At best annual resolution for last 10^4 y. Longest record ~500 000 y (Vostok, Antarctica)	Glaciated regions. Polar and high mountain regions – primarily Greenland/ Antarctica and ice-caps of Peru, Bolivia, China and Tibet
Terrestrial geomorphic evidence	Closed lake basin – lake levels; glacial and periglacial features; aeolian deposits and loess; speleothems; relict soils	Effective/net rainfall, runoff, temperature	Discontinuous records. Low resolution (10^2 y), often poor chronological control. Coverage varies, up to 10^6 y	Continental areas. Lake levels – arid/semi-arid regions; glacial/periglacial features mid-high latitudes
Marine cores	Isotopic composition, geochemistry/mineralogy, terrestrial (aeolian) dust and ice-rafted debris, floral and faunal abundance	Temperature (surface and deep water), ice volume, ocean circulation, aridity of continents, intensity and direction of winds	Continuous records. Resolution varies from about 10^2 to 2500 y depending on rate of deposition. Some coastal basins are varved (annual resolution). Temporal coverage up to 10^8 y.	Data from virtually all oceans and latitudes (70% earth's surface) – integrated global signal

Low-resolution, regional-global scale patterns

Source: Based on Lamb (1995) and Bradley (1999)

information is derived where earth materials accumulate continuously through time – for example, sediments in ocean and lake basins, and snow and ice in polar and alpine ice fields. Analyses of rates of accumulation, physical properties (size, shape) and geochemistry (mineralogy, trace element content and isotopic composition) of the materials all yield information about past climates. The key sources of long-term climate history (thousands to millions of years) come from ocean cores (Bradley, 1999; Ruddiman, 2001). Most geographers, however, given their interest in natural climate variability on societal timescales and the impact of humans on the environment, are concerned with more recent, higher-resolution geological records where changes over years to centuries can be identified. Optimal records for such applications include ice cores (polar and alpine environments) and varved (annually layered) lake sediments (most commonly found in glaciated and arid environments).

Potential sources of historical palaeoclimatic information also include ancient inscriptions, annals and chronicles, government records, private estate records, maritime and commercial records, personal papers (such as diaries and letters) and scientific and/or quasi-scientific writing (such as non-instrumental weather journals). Contained within these, historical (documentary) evidence can be grouped into three main categories (Bradley and Jones, 1995): (1) direct observations of weather, for example, the frequency and timing of frosts or the occurrence of rainfall or snowfall; (2) records of weather-dependent phenomena, such as droughts, floods, river or lake freeze-up and break-up; and (3) phenological records, which describe weather-dependent biological phenomena, for example the flowering of shrubs or trees, the arrival of migrant birds, or crop yields. Stone inscriptions related to the Nile flood levels date back to ~5000 years BP; Arabic chronicles of rainfall through the Middle East (Iraq, Syria) to ~1000 years BP. Documentary records do have a number of advantages: normally, events are precisely dated, providing a high level of temporal resolution from daily to yearly timescales and often they describe events with important consequences for humans (floods, droughts, landslides, etc.). However, many records are discontinuous, very localized and may be strongly biased by individual observers. Moreover, such records often tend to focus on extreme events and long-term trends go undocumented. Thus, it is difficult to subject the data to sophisticated statistical analyses. However, historical records have been very important in documenting conditions throughout Europe, China and Japan for the last millennium and have yielded detailed information about the cooling and warming phases through the Little Ice Age, a period of cooler temperatures from ~1550 to 1900 (Bradley and Jones, 1995; Lamb, 1995).

ANALYSIS OF PROXY RECORDS

Regardless of the proxy used, commonly three steps are employed to extract the climate signal from extraneous noise (the influence of other factors). First, the physical, chemical and/or biological attributes of the system dependent on climate are identified. Secondly, the proxy data source is calibrated with contemporary climate data, often through statistical regression. This approach has the inherent assumptions that the modern relationships have operated unchanged through the period of interest (the principle of uniformitarianism) and that a modern analogue exists (conditions from the past can be found somewhere on earth today). Thirdly, the present-day relationships (often referred to as transfer functions) are then applied to the longer proxy record to infer past environmental/climatic conditions.

To illustrate this approach, the analysis of annual variations in tree-ring widths (dendroclimatology), one of the most extensively used proxy records in mid and high latitudes, is considered (Figure 13.1). The basic premise is that a tree's growth is a function of climatic conditions and this climate dependence is recorded in the width, density or isotopic composition of the tree's annual growth rings. To collect raw data on ring widths, an increment borer is used to take tree cores (~0.5 cm in diameter) from multiple trees of the same species within a stand (Figure 13.1a). Analysis of proximal replicate trees aims to remove the effect of site-specific conditions that might influence growth rates of individual trees. All the individual tree cores are then cross-dated – i.e. all rings of the same age are identified (Figure 13.1b). This is essential so the age of each ring is precisely known. The in-built chronology of tree rings, which allows annual or even seasonal resolution, is one of the greatest advantages of tree-ring analysis. If tree rings are collected from living and dead trees (in the ground, preserved in peat-bogs or incorporated into buildings, for example), the chronology may extend back centuries to thousands of years (see examples cited in Bradley, 1999). Next, growth effects are removed from the time series of ring widths. Smaller (younger) trees tend to have wider widths than larger (older) trees, both because they grow more rapidly and because the circumference of a young tree is smaller, so a given amount of growth results in a thicker ring. This growth trend is usually removed by fitting some sort of mathematical function to the raw data (Figure 13.1c). Modified exponential functions, orthogonal polynomials and cubic splines are all used, depending on whether the tree is growing in isolation or in a closed canopy. The difference between the ring widths and fitted values averaged for the site is the proxy climate signal to be interpreted. Mathematical and/or statistical procedures are used to derive an equation relating the tree-

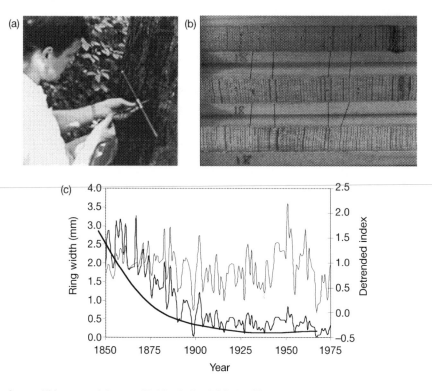

Source: All images and data provided by Dr Henri Grissino-Mayer

FIGURE 13.1 Analysis of proxy records. The example of tree-ring analysis: (a) collecting the raw data – coring a tree; (b) cross-dating cores from multiple trees; (c) removing the growth trend from time series of ring widths (solid line) to generate a detrended index (dashed line). These data are averaged for the stand and calibrated with observed climate data to develop transfer functions to be applied to earlier tree-ring records to infer climate conditions before the period of instrumental records

ring width indices to a climate variable for a given period (referred to as calibration). Different methods, which range from simple to multiple to spatial regression, principal components analysis and canonical regression analysis, have been used and are described by Bradley (1999). Ideally, some climate data outside the calibration period are used to assess the performance of the fitted equation with independent data (verification), before the relations are applied to infer past climates. Clearly, careful attention needs to be directed to what climate data should be used and for what seasons. The best results tend to be obtained at sites where trees are growing close to ecological limits; thus growth rates are sensitive to climate. The two climatic stresses most commonly recognized are moisture (drought or flooding) and temperature (see the Ultimate Tree-Ring web pages for specific examples and further details – web.utk.edu/ ~ grissino).

ISSUES TO CONSIDER WITH PROXY DATA AND RECONSTRUCTIONS

Climatic interpretation

Often there is a tendency to interpret proxy data in terms of single climatic variables, commonly those used in numerical climate models (e.g. average annual temperature, July and/or January temperatures). However, proxies frequently represent the combined influence of climatic and non-climatic factors. For example, the isotopic composition of ice relates not only to local temperature conditions but also to ice volume, the source regions of air masses and an array of elevational/geographic effects. Tree-ring widths are a function of growing season climates and conditions in preceding years, not just annual average temperature in the year they grow.

Spatial coverage

The geographic coverage of proxy data depends both on where the methods are appropriate and where the data have actually been collected (Table 13.2). For example, tree rings, a widely used source of high-resolution climate data in temperate and high latitudes, cannot be used widely in the tropics where trees characteristically do not form annual growth rings. This is a significant issue given over half the surface area of the earth lies in this latitudinal zone. As with the instrumental record, proxy data have not been analysed evenly across the globe. Rather, data which have been analysed are clustered in and around North America and Europe, with a notable paucity in the Southern Hemisphere. Jones et al. (2001) estimate that the number of records for the Southern Hemisphere is nearly two orders of magnitude lower than the Northern Hemisphere – in part because of the small landmass, but more so because less research has been undertaken. This is changing, however, with greater attention directed to the importance of global datasets. Sites where data have been collected are mapped on to the World Data Center for Paleoclimatology (WDC) website (http://www.ngdc.noaa.gov/paleo/).

Spatial representativeness

The spatial scale of climate that the proxy data actually responds to varies: signals can reflect local, regional or global conditions. For example, analyses of different attributes of ice cores provide records of climatic conditions at very different spatial scales: local information can be derived from stable isotope ratios (temperature) and snow accumulation; regional information on winds (intensity and direction) from wind-blown salt and continental dust; and the gas content of

TABLE 13.2 Climate resolution for sediment archives in three environments. Duration of events that can be resolved relate to the rate of influx and accumulation of sediment, and the amount, activity and depth of organisms burrowing in the sediments (bioturbation) which mix sediments that have accumulated, blurring the degree of detail that can be resolved in the permanent record. Numbers are approximate.

	Influx rates	Depth of mixing	Typical resolvable resolution	Length record
Continental lakes	1 mm per year	1–10 cm	10–100 y	1–10^5 y
Continental shelves/shallow seas	10 cm–1 m per 1000 years	10 cm–1 m	100–1000 y	1–10^8 y
Deep ocean	1 cm per 1000 years	1–10 cm	1000–5000 y	1–10^8 y

Source: Adapted from Ruddiman (2001)

trapped bubbles provides insight into global biogeochemical cycles. Increasingly sophisticated statistical techniques of upscaling allow different lines and scales of evidence to be integrated.

Temporal representativeness

The time period for which reconstructions can be made relates both to the nature of the proxy's response and to the nature of the archive in which it is stored. Proxy data can respond to an event or can provide an integrated response to conditions/events over a number of years. Different proxy systems have different levels of inertia with respect to climate: some systems vary essentially in phase with climatic variations, whereas others lag by many centuries. For example, the latitudinal/elevational boundaries of a forest may take hundreds of years to respond to marked changes in climate, whereas one winter with temperatures below a certain threshold will cause a species of beetle to die out (see further examples in Table 13.1). Clearly the nature of the climate change, a rapid/step change versus gradual warming/cooling, and the extent to which the proxy is able to adapt to new conditions are relevant here also. For high-resolution proxy records, it is likely the parameter being used (tree-ring growth, ice-core isotopic value) actually reflects conditions in a particular growing or snowfall season of the year, not an integrative measure for the year. Historical records of variables, such as grape flowering, harvest dates or snow-cover duration are seasonally specific. Interpretations must take this into account. Stating that the seventeenth century was cooler than the twentieth century based on a greater number of days of snow lying on the ground may not be correct if the non-winter temperatures were warmer.

Issues related to the nature of the record are continuity, resolution and temporal coverage. Accumulation of ice, sediments, etc., may be continuous through time, although rates may have varied or be the result of a series of discontinuous events. This is often an issue with terrestrial geological records. For example, moraines represent maximum glacial extents and recessional positions but give little information on conditions during periods of advance. Marine records, in contrast, tend to be continuous and for this reason are commonly the chronological and palaeoclimatic frame of reference for long-term climatic fluctuations. Resolution relates to the smallest time period for which an event/change can be identified. In addition to the nature of the proxy's response (discussed above), this is dictated by the accumulation rate of a system, given that a minimum sample size (sediment, ice, biological material) is needed for analysis and how fixed the proxy signal remains with time. Many properties of sediments undergo changes (diagenesis) once buried. For example, isotopic/geochemical signatures diffuse, smearing the record, while sediments are often subjected to physical mixing or the effects of bioturbation, thereby reducing the resolution of the information that can be derived. A summary of the degree of resolution for sediments in three different environments (continental lakes, coastal shelves and shallow seas, and deep oceans) is provided in Table 13.2. In general, the greater the accumulation rate, the shorter the time period (temporal coverage) the sample represents. Long records (10^6 plus years) tend to have poor resolution (for example, deep ocean cores); high-resolution records (providing information on seasonal/annual timescales) tend to cover only recent earth history (for example, tree rings or historical data) (Table 13.1).

Absolute chronology

Without reliable estimates of the ages of events in the past it is impossible to know if changes occurred synchronously or if certain events led or lagged others. Nor is it possible to assess accurately rates of change. Few proxy data contain an absolute chronology inherently within their records. Notable and very important exceptions are tree rings, varves (annually laminated sediments) and ice cores. A wide array of absolute (radiometric and radiogenic), relative and correlative dating techniques exists. Special issues of the journal *Quaternary Science Reviews* are dedicated to innovations in methods and their application, and reviews are given in such edited volumes as Rutter and Catto (1995). Increasingly, smaller samples can now be dated, resulting in more precise dates and thus higher-resolution records. Today it is almost standard to use AMS (accelerator mass spectrometer) radiocarbon dates (which provide a direct measure of the ^{14}C content of a sample, rather than the former method which measured beta particles, the decay product). AMS analysis requires samples of only fractions of a gram. Thus individual seeds can be dated rather than having to have large samples which resulted in bulk sediment dates integrated over some depth of core. However, the ability to analyse smaller samples means the question of what actually is being dated its stratigraphic context, and how it relates to the event(s) of interest becomes even more important to resolve.

A summary of characteristics of some of the proxy data organized by the key issues outlined above is presented in Table 13.1. Clear differences are evident: ocean sediments, for example, which theoretically can yield data for 70% of the earth's surface and may provide long continuous records, cannot be used under most circumstances to give high-resolution (years/decades) information about recent climates because of the low rate at which they accumulate (1 cm of core may integrate processes of more than a thousand years) and the problem of accurately dating the materials. At the other end of the spectrum, historical records can provide annual (or intra-annual) data for up to a thousand years in some areas. This proxy set tends, however, to provide localized information and, for most areas of the world, its potential has only been realized for the last few centuries.

CONCLUSION

Studies of past environments must begin with an understanding of the types of proxy data available and the methods used in their analysis. At the most qualitative level, proxy data and their analysis and interpretation provide information on wetter–drier and cooler–warmer periods. At the other extreme, precise quantitative predictions are possible. The most common climate variable reconstructed is temperature, although reconstructions of precipitation, evaporation, soil moisture (or some integrated hydrometeorological measure), winds (direction and intensity) and features of the global circulation (e.g. the polar front, inter-tropical convergence zone) have all been developed.

Summary

• Proxy records are historical (documentary) archives or natural (physical, chemical or biological) systems that are dependent on climate and which incorporate in their structure some measure of this dependency so it is preserved through time.

- The challenge in interpreting these data is to separate the effects of climate (the signal) from all the other non-climatic influences (the noise).

- It is important to be aware of the difficulties and assumptions associated with each method. Only then is it possible to select the most appropriate method(s) for a given question and site, or to synthesize the different lines of evidence to provide comprehensive insight into former climatic conditions.

- More, longer sequences of proxy data, combined with a better understanding of the climate dependency in natural phenomena today, are still needed to provide critical insight into the patterns and causes of environmental change, and to provide a backdrop against which the effects of human actions can be assessed.

Further reading

- Bradley (1999) provides an excellent overview of proxy data sources with examples of how climatic variation in the past has been studied. With Jones (Bradley and Jones, 1995), Bradley has edited a collection of papers on climate variations over the last 500 years, emphasizing the 'Little Ice Age' and climate of the twentieth century. The focus is on high-resolution data – documentary (historical), dendroclimatic and ice-core lines of evidence, as well as consideration of forcing factors (causes).

- The Intergovernmental Panel on Climate Change (IPCC) (Houghton et al., 2001) was created by the World Meteorological Organization and the United Nations Environment Program to provide an assessment of all aspects of climate change, including how human activities can cause such change. Chapter 2 provides an excellent, brief summary of high-resolution proxy data and the results derived from their analysis, which provide the context for more recent changes documented by instrumental records.

- Jones et al. (2001) is one of seven papers in a special issue of the journal, *Science*, on the earth's past climate. This summarizes what is known about changes in temperature and two major ocean circulation systems (El-Niño–Southern Oscillation and the North Atlantic Oscillation) over the last 1000 years.

- Lamb (1995) provides a comprehensive introduction to historical (documentary) sources of climate data, while Ruddiman (2001) is a useful introductory climate science textbook which summarizes 550 million years of the earth's climate changes, including the impact by and on humans. Coverage extends to predictions about future climate changes. The text is accompanied by numerous, very useful, colour photographs and illustrations.

- Rutter and Catto (1995) provide a thorough review of dating methods used for proxy data. For more recent developments, start with the journal *Quaternary Science Reviews*.

- Key websites include: NOAA/NGDC Paleoclimatology Program, Boulder, Colorado, USA (world data center for paleoclimatology): http://www.ngdc.noaa.gov/paleo/ – the mirror site in Europe is http://medias.meteo.fr/paleo/www/anglais/activities/donnees – A wealth of links are provided to key sites elsewhere. PAGES, the international Geosphere-Biosphere Programme (IGBP) Core project: http://www.pages-igbp.org/. Two sites, maintained by individual researchers, which provide extensive explanations and links to other sites, are the Ultimate Tree-Ring web pages (web.utk.edu/~grissino) and the World Wide Web Virtual Library: Paleoclimatology and Paleoceanography (www.datasync.com/~farrar/www_vl_paleoclim.html).

Note: Full details of the above can be found in the references list below.

References

Bradley, R.S. (1999) *Paleoclimatology: Reconstructing Climates of the Quaternary* (2nd edn). San Diego, CA: Harcourt Academic Press.
Bradley, R.S. and Jones, P.D. (eds) (1995) *Climate Since AD 1500* (2nd edn). London: Routledge.
Houghton, J.T., Ding, Y., Griggs, D.G., Noguer, M., van der Linden, P.J. and Xiaosu, D. (eds) (2001) *Intergovernmental Panel on Climate Change (IPCC). Climate Change 2001: The Scientific Basis*. Cambridge: Cambridge University Press.
Jones, P.D., Osborn, T.J. and Briffa, K.R. (2001) 'The evolution of climate over the last millennium', *Science*, 292: 662–7.
Lamb, H.F. (1995) *Climate, History and the Modern World* (2nd edn). London: Routledge.
Ruddiman, W.F. (2001) *Earth's Climate: Past and Future*. New York: W.H. Freeman.
Rutter, N.W. and Catto, N.R. (eds) (1995) *Dating Methods for Quaternary Deposits*. St John's, Newfoundland: Geological Association of Canada.

Acknowledgements

As an undergraduate my interest and understanding of this material was inspired by my adviser, Dr Jean M. Grove. This chapter is dedicated to her memory.

14

Making Observations and Measurements in the Field: an Overview

Ian Reid

Synopsis

For an inexperienced physical geographer, the prospect of making observations and measurements in the field can be daunting. Once a hypothesis has been formulated, there then comes the problem of deciding on an area or a set of locations to gather information with which to test or modify the hypothesis. In some cases this may involve opportunism (for example, the opening of a gravel pit to view soil and geological sections), and the choice is both directed and restricted. But in many cases there may be several locations which might be used, each with advantages and disadvantages. In these cases, a structured approach may be adopted to site selection, in which key decisions must be made with respect to the following issues: generality or specificity of the study; suitability of field sites; appropriateness of sampling designs and strategies; and opportunities and limitations to study arising from the availability and/or choice of field instrumentation.

The chapter is organized into the following sections:

- Introduction
- Site selection
- Sampling strategies
- Instrumental kit
- Conclusion.

INTRODUCTION

The search for suitable field sites and the problems associated with establishing a sampling programme may seem daunting, but great rewards lie in store for those who rise to the challenge. There is considerable satisfaction to be gained from having distilled a pattern and process from what, at first sight, appears to be the complexity of nature. In order to choose an appropriate field setting for a study, some may approach the challenge intuitively. Others, perhaps the majority, might be advised to set out a number of criteria for the adoption or rejection of sites before venturing into the field on reconnaissance. If this is done, each criterion can be given a score on an arbitrary scale at each site that is visited. This procedure is important because the eventual choice of site will involve optimal rather than maximal satisfaction of the criteria. Indeed, it is axiomatic that all sites will be rejected if there is an expectation that all criteria should be fully met. Adopting a scoring procedure is also important because it is easy to confuse different sites once several have been visited. In this chapter, some of the conceptual as well as practical issues involved in selecting field sites are introduced. Attention is then given to a discussion about implementing sampling or measurement programmes, and to some of the requirements of field instrumentation and data interpretation.

SITE SELECTION

Hypothesis and test-bed

Which is the chicken and which the egg? This is a question that often exercises an inexperienced researcher setting out to conduct a project on which hangs not only a developing reputation among peers and tutors but also credit that can be gained towards a qualification. In this case, the chicken is a concept or a hypothesis that may involve either a natural or artificial process that manifests itself in space and time, or a sequence of events, or a set of diagnostic properties that are locked in a column of sediment. The egg is the test-bed where data will be gathered. The research 'question' – the hypothesis – has to be the primary driver. Indeed, it might be argued that constructing the hypothesis in the context of a specific location begs the 'question' and can lead to predetermination of the answers. However, in constructing a hypothesis independently of location, the researcher is faced with the 'problem' of finding a field site that offers the prospect of a reasonably rigorous test. This more often than not means eliminating or 'holding constant' factors that might contribute variance and, thereby, cloud the relation between hypothesized cause and effect, so reducing the degree of certainty that a process or a sequence of events has been tested adequately. So, for instance, if the impact of slope aspect on soil thermal regime is under scrutiny, there may be a number of factors, the influence of which would ideally be eliminated at each site of measurement. These will include slope angle; vegetation; soil properties, such as organic matter, horizonation, porosity and water content; differential topographic shadowing; and elevation. In this case, the researcher might start searching for a conical hillock of simple lithology and geological structure, clothed in undifferentiated vegetation and set on a plain apart from other upstanding topographic features! Much time might be spent inspecting prospective locations with inevitable frustration. Compromise is (almost) inevitable and an 'imperfect' site will have to be selected. In this case, the researcher will have to accept that cause and effect are multifactorial, if this has not always been part of the research design, and that the testing of the hypothesis will involve statements of probability about relations between alleged causes and effect.

In reality, the chicken and egg are not easily differentiated, and an idea – a research hypothesis – may spring to mind when observing a field phenomenon. Or it might be that a specific field test-bed springs to mind when formulating a hypothesis. In these cases, the imperfections of the field site may be accepted as contributing to unexplained variance in the relation(s) between effect and cause(s). So, for example, in reconstructing the Holocene development of a floodplain, hypothesizing that climate or land use-induced changes in runoff regime had led to a change in channel pattern from multi- to single-thread, the evidence may come necessarily from sedimentary sections exposed by a gravel extraction company rather than from locations that are aligned best to test lateral or longitudinal gradients in type facies. In this case, extrapolation and inference may become key ingredients in reconstructing environmental change and conclusions will be commensurably tentative. Here, the philosophical approach adopted at the outset might best involve multiple working hypotheses, since there are often too many 'unknowns', and a researcher might be faced with assessing the likelihood that several 'explanations' can be used in understanding the patterns that are found.

Generalist or specifist?

A piece of research may aim to provide an answer to a general question or it may attempt to explain a process, a condition or a sequence that is specific to a particular location or setting. Most projects will attempt to generalize while acknowledging those limitations that are imposed by the choice of field site, the experimental design or the measurement techniques. This is understandable in that there is merit in being able to broaden the applicability of research findings, and it might be argued that the majority of work about which an undergraduate has read will adopt this approach. So, it will be familiar. There may be problems of justification in that the limits on time and testing (often a resource issue if expensive analyses are involved) make it necessary to generalize from the particular or, at best, from an insubstantial number of particular cases. Nevertheless, providing that any claims and conclusions are qualified and that supporting arguments are constructed to acknowledge possible flaws or shortcomings, this is a reasonable stance. So, for example, in attempting to understand the relation between the mass movement of slab slides and rainfall, there may be only one accessible slide and a limit on the number of piezometers that can be installed, the more so if these are continuously recording. The findings will be specific to this slide and, even more confining, to the range of weather-related soil conditions that have been monitored. However, some speculative generalizations would be allowable, even encouraged, providing they are not outrageous.

Avoiding (or acknowledging) legacies and process feedback

In setting out to make field observations, there is a need to be cognizant of boundaries and to acknowledge that what is being measured is more than likely to involve either legacies of processes occurring beyond the area or the period of immediate interest, or process feedback, or indeed 'leakage'. It is virtually certain that no system of natural processes is discrete in any dimension. There is, therefore, a need to take this into account in designing an experiment. So, for instance, in examining the species composition of an ecosystem in the context of testing a hypothetical edaphic control, it would be essential to assess the likelihood of having avoided sampling an ecotone. In the event of not being able to locate a sampling area distant from other ecosystems, this might mean establishing the credentials of a core by assessing if there are spatial gradients of species composition that extend into the periphery. In many instances – perhaps most – discretization is impossible. This is most obvious where the system involves the passage of fluid, whether air or water. So, for instance, in attempting to relate the magnitude of primary and secondary currents to river channel dimensions, it is impossible to avoid some legacy of the channel upstream and the effects that it has imparted to the structure of the flow entering the reach in which the researcher is interested.

    Process feedback can be just as important and has to be allowed for in designing a monitoring programme if it, in itself, is not the subject of investigation. Indeed, in attempting to introduce a means of measurement, the problem of feedback is sometimes – perhaps often –

inadvertently introduced by the researcher. So, for example, in attempting to show that soil interflow is dependent on some variable such as ground slope, interceptors might be introduced after excavating soil pits. But each pit face will act as a macropore such that, unless the soil water immediately upslope has a pressure potential near atmospheric, water that would have moved downslope under a gradient of total potential will back up above the pit and the interceptor will give a false indication of flux. There may be little or no alternative in circumstances such as these. Indeed, many field sciences are faced with the problem of reconciling the fact that a sledge-hammer – however expensive and sophisticated – is the only tool available to crack the nut! Providing that the method adopted is not seriously flawed, it might be justifiably applied. But it is important that the researcher recognizes the limitations of the experimentation, acknowledges these in a suitable fashion and is appropriately circumspect about the results being reported.

Deploying scarce resources of time and/or equipment

Completing a research project successfully requires efficient use of time and equipment. Although it may be considered by some to be pedantic, it may be useful to define specific objectives in detail so that they can be seen to feed the project aims. It is also important to remain flexible. Once in the field, it may become self-evident that some things will be impossible to execute. In this circumstance, a researcher may wish to revisit the objectives and evaluate whether the substitution of new or modified objectives is desirable. For the sake of ego and self-esteem, an inexperienced researcher should be aware that even the most seasoned field scientists change their plans on moving into the field. Nature is, after all, wonderfully unpredictable – hence our interest in finding out how it works!

The way field observations are obtained will depend upon a number of factors. Some involve the nature of the subject under investigation; others are governed by the time a researcher can devote to a field project. There are two basic models. The first involves an uninterrupted period of field data collection, perhaps of two to four weeks or longer. But, before this, an initial reconnaissance visit is highly recommended. This will ensure that sites are accessible and that conditions will permit data acquisition. It should be used to gain permission for access from landowners. In some circumstances, research permits may be required from government departments. In others (e.g. quarries, gravel pits), gaining permission is obligatory in order to satisfy safety regulations. Here, there may be severe restrictions dictated by quarrying operations, and these will affect the amount of data that can be acquired. Permission is vital in order to ensure that the full sampling programme can be executed without eviction. It cannot be stressed enough that the goodwill of landowners is invaluable. Without it, a project may founder; this is all the more tragic if eviction comes after time has been invested in acquiring part of the dataset and a change in direction, even a change in the nature of the project, has to be sought. Under this operational model, following the reconnaissance visit, a researcher moves into the field, establishes a sampling scheme and a schedule and accomplishes the entire data collection within the period. This model favours morphological and stratigraphic studies and those where the process of interest is short-lived (e.g. a flowering season or snow-melt runoff) or where the range of conditions is likely to be met in a finite period (e.g. a tidal cycle). It also favours studies in remote locations where cost precludes repeat visits.

The other operational model involves local sampling over one or two days at either regular or irregular intervals over a period of a number of months. This is likely to be dictated as necessary by temporal changes of longer wavelength, often seasonal, and may be *required* by some hydrological, ecological and weather studies. A number of variants on these two models can be tailored to fit the requirements of individual projects, but experience suggests little variation is required for most undergraduate research.

The sampling scheme – even the very nature of the project – might be determined by the availability of equipment. Many university field science departments make instruments available to students. However, the more sophisticated or expensive the equipment, the more it is likely to be loaned for short periods only, depending on demand. Projects of model type 1 above, involving long periods of resident fieldwork in distant locations, are most likely to be refused loan of rare equipment because they tie it up, reducing its availability to others. The solution to this difficulty will vary according to individual circumstances. Some may be able to secure equipment from elsewhere (e.g. surveyors' levels can be hired). Others will be imaginative and use alternative methods, perhaps constructed from cheap materials, while some may make sensible adjustments to the study, so obviating the need for certain items of equipment. Whatever the solution, it is important to review whether any modification in the type and amount of data that will be acquired will allow a satisfactory assessment of the aims and objectives. If this is impossible, a serious review of the project should be undertaken. But before panic sets in, bear in mind that it is more likely to be the level of significance that can be attached to any statistical test conclusions that will deteriorate rather than the study as a whole.

SAMPLING STRATEGIES

Defining boundaries

It is essential to define the bounds of a project – both philosophical and dimensional. Much will depend on whether the task is part of a single module or a dissertation that forms a large constituent of work towards the qualification. In both cases, there is a need to match the amount of effort with both the nature of the contribution and the aims and objectives of the project. It is self-evident that quantity is no substitute for quality. However, too few data will almost certainly limit the analysis and compromise an ability to reach reasoned conclusions. Too much may mean that time could have been devoted to more productive activity! It is here that the advice and guidance of the project tutor should be sought.

Space and time trade-offs

The sampling strategy of a project is second in importance only to the hypothesis. Making field observations is not difficult. Making appropriate field observations is a challenge. Sampling design involves an incredible number of decisions and choices, many of which are made subconsciously, but many of which involve agonizing. Given that a great deal of effort is about to be invested in obtaining materials, recording locations and measuring processes, there is often considerable anxiety attached to setting out a sampling scheme. There are a number of situations where choices are limited, of course. So, for instance, if interest lies in determining the impact of a sewer outfall on the water quality of a river, the number of sampling locations will be limited and reasonably obvious. However, most studies involve the researcher in defining experimental boundaries in both space and time. For example, in assessing the spatial variation of soil properties in undifferentiated topography, the extent of the sampling grid is going to be determined by the researcher. In contrast, where topography is broken, each slope facet may well be easily identifiable, so directing the researcher's decisions. But here the problem might be that it is impossible to replicate sampling points or plots, so running the risk of infringing the need to demonstrate that processes or patterns under investigation repeat themselves and are, thus, predictable at given levels of probability.

The sensible location of sampling stations, or the appropriate definition of a sampling arena, is a function of experience. An inexperienced researcher should not shy away from decision-making – we all learn from mistakes – but it is advisable to lean on a supervisor more at this early stage than later when the data gathering is under way.

However, harsh decisions may have to be made about the nature of the data gathering. If a project involves assessing processes in both time and space, pragmatism may require a choice between gathering lots of pertinent information over time at a few locations or gathering data at a lot of locations but on few occasions. The two are not infrequently interchangeable. In the first, the element of risk lies in the fact that the few sites chosen may be inappropriately located and, therefore, yield data that do not capture the process under consideration. This may not become apparent until the investigation is under way. (Heavy reliance on expensive or complex instrumentation often forces this mode of operation, the use of the equipment itself making 'mistakes' seem even more expensive if they occur.) In the second, the risk lies in missing the very events that are under investigation. So, for instance, a time series might become akin to a moving average, ironing out the interesting extremes or, just as difficult, a set of extremes might be recorded, preventing an assessment of base conditions from which the extremes protrude.

Another version of trade-off involves the number of environmental parameters that can be investigated. Here, for example, it might be highly desirable to maximize the number of sampling locations; but, given the limits on time and resources, this might require a reduction in the number of environmental attributes that are measured at each site. The risk lies in choosing attributes that may turn out to be inappropriate in attempting to assess cause and effect in process studies, or in assessing environmental change in microstratigraphic studies.

Depending upon the nature of the project, a compromise might be achieved by adopting a 'nested' study strategy. Here, the broad framework may be established with a spatially extensive study of many sites, at each of which comparatively few diagnostics are obtained. Within this framework, at a strategic site or sites, an analysis may be made of many diagnostics and/or the sampling interval in a sedimentary sequence or soil profile may be much reduced, or a detailed temporal study may be undertaken to capture event-based changes which would be missed by the broader framework. Here, strength comes from the juxtaposition of the complementary approaches. There are different expectations of each 'nested' component. The localized studies may indicate processes or patterns at specific places and be 'guilty' of investigating the unique. The geographically extensive component may provide a broad framework and be insensitive to vital detail. But the researcher is seeking corroboration, either direct or through inference, and is hoping to show that each nested study points towards the same set of conclusions.

Harmony between nature's rhythms and sampling schema

Nature's rhythms are a gift to the field researcher and should be exploited to the full if the study allows. Transitions between wet and dry seasons often involve measurable changes that highlight processes which would not manifest themselves at other times. So, for instance, if, in a study of soil-water redistribution mechanisms, it is important to evaluate by-pass flows that are facilitated by cracks between pedons, there may be a limited but valuable window of opportunity in the autumnal transition between summer and winter. These transitional seasons might also be valuable in confirming the magnitude or direction of processes that have been measured during the height of summer or the depth of winter. So, for example, in confirming the pattern and quantity of interception by a plant community, it might be useful to extend the study from summer through an autumn period of senescence, even if time precludes carrying the measurement programme into the winter season of defoliation.

Some systems have comparatively short cyclical wavelengths, and conditions might be considered transitional at any time. In these cases, allowance for non-stationarity will be an essential part of the research design. So, an obvious case involves studies of the marine shoreline where significant diurnal and monthly changes in tide level might mean that a process under study such as wave break migrates shoreward or seaward by several hundreds of metres and, in some cases, by as much as several kilometres. Cruelly! In this particular environment, if the study involves progress towards or away from high water at either neaps or springs, data gathering will inevitably mean a shift in the time of sampling and this may well dictate nocturnal activity in order to complete the analysis of processes that occur at a state of the tide that is first encountered in full daylight!

Exploiting nature's rhythms and the transitions they offer can be invaluable. However, it may be essential not to miss the 'main event'. So, it should be self-evident that if an aim is to measure flash-flood dynamics, fieldwork should avoid the dry season! If there is an interest in discovering the impact of Late Glacial environmental change on local biodiversity, it would be prudent to adopt a facies-orientated microstratigraphic sampling scheme rather than a rigid interval sampling scheme in analysing a sediment sequence.

All this suggests that preparation is important. The acquisition of adequate maps (if available) is a necessary prerequisite. In the absence of suitable maps, there may be air photographs (even in remote locations) or satellite imagery (though the usefulness of this last may depend on the scale of the study and the cost of acquisition). For process studies, there may be available from government agencies, water supply companies, etc., a wealth of information regarding weather and hydrology that can be used both as a backcloth for the study and to determine timing of fieldwork. Here, however, it is as well to bear in mind that epicyclic changes in weather and instability that may be driven by longer-term climate change could affect the incidence of events, producing significant departures from patterns that come out of historical records, even those of very recent periods.

The field and subsequent analytical budget costs

Cost implications need to be assessed before embarking on a study. It is usual to be fully cognizant of the costs of travel to, and subsistence in, the field. What may not be taken into account are the costs of analyses, either in the field or subsequently on returning to the laboratory. Some water, soil and sediment chemical analyses can incur substantial bills for reagents. Access to scanning electron microscopes more often than not involves fees that are substantial. There may also be the costs of insuring against loss of, or accidental damage to, equipment while it is in use in the field.

Some costs can be avoided or minimized by changing the emphasis of a study. However, if a reduction in the number of samples is brought about by a consideration of cost, the question of whether the project aims and objectives can be met should be revisited. As with the overall model of sampling outlined above, it might be acceptable to adopt a 'nested' approach to measurement. So, for example, in assessing the influence of urban runoff on the quality of water in the nearby arterial river, it may be acceptable to interrogate the system for diagnostics such as electrical conductivity, pH and temperature at frequent intervals, since these involve little, if any, analytical costs, and to conduct a full ionic analysis at less frequent intervals.

INSTRUMENTAL KIT

Importance of thinking ahead, especially for work in remote areas

Experience is the best guide to what equipment or tools might be needed to execute fieldwork. For the inexperienced researcher, a discussion with a tutor may be invaluable, but careful thought about the operations that are planned – a rehearsal in the mind's eye – will help to sort out a list of essentials. Fieldwork may be sited far from base or from the nearest hardware store; it may involve an arduous hike off the beaten track; and it might require carrying large weights of soil, sediment or water back to a collection point. All these mean that a researcher should have the variety and number of items required while not overburdening him or herself with kit that might never be used.

Commonality of instruments

Fortunately, much can be achieved with a set of basic equipment, and the best research is often accomplished with simple tools. So, a stock 'tool-kit' might include one (preferably two) 25 m glass-fibre measuring tapes; a folding 1 m ruler; a compass-clinometer (preferably, also, a prismatic compass for large-scale survey if trilateration is impracticable); a camera (digital or conventional; for stratigraphic work, a Polaroid camera can be useful for locating sampling positions within identifiable facies on-site, obviating the need to tie in notes and sketches back at base); a folding trenching tool; a multipurpose pocket tool which may or may not have affiliation with the Swiss military; zipping sample bags; a 'permanent' marking pen; and a water-resistant (if affordable) notebook and pencil (not, as some would have it, a ballpoint pen!). In addition, it might be useful to carry an Abney level (which can double as a spirit level or clinometer) and a 2 m ranging rod.

Complex and costly instruments may be available on loan from a field science department. If this is the case, make the most of them and, in order to preserve goodwill, ensure they are returned in good condition! The borrower may be required to take out insurance, but this is wise anyway if the equipment is not already insured. It is not always appreciated that a seemingly uncomplicated current meter or a basic electrochemical water-test kit can each cost £2000 or more to replace and a surveyor's level may be valued at £500 or more.

Customized instruments

There are many circumstances where improvised equipment provides an adequate substitute for expensive items borrowed from stock. Indeed, this may be the only way of obtaining data. Plumbing materials are particularly useful. So, for example, if the longitudinal slope of a meandering river reach is required, careful use of a garden hose with a transparent water-level site tube fixed to a 1 m ruler at either end may be a better way of establishing the fall in the bed or water surface than attempting to level through thick riparian vegetation, which would, in any case, require a companion. If cash resources are small, many materials – especially timber – can be gleaned (with care and permission!) from builders' skips. So, if there were a need to install 30 flood crest-stage gauges in a river network, an initial look at other people's rubbish might save the purchase of new timber and the money can then be spent on another component – drain-pipe!

CONCLUSION

There is no limit to the number of practical field projects that can be tackled at all levels in physical geography, and there is nothing more satisfying than establishing and executing a field monitoring programme which successfully demonstrates cause-and-effect relations in a process study, or which confirms a conjectured sequence of events in a palaeoenvironmental reconstruction. For those who are inexperienced, setting out to test a hypothesis can seem daunting, given the complexity of the natural world. By adopting a structured approach to research and research design, however, a student minimizes angst and maximizes the prospect of success.

Having taken a structured approach to both site selection and the subsequent measurement programme, it is easier to provide a justification for the veracity and potential generality of the results obtained. All this needs to be done in the knowledge of what subsequent analytical procedures are both available and affordable, and with a weather eye on the possible uses of the data in constructing or illustrating the arguments that may be developed in the project report.

Summary

- The search for suitable field sites and the problems associated with establishing a sampling programme may seem daunting, but great rewards lie in store for those who rise to the challenge.
- This involves considering, first, some very general questions, such as the purpose of the research, the degree to which general conclusions might be derived from one or more sites, and the nature of cause and effect at the site.
- Next come more practical considerations, such as the cost in time or money of using particular sites and the access to them.
- Even when a location has been chosen, decision-making is far from complete because implementing a sampling or measurement programme again involves compromise and choices based upon a mixture of theoretical and practical considerations.
- A researcher needs to be consistent in applying a set of either self-imposed or conventional procedural rules in order to ensure adequate justification.

- The availability of field instrumentation is also an issue, since this determines the kind and range of data which may be collected, and, in turn, the nature of any explanation which is offered.

Further reading

This chapter has been concerned more with approaches to making observations and collecting data in the field rather than with the data gathering itself for a number of reasons. First, geographers faced with a choice of projects in physical and human geography often perceive it to be more difficult to establish a meaningful programme of data gathering in physical geography. This is a misconception that should be dispelled and it is hoped that this chapter goes some way towards accomplishing this whilst acknowledging that getting something worthwhile done always involves a challenge. Secondly, the field of physical geography is huge. Although there are common tools and methods (some, such as those involved in sampling and conducting experiments, being dealt with in other chapters), the range is too large to cover adequately in a single chapter. Rather should students use the messages conveyed here as a guide to establishing a field programme. For the specifics of how to tackle field observations in each of the sub-realms of physical geography, the reader is referred to a number of texts:

- Gilbertson et al. (1985) demonstrate that establishing ecological patterns and processes involves not only measuring the physical and chemical environment but parameters that are peculiar to this branch of physical geography. Some of the targets of study move of their own volition and have 'preferences'! So, if your study involves biota, delve into a book like this to assess ways of measuring and observing the terrestrial living world.

- For riverine projects, Gordon et al. (1992) provide an excellent mix of theory and practice. Even more, as a collaborative venture between an ecologist, a hydraulic engineer and a fluvial geomorphologist, it offers a balanced approach to river studies and removes the divergence of thinking and terminology that not infrequently complicates interdisciplinary areas of study. If you are a fluvial *geomorphologist*, do not be put off by the book's subtitle *An introduction for ecologists*. The book will cover the ground that you need and open your eyes to other ways of approaching river studies.

- Andrew Goudie's many contributions to geomorphology include an extremely useful introduction to the field techniques (Goudie, 1990) commonly deployed in a number of subdisciplines. The book is a set of essays written by experts in each area, which include studies of landform, surface processes and sediments in hillslope, riverine, glacial and littoral contexts. It is a first port of call for many field studies in physical geography.

- Only part of Haynes (1982) – on ground surveying – addresses the question of field measurement, but the remainder provides a wealth of information on treating environmental data mathematically, statistically and graphically that is useful to know before venturing into the field so that sampling design satisfies post-data-collection test criteria.

- Linacre (1992). Meteorological data not only forms the root stock of many studies that are directly concerned with weather phenomena, it is often an essential ingredient for ecological, hydrological, glaciological and geomorphological studies. This book has one chapter on setting out to measure weather. But it also alerts the student to the potential pitfalls encountered in self-made measurements and in climate data obtained from other sources.

- The fundamentals of recording weather are set out in a really useful, extending series of articles published in the Royal Meteorological Society's periodical *Weather*. Of particular interest are:
    Strangeways, I.C. (1995) 'The "met enclosure". Part 1, Its background', *Weather*, 50: 182–88.
    Strangeways, I.C. (1996) 'The "met enclosure". Part 2a, Raingauges', *Weather*, 51: 274–79.
    Strangeways, I.C. (1996) 'The "met enclosure". Part 2b, Raingauges, their errors', *Weather*, 51: 298–303.
    Strangeways, I.C. (1998) 'The "met enclosure". Part 3, Radiation', *Weather*, 53: 43–49.
    Strangeways, I.C. (1999) 'The "met enclosure". Part 4, Temperature', *Weather*, 54: 262–69.
    Strangeways, I.C. (2000) 'The "met enclosure". Part 5, Humidity', *Weather*, 55: 346–52.
    Strangeways, I.C. (2001) 'The "met enclosure". Part 6, Wind', *Weather*, 56: 154–61.

In order to review the growing list (and much more besides), use the World Wide Web URL: www.met.reading.ac.uk/~brugge/index.html and click successively on the links: 'Weather' and 'Back to Basics'.

Note: Full details of the above can be found in the references list below.

References

Gilbertson, D.D., Kent, M. and Pyatt, F.B. (1985) *Practical Ecology*. London: Hutchinson.

Gordon, N.D., McMahon, T.A. and Finlayson, B.L. (1992) *Stream Hydrology*. Chichester: John Wiley.

Goudie, A.S. (ed.) (1990) *Geomorphological Techniques* (2nd edn). London: Unwin Hyman.

Haynes, R. (1982) *Environmental Science Methods*. London: Chapman & Hall.

Linacre, E. (1992) *Climate Data and Resources: A Reference Guide*. London: Routledge.

15

Sampling in Geography

Stephen Rice

Synopsis

Geographers recognize the value of both extensive statistical sampling and intensive 'case-study' sampling for exploring an uncertain world. A benefit of extensive sampling is that a set of techniques known as inferential statistics can be applied to make probabilistic statements about the population from which the sample is drawn. Sampling is therefore a powerful tool, but geographical research frequently engages with very heterogeneous phenomena that require careful sampling in order to maximize the accuracy of conclusions. Thoughtful design of the sampling programme is therefore crucial and is driven by both the research aims and available resources.

The chapter is organized into the following sections:

- Introduction
- Samples and case studies
- What makes a good sample?
- Designing a sampling programme
- Statistical inference
- Sample size
- Conclusion.

INTRODUCTION

Sampling is the acquisition of information about a relatively small part of a larger group or population, usually with the aim of making inferential generalizations about the larger group. Sampling is necessary because it is often not possible, practicable or desirable to obtain information from an entire population. For example, it is essentially impossible to measure all the sand grains on a beach to ascertain their average size; impracticable, in the course of a normal day's work, to question every person on the beach to determine the variety of their views about personal use of public spaces; and undesirable (never mind unethical) to stress the fish community of a seashore rock-pool by catching and examining all its members. Moreover, in quantitative research, a set of procedures known as inferential statistics can be applied to sample data in order to make generalizations, validated by probability statements, about the entire population from which the sample was drawn. Thus, it is not *necessary* to interrogate a whole population to make useful generalizations about it.

In one form or another, sampling is the basis of almost all empirical research in both physical and human geography and is widely relied upon. However, as with so many attractions, such a powerful methodological tool comes with a set of health warnings: samples are only as valuable as they are representative of the larger population; at best, bad sampling leads to imprecision; at worst, bad sampling yields incorrect or prejudiced results. In the extreme, lack of sampling rigour makes inferences meaningless: 'The tendency of the casual mind is to pick out or stumble upon a sample which supports or defies its prejudices, and then to make it the representative of a whole class' (Walter Lipman, journalist, 1929). This remark may be directed at the dangers of using isolated examples to make unfounded inferences in everyday life rather than in academic research, but it nicely highlights the important association of weak, non-rigorous sampling with bias and inaccuracy.

This association, the difficulties that are frequently encountered in order to collect representative samples and the impenetrable nature of 'stats' for some students breed a scepticism about sampling and statistical inference that can run deep. It is fairly common to hear the dismissive statement 'Well, you can prove anything with statistics'. And Benjamin Disraeli's remark that 'There are three types of lies: lies, damned lies and statistics' has been a convenient aphorism of opposition politicians concerned with the scruples of their counterparts for over a century. This scepticism is, however, unfortunate because geographers often seek to understand spatially diverse, highly complex phenomena that can be efficiently accessed by sampling and usefully examined using statistical inference.

While questions about the motivations and affiliations (political, epistemological, etc.) of researchers have a place in the critical assessment of all research, for students of method, the value of statistical inferences must be judged simply against the quality of the sample data and the quality of the analysis applied to them. These issues are the focus of this chapter. It reviews the use of sampling by geographers; considers the criteria by which samples should be judged;

discusses the design and implementation of sampling schemes; introduces some of the basic elements of statistical inference; and suggests some methods for defining sample size. The overall message is that sampling is a powerful tool that most geographers need but, if research methodology aims to be as impartial and free of error as possible, sampling must be done thoughtfully and rigorously.

SAMPLES AND CASE STUDIES

Not all geographers seek to make quantitatively robust inferences about large populations or develop theories and models of universal validity. Nevertheless, a principal aim of most geographical research is to make useful generalizations – that is, to seek out and explain patterns, relations and fluxes that might help model, predict, postdict or otherwise understand better the human and physical worlds around us. Thus, some geographers may restrict their attention to small areas, short periods of time, small groups or even to individual places or people, but the underlying approach remains nomothetic – it focuses on identifying the general rather than the unique, and in turn most geographical research involves some form of sampling.

Geographical systems are complex and affected by historical and geographical contingencies, indeterminism or singularity (Schumm, 1991). That is, while there are general similarities between objects, there is also inexplicable (or at least, as yet unexplained) variation in any population so that each item is a little different from the others. This means that while we may be able to develop generalizations that are valid for a whole population, estimating the behaviour or character of any single individual is difficult and prone to error. This is as relevant to predicting river discharge as it is to commenting on the global reach of large corporations or people's views on street architecture. Thus, the geographical world is an uncertain world.

Geographers have adopted a variety of strategies to search for general understanding while recognizing this uncertainty. A useful distinction can be made between approaches that intensively examine a small number of examples and those that sample extensively (Harvey, 1969; Richards, 1996). In the extreme, geographers have made substantial use of single case studies (samples of one) to learn about both physical and human phenomena. While case studies provide detailed information, a fundamental criticism of the approach is that the generality of the case is unknown. That is, there is no formal basis for substantiating inferences made about a population on the basis of a single sample, such that extrapolating the findings of the case to the population remains merely a matter of intuitive judgement on the part of the investigator. This is problematic because most geographers would and should be sceptical of such subjectivity. In contrast, the collection of large, extensive samples provides the opportunity to utilize statistical methods which, at least when sampling is appropriately conducted, offer a means of assigning objectively derived conditional statements to population inferences.

While recognizing the limitations of case studies and the benefits of statistical inference, it is important not to denigrate the value of intensive investigations, not least because there is no single, simple model that defines how geographical knowledge can or should be obtained and ratified (see Chapters 14, 16 and 17). Case studies may not provide a basis for making wide-ranging inferences about a population, but inferences based on case studies are not necessarily false or unreasonable. Rather, they stand to be substantiated, and the detailed information gathered in a case study may reveal general structures or relations that can be used to generate or modify models or hypotheses (Harvey, 1969). Similarly, case studies may present unique opportunities for understanding the mechanisms that underlie empirical observations. In geomorphology, Richards (1996) argues that as long as the location of field sites is carefully planned, case studies offer important advantages over extensive approaches. This is because case studies provide an opportunity to ask fundamentally different questions in a fundamentally different way. Case studies often aim to explain the mechanisms that generate patterns observed in extensive studies (cf. Blaut, 1959) and, as such, case studies should not be judged by their representativeness (or lack thereof) but by the quality of the theoretical reasoning they generate (Richards, 1996). Similar arguments, based on alternative views of generalization and theory validation, are used by qualitative researchers to explain the use of case studies (e.g. Miles and Huberman, 1994).

While statistical theory guides sample collection in extensive approaches, the selection of case studies for intensive study is based on less well established criteria. Richards (1996) highlights the importance of carefully selecting cases posessing those properties which facilitate rigorous tests of the hypotheses under consideration, and meticulous definition of local conditions. Curtis et al. (2000) review case-selection criteria formulated for qualitative research which include ethical considerations and relevance to a theoretical framework. Using examples from medical geography, they find that it can be difficult to reconcile these two criteria such that ethical considerations are at odds with the selection of the most useful or theoretically relevant cases.

The collection of information within a case study, especially in quantitative research, will typically require extensive sampling *internally*, albeit on relatively small temporal or spatial scales. If useful internal inferences are to be made then this sampling must be rigorous and statistically accountable. In turn, as Richards (1996) points out, the practical distinction between 'case studies' and 'extensive studies' is somewhat fuzzy in geomorphology, and largely semantic. In practice, many geographers are typically engaged in studies that shift between these two styles, using their attendant opportunities to explain geographical phenomena. The literature reviewed above provides a starting point for further consideration of case studies as a form of sampling. In the remainder of this chapter, the focus is on extensive sampling and classical statistical inference.

WHAT MAKES A GOOD SAMPLE?

In the context that sampling is used to make generalizations about a larger population, the aim of sampling is to obtain a 'representative characterization'. Characteristics of a population are referred to as parameters, while those of a sample are called statistics. A good sample may then be defined as one that satisfies two criteria: it must provide an unbiased estimate of the parameter of interest, and it must provide a precise estimate of the parameter of interest.

Accuracy, precision and bias

The very nature of sampling means that, in repeated sampling of the same population, different items (individuals, businesses, households, etc.) will be drawn. Thus, statistics vary between samples and, in each case, differ to a greater or lesser degree from the population parameters they estimate. The difference between sample estimates and the true population value is referred to as the accuracy of the sample. Accuracy is gauged in terms of its two components: bias and precision.

These terms can be illustrated by representing repeated sampling of a population by a darts competition in which each dart represents a single sample estimate and the bull's eye represents the true population value (Figure 15.1). Bias refers to the systematic deviation of sample statistics from the true value. A set of sample statistics that vary around the true value without any discernible pattern is said to be unbiased (Figure 15.1b and d), whereas a tendency to be consistently different (for example, too high or too low) reveals bias (Figure 15.1a and c). Lack of bias ensures the representativeness of a sample and is a fundamental requirement of statistical inference. In principle, lack of bias is achieved by sampling randomly from a population and in practice this becomes the main challenge in sample collection. Precision refers to the size of the deviations between repeated estimates of a given statistic. It describes the degree to which repeated estimates are clustered together: are they tightly bunched (Figure 15.1c and d) or widely dispersed (Figure 15.1a and b)? In real sampling

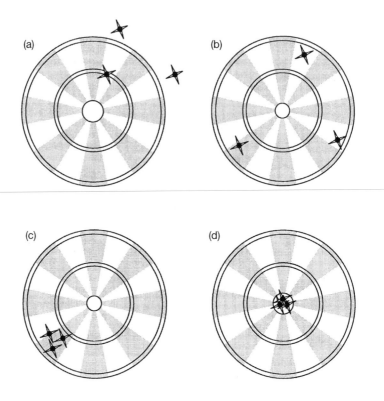

FIGURE 15.1 Precision and bias represented as a game of darts

problems, bias and precision cannot be observed directly because the values of the population parameters are unknown (otherwise there would be no need to sample!) and, typically, we only collect a single sample. When sampling, we should then aim to maximize precision and minimize bias so that we can have faith that the single sample we collect is accurate (Figure 15.1d).

Minimizing bias

Bias is introduced into a sample in one of three main ways. First, the actions of the person collecting the data may introduce operator bias. Two different individuals asked to sample the same population may produce samples that are not just different (this is expected), but systematically different from one another. For example, in approaching passers-by to interview on a city street, an operator may knowingly or inadvertently select individuals of a certain age, gender or ethnicity, thereby over-representing those groups in the sample. In collecting pebbles from a beach to determine their average size an operator may tend to pick up pebbles that are more distinctive in colour and more easily seen. Since colour depends on lithology and lithology affects size, the sample could be unrepresentative of the range of sizes present. Operator bias is reduced by careful training and adherence to a standardized set of procedural rules consistent with the sampling method (see below). Thus, the street interviewer may be asked to minimize selection-based bias by approaching every tenth person who passes by and the pebble picker may be instructed to collect the pebble that lies beneath the metre marks on a measuring tape laid down on the beach surface. However, the propensity for humans to ignore rules or simply to make mistakes means that operator bias is difficult to remove entirely. Thus, looking down at the correct spot on the tape, the pebble picker can choose between that pebble that lies to the right or left of the metre mark, and it is common for operators to select pebbles preferentially that fit comfortably in their hand, potentially ignoring large or small clasts and thereby biasing the sample (see Marcus et al., 1995, for a fuller discussion of this case). Additional procedural rules can be made to try to minimize such errors, but ultimately it may be difficult to eradicate them entirely.

Secondly, bias can be introduced by faulty or misused measurement devices that systematically misrepresent the characteristic of interest. This is fairly clear in the case of instruments that measure some chemical, physical or biological property and have the potential to be miscalibrated or broken. Miscalibration is a problem because many measurement devices do not measure the property of interest directly but via some more easily assessed quality that is quantitatively related to the target property. For example, stream velocity is routinely measured using devices that generate a small magnetic field in the flowing water and detect changes in the electrical current that is produced as water flows through it. The current is proportional to the water's velocity so that velocity can be determined by measuring the output current. However, care must be taken to calibrate and set up such instruments correctly to ensure that no systematic error produces consistently deviant velocity estimates. In the widest sense, a question on a questionnaire, for example, is also a type of measurement device that can return a biased response by asking leading or biased questions (see Chapter 6).

Thirdly, and perhaps most commonly, bias can be introduced during the design of the sampling programme, particularly poor definition of the population or choice of an inappropriate sampling method by which individual sample items are drawn from the population. In geographical research, the content of the population is apt to vary in space, in time and with the scale of interest, often in a systematic way. Unfortunately, this means it is relatively easy to collect a sample that is not representative of the intended population (these issues are discussed at length below).

Maximizing precision and non-systematic measurement errors

Precision is largely a function of three things: the number of observations that make up a sample; the heterogeneity (variability) of the characteristic of interest within the population; and non-systematic errors that arise from the technical limitations of the measuring procedure. We shall return to the first two below; suffice to say here that the larger the sample size and lower the population heterogeneity, the more precise sample estimates will be.

Measurement errors, sometimes called pure errors, exist in all measurement operations because of technical limitations within the measuring system (the people and instruments involved). These are the irreducible errors that one has to be willing to accept in making any set of observations. Wherever possible, every effort should be made to minimize them and, of course, one must endeavour to ensure that the errors are free of bias. A brief example from geomorphology illustrates several relevant issues.

In the field, we use an instrument called a total station to survey topography – for example, exposed gravel bars in an alpine river channel. The instrument is set up at a base station and measures distances, declinations from horizontal and directional angles to survey points across the bar surfaces. A target prism attached to a pole of known length is used to mark each survey position and its reflective properties allow distance and angle measurements to be made to it. These measurements are used with simple trigonometry to derive the elevation of each point and its position within a Cartesian co-ordinate system. The instrument measures declinations from horizontal with an accuracy of ±5 seconds (where 1 second is $1/3600^{th}$ of a degree), and distances of over 1 km with an accuracy of ±3 mm. This kind of instrument is virtually state of the art, it is expensive and the measurements it makes are incredibly refined. Nevertheless, a number of measurement errors can be identified:

- As indicated by the manufacturer's specifications, repeated measurements of exactly the same target position return declination values that vary by as much as 5 seconds either side of the true value. In practice this error is very small, introducing a deviation of no more than 5 mm into the calculation of the elevation of a position 1 km away.
- Holding the target pole still, especially in cold, windy weather, is difficult. Despite every effort it is common for the target to move a few millimetres back and forth over a period of seconds. In turn,

repeated measurements with the pole in the same position will yield very slightly different distance estimates.
*   The survey aims to characterize bar topography but measurements are affected by the smaller scale, gravelly surface texture of the topography being surveyed. Thus, choosing to place the target pole 1 cm to the left or 1 cm to the right of a particular spot can mean measuring the elevation of a hole between two pebbles, or the elevation of the top of a pebble. In this case, the estimate of bar elevation at that position can vary by tens of millimetres.

In practice, the first two sources of error are of little concern because they introduce only a very small amount of uncertainty into the results. They could be reduced further, for example, by moving the base station closer to the survey positions and using a tripod rather than a student to hold the target pole. The third error is more worrying because it is slightly larger. However, there is no bias involved because the student selecting each target position is instructed to place the pole randomly rather than, for example, consistently selecting pebble tops. This ensures that, while individual points may be a few centimetres higher or lower than the average bed elevation in their vicinity, the overall surveyed surface is neither consistently above nor below the average. Most importantly, all the errors combined are very small relative to the variations in bar topography that we aim to characterize (millimetres compared with metres). This means that we can be confident that what little uncertainty the errors do introduce does not affect our ability to describe the overall shape of the bar surfaces. Clearly this kind of decision-making is dependent on the job at hand, its aims and whether or not the overall precision achieved is sufficient to meet the objectives. Making such a decision always depends on appreciating the measurement errors involved so that every effort should be made to characterize them.

DESIGNING A SAMPLING PROGRAMME

In any project, there are two main controls on the design of the sampling programme: the research aims and the resources available (time, money, person power). While the research objectives should drive the sampling design, more often than not it is resource issues which limit the sampling programme, and compromises have to be made. Two key issues which follow from this are the definition of the population of interest and the choice of sampling method.

Defining the target population

Defining the target population is a critical step and begins with a clear definition of the unit of study (the items about which generalizations are to be made and that will be sampled). In social geography this might be an individual, a household or an organization. In fluvial geomorphology it might be a channel cross-section, a bar, a hydrological link or a river basin. As a result of geographical variations, defining the population is an iterative process in which several questions are asked: how do the population characteristics of interest vary spatially and temporally? Do the aims of the research necessitate their inclusion or exclusion? Is it possible to include them given resource limitations? Can the research aims be modified to accommodate practical difficulties? Answering these questions depends on careful investigation of published research, consideration of what one might expect to be the case and a clear understanding of the research aims. In turn, the spatial extent and temporal character of the intended population should be stated and used to guide design of the sampling programme.

Failing to accommodate temporal and spatial variability, by targeting too narrow a slice of the possible population, will produce a sample that is unrepresentative. For example, questioning households about their leisure activities in only one district of a city is unlikely to yield results that are applicable to the city as a whole because one district is unlikely to encapsulate the range of economic, ethnic and age-related factors that influence use of leisure time across the city. Similarly, sampling suspended sediment concentration in a stream only during the rising limb of a flood is likely to yield an average value that is too high for the flood as a whole because of temporal variations in sediment availability over the course of the event (the hysteresis effect). Equally, it is possible to target too much of the possible population in terms of its spatial extent, temporal boundaries or internal structures. This not only spreads precious resources thinly with implications for sample precision (see below) but may also add sources of variability that are not of direct interest and that obfuscate or dilute critical information. Thus, it may be necessary to exclude sources of variability from a sampling programme and focus attention on particular objects, places, times or patterns.

Choice of sampling method

Having clarified the spatial, temporal and structural dimensions of the target population, the next problem is to determine the best way of sampling from this. A variety of sampling methods are used by geographers. These fall into two basic groups: non-probability methods and probability-based methods. Non-probability methods cannot be used to make statistical inferences about the population from which they are drawn. In choosing to adopt non-probability methods one must therefore accept that statistically rigorous representativeness is not a primary issue in the research design (which may be the case, for example, in some research utilizing case studies). If the intention is to make generalizations about a larger population then non-probability methods should only be used with extreme caution and it is in this context that such methods are briefly reviewed here.

In *accessibility sampling*, units are selected on the basis of convenience, such that one selects the most accessible units from the population. For example, if interviewing about leisure activities, one might question one's friends. Such samples are unlikely to yield an unbiased sample of even the most limited target population. In *judgemental* (also referred to as *purposive*) *sampling*, units are selected subjectively by the researcher on the basis of prior experience. This is problematic because the researcher's previous experience may be limited and his or her own prejudices, derived from his or her expectations and viewpoint, become an integral part of the selection process. Even if, by chance or skill, a judgemental approach yields an unbiased sample, it is difficult to prove that this is the case and therefore difficult to convince critics of the value of any generalizations that are made. *Quota sampling* aims to be more representative by attempting to produce a sample that replicates the general structure of the population. Predefined quotas based on such factors as age, gender and class are filled, thereby imposing some useful control on the selection of units, but the choice of individual items within each quota group is still subjective. Kitchin and Tate (2000) suggest that this method can yield representative samples, but it should only be used in situations where prior work has shown this to be the case.

By contrast, *probability-based sampling* methods aim to preclude bias and produce representative samples. Their common characteristics is that the sampling units are selected by chance and the probability of any unit being selected can be determined. Probability-based methods must be used if one intends to use inferential statistics to generalize from the sample to the population. These methods require that a sampling frame exists or can be developed. A sampling frame is a list or other representation of the target population from which units can be drawn (for example, an electoral roll, a catalogue of discharge gauging stations, an aerial photograph, a map or a street directory).

Table 15.1 illustrates several probability-based methods. The two basic methods are *simple random* and *systematic* sampling. Their common feature is that there is an equal probability of selecting each and every unit within the sampling frame. Two issues are worth considering when adopting these methods. First, if systematic sampling is applied within a sampling frame that includes a repetitive structure and the sampling interval that is chosen coincides with that structure, bias will be introduced. For example, many alluvial rivers

TABLE 15.1 Basic sampling methods

	Description	Physical illustration	Human illustration
		Sediment size *Aim*: ascertain average size of the sediment particles on a river bar. *Population*: all particles on the river bar. *Unit*: a sediment particle. *Frame*: a map of the bar surface located in an arbitrary Cartesian space. *Measurement*: using a size template.	**Campus safety** *Aim*: ascertain views of university. students on campus safety. *Population*: all students at the university. *Unit*: an individual student. *Frame*: a list of students and their addresses. *Measurement*: by questionnaire.
Simple random	Within the sampling frame each unit is assigned a unique number or position. Numbers and thence units are selected at random from the sampling frame.	A random number generator is used to pick *x* and *y* co-ordinates. These co-ordinates locate particles for measurement.	Each student on the list is assigned a unique number. A random number generator is used to pick numbers and the corresponding people are sent questionnaires.
Systematic	A sampling interval is defined (e.g. every 10 m, every fourth individual, every 60th second). The first unit is randomly selected as above and subsequent units are selected systematically according to the sampling interval.	The bar is approximately 40 m^2 and a sample of 100 is required. A sampling interval of 2 m is defined. From an arbitrary origin, a grid of 2 m squares is projected on to the sampling frame map. Grid intersections locate particles for measurement.	The list contains 500 names and a sample of 100 students is required. An interval of 4 units is defined. One name is randomly selected as above. Subsequently, every fourth student is selected. If the end of the list is reached, counting continues at the beginning.
Stratified	Mutually exclusive subgroups (strata) are identified and sampled randomly or systematically in one of two ways. *Proportionate*: each stratum is sampled in proportion to its true population proportion. This is necessary if the sampling frame is inadequate. *Disproportionate*: an equal number of units is sampled from each stratum irrespective of its true population proportion. This is necessary when comparisons between strata are required.	Four strata corresponding to distinct facies (areas of homogeneous sedimentary character) are evident. In this case the frame is adequate and there have been no measurement problems. Simple random and systematic sampling are adequate. Suppose one wishes to compare size in facies 1 (a small area) and facies 4 (a larger area). An equal number of particles should be selected from each. Thus, disproportionate sampling is necessary (note this will yield a biased sample of the population so weighting is required).	It is suspected that gender is an important factor in determining views on campus safety. Suppose the supplied list is for students in only one faculty. Different faculties typically exhibit distinct gender distributions. In this case the list is not representative of gender distribution across the university. Proportionate sampling is required: stratify (male, female) and randomly sample in each group to obtain numbers that yield the female:male ratio for the university as a whole.

exhibit repetitive pool-riffle bar from morphology in which the spacing between bar units is typically five to seven times the channel width. If water depth or grain size or stream velocity are systematically sampled using a similar interval, it is possible that measurements will be biased towards the characteristics of pools or riffles. Secondly, with a target population where the characteristic of interest is heterogeneous but also exhibits some internal pattern, it is important to obtain uniform coverage of the sampling frame without any gaps. Simple random sampling may not do this as well as systematic sampling because it is possible for sampled units to be unevenly distributed, as illustrated for the case of river sediment characterization by Wolcott and Church (1991).

In a *stratified* sample (Table 15.1) a number of homogeneous subgroups or strata, differentiated by some relevant characteristic, are recognized within the population. In contrast to the simple and systematic methods the probability of selecting an individual unit from the sampling frame varies, depending upon the stratum the unit belongs to. Three common reasons for utilizing stratified sampling illustrate its value. First, it can be used to ensure that the number of units drawn from distinctive strata is in proportion to their true size in the population. This is known as *proportionate stratified sampling* (Table 15.1). Simple random and systematic sampling will achieve this by default if the sampling frame is appropriate, comprehensive and accurate, which should be the case if the sampling frame is developed for the research project. However, it is not uncommon for the sampling frame to be obtained from a source that compiled the frame for purposes other than those for which it is now intended. Such frames may be biased in favour of one or other strata. Similarly, instrument malfunction at a particular time or place or non-responses to questionnaire surveys may yield a sample that is known to be biased. In either case, if the true population proportions are known then each strata can be randomly subsampled in those proportions to obtain an unbiased sample. Secondly, it may be uneconomical or unfeasible to sample strata of very different sizes in proportion to their size (total area, number of units, etc.). A more efficient method is often to collect a random sample of common size within each stratum, then weight the statistics obtained for each strata according to the stratum's size within the population and combine them appropriately in order to generate population estimates. Sampling the same number of units from strata of different size is referred to as *disproportionate stratified sampling* (Table 15.1). Thirdly, individual research projects may ask questions about the strata, often requiring that comparisons are made between them. In this case it is necessary to obtain equally precise samples for each stratum, which means selecting a similar number of units from each. Simple random or systematic sampling does not do this but, rather, selects a number of units from each stratum that is in proportion to the stratum's size. With disproportionate stratified sampling this problem is overcome by randomly selecting the same number of units from each stratum, irrespective of their true relative sizes. In using this method, it is important to remember that, as far as the population as a whole is concerned, one has created a biased sample so that if estimates of population parameters are required, strata estimates must be combined using appropriate weighting techniques.

A final example of a probability-based method is the multi-stage or *hierarchical sample* in which the sample is selected in several stages that usually relate to spatial or temporal scale. For example, if the campus safety study (Table 15.1) were extended to a global scale the aim might be to sample 100 universities from around the world. First, ten countries might be randomly selected, then within each country five cities and, ultimately, within each city, two universities. Multi-stage surveys are an efficient method when faced with a very large (in space or time) population.

Choosing between these various probability-based methods (and the many others that have been suggested) requires some prior knowledge or reasoned judgement concerning any spatial or temporal structures within the population, a thorough understanding of the sampling frame and a clear set of aims. Without a good appreciation of these, it is possible inadvertently to choose a sampling method that systematically favours some parts of the population over others, in which case the characteristics of interest are not properly represented. This basic point has been stressed by several authors who have considered the specific details of applying standard sampling methods to spatial data (e.g. Berry and Baker, 1968; Harvey, 1969). Haining (1990) suggests that systematic sampling is superior where the underlying spatial variation is random. Wolcott and Church (1991) find that a particular combination of grid and random sampling known as stratified systematic unaligned sampling (cf. Smart and Grainger, 1974; Taylor, 1977) performs well for areally structured data. They point out that it avoids the primary problems with each of random and systematic sampling: the possibility that random sampling is unevenly distributed thereby missing small spatial structures, or that the data contain spatial structures that have the same spacing as the grid spacing, thereby introducing bias.

In summary, non-probability methods are less desirable than their probability-based counterparts and certain probability methods are more appropriate than others in certain circumstances. Nevertheless, it is important to recognize that the vagaries of empirical research often make meeting the ideal difficult (if not impossible) with the result that the target population and the sampled population differ (cf. Krumbein and Graybill, 1965). This might be because the resources necessary are not forthcoming. It may be that accurate information about the population is not available to guide programme design or that there are unknown and hidden sources of variation within the population. It may be that an appropriate sampling frame does not exist or that we are forced to accept an accessibility sample because only certain people will talk with us or only certain places can be reached. In cases like these it is incumbent on the researcher to make it very clear exactly how sampling was conducted and for him or her to interpret his or her results in the light of suspected sampling weaknesses.

Analytical requirements

Finally, in designing a sampling programme it is important to think ahead to the analytical stage of the research and identify any restrictions or requirements that the intended analysis imposes on the sampling strategy. For example, it may be that the inferential statistics used require a minimum number of samples, or that a laboratory machine requires individual samples to be of a particular mass. It is certainly the case that any hypothesis being tested will require the data to be collected in a particular manner. In experimental and some observational projects, the experimental design will be an integral part of designing the correct sampling programme. It is therefore crucial to identify the analytical procedures you intend to use in the laboratory or at your desk *before* you set out with your clipboard or shovel.

STATISTICAL INFERENCE

We have already noted that geographical inquiry must deal with uncertainties. Hicks (1982: 15) defines inferential statistics as 'a tool for decision making in the light of uncertainty', and geographers have certainly found inferential statistics to be a valuable tool. Inferential statistics use sample data to make probabilistic statements about the population from which they are drawn. Statements can be made about (1) the characteristics of the population, which is referred to as parameter estimation; and (2) whether a particular supposition about the population is true or false, which is referred to as hypothesis or significance testing.

Numerous textbooks are available that explain the principles and practical application of the great array of inferential statistical techniques used by geographers. These include spatial techniques that extend statistical analysis to the examination of patterns in space (e.g. Norcliffe, 1977; Williams, 1984; Haining, 1990; Shaw and Wheeler, 1994; Fotheringham et al., 2000; Rogerson, 2001). Particular attention should always be paid to the data assumptions that these procedures have and whether so-called parametric or non-parametric techniques are most appropriate. There are also some specifically geographical issues to be aware of too, particularly spatial autocorrelation. This refers to the propensity for the value of a variable at one location to be related to the value of that same variable in a nearby location (Rogerson, 2001). It is problematic because inferential statistical techniques often require that each sample measurement is independent of all others. In spatial data, autocorrelation is common (otherwise location would not matter) such that the performance of standard methods may be degraded and there is the potential for misinterpretation. It is possible to measure the significance of spatial autocorrelation in a dataset (see, for example, Kitchen and Tate, 2000), and standard inferential procedures can be adapted to minimize its impact (see, for example, Cliff and Ord, 1975; Fotheringham et al., 2000; Rogerson, 2001).

There are many introductory texts that can provide a detailed step-by-step introduction to inferential statistical methods. The aims of this section are limited to explaining the apparent incongruity of statistical inference – how can one make statements about a population based on a single sample drawn from it, even though one knows that no two samples would ever be exactly the same? – an apparent leap of faith that brings to mind Jean Baudrillard's comment (1990: 197) that 'Like dreams, statistics are a form of wish fulfilment'. The simple answer is that, although we know our sample to be unique, statistical theory allows us to assess the reliability of sample estimates (called statistics) such as the sample mean. It is, therefore, possible to ascertain the likely difference between a sample statistic and the equivalent population parameter, *without knowing* the value of the population parameter. In turn, the differences between sample statistics, for example mean values from different groups, can be compared with one another to test the hypothesis that they come from different populations. The following exposition of these ideas is necessarily very brief and non-technical and focuses on ascertaining reliability rather than hypothesis testing. The reader is directed to one of the above texts (Shaw and Wheeler, 1994; Rogerson, 2001) for a fuller account.

Probability and the 'normal' distribution

A basic understanding of probability distributions is necessary before continuing. A probability distribution describes the changing frequency with which particular values of a variable of interest are measured. It is commonly visualized as a histogram in which the ordinate shows the number of occurrences (the frequency) with which groups of values occur. For example, one can describe the frequency distribution of beach pebble sizes in a 100-pebble sample by indicating the number of particles in each of several consecutive 10 mm grain-size classes. Frequencies can be represented as absolute numbers or as relative proportions, in which case they represent the empirical probabilities of measuring a value in each class. Thus, if 35 of the 100 pebbles were found to be between 40 and 50 mm in diameter, it follows that there was a probability of 0.35 (a 35% chance) of finding a pebble in that size range on the beach. Probability distributions for measured phenomena take a wide variety of forms, but a typical situation is that values close to the mean are common and those further away are proportionately less common. Specifically, many phenomena exhibit an approximately 'normal' distribution (sometimes referred to as a Gaussian distribution after the mathematician who first defined it) with its characteristic bell-shaped curve, centred on the mean.

The properties of the normal distribution and, in turn, empirical probability distributions that approximate it are at the heart of basic statistical inference. Any normal distribution can be described in a standardized form in which raw empirical data are transformed into so-called z values. These numbers express changes in the measured values as multiples of the sample's standard deviation. In standardized form the mean of the distribution is zero and the standard deviation of the distribution is 1.0. Because the mathematical form of the standardized distribution is known, the probabilities of observations falling within any given range of z values can be calculated and most statistical textbooks contain tables that give the probability associated with specific z ranges. Thus, there is a 0.68 probability (0.34 either side of the mean) of an event falling in the range $z = -1.0$ to $z = 1.0$ – i.e. within one standard deviation of the mean. Similarly 95.45% of events will be within two standard deviations and 99.73% within three standard deviations of the mean. This is true of any normally distributed variable which means we can apply such reasoning to a wide variety of phenomena in physical and human geography. Using such tables in reverse, the values of z that are associated with selected probabilities can be ascertained. For example, 95% of the values (47.5% either side of the mean) in a normally distributed phenomenon will have z values that are in the range ±1.96. Equally, we can be sure that sampled values with z values outside this range have a probability of being observed 5% of the time or less.

Confidence statements about sample statistics

If repeated samples are drawn from the same population and in each case the mean is calculated, the mean values will vary from sample to sample but will tend to cluster around the true mean of the population. Such a collection of sample means (or indeed any other sample statistic) is called a sampling distribution. A piece of mathematical theory called the Central Limit Theorem (CLT) proves that sampling distributions are normal with a mean value equal to the value of the true population parameter (e.g. the true population mean) and that this holds irrespective of the population distribution. Thus, even for a phenomenon that does not exhibit a normal distribution, the sampling distribution of the mean is normal. The standard deviation of a sampling distribution is known as the standard error and it has the same general properties as the standard deviation of any normal distribution so that, for example, 95% of the sampling distribution lies within 1.96 standard errors of the true population mean.

Standard errors can be determined empirically by repeated sampling of a given population, but this is rarely plausible. It is of significant consequence, then, that standard errors can be calculated on the basis of collecting only a single sample. For example, the standard error of sample means (σ_x) can be calculated as:

$$\sigma_x = s/\sqrt{n}$$

where s is the standard deviation determined from a single sample and n is sample size. Armed with this value and our knowledge of the normal distribution it is possible to make statements about the reliability of the sample mean – that is, to say how confident we are that the true population mean is within a given interval about the sample mean. Remembering that the probabilities in a z table indicate that 95% of a normal distribution lies within 1.96 standard deviations of the mean, we can say that there is a 95% chance that the sample mean lies within 1.96 standard errors of the true population mean. This is equivalent to saying that there is a 95% chance that the population mean lies within 1.96 standard errors of the sample mean.

For a given case, the interval can be specified in the original data units and is known as a confidence interval. So, for example, for a sample of pebble diameters with a standard deviation of 20 mm and $n = 100$, there is a 95% chance that the sample mean lies within $1.96 \times \sigma_x = 1.96 \times (20/\sqrt{100}) = \pm 3.9$ mm of the population mean. This is commonly interpreted as meaning that in 95 samples out of 100 the sample mean would lie within ± 3.9 mm of the population mean, although more precisely it says that if 100 samples were used to construct 100 confidence intervals the true population mean would lie within 95 of them. Confidence intervals for any probability can be constructed using the appropriate z value, so that at 0.99 probability the confidence intervals in the above examples are $2.58 \times (20/\sqrt{100}) = \pm 5.2$ mm. An important caveat for the reader to investigate further is that while large samples always have normal sampling distributions, irrespective of the population distribution, small samples ($n < 30$) have distorted distributions with a form that is a little different from 'normal'. Small samples tend to yield statistics that are distributed according to the t-distribution, sometimes known as 'student's t-distribution'. This has similar properties to the normal distribution and it is used in the same way to determine the reliability of sample estimates, except that probability values from published t tables, rather than z tables, are used.

The CLT and standard errors are so important because they allow us to make rigorous statements about the reliability of the statistics we derive from sample data – that is, to quantify accurately the uncertainty that is inherent in a sample. In turn they provide a basis for making rigorous comparisons between samples and thence for testing hypotheses. Just as confidence intervals are used in assessing reliability, so-called significance levels, denoted by α, are used to attach probability statements to the decisions made in hypothesis testing. There is always the chance that a given decision is incorrect and levels of significance define the probability that one incorrectly rejects a true hypothesis. Significance levels are set by the researcher as part of the testing procedure. Usually we are only willing to accept low levels of error, so significance levels are set to 5 or 1%, though smaller, more stringent values can be used. The important point to make at the end of this section is that statistical inference, beyond the mathematical formulation of the various procedures and tests, involves commonplace ideas of confidence and significance *not* certainty. It allows us to attach probability statements to estimates and decisions but, crucially, statistical techniques do not provide binary, 'black and white', yes and no answers. It is always down to the researcher to choose levels of confidence and significance, and to interpret results thoughtfully in the light of these choices.

SAMPLE SIZE

A frequently asked question is 'How big should my sample be?' The answer reflects a compromise between the desired precision of the sample estimates and the resources available, because maximizing precision and thence the significance that can be attached to statistical inferences (see above) requires the collection of large samples, but also demands greater resource expenditure.

In general, precision improves with sample size in a curvilinear fashion. As sample size increases, precision improves rapidly to begin with but then more slowly (Figure 15.2). In this case, we can imagine that in any sampling procedure there is an optimal sample size

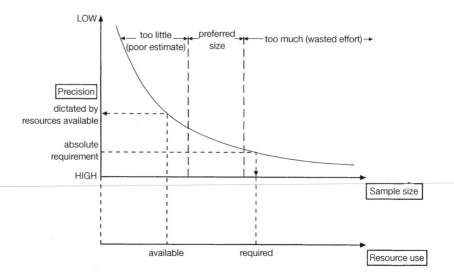

FIGURE 15.2   The relation among estimate precision, sample size and sampling resources

corresponding to that region where the curve begins to flatten out. Beyond this point additional gains in precision are small and do not warrant the additional sampling effort (resources) required. Before this point, the sample size is too small to yield reasonable estimates of the population characteristics and it is worth expending small amounts of effort to obtain significantly better estimates. Ideally, we want to obtain samples of a size somewhere in this optimal zone. Several ways of doing this are discussed below, but it is worth noting two cases where the option of seeking the optimal solution does not arise.

First, a common situation is that the resources needed to collect a sample of the desired size are simply not available, so that low sample precision is inevitable. Similarly, where secondary rather than primary data are being used, the secondary data may be less voluminous than that desired. When designing a sampling programme it is always necessary to consider the allocation of resources in the light of this problem. It may be that sacrifices can be made in one part of the programme in order to improve precision elsewhere. For example, rather than obtaining low-precision estimates of water quality in 20 lakes, it may be advisable to seek quality estimates for five lakes, especially if these 'retained' lakes are carefully selected to test one or more hypotheses.

Secondly, it may be that a specified level of precision is required by the research aims, in which case there may be no option but to collect inefficient, large samples. Work with legal or health implications often has an absolute level of precision that is required by the project objectives. A related issue that constrains sample size is the possibility that a particular physical or statistical technique used to analyse the sample data has sample size requirements. Again, it is vital to know how the data will be analysed before the sampling programme is finalized.

In any given study, the relation between sample size and precision is driven by the variability (heterogeneity) of the characteristic of interest within the population. For a given sample size, precision is worse for populations that exhibit greater spread or variability and the more heterogeneous a population, the greater the sample size required to obtain a given level of precision.

Formulae exist for calculating the sample sizes needed to obtain specified levels of precision for a given statistic. For example, in the case of estimating the population mean μ, precision can be thought of as the error, δ, that we are willing to accept – that is, the acceptable difference between a sample mean, x and μ ($\pm \delta$ units). δ is equivalent to half a confidence interval and a confidence interval has width $2.(z.\sigma_x)$, where σ_x is the standard error of the mean and z is the tabled value associated with the chosen significance level α. Thus:

$$\delta = z.\sigma_x$$

therefore:

$$\delta = z.(s/\sqrt{n})$$

where s is the sample standard deviation and

$$n = (z^2.s^2)/\delta^2$$

This gives the sample size n needed to obtain an estimate of the mean that is within δ units of the population mean with a $100.(1 - \alpha)\%$ level of confidence.

Similar formulae can be developed for estimating other statistics or for use in hypothesis testing (a device known as an operating characteristic curve can also be used to determine optimal sample sizes in hypothesis testing). The operational problem with these methods is quite simply that usually we do not know the sample standard deviation beforehand. This can be overcome by a two-phase sampling procedure or by estimating the standard deviation from previous published work. An additional problem is that researchers seldom find it easy to define an acceptable error, δ.

Empirical approaches to sample size determination may then be useful. As sample size increases from one, the value of any statistic will vary significantly as successive population values are added, but will gradually achieve a degree of stability. This indicates that the sample has incorporated most of the variance evident within the population (see again Figure 15.2). If it is possible to monitor the value of the statistic of interest as the sample is collected, sampling can be curtailed when values for successive n become relatively stable. This can be an especially effective method if the same type of sample is to be obtained from a number of strata or discrete sampling frames where it is anticipated that there is little change in the population variance between those strata or frames. A pilot exercise conducted in one case can then be used to inform sample size for the whole programme. For example, in the case of an insect survey consisting of many discrete quadrat samples where the aim is to examine variations in number of taxa present, it may be worth while to conduct a pilot exercise in which one monitors the changing number of taxa as the size of the quadrat is gradually increased. A graph can be plotted of area against number of taxa and the stabilization point will reveal the optimal quadrat size, to be utilized throughout the survey, for obtaining a reasonable estimate of taxa number (e.g. Chutter and Noble, 1966; Elliot, 1977: 128). More elaborate empirical methods can also be used to examine sample precision and identify optimal sample sizes, for example a technique called bootstrapping (e.g. Rice and Church, 1996), though these require very large datasets and the ability to invest resources in a significant pilot study.

CONCLUSION

A principal aim of most geographical research is to make useful generalizations – that is, to seek out and explain patterns, relations and fluxes that might help model, predict, postdict or otherwise understand better the human and physical worlds around us. Because it is usually impossible or impractical to observe all instances of variation, a smaller number of instances (the sample) is used, from which the 'population' characteristics can be estimated. Achieving this in a reliable and reproducible fashion is the basis of sampling theory and sampling design.

Geographers recognize the value of both extensive statistical sampling and intensive 'case study' sampling for exploring an uncertain world. A benefit of extensive sampling is that a set of techniques known as inferential statistics can be applied to make probabilistic statements about the population from which the sample is drawn.

Sampling is therefore a powerful tool, but geographical research frequently engages with very heterogeneous phenomena that require careful sampling in order to maximize the accuracy of inferential conclusions. Thoughtful design of the sampling programme is crucial and is driven by both the research aims and available resources. The overall message is that sampling is a powerful tool that most geographers need, but if research methodology aims to be as impartial and free of error as possible, sampling must be done thoughtfully and rigorously.

Summary

The most important aspects of sampling are as follows:

- Ensuring the quality of the sample by maximizing the precision and minimizing bias in any measurements or observations.
- Relating individual observations and sample sets to the observed or expected geographical patterns forming the population by the correct sampling design (choosing the sampling method and the sample size).
- Assessing the significance of sample estimates using graphical and statistical means, including the use of inferential statistical hypothesis testing.

Further reading

- Most basic textbooks on statistical analysis include an introductory section on sampling. Two that are not overly technical are Shaw and Wheeler (1994) and Rogerson (2001). The latter is useful in terms of addressing autocorrelation issues.

- Kitchin and Tate (2000) is a general book on research methods in human geography that contains some useful examples of sampling schemes and a lot more besides, and can be recommended to physical geographers for its coverage of basic statistical techniques.

- The book by Haines-Young and Petch (1986) contains a brief but lucid overview of measurement errors and statistical inference in Chapter 11. Harvey (1969) and Richards (1996) provide useful, and in the latter case advanced, discussions of the role of case studies in geography.

- For a detailed consideration of the benefits of alternative spatial sampling techniques, look at Wolcott and Church (1991). Several hypothetical examples used here have been drawn from my own experience of sampling sediments. For anyone embarking on a project involving sediment sampling, Bunte and Abt (2001) provide a plethora of valuable information.

- Finally, a computer-assisted learning package is available as part of the GeographyCAL suite distributed by the CTI Centre for Geography, Geology and Meteorology at Leicester

University (www.geog.le.ac.uk/cti/Tltp/index.htm). The module 'Social Survey Design' considers probability and non-probability methods.

Note: Full details of the above can be found in the references list below.

References

Baudrillard, J. (1990) *Cool Memories* (English translation by Chris Turner). London: Verso.

Berry, B.J.L. and Baker, A.M. (1968) 'Geographic sampling' in B.J.L. Berry and D. Marble (eds) *Spatial Analysis: A Reader in Statistical Geography*. London: Prentice Hall.

Blaut, J.M. (1959) 'Microgeographic sampling', *Economic Geography*, 35: 79–88.

Bunte, K. and Abt, S.R. (2001) *Sampling Surface and Subsurface Particle-size Distributions in Wadable Gravel- and Cobble-bed Streams for Analysis in Sediment Transport, Hydraulics and Streambed Monitoring*. General Technical Report RMRS-GTR-74. Fort Collins, CO: USDA Forest Service, Rocky Mountain Research Station.

Chutter, F.M. and Noble, R.G. (1966) 'The reliability of a method of sampling stream invertebrates', *Archive Hydrobiologia*, 62: 95–103.

Cliff, A. and Ord, J.K. (1975) 'The comparison of means when samples consist of spatial autocorrelated observations', *Environment and Planning A*, 7: 725–34.

Curtis, S., Gesler, W., Smith, G. and Washburn, S. (2000) 'Approaches to sampling and case selection in qualitative research: examples in the geography of health', *Social Science and Medicine*, 50: 1001–14.

Elliot, J.M. (1977) *Some Methods for the Statistical Analysis of Samples of Benthic Invertebrates*. Freshwater Biological Association, Scientific Publication 25. Windermere: Freshwater Biological Association.

Fotheringham, A.S., Brunsdon, C. and Charlton, M. (2000) *Quantitative Geography: Perspectives on Spatial Data Analysis*. London: Sage.

Haines-Young, R. and Petch, J. (1986) *Physical Geography: Its Nature and Methods*. London: Harper & Row.

Haining, R. (1990) *Spatial Data Analysis in the Social and Environmental Sciences*. Cambridge: Cambridge University Press.

Harvey, D. (1969) *Explanation in Geography*. London: Edward Arnold.

Hicks, C.R. (1982) *Fundamental Concepts in the Design of Experiments* (3rd edn). New York: CBS College Publishing.

Kitchin, R. and Tate, N.J. (2000) *Conducting Research into Human Geography: Theory, Methodology and Practice*. Harlow: Prentice Hall.

Krumbein, W.C. and Graybill, F.A. (1965) *An Introduction to Statistical Models in Geology*. New York: McGraw-Hill.

Marcus, W.A., Ladd, S. and Stoughton, J. (1995) 'Pebble counts and the role of user-dependent bias in documenting sediment distributions', *Water Resources Research*, 31: 2625–31.

Miles, M. and Huberman, A. (1994) *Qualitative Data Analysis*. London: Sage.

Norcliffe, G.B. (1977) *Inferential Statistics for Geographers*. London: Hutchinson.

Rice, S.P. and Church, M. (1996) 'Sampling surficial fluvial gravels: the precision of size distribution percentile estimates', *Journal of Sedimentary Research*, 66: 654–65.

Richards, K. (1996) 'Samples and cases: generalisation and explanation in geomorphology', in B.L. Rhoads and C.E. Thorn (eds) *The Scientific Nature of Geomorphology*. Chichester: Wiley, pp. 171–90.

Rogerson, P.A. (2001) *Statistical Methods for Geography*. London: Sage.

Schumm, S.A. (1991) *To Interpret the Earth: Ten Ways to be Wrong*. Cambridge: Cambridge University Press.

Shaw, G. and Wheeler, D. (1994) *Statistical Techniques in Geographical Analysis* (2nd edn). London: David Fulton.

Smart, P.F.M. and Grainger, J.E.A. (1974) 'Sampling for vegetation survey: some aspects of the behaviour of unrestricted, restricted, and stratified techniques', *Journal of Biogeography*, 1: 193–206.

Taylor, P.J. (1977) *Qualitative Methods in Geography*. Boston, MA: Houghton-Mifflin.

Williams, R.B.G. (1984) *Introduction to Statistics for Geographers and Earth Scientists*. London: Macmillan.

Wolcott, J. and Church, M. (1991) 'Strategies for sampling spatially heterogeneous phenomena: the example of river gravels', *Journal of Sedimentary Petrology*, 61: 534–43.

16 The Critical Role of 'Qualitative Thought' in Physical Geography and Geomorphological Research

Colin E. Thorn

Synopsis

Natural and physical scientists are frequently confronted with numerous intellectual challenges, including issues of spatial and temporal scale, system boundary demarcation or closure and contingency (unique history). In the earth and environmental sciences, these issues are inevitable but they may not be amenable to quantitative resolution alone. As a result, researchers must rely, at least in part, on qualitative judgement. This, in turn, must be informed by philosophical reflection as well as by practical experience. Through ideas such as 'systems' and 'models', this chapter explores the interdependence of qualitative and quantitative modes of expression in arriving at 'explanations' in physical geography, and the nature of 'science' and the importance of studies of the philosophy of science. It then goes on to examine questions stemming from the philosophical scrutiny of science, such as absolute truth, the independence of observation and theory, natural kinds and the nature of scientific laws. Finally, it looks at scale linkages, 'closure' and contingency in environmental systems and their implications for modelling and explanation.

The chapter is organized into the following sections:

- Introduction
- Philosophy of science
- Science
- Systems
- Geomorphological models
- Conclusion.

INTRODUCTION

While 'mathematics is the language of science' is a common aphorism, it is one that discounts the profoundly important role played in science by qualitative thought and expression. Indeed, it is equally justifiable to claim that 'qualitative thought is the foundation and frontier' of science because many foundational concepts defy quantification, and many insights are initially purely qualitative. Although focused upon the role of qualitative thinking, this chapter is predicated upon a belief in the complementary nature of qualitative and quantitative expression in science generally, and physical geography, earth science and geomorphology specifically.

A brief chapter permits only a sketch of the role of qualitative thought in science, illustrated briefly by examples drawn from physical geography and earth science but primarily from geomorphology. Unfortunately, many worthwhile topics will receive only scant (and undoubtedly in some instances no) mention. The approach will be to move from the philosophical to the scientific and on to the geomorphological, although emphasis will be on the first two. As a starting point for discussion, it is assumed, first, that the purpose of science is to generate new, and tested, knowledge of the real world; and that, secondly, geomorphology is the scientific study of landforms, their development and their interactions (Rhoads and Thorn, 1993).

Geomorphology (particularly in its association with geology) has often been viewed as historical rather than scientific. However, the choice between history and science is not mutually exclusive. Generalizing from the lead proffered by Schumm (1977), Rhoads and Thorn (1993: 288) suggested that:

> A scientific approach to the discipline embraces the view of geomorphic systems as physical systems with a history . . . In short, geomorphology can be viewed as a science in which time is a focal issue. A scientific approach to any subject is inherently permeated with theory. Thus, geomorphology as a science must strive for explanations infused by theory . . .

A second prefatory step is to identify 'theory' as a construct and 'mathematics' as a language. This distinction is necessary because conflation of these two ideas has been common in geomorphology and other branches of physical geography. Mathematics is a language: it has no quintessential thematic content. Like English or any other spoken language, it is a set of symbols and associated rules that may be applied to communicate ideas about any kind of content. The power of mathematics is its precision, but this precision comes at the cost of necessarily having great knowledge of the matter in hand – a fuzzy grasp of a topic does not permit precise and/or elegant mathematical expression of it. On the other hand, theory is a web of linked, hierarchical ideas; it is the ideas and their inter-relationships that form theory which may subsequently be expressed in any language whether natural (spoken) or artificial.

PHILOSOPHY OF SCIENCE

Philosophers of science contribute to the development of science by probing the strengths and weaknesses of the scientific enterprise. They remind scientists that they are part of, not apart from, the world they are investigating (Rhoads and Thorn, 1994). For many decades, the focus of the philosophy of science was upon instructing practising scientists on the manner in which scientific research should appropriately be conducted. Not surprisingly, this often generated resentment. In recent decades there has been a radical shift in philosophy of science and it is now overwhelmingly committed to the study of how practising scientists pursue their research – essentially 'a science of science'. Generally, physical geographers and geomorphologists have shown little interest in this topic, unlike their colleagues in human geography. However, it is profoundly important to have at least a basic familiarity with the philosophical underpinnings of our day-to-day scientific activities. Those wishing to begin such an investigation are referred to Rhoads and Thorn (1993; 1994; 1996a; 1996b; 1996c). Here, just four critical points will be discussed briefly by way of illustration.

First, is science closing in on some sort of absolute truth (i.e. improving) or is it merely examining the world from a series of different, but non-improving, perspectives as scientific fashions change? This fundamental question intrudes into physical geography in many ways. Resolution of it might justify some research approaches (e.g. a mathematical one) while invalidating others (e.g. a historical one); alternatively, it might vindicate multiple approaches. Even simple mistakes, such as misuse of the term 'paradigm', hamper progress within physical geography on this topic. Technically, the term implicitly embraces the notion of different but non-improving perspectives – yet many scientists use it widely shorn of this philosophically important distinction (Rhoads and Thorn, 1994). It seems probable that most scientists would prefer to think that we are actually improving, rather than just changing, our scientific perspectives when they invoke a 'paradigm shift'.

Secondly, many physical geographers cleave to an outdated philosophical notion that observations and theory are separate entities. Virtually all modern schools of philosophy of science view observations as theory laden (Rhoads and Thorn, 1994). This does not eliminate all problems; rather, it generates different topics of contention.

Nevertheless, it does have a profound impact on geomorphology, for example. The frequently invoked notion that a geomorphologist should go into the field unfettered by theoretical pretensions and simply 'observe' (that is, make 'theory-free observations') is simply fallacious. Observation is always theory laden – the true distinction is between those who know it and those who don't (Chorley, 1978)!

A third, critical philosophical issue, is that of natural kinds. Natural kinds are an ancient concept and have been debated since classical Greek times. The central issue is that, if the 'real' essence of things can be determined, it will be possible to identify causal mechanisms, powers or processes. It will also be possible to identify natural categories that are independent of the context of inquiry. While such arguments have been best developed in biology (particularly with respect to species), the fundamental question for geomorphologists is 'is there really such a thing as a landform, or is it purely an artifact of the human mind?' What are the necessary and sufficient properties to specify a landform? No issue could be more central to geomorphology, yet convincing definitions are absent from the geomorphic literature. This would not be quite so problematic but for the fact that the concept of natural kinds is hotly debated within the philosophy of science (Rhoads and Thorn, 1996c) and, consequently, sound guidelines for making the decision are unavailable. The geomorphological notion of equifinality or convergence (Haines-Young and Petch, 1983; Thorn, 1988; Rhoads and Thorn, 1996c) touches upon this important topic. About all that can be said with certainty at this point is that geomorphologists have not found a scientifically sound way to determine an unvarying relationship between external form and process. If landforms can be established as natural kinds, there is a strong basis for uniting geomorphology around a set of objects, and not around prescribed methodologies (Rhoads and Thorn, 1996c), perhaps thereby reducing methodological contention and opening the door to more substantive debate. Undoubtedly similar problems occur within other subfields of physical geography.

A fourth highlight is generation of 'laws' within science (Rhoads and Thorn, 1996c). A useful starting point here is Goodman's (1967: 96) telling comment about scientific laws: 'Whatever made the world and whatever makes it go, the scientist writes its laws. And whether or not nature behaves according to law depends entirely upon whether we succeed in writing laws that describe its behavior.' In fact, Goodman's remarks merely hint at a forceful debate among philosophers of science about the nature of scientific laws. Are physical geographers condemned merely to expand the domain of chemical and physical laws, or are there, for example, laws to be written about geomorphological kinds (landforms)? The answer to this question, if affirmative, has the potential to be the quintessential intellectual underpinning legitimizing geomorphology as a scientific discipline, as do similar questions and answers in other portions of physical geography.

Even this cursory selection of topics in the philosophy of science demonstrates that philosophical and purely conceptual or qualitative concepts are critical to the pursuit of rigorous science. In short, the quality control of science must come from a largely external, philosophical scrutiny, as well as from within science itself. It certainly cannot come alone from within any single discipline such as geomorphology. The next step is a brief consideration of some general concepts within science itself.

SCIENCE

At root, science is a belief system developed on an ad hoc basis. At any moment, the scientific community (at large and within any individual discipline) subscribes to a set of interwoven beliefs that appear to be an adequate explanation for the past, present and likely future development of what interests them (e.g. landforms and landscapes in geomorphology). Such a set of explanatory beliefs inevitably fails, or appears inelegant and/or inadequate in some areas. Consequently, there are always challenges facing the belief system and it must be modified and/or expanded to explain them. In most instances, there are also individuals who are unconvinced by the opinions held by the majority and who challenge the prevailing ideas. Such individuals or groups are critical to scientific progress. Indeed, it is generally considered problematic when such challenges do not exist. For example, the present lack of an alternative to plate tectonics has been noted with alarm by some (Saull, 1986). However, it is important to note that the stability, integrity or pervasiveness of the belief system varies within science in a hierarchical manner, both among individual disciplines and within any one discipline. Von Engelhardt and Zimmermann (1988: 314–30) have dubbed our most resilient and entrenched guiding beliefs 'regulative principles'. Within science, we might cite the conservation of energy and mass as important regulative principles – in their general form both are unverifiable. Within the earth sciences, von Engelhardt and Zimmermann suggest that uniformitarianism and actualism, catastrophism and evolutionism are among the most fundamental regulative principles.

The deep and pervasive scientific perspective that individual parts of science must interface with other parts of science, and that there is a hierarchy of importance, is a critical one. Physical geographers do not seek to challenge the laws of physics or chemistry. Rather, these are considered foundational. Thus, geomorphologists, for example, seek to explain the development of the surface of the earth using the principles of physics and chemistry. Findings in physical geography and earth science that appear to conflict with physical and/or chemical principles are presumed to be erroneous. Any attempt to modify physics or chemistry in accordance with geomorphological findings would have ripple effects throughout science and is considered unacceptable. One way of characterizing this kind of description of science is to say that scientists view science as a system; in fact, systems are an integral part of scientific thinking in many domains and at virtually all scales.

SYSTEMS

'A system is a set of objects together with the relationships between the objects and between their attributes' (Hall and Fagen, 1956: 18). This classic definition of a system captures not only the essence of ideas about systems but also much of the essence of a scientific view of the world (although it says nothing about scientific motivation). The reason for this overlap is that modern science is absolutely permeated with systems thinking. In fact, this is developed to such a high degree that it frequently passes unnoted. The oversight is an important one. It is a fatal flaw to overlook the reality that systems are simply a human creation – as Strahler (1980) has emphasized, belief in the existence of systems is largely an act of faith and not susceptible to quantification or verification. On the other hand, systems have also received extensive formal treatment such as von Bertalanffy's (1951; 1956) creation and introduction of a 'content-free' discipline that he called General System Theory (GST).

The crux of systems thinking is that scientists are interested not only in objects but also in the origin and development of the objects, as well as the interaction among them. The attributes of an object are simply smaller-scale phenomena, and the objects under study constitute the attributes of larger-scale phenomena (e.g. individual landforms constitute a landscape). At any scale, objects interact among themselves as well as with objects that are both smaller and larger. So immediately the scientist is thrust into a world of scale and scale linkages.

Scale linkage (Haggett, 1965) implies that things of one size and/or timespan are composed of objects and time periods that are smaller and, in turn, are also the formative components of larger assemblages and/or time periods. However, while scale linkage is hierarchical (Haigh, 1987; de Boer, 1992), there is also a myriad of same-scale links. Scale linkage provides all earth scientists with one of their greatest intellectual challenges. There is no right or wrong spatial or temporal scale at which to study the earth. This is a matter of personal choice and scientific objective. Two scale problems, one spatial the other temporal, appear to be most vexing for geomorphologists specifically. In the spatial domain, there is the difficulty of generating sufficient data, given the expense of modern instrumentation. Consequently, generalization from limited spatial data becomes an enormous challenge. Both its importance and potentially contentious nature are seen in such concepts as the catena in pedology and facies in stratigraphy. Satellite data provide the inverse spatial problem – how to cope conceptually with an overwhelming flood of data. While it represents the inverse kind of data problem, its solution also rests in conceptual generalization. Nevertheless generalizations by extrapolation (i.e. from a few to many) and amalgamation (i.e. by the reduction of many to general groupings) are never likely to produce identical results, even if the stipulated categories are allegedly identical. On the temporal front, it is the slowness of geological change that provides the most serious intellectual challenge. This problem is accentuated when large features are studied because it appears that the larger a phenomenon is, the slower is its response time. Such problems are not readily reduced to purely quantitative formulations because development of an adequate temporal data set is infeasible. Conceptual temporal frameworks depicting the possible behaviour of landforms or landscapes over time (e.g. Thornes, 1983; Huggett, 1990) are essential tools for a geomorphologist, but they are ones that are ultimately untestable in a quantitative fashion because of the inaccessibility of adequate timespans. To counter this, geomorphologists have attempted to invoke ergodicity to explain landform development over time by examining the present spatial distributions of the same landform. (Physicists needed to address the pathways of individual, fast-moving molecules. They invoked the concept of 'ergodicity' that states that the mean position of an individual molecule over time is matched by the spatial distribution of many molecules at any one moment.) Unfortunately, the very precise statement of ergodicity by physicists has been transformed into a fairly 'rough and ready' format by geomorphologists that has yet to be tested rigorously. Thorn (1988: 45–50) provides a preliminary discussion of the topic, as well as references to more extensive treatments.

One way in which scientists have long attempted to deal with large-scale problems is by reductionism. Such science scrutinizes things at the smallest scale feasible with the prospect of 'multiplying up' the results to address larger-scale issues. However, this has proven to be a flawed approach. In some instances, computational limitations block such an approach: molecular knowledge of mineral behaviour affords little or no opportunity to understand cliff development – the scale span is simply too great computationally. An even greater problem is the presence of 'emergent variables' (Haff, 1996). At any scale, the researcher must confront and deal with phenomena that are present ('emerge') only at that scale and which are not discernible at smaller scales. Consequently, 'multiplying up' simply misses such issues and their by-products. Reductionist science has the seductive power of seeming precise, and consequently more scientific, but sound science depends on intellectual rigour, not scale.

The scientist is always confronted with the reality that everything is linked to everything else at every scale. While this may be so, it isn't manageable. Because science is ultimately about problem-solving, the unmanageable real world must, therefore, be reduced artificially to ensure manageability. The need for manageability leads to one of the other critical characteristics of systems thinking – the necessity to create a system boundary – often called 'closure' (Massey, 1999; Lane, 2001). The boundary demarcates the end of what will be examined. Beyond it is the 'rest of the world' which, for the purposes of the immediate scientific exercise, will be ignored or, at least, taken as given. Unfortunately, there is no quantitative procedure for determining where boundaries should be placed. The result is that this becomes purely a matter of personal judgement and its validity is dependent on the quality and experience of the scientist making it. Obviously, optimal boundary location should include all that is essential and exclude that which is marginal (as nothing can be truly irrelevant). Nothing attests more to the frequent failure of the scientific community (and humanity in general) to make this judgement adequately than the numerous crises facing animal and plant species stemming from human pressures. Furthermore, recognition of the significance of systems in environmental issues is reflected in the increasing abandonment of attempts to save individual species, and a growing preoccupation with saving habitats (i.e. systems).

While we may believe in the reality of systems, they are too complicated for us ever to describe fully; consequently, we must simplify them (Strahler, 1980). A common approach in these circumstances is to use a model, which may be defined as 'any simplification of a system (that permits prediction)'.

GEOMORPHOLOGICAL MODELS

Models come in many forms, ranging from classic playthings, such as dolls and trains, to the inordinately complex, such as the multiplicity of linked equations comprising a GCM (general circulation model). It is vital to remember that models are human simplifications, but that they are also our only access to all the future and most of the past (Rhoads and Thorn, 1993). Whatever their limitations, it behoves us never to lose sight of them. A model must be created for a specific purpose: while a model may be judged as doing its particular job well or badly, it cannot simply be applied to any job. A 'general' model is really something of an oxymoron – inevitably condemned to do nothing well. This does not mean, however, that there cannot be general principles of model building (e.g. von Bertalanffy, 1951; 1956).

The success of Chorley's (1962) introduction of GST into geomorphology, followed by a comprehensive introduction of the then cutting-edge ideas in Chorley and Kennedy (1971), has had the tendency to leave many geomorphologists believing that GST represents the fundamental set of systems notions. As Smalley and Vita-Finzi (1969) pointed out, systems thinking is much older than GST. While Chorley and Kennedy's (1971) presentation and developments of GST have been criticized (e.g. Kennedy, 1979), this kind of modelling is, rightfully, here to stay in all branches of the earth and natural sciences, including geomorphology. Important decisions include the choice between highly detailed modelling of a few cases or more generalized modelling of many instances (Richards, 1996). Both approaches have utility. Among the many problems facing modellers is that, in many instances, a variety of parameter sets will produce comparable (often good) fits to the systems being modelled (Beven, 1996).

It is perhaps instinctive to judge models by the yardstick of how faithfully they seem to mimic reality. However, while critical in some instances, models fulfil yet another vital role. Models teach us! However simple (even unrealistic) the model, careful manipulation of it shows how things may behave, develop and interact. It is true that we will never remove the 'may'; invariably there will be multiple ways of mimicking reality (Beven, 1996). Nevertheless, as societal demands for management and planning increase, there are always greater demands for knowledge of the future, and environmental and earth scientists have nowhere to turn but to models. Much will be gained by quantitative and technical fine-tuning, particularly turning black boxes to grey boxes to white boxes, as well as by sensitivity analysis. However, technical improvement is always harnessed to conceptual rigour and progress is constrained by whichever of these two components lag.

Like all tools, models are subject to abuse. Two powerful illustrations will suffice to substantiate this point. First, William Morris Davis's (1899) model of landscape evolution, 'the geographical cycle', became in Davis's hands a regulative principle rather than a model (Rhoads and Thorn, 1996b). Davis sought not to test his model but to fit his landscape observations and those of others to his preconceived, irrefutable belief in the geographical cycle. Here, failure to test and sustain scepticism was abuse. Secondly, equilibrium is also essentially a regulative principle in that it ultimately depends on a belief in conservation of energy or mass which, in their general formulation, cannot be tested quantifiably. It has been, and sometimes still is, widely used in geomorphology as a qualitative concept. However, used qualitatively, equilibrium is a blunt instrument; but when specified precisely and manipulated quantitatively it is a powerful (though limited) research tool (Kennedy, 1994; Thorn and Welford, 1994a; 1994b), particularly in quantified modelling. The multiple forms of equilibrium invoked in dynamical systems (models) are the foundation of a rich diversity of geomorphological behaviours (e.g. Phillips, 1999), many of which are sensitively adjusted to initial conditions (contingency raises its head!). The potential abuse with equilibrium rests in imprecise, qualitative use producing a false aura of understanding, and the not uncommon assignment of equilibrium conditions to landforms and landscapes where none has been established quantitatively.

One insurmountable problem for modellers, and geomorphologists in general, is the role of contingency in landscape history (Gould, 1989). The unique sequence of events that make up earth history, and will make up earth's future, cannot be fully excluded from what we see as landform and landscape development. As a result, our interpretation must embrace not only the timeless scientific rules we develop but also a purely 'historic' component that stems from our geomorphological sensitivity. Unarguably, the landscape is a complex physical and biological system which also has a history. A sound interpretation of it, therefore, requires geomorphologists to be rigorous scientists with a strong sense of both scientific rules and history, and a willingness to demean or disparage neither.

CONCLUSION

To end where we began – this is not intended as a polemic in favour of substituting qualitative thought for quantitative expression; rather, it is a plea to see their symbiotic nature. In practical terms, the fundamental suggestion to the student is to view the disciplines of physical geography, earth and environmental science as much as 'worlds of ideas' as they are a world of fieldwork and/or mathematical or laboratory techniques.

Our views of the world, whether our field observations, our laboratory analyses or our subsequent computations, are infused with, and shaped by, our theories. Furthermore, our curiosity is invariably piqued by possibilities first emerging as analogs (Gilbert, 1886), homologues and isomorphs (von Bertalanffy, 1951), or by observations that seem to conflict with our hypotheses or be beyond their reach. In each of these encounters, theory (knowingly or unknowingly) is our reference point.

Our task as physical geographers, or natural scientists in general, is to bring the 'unknowns' or aberrations we encounter into the known and explained world as deeply or firmly as possible. Initially, and fundamentally, this may be a qualitative task. As we progress, I believe that we are constrained by some 'real' phenomena with which we interact. Because we can never interact with them independently of ourselves, we must give profound thought to the nature of that interaction.

As humans we are two footed, so metaphorically are we as scientists, and as geomorphologists specifically: it is true we can hop through life, but we do so inelegantly. It is also true that we can hop through science, depending either on our theoretical foot, unfettered by the real world, or upon an observational foot, naïvely believed to be theory free. Again, we do either inelegantly. The trick is to stride purposely forward balanced on both feet so that the real world informs us wittingly rather than unwittingly – this is human and, therefore, not perfect, but it is also us at our most elegant!

Summary

- This chapter is predicated upon a belief in the complementary nature of qualitative and quantitative expression in science generally, and physical geography and geomorphology specifically.
- It is profoundly important to have at least a basic familiarity with the philosophical underpinnings of our day-to-day scientific activities as a kind of 'quality control'.

Further reading

This listing places heavy emphasis on sources of ideas central to geomorphology:

- Ager (1993) is an excellent and stimulating introduction to the profound importance of ideas in earth science. Don't worry that the topic is stratigraphy.

- The philosophical position in Haines-Young and Petch (1983) is outmoded, but this book is an excellent introduction to the powerful relationship between philosophy of science and quality research.

- Rhoads and Thorn (1996a) is the proceedings of a conference on philosophy, methods, modelling and disciplinary contexts in geomorphology. The entire text is available on the web (www.staff.uiuc.edu/ ~ b-rhoads/book/book.htm).

- Thorn (1988) is an attempt to survey the conceptual issues underpinning modern geomorphology.

- Weston (1992) is a little textbook gem for those seeking to make sound and forceful qualitative arguments.

Note: Full details of the above can be found in the references list below.

References

Ager, D.V. (1993) *The Nature of the Stratigraphical Record*. Chichester: Wiley.

Beven, K. (1996) 'Equifinality and uncertainty in geomorphological modelling', in B.L. Rhoads and C.E. Thorn (eds) *The Scientific Nature of Geomorphology*. Chichester: Wiley, pp. 289–313 (also available at www.staff.uiuc.edu/~b-rhoads/book/book.htm).

Chorley, R.J. (1962) *Geomorphology and General Systems Theory. US Geological Survey Professional Paper* 500-B.

Chorley, R.J. (1978) 'Bases for theory in geomorphology', in C. Embleton et al. (eds) *Geomorphology: Present Problems and Future Prospects*. Oxford: Oxford University Press, pp. 1–13.

Chorley, R.J. and Kennedy, B.A. (1971) *Physical Geography: A Systems Approach*. London: Prentice Hall.

Davis, W.M. (1899) 'The geographical cycle', *Geographical Journal*, 14: 481–504.

de Boer, D.H. (1992) 'Hierarchies and spatial scale in process geomorphology: a review', *Geomorphology*, 4: 303–18.

Gilbert, G.K. (1886) 'The inculcation of scientific method by example with an illustration drawn from the Quaternary geology of Utah', *American Journal of Science*, 31: 284–99.

Goodman, N. (1967) 'Uniformity and simplicity', in C.C. Albritton Jr (ed.) *Geological Society of America Special Paper* 89, pp. 93–9.

Gould, S.J. (1989) 'Response by Stephen Jay Gould', *Bulletin Geological Society of America*, 101: 998–1000.

Haff, P.K. (1996) 'Limitations on predictive modeling in geomorphology', in B.L. Rhoads and C.E. Thorn (eds) *The Scientific Nature of Geomorphology*. Chichester: Wiley, pp. 337–58 (also available at www.staff.uiuc.edu/~b-rhoads/book/book.htm).

Haggett, P. (1965) 'Scale components in geographical problems', in R.J. Chorley and P. Haggett (eds) *Frontiers in Geographical Teaching*. London: Methuen, pp. 164–85.

Haigh, M.J. (1987) 'The holon: hierarchy theory and landscape research', *Catena Supplement*, 10: 181–92.

Haines-Young, R. and Petch, J. (1983) 'Multiple working hypotheses: equifinality and the study of landforms', *Transactions Institute of British Geographers NS*, 8: 458–66.

Hall, A.D. and Fagen, R.E. (1956) 'Definition of system', *General Systems Yearbook*, 1: 18–28.

Huggett, R. (1990) *Catastrophism Systems of Earth History*. London: Edward Arnold.

Kennedy, B.A. (1979) 'A naughty world', *Transactions, Institute of British Geographers*, 4: 550–8.

Kennedy, B.A. (1994) 'Requiem for a dead concept', *Annals of the Association of American Geographers*, 84: 702–5.

Lane, S.N. (2001) 'Constructive comments on D. Massey space-time, "science" and the relationship between physical geography and human geography', *Transactions, Institute of British Geographers*, 26: 243–56.

Massey, D. (1999) 'Space-time, "science" and the relationship between physical geography and human geography', *Transactions, Institute of British Geographers*, 24: 261–76.

Phillips, J.D. (1999) *Earth Surface Systems Complexity, Order, and Scale*. Oxford: Blackwell.

Rhoads, B.L. and Thorn, C.E. (1993) 'Geomorphology as science: the role of theory', *Geomorphology*, 6: 287–307.

Rhoads, B.L. and Thorn, C.E. (1994) 'Contemporary philosophical perspectives on physical geography with emphasis on geomorphology', *Geographical Review*, 84: 90–101.

Rhoads, B.L. and Thorn, C.E. (eds) (1996a) *The Scientific Nature of Geomorphology*. Chichester: Wiley (also available at www.staff.uiuc.edu/~b-rhoads/book/book.htm).

Rhoads, B L. and Thorn, C. E. (1996b) 'Observation in geomorphology', in B.L. Rhoads and C.E. Thorn (eds) *The Scientific Nature of Geomorphology*. Chichester: Wiley, pp. 21–56 (also available at www.staff.uiuc.edu/~b-rhoads/book/book.htm).

Rhoads, B.L. and Thorn, C.E. (1996c) 'Toward a philosophy of geomorphology', in B.L. Rhoads and C.E. Thorn (eds) *The Scientific Nature of Geomorphology*. Chichester: Wiley, pp. 115–43 (also available at www.staff.uiuc.edu/~b-rhoads/book/book.htm).

Richards, K. (1996) 'Samples and cases: generalisation and explanation in geomorphology', in B.L. Rhoads and C.E. Thorn (eds) *The Scientific Nature of Geomorphology*. Chichester: Wiley, pp. 171–90 (also available at www.staff.uiuc.edu/~b-rhoads/book/book.htm).

Saull, V.A. (1986) 'Wanted: alternatives to plate tectonics', *Geology*, 14: 536.

Schumm, S.A. (1977) *The Fluvial System*. New York: Wiley.

Smalley, I.J. and Vita-Finzi, C. (1969) 'The concept of "system" in the earth sciences, particularly geomorphology', *Bulletin Geological Society of America*, 80: 1591–4.

Strahler, A.N. (1980) 'Systems theory in physical geography', *Physical Geography*, 1: 1–27.

Thorn, C.E. (1988) *An Introduction to Theoretical Geomorphology*. Boston, MA: Unwin Hyman.

Thorn, C.E. and Welford, M.R. (1994a) 'The equilibrium concept in geomorphology', *Annals of the Association of American Geographers*, 84: 666–96.

Thorn, C.E. and Welford, M.R. (1994b) 'No dirge, no philosophy, just practicality', *Annals of the Association of American Geographers*, 84: 706–9.

Thornes, J.B. (1983) 'Evolutionary geomorphology', *Geography*, 68: 225–35.

von Bertalanffy, L. (1951) 'An outline of general system theory', *British Journal for the Philosophy of Science*, 1: 134–65.

von Bertalanffy, L. (1956) 'General system theory', *General Systems Yearbook*, 1: 1–10.

von Engelhardt, W. and Zimmermann, J. (1988) *Theory of Earth Science* (translated by L. Fischer). Cambridge: Cambridge University Press.

Weston, A. (1992) *A Rulebook for Arguments* (2nd edn). Indianapolis, IN, and Cambridge: Hackett Publishing.

17 Numerical Modelling in Physical Geography: Understanding, Explanation and Prediction

Stuart N. Lane

Synopsis

This chapter introduces the use of numerical modelling for understanding environmental systems. Numerical models are used extensively throughout society, and hence by physical geographers, for dealing with situations where the researcher is 'remote' from what her or she is studying: where the events that are of interest have occurred in the past (e.g. reconstruction of past climates); may occur in the future (e.g. patterns of inundation associated with future flood events); or where they are occurring now but cannot be measured or studied using other methods. The potential of numerical models aside, the problems of modelling environmental systems mean that models are commonly scientifically wrapped crystal balls whose predictions must be treated with caution at best and scepticism at worst. The chapter begins by introducing the various stages of the modelling process and the criteria for success in modelling. It goes on to discuss the implications of feedback in environmental systems and empirical and deterministic (or physically based) modelling approaches. Finally, it explores what models can and cannot do in relation to the environment and the possible role of modelling as a form of scientific experiment.

The chapter is organized into the following sections:

- Introduction: why model?
- Fundamental aspects of environmental modelling
- What a model can and cannot do
- Conclusion.

INTRODUCTION: WHY MODEL?

In Winnie-the-Pooh, A.A. Milne demonstrates clearly the basic reason why we need to consider using numerical models. Pooh Bear finds a jar labelled 'hunny' with a yellow substance in it. Being a 'good' scientist, he cannot be sure that it is 'hunny' until he has done a proper scientific experiment to test that his hypothesis is valid. This has to be grounded in direct observation of what is in the jar and involves him tasting it. However, he cannot be sure that it is 'hunny' until he has tasted all of it, right to the bottom of the jar as, according to Pooh Bear, his uncle had once said that someone once put cheese in a 'hunny' jar for a joke. The view that *all* jars labelled 'hunny' contain 'hunny' has already been proven false, and so each individual jar that Pooh Bear finds has to be subject to Pooh-type assessment, in which the entire contents of the jar are consumed. If we apply this to environmental systems, we find ourselves visiting a series of great environmental disasters that have been created because, as a society, we have been unwilling to accept certain evidence (e.g. that a jar labelled 'hunny' with a yellow substance in it is honey) until there is definitive observational evidence that confirms that a partial observation or a theory is correct. The best example of this is provided by depletion of stratospheric ozone concentrations. The potential that chlorofluorocarbons might lead to long-term ozone depletion was demonstrated in the early 1970s (Molina and Rowland, 1974). It wasn't until the direct observation of a spring ozone hole over Antarctica in 1985 (Farman et al., 1985) that this theoretical notion was accepted and subsequent environmental policy was developed. The same scenario has emerged in relation to global climate change: we have a theoretical basis to expect that atmospheric greenhouse gas accumulation has the potential to change climate, and much of the critique of this theory is based around the fact that there is no demonstrable evidence to confirm this hypothesis (e.g. Michaels, 1992). We won't do anything about it until we have actually observed it.

The problem of this view is also illustrated by Milne. When Piglet goes to visit his Heffalump trap, he finds what looks like a Heffalump. Being only a small animal, and with the trap being very deep, he needs to get closer to see if the Heffalump (Pooh Bear with an empty 'hunny' jar stuck on his head) is indeed a Heffalump. This mirrors classic observation-based science in which we search for a better understanding of apparent phenomena through more in-depth observation. However, if Piglet discovers that what he thinks is a Heffalump is actually a Heffalump then, being a small animal, he is likely to be in a lot of trouble as the Heffalump may attack him. We don't want, through more intensive investigation based upon observation, to discover things (e.g. Heffalumps, ozone holes, global warming, severe species loss, serious organic pollution). However, sometimes, we can only confirm that what we think is a matter of concern is actually so through more in-depth assessment.

The numerical model is one of the tools the geographer might use to break out of the circularity that Pooh Bear and Piglet find themselves in. It provides a tool for investigating things that might occur in inaccessible places. Inaccessibility arises because we are commonly interested in (1) environments in the past, before records began or where environmental reconstruction is unreliable or impossible; (2) environments in the present from where we cannot obtain measurements, perhaps because those environments are inaccessible, whether because they are remote, too large to measure or too small to measure; and (3) environments in the future, where we are concerned about the possible impacts of decisions made now for future generations. This chapter seeks to introduce the basics of environmental modelling as part of geographical inquiry but also to reflect upon the challenges and problems that result. Thus, the first section of the chapter introduces the fundamentals of environmental modelling in terms of conceptual, empirical and physically based approaches. The second section seeks to evaluate models critically by considering what a model can do and what it can't do.

FUNDAMENTAL ASPECTS OF ENVIRONMENTAL MODELLING

There are two distinct approaches to mathematical modelling: empirical and physically based. However, as this section will note, these two distinct approaches are not actually that distinct and both of them draw fundamentally upon the idea that there is a resilient and defensible conceptual model of the system that is being considered.

The conceptual model

Without a conceptual model of the system that is being modelled it is impossible to develop either an empirical or a physically based mathematical model. Conceptual models involve a statement of the basic interactions between the components of a system. If we think about the understanding of past climates we can start to think about a simple conceptual model for climate change. Empirical evidence has shown that the earth's climate has fluctuated between periods that were much colder than present (glacial periods) and periods that were slightly warmer than present (Figure 17.1). Why has this occurred? The initial conceptual model might assume that it is due to an external forcing, and we know that a good candidate in this respect is the nature of the earth's rotation around the sun (see Imbrie and Imbrie,

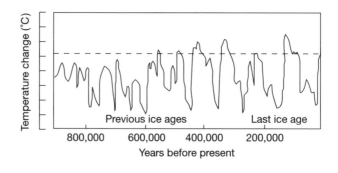

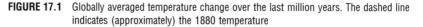

Source: Houghton et al. (1990)

FIGURE 17.1 Globally averaged temperature change over the last million years. The dashed line indicates (approximately) the 1880 temperature

1979), which varies over a number of different timescales. There is good support for some of this cyclical behaviour being explained by orbital forcing (e.g. Imbrie and Imbrie, 1979), but there is also much evidence to question the conclusion that this is the only explanation (e.g. Broecker and Denton, 1990). Thus, researchers have begun to recognize that processes internal to the earth–atmosphere system may exert an important conditioning role upon the way in which these external forcing factors affect climate. This is where we can develop a simple conceptual model to illustrate what form this system might take.

A system is made up of components that are connected together by links. Flows between components are driven by processes in a way that depends upon both the components and the nature of the links. Thus, in the case of glacial cycles, we may start to build a system by considering three components: (1) albedo, which relates to the way in which a surface reflects incoming solar radiation; (2) temperature; and (3) ice-sheet growth. As a first approximation, these may be linked together (Figure 17.2). The links are specified very simply as positive and negative: (1) a negative link between albedo and temperature, which reflects the fact that as albedo (earth surface reflectivity) goes up, more shortwave radiation will be reflected and hence temperature will go down; (2) a negative link between temperature and ice-sheet growth, which reflects the fact that as temperature goes down, ice-sheet growth will go up; and (3) a positive link between ice-sheet growth and albedo, which reflects the fact that as ice-sheet growth goes up, so albedo goes up as ice cover tends to reflect more shortwave radiation than other types of land cover. Put more generally, a positive link involves the effect variable responding in the same direction as the cause variable; a negative link involves the effect variable responding in the opposite direction to the cause variable.

Figure 17.2a illustrates a very important property of feedback which occurs in all environmental systems and which arises from the combined interaction of these links between components: as temperature goes down, ice-sheet growth goes up; as ice-sheet growth goes up, albedo goes up; and as albedo goes up, temperature will go down. Thus, the net effect of the system in Figure 17.2a is positive feedback, where an initial reduction in temperature would be amplified through the ice-sheet growth and albedo links to result in further temperature reduction. Positive feedback causes a system to change or to evolve.

In Figure 17.2b, precipitation is introduced as an additional and important component of glacier growth and decay: precipitation is required to add mass to an ice sheet. Precipitation levels will be governed by temperature as this controls evaporation. For ice-sheet

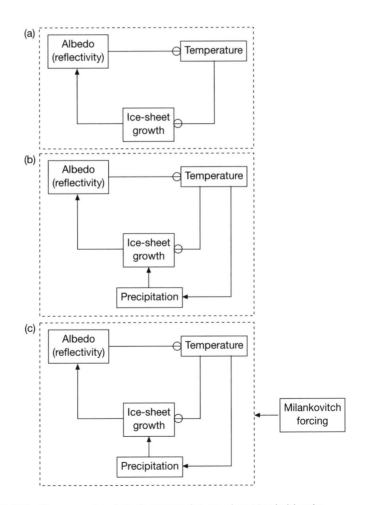

FIGURE 17.2 Three examples of simple systems that are relevant to glacial cycles

growth, the global oceans act as the major water source. As temperature goes down, precipitation will go down due to less evaporation (i.e. the link is positive as both changes are in the same direction). If precipitation goes down, ice-sheet growth will go down. Thus, if we examine the links from ice-sheet growth to albedo to temperature to precipitation and back to ice-sheet growth we have a negative feedback: as ice sheet growth goes up, albedo goes up, temperature goes down, precipitation goes down and, hence, ice-sheet growth goes down. Negative feedback is a self-limiting feedback which causes a system to resist change.

With these basic ideas about positive and negative feedback, we have a means of understanding the dynamics of glacier growth and decay in the system in Figure 17.2b. If negative feedbacks dominate in a system, the system will be maintained in the state it is in, either as a glacial period or as an interglacial period. If positive feedbacks dominate, we may have change, which may be rapid and which may cause a system to evolve, either from a glacial to an interglacial or from an interglacial to a glacial. Thus, if there is some sort of external forcing, the effects of that forcing will depend upon whether the system displays positive feedback and hence the forcing causes change, or negative feedback, in which the forcing is absorbed by the system. Thus, we have a basic conceptual model for considering the effects of Milankovitch forcing: the orbital driving factors are either enhanced or reduced by the system to which they are applied. In practice, the system is much more complex than this. Feedbacks may be delayed, and simple components like 'temperature' and 'precipitation' need to be disaggregated in relation to glacier growth (e.g. enhanced winter snowfall (increased precipitation) and reduced summer melt (lower temperatures) are generally considered to be the most conducive to glacier growth).

Thus far we have only considered the environment as changing through time. However, the environment has a critical spatial and vertical dimension. Thus, feedbacks don't simply operate through time but also through space: if you change one part of the environment, feedbacks may not be limited to just that part of the environment but may also affect processes operating in other places in the environment. For instance, in the case of glacial cycles, it is well established that ice-sheet growth in the high latitudes can have major environmental impacts at both the mid and the low latitudes. As the ice sheets extended to lower latitudes than present, so cold polar air extended to lower latitudes, causing the mid-latitudinal zones where warm and cold air mix and the westerly jet stream to move equatorwards. In the American south west, this caused substantial increases in precipitation during the last glacial maximum (e.g. Spaulding, 1991). Similarly, as more water was locked up in the ice sheets during the last glaciation, and with changes in atmospheric and oceanic circulation, the tropical latitudes were somewhat cooler and drier than today. This has resulted in suggestions that tropical plant and animal communities retreated into refugia, although this is contested, with suggestions that continuous forest cover was maintained throughout the Pleistocene (e.g. Colinvaux et al., 2000). This provides an important reminder of the connected nature of the environment and the way in which changes in one part of the environment can affect other parts of the environment.

It follows that the way the environment is modelled needs careful consideration so that these feedbacks are properly incorporated. This allows us to note what an effective model of the environment must allow for: (1) the proper representation of the spatiality of processes, especially because many processes are driven by gradients that are spatial (e.g. the slope of a hill controls the velocity with which water moves over it); (2) vertical gradients of process (e.g. flow at the bed of a river is generally faster than at the water surface); (3) the system to evolve through time in response to the operation of these two- and three-dimensional processes; and (4) feedbacks, manifest as change in the process drivers in response to the operations of processes themselves (e.g. as water moves over a hillslope, it may cause erosion (or deposition) and so change the slope so that processes at future times operate in different ways). The conceptual model allows us to specify (1) through (4) and hence: (a) what is included explicitly in the model; (b) what is included but represented in a simple way; and (c) what is excluded. As we will see later, this means that models are rarely generic representations of whole systems but truncated representations of reality, where model predictions are partly defined by how the modeller builds the model (Lane, 2001). Regardless of the conceptual model chosen, it is possible to conceive of a continuum from models that are largely built around a set of observations (empirical models) to models that are largely built around a set of laws (physically based models). This is an artificial continuum as most empirical models have an implicit theoretical justification and most physically based models have an explicit need for empirical relationships. However, it provides a useful categorization for exploring the nature of environmental models.

Empirical approaches

Mathematical models based upon empirical approaches involve making a set of observations of a number of phenomena and then using these observations to construct relationships among them. Thus, this approach is heavily dependent upon either field or laboratory measurements to provide the data necessary to construct the models. Central to empirical approaches are statistical methods. The key assumption is that one or more forcing (independent) variables cause changes in a response (dependent) variable. A good example of this is the eutrophication of shallow lakes. Eutrophication is a natural process associated with a progressive increase in lake primary productivity as lakes are ineffective at removing accumulated nutrients. Research in the 1970s (e.g. Schindler, 1977) established that the key limiting nutrients for eutrophication are phosphates and not nitrates, as is commonly assumed, because certain primary producers are capable of fixing atmospheric nitrogen. Figure 17.3 illustrates this for Barton Broad, using a dataset when no submerged plants (macrophytes) are present: as the total phosphate concentration of the water column rises, so the level of primary productivity increases. Thus we have a conceptual model in which we assume that levels of eutrophication are determined by levels of phosphate in a given lake ecosystem. By synthesizing a set of data from different lake ecosystems, Cullen and Forgsberg (1988) were able to build simple statistical rules that allow levels of chlorophyll-a (an indicator of the level of primary productivity in the system) to be predicted from phosphate concentrations. In practice, eutrophication is also affected by predator–prey interactions and seasonality effects, and more complex models can be developed for predicting the level of eutrophication in a lake by using more than one forcing variable (e.g. Lau and Lane, 2002). This is not a straightforward task if the forcing variables are themselves related to one another or,

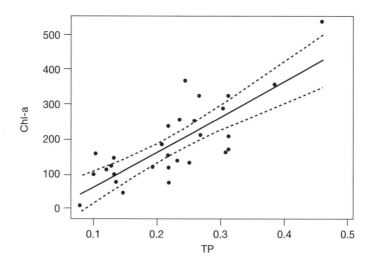

Source: Lau (2000)

FIGURE 17.3 The relationship between chlorophyll-a loading (a measure of the level of eutrophication) and total phosphate concentration for periods of macrophyte absence in Barton Broad, Norfolk

as is commonly the case with ecological data obtained for a single ecosystem (e.g. a lake), individual observations are correlated through time or across space. Additional information on empirical modelling of quite complex environmental systems is provided in Pentecost (1999).

The use of an empirical approach to modelling requires a number of assumptions. First, the empirical relationship must have a justifiable theoretical basis in the form of an appropriate conceptual model. While the use of statistical methods allows assessment of how good the model is (e.g. by assessing the goodness of fit of the empirical relationship) this is not normally a sufficient test of a model: empirical relationships may be spurious and, hence, have a poor predictive ability. They may also be strongly affected by individual observations, especially at the extremes of a variable, which can affect the model's predictive ability in relation to extreme situations. Unfortunately, many of the environmental issues of most concern are those that are connected with extremes (e.g. floods and droughts). Secondly, many empirical models perform poorly when used to predict beyond the range of observations upon which they are based. For instance, Cameron et al. (2002) compare traditional (using probability distributions and kinematic routing) and neural network approaches to flood forecasting. The latter is based upon using artificial intelligence methods to construct an empirical relationship between forcing parameters (e.g. upstream rainfall) and the key response variable (e.g. water level). Cameron et al. (2002) found that these models are only good at predicting water levels associated with patterns of forcing parameters that have happened before. Questions over temporal validity also appear in relation to spatial validity. In the case of lake eutrophication, there is a good reason to assume that lakes are generally phosphate limited, but research has shown (Kilinc and Moss, 2002) that in certain situations lakes can be nitrate limited. Thus, a generalized model (e.g. Figure 17.3) will not hold everywhere. There are two important implications of this problem: empirical models don't always hold through time and they often transfer poorly to different places. At the root of these problems is the poor generalizability of empirical models. It is often argued that this is because they don't necessarily have a good physical basis, as they are grounded upon statistical interactions rather than fundamental physical processes.

Thirdly, however, the relationship is constructed, it is important to remember that empirical relationships involve statements of uncertainty. The scatter around the line in Figure 17.3 means that for a given phosphate loading there is a range of possible levels of eutrophication. The greater the scatter, the greater the uncertainty, and it is necessary to provide predictions with an uncertainty attached to them, normally in the form of a standard deviation (i.e. prediction ±

standard deviation). When predictions from more than one empirical model are combined, it becomes especially important to propagate the uncertainty associated with each prediction. For instance, it is common to estimate the sediment load produced by a catchment from the product of an estimated discharge (predicted on the basis of an empirical relationship between discharge and the more readily measured parameter, water level) and an estimated suspended sediment concentration (predicted on the basis of an empirical relationship between suspended sediment concentration and discharge). Both these have uncertainties that will tend to magnify error when those uncertainties are combined. In some situations, this uncertainty can be propagated mathematically (e.g. Taylor, 1997). However, as the assumptions required for mathematical propagation of error are commonly violated, it is more common to propagate error statistically using techniques such as Monte Carlo analysis.

Physically based numerical models

Empirical approaches have a physical basis in so far as they have some sort of conceptual model that justifies the form of the relationship developed. Physically based numerical models take this one stage further by using the conceptual model to define links among fundamental physical, chemical and occasionally biological principles, which are then represented mathematically in computer code. At the scale of the natural environment, we are fortunate in having a number of key principles, largely deriving from Newtonian mechanics: (1) rules of storage; (2) rules of transport; and (3) rules of transfer. Rules of storage are based upon the law of mass conservation: matter cannot be created or destroyed but only transformed from one state into another. Thus, in relation to the prediction of flood inundation extent, and assuming that evaporation and infiltration losses are negligible, an increase in river discharge must result in one or more of (1) an increase in flow velocity; (2) an increase in water level; or (3) transfer of water on to the floodplain. Mass conservation underpins almost all numerical models in some shape or form. However, it is rarely sufficient to predict how a system behaves. Thus, in the case of flood inundation, how an increase in discharge is divided among changes in flow velocity, water level and floodplain transfer depends upon what is conventionally labelled a force balance, and which is normally based upon further Newtonian mechanics – e.g. every body continues in its same state of rest or uniform motion unless acted upon by a force. In the case of a river, this partitioning involves consideration of the pressure gradients and potential energy sources which drive the flow and the loss of momentum, due to friction at the bed and due to turbulence, which slows the flow. For instance, in the simplest of terms, with a rougher bed and subject to the shape of the river channel, an increase in discharge is less likely to lead to an increase in flow velocity than an increase in water level than with a smoother bed. Finally, rules of transfer allow for the possibility that chemical reactions cause a change in the state of an entity. For instance, phosphate bound to aluminium or iron may become soluble in a eutrophic lake, and hence available for fuelling eutrophication if there are oxygen deficits sufficient for reducing conditions to develop.

Following from the discussion of conceptual models above, the key to the operation of these rules is to allow feedback between them and the parameters that describe them. For instance, in the case of the river example above, an increase in discharge was noted to lead to more of an increase in water level for a rougher bed than a smoother bed. However, if water level rises, the effect of bed friction will be reduced, making it more easy to translate increases in discharge into increases in velocity. Here we see one of the fundamental advantages of numerical models over empirical models: they are more likely to capture the dynamics of the system through incorporating feedbacks between system parameters, something that our consideration of conceptual models considered as being crucial.

Stages in the development of a physically based numerical model

Stages in the development of a mathematical model are shown in Figure 17.4 for the case of a model of eutrophication processes in a shallow lake (from Lane, 2002). This introduces a number of important components in model building. First, it demonstrates the dependence upon a proper conceptual model. As noted above, the conceptual model is 'closed' as the boundaries of the system that the model will address must be defined. Ideally, all relevant processes will be included, and the closure will not exclude processes that might matter. In practice, processes are excluded for two reasons: (1) if they don't matter for the particular system being studied; or (2) if there are limitations on the possibility of including a particular process. Situation (1) can arise for a number of reasons. First, the geography or the history of the problem may allow processes to be excluded (i.e. they may not be occurring in a particular place or at a particular time, and so can be ignored). Secondly, there may be timescales or space scales that are not relevant to the system that is being considered. For instance, if we wish to model the spatial patterns of flood inundation over large lengths of river, it is not necessary to include a sophisticated treatment of turbulence, a short timescale aspect of the process, as uncertainties in other aspects of the model will tend to dominate. This reflects the common situation that time and space are coupled (e.g. Schumm and Lichty, 1965), where processes of interest over large

(a)

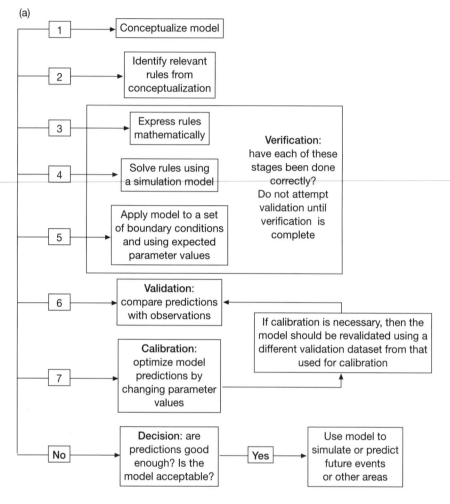

Key to Figure 17.4a

1 Is the model properly conceptualized? Are there enough processes represented? Are there too many processes represented? Are the relevant processes represented?

2 Have the correct rules been identified? Are the rules properly specified? What asumptions have been introduced here? Are they accurate?

3 Is the mathematical expression of the rules correct? Have any unsupportable assumptions been introduced during expression?

4 Are the rules solved correctly? Is the computer program free of programming error? Is the numerical solver sufficiently accurate? Is the model discretization robust?

5 Are the boundary conditions properly specified? Is there error in the boundary conditions that is affecting the model? Is the poor model performance due to a lack of necessary boundary conditions?

6 Are the observations being used to validate the model correct? Are the observations equivalent to model predictions? Are they representative of model predictions? Where (in space and time) is the model going wrong when compared with observations and predictions?

7 Has the calibration process produced realistic parameter values? Does the model produce more than one set of parameter values that are equally good at predicting when revalidating the model? Is the model overly sensitive to parameterization?

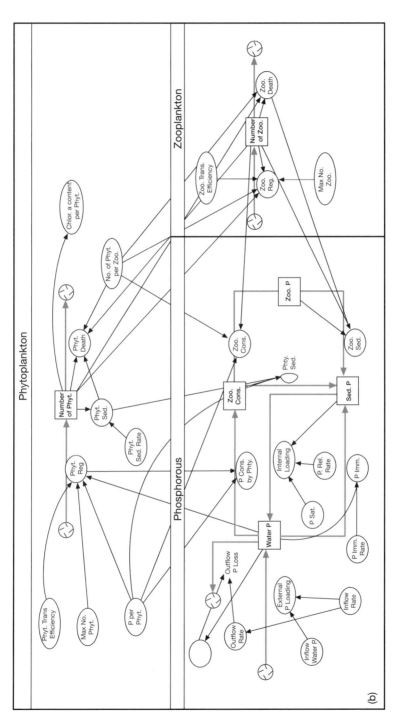

FIGURE 17.4 A general approach to model development and the conceptual model applied to Barton Broad, Norfolk (Lau, 2000). This shows that the model has three major components: the algal component (phytoplankton), a nutrient component (phosphorous) and an algal grazing component (zooplankton). There were then interactions both within and between these three major components. For instance, phytoplankton are regenerated according to nutrient availability and die naturally. They are also grazed by zooplankton. As phytoplankton are generated, die and are grazed, so phosphorous is moved around the various components in which it can be stored

spatial scales are commonly associated with longer timescales. Unfortunately, numerical modelling faces a serious challenge when processes couple across time and space scales, and a fine resolution of process representation (in space and time) is required in order to get an adequate system representation. Situation (2) follows from (1) when there are limits placed upon process representation because of limits to computation, but also more generally where either the process knowledge is missing or process representation is difficult. For instance, in eutrophication modelling, there is a major difficulty in coupling the aggregate behaviour of phytoplankton to the species-specific behaviour of an individual fish, in relation to its life-cycle. This situation is commonly dealt with by using a more simple version of a process's affects. In the case of lake eutrophication, the system may be driven by both nutrient limitations and food-chain interactions. The latter are controlled by the presence of bottom-growing vegetation which act as refugia for zooplankton that graze upon algae. It is not necessary to model plant life-cycles in most shallow lakes as these are relatively straightforward functions of seasonality. Thus, they may be dealt with using simple parameterizations (see below).

The second component of model building involves taking the conceptual model, identifying appropriate process rules and transforming these into a simulation model that can solve the equations. This can be the most difficult stage of model development as many equations do not lend themselves to easy solution. This is well illustrated for the case of predicting the routing of flood waters through a drainage network. The discharge entering the network varies as a function of time. In simple terms, the rate at which it moves through a part of the network depends upon the water-surface slope in that part of the network (steeper slopes mean faster flows). Thus there is a spatial dependence. The water surface itself will evolve as the water moves through the network. This leads to two basic problems. First, the combined space–time dependence of flood routing means that the dominant equations are partial differentials (as they contain derivatives in time and space). This is common to almost all environmental models, as we are interested in how things move through space. Moving through space takes time and hence all models should contain both space and time. Partial differentials are very difficult to solve. Secondly, all models require some form of initial conditions. In this case we need starting values for water level and discharge throughout the drainage network. We will also need to know some combination of discharge, velocity and depth at the inlet and/or the outlet. Hence, models have a crucial dependence upon the availability of data to initialize them.

Solution of the governing equations commonly requires us to

introduce parameters. This may be because, during the conceptualization process, we chose either to exclude certain processes or to represent them in a simplified way. In algal modelling, lake vegetation effects were treated in a simple way (a presence or absence conditioned by seasonality). Flood routing commonly ignores the lateral and vertical movements of water and flow turbulence. These affect the routing of discharge but their inclusion in a model may make solution impossibly time-consuming if we are interested in drainage network-scale flood routing. However, parameterization also results from the fact that, while equations can have a good physical basis, as they are simplified, new terms can appear whose physical basis is less certain and, most commonly, whose field or laboratory measurement is especially difficult. Parameterization can also be required in situations where a process has been excluded from a model, with its effects being represented through one or more parameters. A good example of this in flood routing studies is the use of a roughness parameter which not only represents the effects of friction at the bed upon flow hydraulics but also turbulence. The process of parameterization is rarely conducted independently from check data in which the model is optimized by changing parameter values such that the difference between check data and model predictions is minimized. This can result in parameters taking on values that are very different from what they might appear to be if measured in the field. Importantly, this causes us to question the supposed generality of models as the results of parameterization may not be guaranteed to hold beyond the range of conditions for, and the location at, which parameterization has been undertaken.

Model assessment involves two important stages: verification and validation. Verification is the process by which the model is checked to make sure that it is solving the equations correctly. This may involve debugging the computer code, doing checks upon the numerical solution process and undertaking sensitivity analysis. The latter may be used to make sure the model behaves sensibly in response to changes in boundary conditions or parameter values. Validation is the process by which a model is compared with reality. This normally involves the definition of a set of 'objective functions' that describe the extent to which model predictions match reality (see Lane and Richards, 2001). This is where there can be confusion with regard to validation and parameterization. Once an objective function has been determined, a model may be developed to reduce the magnitude of the error defined by the objective function. This may involve changing parameters or checking boundary conditions through the process of optimization described above. It may involve a more radical re-development of the model through the incorporation of new processes or alternative treatment of existing processes. Figure 17.5 shows default and optimized predictions of chlorophyll-a concentrations obtained for the eutrophication model described above, as applied to Barton Broad. Figure 17.5a shows the default predictions and Figure 17.5b shows the optimized predictions, in which parameterization was used to maximize the goodness of fit between the model and

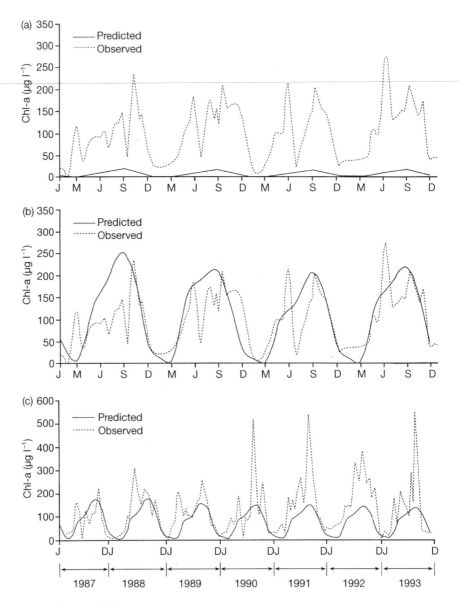

Source: After Lau (2000)

FIGURE 17.5 The default predicted (a), optimized using 1983–6 data (b) and predicted using 1987–93 data (c) chlorophyll-a concentrations for Barton Broad, Norfolk

independent data. Unfortunately, Figure 17.5b is not an example of validation as we have no evidence that the fit between the model and the measurements is due to anything other than the fact that we have forced the model to fit the measurements. To claim, from Figure 17.5b, that the model is validated is tantamount to cheating. Modellers get around this problem through one of a number of means. First, it may be possible to do a split test, in which some data are not used for model parameterization but held back for independent use for validation. An example of this is shown in Figure 17.5c, in which the optimized model (Figure 17.5b for 1983–6 data) is used to predict for 1987–93. This demonstrates why the optimization undertaken for 1983–6 does not constitute validation: when the model is applied in a predictive mode, it is clear that there is a progressive divergence between observations and model predictions and the model is not capturing some part of the system. An optimized model is not necessarily a validated model. Secondly, validation can be undertaken by taking a model that has been parameterized for one location and, under the assumption that the parameters can be transferred, applying it to a second location in order to provide independent validation data.

The main purpose of validation is to place confidence in the extent to which a model can be used to simulate or to predict beyond the range of conditions for which it is formulated. As we noted in the introduction to this chapter, this is the core purpose of a model: to extend the bounds of space and time. Here we face some serious difficulties, and these cause us to look very critically at deterministic mathematical models and what they can and cannot do.

WHAT A MODEL CAN AND CANNOT DO

The above examples illustrate four basic reasons why we might wish to develop numerical models and, connected to each of these, four reasons why models might not deliver what we hope. The first relates to the need to understand whole systems. This relates to the idea that there is a potentially large number of processes that might operate in a system, that these connect with one another through both positive feedbacks (which amplify change) and negative feedbacks (which slow or even stop change), and that these processes operate over space and through time. Designing either field or laboratory experiments to understand such systems is a serious challenge. The numerical model provides the opportunity to integrate processes together in order to understand whole-system behaviour. The problem here is that numerical models are no different from field or laboratory experiments in that they have been subject to closure. Models require assumptions to be made about what does and does not matter. For instance, climate models vary in the extent to which they include feedbacks associated with core components of the earth system (e.g. vegetation feedbacks). And noted above, they also involve simplifications over how processes are included. Also in climate models, the effects of clouds upon the energy and water balance of the atmosphere have to be dealt with using simplifications as the spatial scale of cloud physics is typically smaller than that at which models are operating. Thus models are closed representations of reality and a model is partly dependent upon how the modeller sets up the model (i.e. defines the closure). This is similar to the way in which the experimental design associated with fieldwork (e.g. choice of field site, measurement methods) and laboratory work (e.g. definition of controlled experimental conditions) partly determines the results that are obtained (Lane, 2001).

The second important role of a model relates to system understanding through sensitivity analysis. This is where the model is used in the same way as a laboratory experiment: model predictions derived from models with different structures, process representations or parameter values are compared in order to understand which processes and parameters have most effect upon the system. This tells us which aspects of a system are most important for further research, for fieldwork or for laboratory analysis. In a three-dimensional numerical model of flow over a rough gravel bed, model predictions were highly insensitive to changes in bed roughness, but highly sensitive to bed-surface structure (Figure 17.6; Lane et al., 2002). Bed roughness affects the momentum balance at the bed. Bed-surface structure affects both the momentum balance and mass conservation (i.e. flow blockage). The poor sensitivity to bed roughness demonstrates that the effect of bed-surface structure will not be properly represented through a bed roughness term and requires us to reconsider how bed-surface structure is represented in models of flow over gravely surfaces. The problem with this type of sensitivity analysis is similar to the whole-system problem. Do the results hold in a more general sense, or are they a product of the nature of the model as it has been developed and as it is being applied? Unfortunately, this is not a simple question to answer. In the gravel-bed river-flow model, the treatment of turbulence is known to be inappropriate for this kind of flow. Turbulence has a major effect upon momentum transfer at the bed. Thus, while the conclusion that blockage effects associated with bed-surface structure matter does hold, the extent to which this is the case must remain uncertain as turbulence effects are not being fully included.

The third role of a model is the one that is of most interest to environmental managers: the prediction of system properties either at future time periods or for locations where reliable measurements

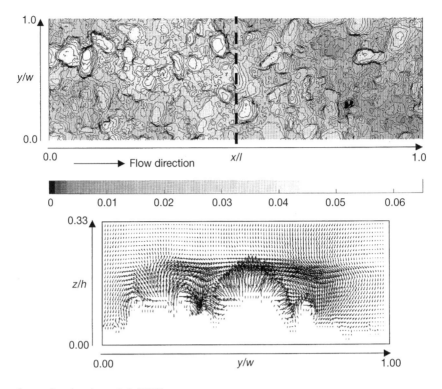

Source: Based on Lane et al. (2002)

FIGURE 17.6 Model predictions of flow over a rough gravel-bed surface: (a) shows the digital elevation model and (b) shows velocity vectors in a cross-section taken along the dashed line in (a) (i.e. vectors show the cross-stream and vertical components of velocity. *w* is width, *h* is depth and *l* is channel length. *y* is position along the section, *x* is position downstream and *z* is position in the vertical)

cannot be made. This rationale allows us to break out of the 'hunny' jar and Heffalump problems described in the introduction to this chapter by using technology (a properly validated numerical model) to try to make statements about what might happen before they actually happen. We live with this type of modelling on a day-to-day basis. For instance, short-term weather forecasting is based upon regional predictive models of weather systems. Flood warning systems are commonly driven by a predictive model that indicates where water is expected to reach for a range of river discharges, and which can be used to issue warnings to potentially affected properties when a particular discharge occurs. The main issue here is the extent to which trust can be placed in a model's predictions as a result of uncertainties in model predictions. These uncertainties can be classified into seven broad headings (Table 17.1). Table 17.1 identifies a number of core issues that we must evaluate when considering

TABLE 17.1 Model uncertainties

Type of uncertainty	Explanation
Closure uncertainties	This relates to uncertainties that arise because certain processes have been included or excluded during model development. A common strategy to deal with this is to include processes, test their effects and to ignore them or to simplify them if they seem to have only a small effect upon model predictions. The problem with this is that the effect of a process can often change as the effects of other processes (i.e. system state) change. Thus, it is impossible to be certain that a given process will always be unimportant. Note that problems of closure are not just associated with models, but are an inherent characteristic of all science (see Lane, 2001). In general, closure uncertainties are easy to spot in any model application. We often forget that the same criticism of closure applies equally to almost all other aspects of scientific method
Structural uncertainties	These arise because of uncertainties in the way in which the model is conceptualized in terms of links between components. A good example of this is whether or not a component is an active or a passive component of a model. For instance, in a global climate model, the ocean may be assumed to be a passive contributor to atmospheric processes: it acts as a heat and moisture source, but does not itself respond to atmospheric processes. This type of structural uncertainty will define limits to model applicability. For instance, the ocean has a high specific heat capacity, and so will respond slowly to atmospheric processes. Thus, specification of the ocean as a passive source of heat and moisture is acceptable if the model is being applied over a timescale (e.g. days) when the ocean can assume to be in steady state. As with closure uncertainties, structural uncertainties limit a models applicability and remind us of the importance of evaluating a model in relation to the use to which it will be put. Any model can be criticized on the grounds of structural uncertainties without a purpose-specific evaluation
Solution uncertainties	These uncertainties arise because most numerical models are approximate rather than exact solutions of the governing equations. Commonly, numerical solution involves an initial guess, operation of the model upon that initial guess and its subsequent correction. This continues until operation of the model on the previous corrected guess does not alter by the time of the next guess. This process can result in severe numerical instability in some situations, which is normally easy to detect. However, more subtle consequences, such as numerical diffusion associated with the actual operation of the solver, can be more difficult to detect. Following guidelines in relation to good practice can help in this respect
Process uncertainties	These arise where there is poor knowledge over the exact form of the process representation used within a model. A good example of this is the treatment of turbulence in models of river flow. Turbulence can have an important effect upon flow processes as it extracts momentum from larger scales of flow and dissipates it at smaller scales. Most river models average turbulence out of the solution but then have to model the effects of turbulence

TABLE 17.1 *continued*

Type of uncertainty	Explanation
	upon time-averaged flow properties. Turbulence models vary from the simple to the highly complex and it can be shown that different turbulence treatments are more or less suitable to different model applications. Thus, the process representation in a model, as with structural aspects of a model, needs careful evaluation with reference to the specific application for which the model is being used
Parameter uncertainties	These arise when the right form of the process representation is used but there are uncertainties over the value of parameters that define relationships within the model. Particular problems can arise when parameters have a poor meaning in relation to measurement. This can arise in two ways. First, some parameters are difficult to measure in the field as they have no simple field equivalent. Secondly, during model optimization, parameters can acquire values that minimize a particular objective function, but which are different from the value that they actually take on the basis of field measurements. A good example of this is the bed roughness parameter used in one-dimensional flood routing models. It is common to have to increase this quite significantly at tributary junctions, to values much greater than might be suggested by the shape of the river or the bed grain-size. In this case, there is a good justification for it, as one-dimensional models represent not only bed roughness effects but also two- and three-dimensional flow processes and turbulence through the friction equation. Roughness is therefore representing the effects of these other processes in order to achieve what Beven (1989) labels the right results but for the wrong reasons
Initialization uncertainties	These are associated with the initial conditions required for the model to operate. They might include the geometry of the problem (e.g. the morphology of the river and floodplain system that is being used to drive the model) or boundary conditions (e.g. the flux of nutrients to a lake in a eutrophication model)
Validation uncertainties	Given the above six uncertainties, a model is unlikely to reproduce reality exactly, and validation is required to assess the extent to which there is a reasonable level of agreement. However, validation data themselves have an uncertainty attached to them. This is not simply due to possible measurement error, but also when the nature of model predictions (their spatial and temporal scale, the parameter being predicted) differs from the nature of a measurement. A commonly cited example of this is validation of predictions of soil moisture status in hillslope hydrological models, when point measurements of soil moisture status (in space and time) are used to validate areally integrated predictions. This creates problems for modelling in two senses. First, apparent model error may actually be validation data error. Secondly, if validation data are then used for model optimization (and note that data used for model optimization should not then be used for validation), uncertainty will be introduced into model predictions as the data that the model are optimized to may be incorrect. Following Beven (1989) this means that we may get the wrong results for the wrong reasons

models as a strategy within physical geography. First, uncertainty will be an endemic characteristic of all modelling efforts. Attempts to reduce uncertainty are tempting through the adoption of more sophisticated modelling approaches or the improved specification of boundary conditions. Unfortunately, research has shown that attempts to deal with uncertainty tend to introduce yet more uncertainty, in the way that Wynne (1992) conceptualizes the dynamics of science as being the generator of uncertainty rather than the eliminator of it. For instance, Lane and Richards (1998) argue that three-dimensional models of river flow are required in tributary junctions in order to represent properly the effects of secondary circulation upon flow processes. Lane and Richards (2001), having used a three-dimensional model for this purpose, demonstrate the significant new uncertainties that have emerged from (1) the difficulty of specifying inlet conditions in each of the tributaries in three-dimensions; (2) problems of designing a numerical mesh that provides a stable numerical solution that minimizes numerical diffusion; (3) uncertainties over the performance of a roughness treatment in three dimensions; and (4) problems with finding an appropriate turbulence model. The continual creation of uncertainty keeps the science of this sort of modelling alive. However, when models need to be put to practical use for forecasting, or process understanding, obvious questions emerge as to whether or not numerical models are little more than computationally intensive crystal balls.

Secondly, Table 17.1 emphasizes that when models are to be used in an applied sense, it is vital that the associated uncertainty is communicated along with those model predictions. This is necessary to avoid false faith being placed in model predictions. This requires methods for (1) determining what the uncertainty is; (2) representing the uncertainty in a manner that has meaning to the user of a model's predictions (as well as the modeller themselves); (3) communicating uncertainty; and (4) persuading those who use model predictions to accept both uncertainties in model predictions and uncertainties in determining the uncertainty of model predictions. Before commenting on these, it is worth remembering the four-fold division introduced by Wynne (1992) of different types of uncertainty. Wynne argues that all uncertainties can be given one of four labels: (1) risks or quantifiable uncertainties; (2) unquantifiable uncertainties; (3) uncertainties that we are ignorant about but which we may find out about through further experience or investigation; and (4) indeterminacies or uncertainties that cannot be determined through any form of investigation prior to them happening. Thus, the determination of uncertainty is largely about the determination of risk, or quantifiable uncertainty, and the representation of that uncertainty accordingly. However, as uncertainty is communicated, it is vital to include unquantifiable uncertainty, ignorance and indeterminacy. As an example, consider estimation of the uncertainty in patterns of floodplain inundation. Sensitivity analysis can be extended to uncertainty analysis through assessing the implications of model uncertainties (e.g. Table 17.1) for model predictions. As Binley et al. (1991) have demonstrated, we need formal methods for doing this (e.g. generalized likelihood least-squares uncertainty estimation, or GLUE). This is not a straightforward task as a result of the large number of parameters that we are typically uncertain about. None the less, uncertainty bands can be determined in the form of probabilities of floodplain inundation for events with different return periods (e.g. Romanowicz et al., 1996). These provide a partial account of uncertainty. Actual predictions of flood inundation will be affected by unquantifiable uncertainties (e.g. the difficulties of determining likely runoff generation given a particular combination of antecedent moisture conditions and rainfall patterns), ignorance (e.g. incorrect assumptions built into the model, such as that extreme flood events only occur when a catchment is fully saturated) and indetermi-nacy (e.g. aspects of floodplain management, such as culvert main-tenance and repair, that cannot be determined in any realistic way before an event occurs). Unfortunately, we are not good at commu-nicating these uncertainties *and* accepting these uncertainties as a natural aspect of the science of environmental management. The lat-ter is compounded by the problem that uncertainty is measured on a continuous probability scale whereas decisions over floodplain man-agement have to be made on a discrete scale involving action (e.g. improve floodplain defences, refuse flood-insurance protection) or no action (do not improve floodplain defences, allow flood-insurance protection).

The final role for modelling is for simulation in which we ask 'what if?' questions about environmental behaviour. In many senses, this is the combination of the above three reasons for using numerical models: use a representation of the whole system (reason 1), in which we vary system parameters (sensitivity analysis, reason 2) to make predictions (reason 3) that allow us to improve environmental man-agement. The need to do this was emphasized in the introduction to this chapter: waiting for confirmed observational evidence from a noisy system means that significant damage may have occurred before action can clearly be justified.

CONCLUSION

Harré (1981) identified three different roles for scientific investigation: (1) as formal aspects of methodological investigation; (2) to develop the content of a theory; and (3) in the development of technique. Table 17.2 applies these to numerical models in an attempt to demonstrate

TABLE 17.2   What models can and cannot do

Models as formal aspects of method	Can a numerical model help?
A1 To explore the characteristics of a naturally occurring process	✔ The classic role of numerical simulation – asking 'what if?' questions
A2 To decide between rival hypotheses	✔ Used as part of sensitivity analysis in which rival hypotheses are tested using a model
A3 To find the form of a law inductively	✖ Generally not possible as laws are required to get a model to work, so generating laws from model predictions runs the risk of circular argument (but what is a law?)
A4 As models to simulate an otherwise unresearchable process	✔ A critical function of models and perhaps where they are most powerful but also most problematic (are a model's results a reflection of reality or a reflection of the way the model has been set up?)
A5 To exploit an accidental occurrence	✔ But more to understand possible accidental occurrences that might occur (e.g. what if there was a dam break?)
A6 To provide negative or null results	✖ A major problem for models – is the negative or null result a true property of the system, or simply because a model has been used?

Models in the development of the content of a theory	Can a numerical model help?
B1 Through finding the hidden mechanism of a known effect	✖ Can't find hidden things if they are not included in the model
B2 By providing existence proofs	✔ Models can provide corroboration of things observed using other methods
B3 Through the decomposition of an apparently simple phenomenon	✖ Not possible
B4 Through demonstration of underlying unity within apparent variety	✔ Identification of general patterns

In the development of technique	Can a numerical model help?
C1 By developing accuracy and care in manipulation	✖ Not possible
C2 Demonstrating the power and versatility of apparatus	✖ Not possible

Source: After Harré (1981)

what models can and cannot do. In practice, this sort of classification should be treated as a fuzzy one, and this table is included as a basis for discussion, perhaps in a tutorial. For instance, following A1 in Table 17.2, models may be used to explore the characteristics of a naturally occurring process, but the extent to which this can be done will depend upon the confidence that can be placed in a model and, as noted above, there will always be uncertainty in the 'natural' characteristics that the results from a model may suggest. Similarly, strictly, models may not be used to provide negative results (A6), as we never know whether or not the null result is due to the model as constructed or a real characteristic of the system that the model is being used to represent. However, a model may provide an indication of a null result that causes us to look elsewhere in order to find additional supporting evidence. This discussion of both A1 and A6 emphasizes a key theme in the use of models: a model's effectiveness largely depends upon the ability of a modeller to engage with a broad spectrum of methods, including those that are field and laboratory based. This is where Winnie-the-Pooh again provides us with key guidance. During the flood, Pooh Bear finds a 'missage' in a bottle. Being unable to read, he has to get to Owl. Surrounded with water he uses a classic piece of reasoning (i.e. a model): if a 'missage' in a bottle can float, then a bear in a 'hunny jar' can float. Having got the skeleton for his model, he develops his model by trying various positions on the jar until he finds one that is stable. Here we see the crucial iteration between model development and empirical observation. He now has an optimized model that he proceeds to validate by successfully floating off to Owl. Finally, he demonstrates the transferability of his model by going to rescue Piglet, with Christopher Robin, in an upside-down umbrella. This requires much less development and represents the common transition of a model, through time, from being a developmental piece of science to a practical part of technology. Unfortunately, Pooh Bear gives us little guidance as to when a model of a piece of science is sufficient for us to make it a practical piece of technology. That is a decision that society has to take.

Summary

- Numerical models are used by physical geographers for dealing with situations where the researcher is 'remote' from what her or she is studying: where the events that are of interest have occurred in the past (e.g. reconstruction of past climates); may occur in the future (e.g. patterns of inundation associated with future flood events); or where they are occurring now but cannot be measured or studied using other methods.

288

- Without a conceptual model of the system that is being modelled, it is impossible to develop either an empirical or a physically based mathematical model. Conceptual models involve a statement of the basic interactions between the components of a system.
- The way the environment is modelled needs careful consideration so that feedbacks are properly incorporated.
- Empirical approaches have a physical basis in so far as they have some sort of conceptual model that justifies the form of the relationship developed. Physically based numerical models take this one stage further by using the conceptual model to define links among fundamental physical, chemical and occasionally biological principles, which are then represented mathematically in computer code.
- Model assessment involves two important stages: verification and validation. Verification is the process by which the model is checked to make sure that it is solving the equations correctly. Validation is the process by which a model is compared with reality.
- Modelling has an important role in understanding the operation of environmental systems. This is achieved by using models to integrate multiple processes over many timescales; to undertake experiments of sensitivity; to predict outcomes; and (through simulation) to ask 'what if?' type questions.
- Models and the modelling process demonstrate many of the fundamental aspects of scientific research. The transition from scientific model to a usable technology for environmental management requires that it is transferrable between case studies, and reflects the need for qualitative as well as quantitative judgements.

Further reading

- Beven (1989) is a very useful introduction to critical thinking in relation to numerical models. Beven introduces a set of ideas that challenged an emerging paradigm regarding the power of numerical models and provides a critical framework for evaluating the role of modelling, in this case for the hydrological sciences but with implications for the modelling of environmental systems more generally.

- Rather like the Beven paper, the book by Anderson and Bates (2001) is useful because it brings together a very wide range of theoretical and methodological perspectives in relation to numerical modelling, albeit around the hydrological sciences.

- Kirkby et al. (1992) provide a good general introduction to numerical modelling in physical geography. This book is especially strong on the way modelling is done and has some easy but useful examples of models that can be coded to illustrate principles of model building.

- Beven (2000). This is definitely the best book around on modelling in hydrology in general, with excellent coverage of material specific to hydrology but illustrating environmental modelling in general. Likewise, Huggett (1993) is very effective on conceptual modelling across the environmental spectrum and how to apply conceptual models using a range of modelling techniques.

- Jakeman et al. (1993). This is useful for understanding modelling over a range of spatial scales and especially at the global scale.

Note: Full details of the above can be found in the references list below.

References

Anderson, M.G. and Bates, P.D. (eds) (2001) *Model Validation: Perspectives in Hydrological Science*. Chichester: Wiley.

Beven, K.J. (1989) 'Changing ideas in hydrology: the case of physically-based models', *Journal of Hydrology*, 105: 157–72.

Beven, K.J. (2000) *Rainfall-runoff Modelling: The Primer*. Chichester: Wiley.

Binley, A.M., Beven, K.J., Calver, A. and Watts, L.G. (1991) 'Changing responses in hydrology: assessing the uncertainty in physically-based model predictions', *Water Resources Research*, 27: 1253–61.

Broecker, W.S. and Denton, G.H. (1990) 'What drives glacial cycles?' *Scientific American*, 262: 42–50.

Cameron, D., Kneale, P. and See, L. (2002) 'An evaluation of a traditional and a neural net modelling approach to flood forecasting for an upland catchment', *Hydrological Processes*, 16: 1033–46.

Colinvaux, P.A., De Oliveira, P.E. and Bush, M.B. (2000) 'Amazonian and neotropical plant communities on glacial time-scales: the failure of the aridity and refuge hypotheses', *Quaternary Science Reviews*, 19: 141–69.

Cullen, P. and Forgsberg, C. (1988) 'Experiences with reducing point sources of phosphorous to lakes', *Hydrobiologia*, 170: 321–36.

Farman, J.C., Gardiner, B.G. and Shanklin, J.D. (1985) 'Large losses of total ozone in Antarctica reveal seasonal CLOx/Nox interaction', *Nature*, 315: 207–10.

Harré, R. (1981) *Great Scientific Experiments*. New York: Dover Publications.

Houghton, J.T., Jenkins, G.J. and Ephramus, J.J. (eds) (1990) *Climate Change: The IPCC Scientific Assessment*. Cambridge: Cambridge University Press.

Huggett, R.J. (1993) *Modelling the Human Impact on Nature*. Oxford: Oxford University Press.

Imbrie, J. and Imbrie, K.P. (1979) *Ice Ages: Solving the Mystery*. London: Macmillan.

Jakeman, A.J., Beck, M.B. and McAleer, M.J. (1993) *Modelling Change in Environmental Systems*. Chichester: Wiley.

Kilinc, S. and Moss, B. (2002) 'Whitemere, a lake that defies some conventions about nutrients', *Freshwater Biology*, 47: 207–18.

Kirkby, M.J., Naden, P.S., Burt, T.P. and Butcher, D.P. (1992) *Computer Simulation in Physical Geography*. Chichester: Wiley.

Lane, S.N. (2001) 'Constructive comments on D. Massey space-time, "science" and the relationship between physical geography and human geography', *Transactions, Institute of British Geographers*, 26: 243–56.

Lane, S.N. (2002) 'Environmental modelling', in A. Rogers and H. Viles, *The Student's Companion to Geography*.

Lane, S.N., Hardy, R.J., Elliott, L. and Ingham, D.B. (2002) 'High resolution numerical modelling of three-dimensional flows over complex river bed topography', *Hydrological Processes*, 16: 2261–72.

Lane, S.N. and Richards, K.S. (1998) 'Two-dimensional modelling of flow processes in a multi-thread channel', *Hydrological Processes*, 12: 1279–98.

Lane, S.N. and Richards, K.S (2001) 'The "validation" of hydrodynamic models: some critical perspectives', in P.D. Bates and M.G. Anderson (eds) *Model Validation: Perspectives in Hydrological Science*, pp. 413–38.

Lau, S.S.S. (2000) 'Statistical and dynamical systems investigation of eutrophication processes in shallow lake ecosystems'. PhD thesis, University of Cambridge.

Lau, S.S.S. and Lane, S.N. (2002) 'Biological and chemical factors influencing shallow lake eutrophication: a long-term study', *Science of the Total Environment*, 288: 167–81.

Michaels, P.J. (1992) *Sound and Fury: The Science and Politics of Global Warming*. Washington, DC: Cato Institute.

Molina, M.J. and Rowland, F.S. (1974) 'Stratospheric sink for chlorofluoromethanes: chlorine atom-catalysed destruction of ozone', *Nature*, 249: 810–12.

Penetcost, A. (1999) *Analysing Environmental Data*. Harlow: Longman.

Romanowicz, R, Beven, K.J. and Tawn, J. (1996) 'Bayesian calibration of flood inundation models', in M.G. Anderson et al. (eds) *Floodplain Processes*. Chichester: Wiley.

Schindler, D.W. (1977) 'Evolution of phosphorous limitation in lakes', *Science*, 195: 260–2.

Schumm, S.A. and Lichty, R.W. (1965) 'Time, space and causality in geomorphology', *American Journal of Science*, 263: 110–19.

Spaulding, W.G. (1991) 'Pluvial climatic episodes in North America and North Africa – types and correlations with global climate', *Palaeogeography, Palaeoclimatology and Palaeoecology*, 84: 217–27.

Taylor, J.R. (1997) *An Introduction to Error Analysis: The Study of Uncertainties in Physical Measurements* (2nd edn). Sausalito, CA: University Science Books.

Wynne, B. (1992) 'Uncertainty and environmental learning: reconceiving science and policy in the preventive paradigm', *Global Environmental Change*, 2: 111–27.

18 Using Remotely Sensed Data

Paul Aplin

Synopsis

The aim of this chapter is to provide a broad introduction to the field of remote sensing. To understand remotely sensed data and their uses fully, it is necessary to appreciate the underlying scientific principles facilitating remote sensing, to know something of the historical background to remote sensing and to consider the uses to which remote sensing is put. This chapter defines the term 'remote sensing' and summarizes the historical development of remote sensing. It then goes on to explain the key role of electromagnetic radiation in providing a measurable parameter for remote sensing and reviews different remotely sensed data sources, formats and characteristics. Methods of data analysis are described next, focusing in particular on digital image processing techniques. Finally, a brief discussion on the various uses of remotely sensed data is provided, identifying the key advances in imaging and air/space travel that affect the field.

The chapter is organized into the following sections:

- Introduction
- Principles of remote sensing
- Sources of remotely sensed data
- Data characteristics
- Image analysis
- Conclusion.

INTRODUCTION

Like many developing fields, remote sensing has been defined in numerous ways. Definitions have ranged from the very technical ('the interpretation of measurements of electromagnetic energy reflected from or emitted by a target from a vantage point that is distant from the target' – Mather, 1999: 1), to the very 'honest' ('staying as far away from the problem as possible' – Wright, pers. comm., 1994). To clarify the issue, broadly speaking, remote sensing simply involves the collection of information about distant objects. The first point to raise, therefore, is that remote sensing is not as abstract a concept as may be imagined. For instance, human vision is a form of remote sensing since objects are seen at a distance. Similarly, to take a photograph is to acquire a remotely sensed image. However, despite these general principles, in the context of geographical study, remote sensing refers specifically to the collection of images of parts of the Earth's surface using specialized instruments, commonly aerial cameras and satellite sensors (Figure 18.1).

The history of remote sensing stretches back nearly 150 years to when pioneer photographers ascended in hot air balloons to take snapshots of the landscape. For around a century remote sensing was almost exclusively the domain of aerial photography (Lillesand and Kiefer, 2000). Then in the 1960s satellite reconnaissance emerged, providing a powerful new platform for remote-sensing activities. Around this same period, a second revolutionary development in remote sensing occurred as electronic sensors were developed to generate 'digital' images, in contrast with traditional analogue photographs (Campbell, 1996). With these innovations remote sensing entered the computer age, and subsequent development, as with almost any ICT-related field, has been rapid. However, it is not only technological progress that has shaped recent development in remote sensing. Political change has also played a part. While the foundations of remote sensing were principally in the military and meteorology (television weather forecasters still prize their 'satellite pictures'), the last three decades has seen widespread adoption of remote sensing by civilian users. Most recently, in the 1990s, ex-Cold War 'spy satellite' technology was declassified for general, public sector use (Aplin et al., 1997). Nowadays, remote sensing is used as commonly by house buyers to view likely locations (e.g. using http://www.multimap. co.uk) as it is by military strategists to deploy resources. The implications for geographers are immense. Never has there been such a ready supply of spatial data, the central component of so much geographical analysis.

PRINCIPLES OF REMOTE SENSING

It has been stated above that remote sensors generate images of the earth's surface. What has not been explained is how they do this, so a simple explanation follows. For a sensor to generate an image, it must

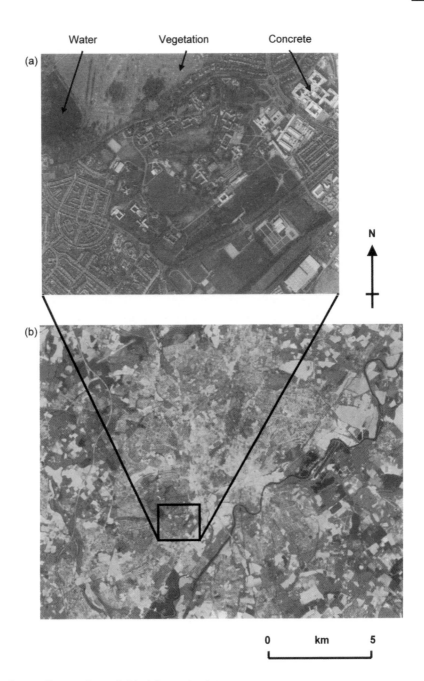

Sources: Photograph supplied by Infoterra; Landsat

FIGURE 18.1 (a) Black and white aerial photograph of The University of Nottingham campus; and (b) false colour composite satellite sensor (Landsat-7 Enhanced Thematic Mapper) image of Nottingham

measure some parameter related to the feature being represented (Gibson, 2000). The parameter used is electromagnetic radiation, defined as a 'system of oscillating electric and magnetic fields that carry energy from one point to another' (Rees, 1999: 55). Although this terminology may be a little perplexing to non-physicists, the most common example of electromagnetic radiation is blindingly straight-forward: solar radiation or sunlight. Put simply, a remotely sensed image is a representation of reflected sunlight (Figure 18.2). In fact, this explanation is perhaps a little 'too' simple since it infers that 'all' sunlight is reflected by the earth's surface and received by the remote sensor, and that reflected sunlight is the 'only' source of electro-magnetic radiation used by remote sensors. Neither of these infer-ences is true, as will be shown below.

The most important point to raise in objection to the preceding explanation of remote sensing is that sunlight occurs over a range of wavelengths (the electromagnetic spectrum), but not all these are useful for remote-sensing purposes. (It should be borne in mind that sunlight does not refer simply to light that is sensitive to human vision (visible light). On the contrary, visible light accounts for only a tiny portion of the electromagnetic spectrum.) That is, while the electromagnetic spectrum extends from short wavelength gamma rays to long radio waves and includes X-rays and ultraviolet waves, only

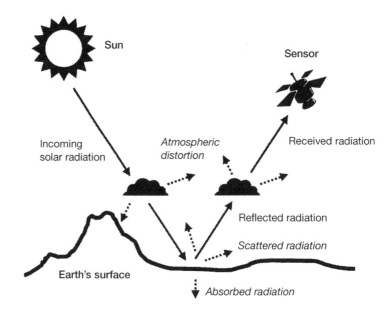

FIGURE 18.2 Reflected sunlight is the most common source of electromagnetic radiation received by remote sensors to generate images of the earth's surface, although some radiation is absorbed or scattered at the surface and in the atmosphere

parts of the visible, infrared and microwave wavelengths are commonly used in remote sensing.

Of the sunlight that 'is' used by remote sensing, a considerable portion never reaches the sensing instrument. First, sunlight is affected by the earth's atmosphere (light waves are scattered or absorbed at various wavelengths depending on atmospheric conditions), distorting the remotely sensed image of the earth's surface (Figure 18.2). To help understand this concept, consider the colour of the sky. During the day, when the sun is directly overhead and sunlight has to travel only a short vertical path through the atmosphere, short wavelength (blue) light is scattered, resulting in a blue sky. In the evening, when the sun is near the horizon and sunlight has to travel a longer oblique route through the atmosphere, longer (red) wavelengths are scattered, causing a red sunset.

Where sunlight does pass through the atmosphere and reach the Earth's surface, a certain proportion of the incoming radiation is absorbed or scattered rather than reflected, depending on the surface feature (Figure 18.2). In fact, this is the basis for discriminating among features using remotely sensed images and is, therefore, a very useful property. For instance, water may appear dark on a black and white aerial photograph because most of the sunlight is absorbed. In contrast, concrete may appear light because most sunlight is reflected, and vegetation may appear mid-tone (medium grey) because some sunlight is reflected but some is scattered due to the rough surface texture of vegetation (see Figure 18.1).

Finally, sunlight is not the only source of electromagnetic radiation used in remote sensing. Certain 'active' remote sensors emit electromagnetic radiation and record the amount of radiation scattered back from the earth's surface. The most common of these are 'radio detection and ranging' (radar) devices, which emit and receive microwave radiation. (Radars for tracking ships, aircraft and so on operate in the same way.) In contrast, 'passive' remote sensors, by far the majority of remote-sensing instruments, simply receive electromagnetic radiation from an external source (primarily reflected sunlight, as shown in Figure 18.2).

SOURCES OF REMOTELY SENSED DATA

So far, only two sources of remotely sensed data have been mentioned: aerial cameras and satellite sensors. Although these are important, many other data sources exist. Furthermore, the neat distinction between aerial photography and satellite sensor imagery is a little misleading since both cameras and digital sensors can be operated from either aircraft or satellites. Perhaps the clearest way of describing remotely sensed data sources is to separate the remote-sensing technology into two components: the 'platform' on which the sensing instrument is installed and the 'sensing instrument' itself.

There are three main categories of remote-sensing platforms (Figure 18.3): satellites (which also include manned space shuttles), aircraft (including aeroplanes, helicopters and even balloons) and ground-based devices (e.g. hand-held structures and hydraulic platforms). Ground-based remote sensing is performed for various reasons, but commonly it is used to measure the spectral properties of surface

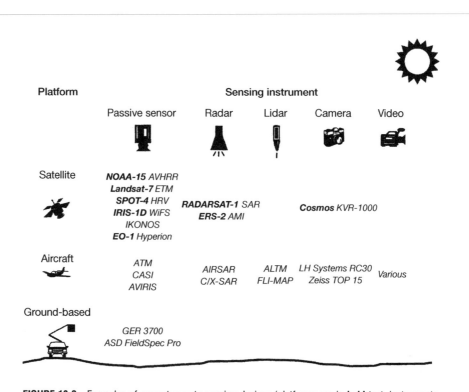

FIGURE 18.3   Examples of current remote-sensing devices (platforms are in **bold** text, instruments are in normal text; IKONOS is the name of both the satellite and the sensor). Please note that these are only a few of the more commonly used sources of remotely sensed data. 'Many' others exist. NOAA = National Oceanic and Atmospheric Association, AVHRR = Advanced Very High Resolution Radiometer, ETM = Enhanced Thematic Mapper, SPOT = Satellite Pour l'Observation de la Terre, HRV = High Resolution Visible, IRS = Indian Remote Sensing Satellite, WiFS = Wide Field Sensor, EO-1 = Earth Observing-1, SAR = Synthetic Aperture Radar, ERS = European Remote Sensing Satellite, AMI = Active Microwave Instrument, ATM = Airborne Thematic Mapper, CASI = Compact Airborne Spectrographic Imager, AVIRIS = Airborne Visible-Infrared Imaging Spectrometer, AIRSAR = Airborne Synthetic Aperture Radar, C/X-SAR = C & X Bank SAR, ALTM = Airborne Laser Terrain Mapper, FLI-MAP = Fast Laser Imaging and Mapping Airborne Platform, GER = Geophysical and Environmental Research Corporation, ASD = Analytical Spectral Devices

features or to 'ground truth' airborne or satellite-based imagery to correct for atmospheric distortion.

Remote-sensing instruments include cameras and digital sensors as described above, but also video and recently developed 'light detection and ranging' (lidar) devices (Figure 18.3). Lidars operate in the same way as radars but use lasers instead of microwaves as the source of electromagnetic radiation. Another current trend is towards the use of digital cameras and video (another form of digital sensing).

At this point it would be useful to describe specific remote sensors in detail, but they are simply too numerous to cover here. Instead, Figure 18.3 lists a small selection of current remote-sensing devices, and certain key instruments will be mentioned with reference to their data characteristics in the following section. For fuller discussions of remote-sensing devices, see Lillesand and Kiefer (2000) for reviews of past technology and Donoghue (2000) for a review of current and future technology.

DATA CHARACTERISTICS

Remotely sensed data are produced in many forms depending on the specifications of the sensing instrument. Certain characteristics of remotely sensed data are particularly important since these can influence how the data are used. Four main data characteristics will be discussed here: data format, spectral properties, spatial resolution and temporal aspects.

Data format

The first major distinction that can be used to categorize remotely sensed data is data format. In general terms, data are either analogue or digital. Analogue data are generated in hard copy form and features are distinguished on the basis of colour, texture, context and so on (Avery and Berlin, 1992). The most obvious example here is photography. While photographs are useful for manual interpretation (e.g. an observer can 'see' that fields are regular and roads are linear), they are not directly usable in digital analysis (Figure 18.4), although they can be scanned electronically. In contrast, digital image data comprise a regular grid of numbers where each square grid cell (known as a picture element or 'pixel') corresponds to a certain area on the earth's surface and has a number corresponding to the amount of sunlight reflected from that surface area (Jensen, 2000). Features are distinguished, therefore, on the basis of numerical differences. These numerical grids can be displayed as images (e.g. on computer monitors) through digital processing, where pixel numbers are assigned colours (Figure 18.4).

Analogue data (photograph)   **Digital data (image)**

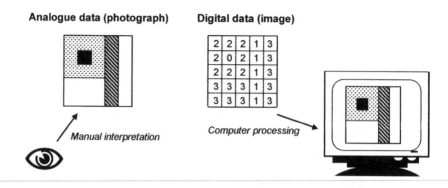

FIGURE 18.4   A simple scene is represented in both analogue (picture) and digital (numeric) format. The black square feature (e.g. building) in the analogue representation is indicated by the number 0 in the digital representation, the striped linear feature (e.g road) is indicated by 1 and so on. Analogue data require manual interpretation by an observer, while digital data can be used for computer-based display and analysis

Spectral properties

The spectral properties of remotely sensed data are perhaps the most significant characteristic in terms of recognizing features from images (spectral properties refer to the part or parts of the electromagnetic spectrum used to generate an image). The important concept to grasp is that features reflect electromagnetic radiation in different ways at different parts of the spectrum (Sabins, 1997). Therefore the representation of features in remotely sensed data varies according to which part of the spectrum (spectral waveband) is used. For instance, cities may appear dark in an image generated using a 'visible light' waveband (due to grey roads, roofs, etc.), but bright using a 'thermal infrared' waveband (since they are relatively warm features).

Generally, remote-sensing devices use either a single spectral waveband (panchromatic images), several wavebands (multispectral images) or many wavebands (hyperspectral images). The benefit of increasing the number of wavebands is that, since features appear different at different parts of the spectrum, an increasing amount of information is provided to help distinguish between features. However, each waveband produces a layer of data, so increasing the number of wavebands also increases data volume and computer processing time.

The defining feature of panchromatic images is their black and white (or, more correctly, 'grey scale') representation. Much aerial photography is black and white (e.g. Figure 18.1a), and some digital sensors generate panchromatic images. Although useful for basic remote-sensing analysis, panchromatic images have a low information content compared to other types of remotely sensed image. Multispectral images, on the other hand, have a relatively high information content and can be represented in numerous colour combinations (e.g. Figure 18.1b). Essentially, multispectral images are multilayer data-sets where each layer corresponds to a specific spectral waveband and all layers share a common grid of pixels. Typically, multispectral remote-sensing devices use between three and seven spectral wave-bands, from the visible and infrared parts of the electromagnetic spectrum. For instance, the Landsat-7 satellite's Enhanced Thematic Mapper (ETM) sensor (the latest instrument in the widely used series of Landsat satellite sensors) uses three visible wavebands, one 'near' infrared waveband, two 'mid' infrared wavebands and one 'thermal' infrared waveband. (It is a common misconception that the term 'infrared' is associated singularly with heat. In fact, there are three main categories of infrared radiation, only one of which provides an indication of temperature. The other two represent different forms of electromagnetic radiation.)

Automated remote-sensing analysis can be performed simultane-ously on 'all' layers of a multispectral image. However, only 'three' layers or less can be used for visual display, either on screen or printed out. This is because computer-based display involves only three basic colours (red, green and blue) which are combined to create a multi-colour presentation. Since any layer of a multispectral image can be displayed using any of these three colours, different representations (colour combinations) are possible for images with more than three layers. At first glance, Figure 18.1b may seem to provide an 'unnatu-ral' representation of the landscape around Nottingham since rural fields appear red or blue, rather than the more familiar green or brown. In fact, this is simply the result of displaying an infrared layer with two visible layers in a combination commonly known as a 'false colour composite' (Gibson and Power, 2000). The inclusion of an infrared layer enables certain characteristics of the landscape to be depicted that are normally invisible to the human eye. For instance, the red and blue fields (e.g. corresponding to irrigated and non-irrigated crops, respectively) may all appear green on a visible 'true colour composite' image.

It is clear from comparing a grey-scale panchromatic image (Figure 18.1a) and a colour multispectral image (Figure 18.1b) that the multi-spectral image enables more accurate 'spectral' discrimination between features. (Spatial discrimination is discussed below.) How-ever, despite fundamental differences between panchromatic and multispectral remote sensing, they are often used in combination. In fact, several remote-sensing devices generate coincident panchromatic and multispectral images, such as the Satellite Pour l'Observation de la Terre (SPOT) satellite's High Resolution Visible (HRV) sensor and the IKONOS satellite sensor.

In contrast to multispectral remote sensors that use a few broad spectral wavebands, hyperspectral instruments use tens or hundreds of very fine wavebands. The benefit of using hyperspectral devices, such as the EO-1 satellite's Hyperion sensor, is that certain features can be identified very precisely according to their specific spectral properties. For instance, hyperspectral images are commonly used for identifying minerals (Drury, 1993) or determining water quality (Koponen et al., 2002).

Finally, while most passive remote-sensing instruments use spectral wave*bands* as described above, active remote sensors use individual wave*lengths*. For instance, radars emit and receive electro-magnetic radiation at an exact wavelength, not over a range of wavelengths (i.e. a waveband). The Radarsat satellite's Synthetic Aper-ture Radar (SAR) instrument, for example, uses a wavelength of 5.6 cm. The product of radar imaging is a single layer of data similar to a grey-scale panchromatic image. As such, radar images are fairly limited in terms of spectral information.

Spatial resolution

Spatial resolution refers to the size of features discernible from remotely sensed data. Various different technical definitions exist, but spatial resolution is often used to refer simply to the 'pixel size' of remotely sensed images (Fisher, 1997). For example, the spatial resolu-tion of Landsat ETM is 30 m since its image pixels measure 30 m by 30 m. The importance of spatial resolution for image interpretation is obvious. High (or fine) spatial resolution images enable relatively small features to be identified, whereas low (or coarse) spatial resolu-tion images can only be used to identify fairly large features. For instance, the aerial photograph presented in Figure 18.1a has a spatial resolution of approximately 0.5 m, enabling the identification of small features such as cars and trees. In contrast, in the 30 m spatial resolution Landsat ETM image (Figure 18.1b), only relatively large or general features such as agricultural fields or urban areas can be distinguished accurately.

The spatial resolution of remotely sensed data varies widely, depending on the data source. At the finest scale, images with a spatial resolution of a few centimetres can be generated by low-altitude airborne instruments. In contrast, certain satellite sensors orbiting hundreds of kilometres above the Earth's surface generate images with a spatial resolution of several kilometres. For instance, the 15th National Oceanic and Atmospheric Association satellite's (NOAA-15) Advanced Very High Resolution Radiometer (AVHRR) has a spatial resolution of 1.1 km. However, other satellite instruments are able to generate relatively fine spatial resolution imagery. For instance, panchromatic IKONOS imagery has a spatial resolution of 1 m.

Associated with the spatial resolution of an image is its 'area of coverage' (the area on the earth's surface represented by the image). Generally, the finer the spatial resolution of an image, the smaller the area covered. Therefore, while an aerial photograph may offer fine spatial detail (e.g. Figure 18.1a), it can have fairly limited areal coverage. In contrast, a satellite sensor image may have relatively poor spatial detail (e.g. Figure 18.1b) but can cover a large area. In fact, Figure 18.2b represents only a small section of a Landsat ETM image. The original image measures 185 km by 185 km. At an even wider scale, an entire NOAA-15 AVHRR image can measure 2600 km by 2600 km.

Temporal aspects

Time is an important factor in much remote-sensing analysis. Many studies use images acquired at different times (multi-temporal images) to investigate dynamic features on the earth's surface. For instance, meteorologists require images every few hours to update weather predictions (Bader et al., 1995) or, at a longer timescale, environmentalists need images every few months to monitor tropical deforestation (Franklin, 2001). The availability and nature of multi-temporal images depend on the 'temporal resolution' (the frequency with which an image is acquired at a constant location) of a remote-sensing device.

Satellites have a significant advantage over aircraft when it comes to acquiring multi-temporal images. Satellite sensors are in a constant orbit and therefore provide a 'continuous' data supply. In contrast, airborne remote sensing occurs sporadically (as and when a decision is made to fly an aircraft) such that data are available on an occasional basis. Thus, while the temporal resolution of airborne sensing instruments is highly irregular, it is routine for satellite sensors. However, that is not to suggest that all satellite sensors have the same temporal resolution. On the contrary, the temporal resolution of such instruments varies considerably, depending on each instrument's area of coverage and tilting capabilities. Generally, sensors with a wide area of coverage acquire images of a constant location more frequently than those with a narrow area of coverage. However, some sensors can tilt obliquely, enabling them to target locations more frequently than sensors pointing vertically downwards. For instance, the temporal resolution of Landsat ETM, which covers an area 185 km wide but has no tilting capabilities, is 16 days. In contrast, the temporal resolution of IKONOS, which only covers an area 11 km wide but can tilt to ±45°, is less than 4 days.

IMAGE ANALYSIS

There are many different types of remotely sensed images and these are used for many purposes (Narayan, 1999). Consequently, there are numerous methods of image analysis. Generally, analysis is either manual or digital. Manual analysis involves an observer interpreting hard-copy or on-screen displays, identifying features according to their shape, size, colour, texture, location and context (Philipson, 1997). However, this type of analysis is prone to error due to its subjective nature and it can be very time-consuming. Perhaps more useful in today's ICT-led working environment is digital analysis. Computers provide the opportunity for advanced, accurate and very fast image analysis, and a wide range of 'digital image processing' (DIP) techniques has been developed. Specialized DIP systems are available for use, but many geographical information systems (GIS) also provide capabilities for DIP.

Remote-sensing studies have varying DIP requirements, depending on their objectives. Common to many studies, however, is the need to 'preprocess' the data prior to the main part of analysis (Jensen, 1996). Preprocessing is performed to correct data distortions or to standardize images. For instance, atmospheric correction may be carried out to remove 'haze' from images (Du et al., 2002). Alternatively, where several images are to be combined for analysis (e.g. multi-temporal images), geometric registration may be necessary to align the images to a known map co-ordinate system (Mao et al., 2001).

Following preprocessing, it is often useful to 'enhance' the presentation of an image to aid interpretation. Commonly, 'contrast stretching' is performed to increase contrast throughout an image, making features more distinct and identifiable. To illustrate this, the Landsat ETM image displayed in Figure 18.1b was subsequently contrast stretched and redisplayed in Figure 18.5a. On comparing the two images, it is clear that features are displayed in brighter tones (such that they are more distinct) in the contrast stretched image than the original image.

Image enhancement is a simple means of altering the visual display of images without affecting pixel values. More advanced analytical techniques involve some operation altering pixel values to generate useful information. Such techniques are too numerous to cover in detail here so three example procedures will be described: 'image filtering', 'image arithmetic' and 'classification'. For further information on digital image processing, see Gibson and Power (2000) in the first instance, and Mather (1999) for more advanced instruction.

Image filtering is used selectively to stress or suppress information within an image. For instance, by passing an 'edge detection' filter over an image, a new image is created that emphasizes linear features (e.g. roads or field boundaries). Alternatively, where general patterns are sought rather than fine detail, a 'smoothing' filter may be used to blur sharp features (Liu, 2000). For instance, the original Landsat ETM image of Nottingham (Figure 18.1b) was smoothed and the new image displayed in Figure 18.5b. Broad patterns, such as the differentiation between urban and rural land, are clearly more visible in the smoothed image than the original image.

Image arithmetic is used to generate new, useful images from one or more multispectral and/or multi-temporal image layers. Essentially, simple arithmetic operations such as addition, subtraction and

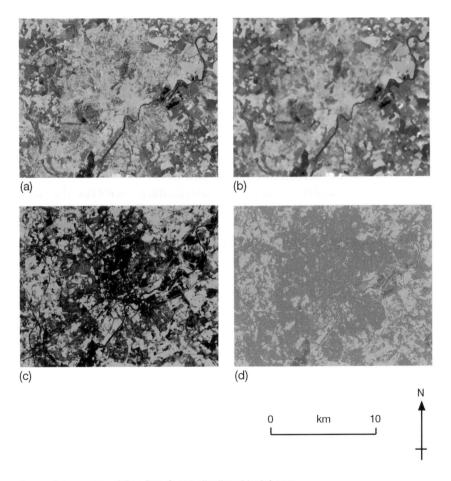

(a)

(b)

(c)

(d)

N

0   km   10

Source: Data courtesy of Eros Data Centre, distributed by Infoterra

FIGURE 18.5   Landsat-7 Enhanced Thematic Mapper image following (a) contrast stretching; (b) smoothing; (c) calculation of normalized difference vegetation index (vegetation is displayed in green); and (d) land-cover classification (green = vegetation, brown = bare soil, blue = water and grey = urban areas

so on are performed by associating corresponding pixels in different data layers. Various types of image arithmetic are possible, but one common example is image differencing (subtraction) for change detection. By subtracting an image acquired at a certain time from another image acquired at a different time, areas of change are highlighted (Sunar, 1998). Other more complex arithmetic procedures can also be performed. One particularly useful example is the 'normalized difference vegetation index' (NDVI), which highlights areas of healthy vegetation using an arithmetic combination of the near infrared and red wavebands of a multispectral image:

$$\mathrm{NDVI} = \frac{NIR - R}{NIR + R}$$

where NIR = near infrared waveband and R = red waveband (Oindo and Skidmore, 2002). The NDVI is particularly useful for this purpose because healthy vegetation has a predictable spectral response at the infrared and red parts of the electromagnetic spectrum. To illustrate this, the near infrared and red wavebands of the original Landsat ETM image of Nottingham (Figure 18.1b) were used to generate a NDVI (Figure 18.5c). Healthy vegetation is indicated in green. Bright green patches are likely to indicate tended vegetation (e.g. irrigated and/or fertilized crops or pasture).

Finally, 'land-cover classification' involves the segmentation of images into meaningful categories. To achieve this, pixels with similar spectral properties are grouped, forming general classes (Tso and Mather, 2001). Classification has many uses, including generating environmental inventories and identifying areas suitable for development. To illustrate the process, the original multispectral Landsat ETM image of Nottingham (Figure 18.1b) was classified into four general classes: vegetation, bare soil, water and urban features (Figure 18.5d).

CONCLUSION

The field of remote sensing has undergone significant development since the advent of aerial photography 150 years ago, and even since the emergence of satellite reconnaissance in the 1960s and 1970s. Thirty years or so ago, remotely sensed data were scarce and lacked detail (both spectral and spatial), and image analysis was limited by technological constraints. Today, in contrast, an extensive range of detailed data and advanced analytical techniques are available for geographical study. To conclude, it is useful to consider the range of applications for which remotely sensed data are currently used.

The implications of remote sensing for basic topographic mapping are obvious. It is considerably cheaper and quicker to generate topographic maps from remotely sensed images than from conventional field survey data. Traditionally, topographic mapping has been performed using aerial photographs since a high level of spatial detail is required. (In fact, this is a major focus of the affiliated field of 'photogrammetry'. See the Remote Sensing and Photogrammetry Society's website (http://www.rspsoc.org) for a definition and further details.) Recently, however, fine spatial-resolution satellite sensor images have been adopted for this purpose, following the launch of IKONOS in 1999. Similarly, many urban remote-sensing studies (e.g. land-use planning), which also require fine spatial-resolution data, are now feasible with satellite sensor images (Donnay et al., 2001). Other forms of mapping (in particular, elevation) are being revolutionized by new remote-sensing developments such as 'interferometric' radar analysis (Lanari et al., 1996) and the availability of lidar data (Flood, 2001).

Further uses of remote sensing include environmental applications, such as monitoring water resources (Neville et al., 2000) and measuring atmospheric pollutants (Carleton, 1991). Remote sensing also holds much practical value for mineral exploration (Gupta, 1991), forestry (Coppin et al., 2001) and agriculture. Recent developments in 'precision farming', for instance, involve the use of remotely sensed data to distribute fertilizer efficiently (Wood et al., 2000). Finally, traditional military and meteorological remote-sensing applications remain common (Schaber, 1999; Yao et al., 2001). However, this is by no means a comprehensive list of the uses of remotely sensed data. For a fuller discussion of applications, see Drury (1998). In the current technological climate dominated by ICT, there is an insatiable demand for data. It has been recognized widely that the 'spatial' domain provides an excellent means of integrating data, due at least in part to the growing popularity of GIS. Consequently, by providing an extensive, continuous and accurate supply of spatial data, the field of remote sensing appears set to prosper.

Summary

- In the context of geographical study, remote sensing refers to the collection of images of parts of the earth's surface using specialized instruments, commonly aerial cameras and satellite sensors.
- Remote-sensing technology may be separated into two components: the 'platform' on which the sensing instrument is installed and the 'sensing instrument' itself.

- There are three main categories of remote-sensing platforms: satellites, aircraft and ground-based devices.
- Remote-sensing instruments include cameras and digital sensors, video and recently developed 'light detection and ranging' (lidar) devices.
- Generally, remote-sensing devices use either a single spectral waveband (panchromatic images), several wavebands (multispectral images) or many wavebands (hyperspectral images).
- The spatial resolution of remotely sensed data varies widely, depending on the data source. At the finest scale, images with a spatial resolution of a few centimetres can be generated by low-altitude airborne instruments. In contrast, certain satellite sensors orbiting hundreds of kilometres above the earth's surface generate images with a spatial resolution of several kilometres.
- Following preprocessing, it is often useful to 'enhance' the presentation of an image to aid interpretation. Commonly, 'contrast stretching' is performed to increase contrast throughout an image, making features more distinct and identifiable. Other procedures include 'image filtering', 'image arithmetic' and 'classification'.
- Computers provide the opportunity for advanced, accurate and very fast image analysis, and a wide range of 'digital image processing' (DIP) techniques has been developed. Specialized DIP systems are available for use, but many geographical information systems (GIS) also provide capabilities for DIP.

Further reading

- Lillesand and Kiefer (2000). This is the classic textbook on remote sensing but, because of its origins in the 1970s (first edition), it is a little dated in parts. Comprehensive discussions are provided on aerial photography and early satellite sensing.

- Drury (1998) is a good, readable introduction to remote sensing that focuses on the 'uses' of remote sensing rather than the technology. Excellent and extensive use is made of colour illustrations to demonstrate points raised in the text.

- Gibson (2000) is a thorough but simple introduction to the subject. It is a good starting point for beginners to remote sensing. The companion to it is Gibson and Power (2000). This book offers another straightforward introduction, this time to digital image processing. A CD comprising image-processing software and exercises is provided.

- Mather (1999) provides an excellent, thorough overview of digital image processing, although it may be a bit too technical for complete beginners. A CD of image-processing software and exercises is provided.

Note: Full details of the above can be found in the references list below.

References

Aplin, P., Atkinson, P.M. and Curran, P.J. (1997) 'Fine spatial resolution satellite sensors for the next decade', *International Journal of Remote Sensing*, 18: 3873–81.

Avery, T.E. and Berlin, G.L. (1992) *Fundamentals of Remote Sensing and Airphoto Interpretation* (5th edn). New York: Maxwell.

Bader, M.J., Forbes, G.S., Grant, J.R., Lilley, R.B.E. and Waters, A.J. (eds) (1995) *Images in Weather Forecasting: A Practical Guide for Interpreting Satellite and Radar Imagery.* Cambridge: Cambridge University Press.

Campbell, J.B. (1996) *Introduction to Remote Sensing* (2nd edn). London: Taylor & Francis.

Carleton, A.M. (1991) *Satellite Remote Sensing in Climatology.* London: Belhaven.

Coppin, P., Nackaerts, K., Queen, L. and Brewer, K. (2001) 'Operational monitoring of green biomass change for forest management', *Photogrammetric Engineering and Remote Sensing*, 67: 603–12.

Donnay, J.-P., Barnsley, M.J. and Longley, P.A. (eds) (2001) *Remote Sensing and Urban Analysis.* London: Taylor & Francis.

Donoghue, D. (2000) 'Remote sensing: sensors and applications', *Progress in Physical Geography*, 24: 407–14.

Drury, S.A. (1993) *Image Interpretation in Geology* (2nd edn). London: Chapman & Hall.

Drury, S.A. (1998) *Images of the Earth: A Guide to Remote Sensing* (2nd edn). New York: Oxford University Press.

Du, Y., Guindon, B. and Cihlar, J. (2002) 'Haze detection and removal in high resolution satellite image with wavelet analysis', *IEEE Transactions on Geoscience and Remote Sensing*, 40: 210–16.

Fisher, P. (1997) 'The pixel: a snare and a delusion', *International Journal of Remote Sensing*, 18: 679–85.

Flood, M. (2001) 'Laser altimetry: from science to commercial lidar mapping', *Photogrammetric Engineering and Remote Sensing*, 67: 1209–18.

Franklin, S.E. (2001) *Remote Sensing for Sustainable Forest Management.* Boca Raton: Lewis Publishers.

Gibson, P.J. (2000) *Introductory Remote Sensing: Principles and Concepts.* London: Routledge.

Gibson, P.J. and Power, C.H. (2000) *Introductory Remote Sensing: Digital Image Processing and Applications.* London: Routledge.

Gupta, R.P. (1991) *Remote Sensing Geology.* Berlin: Springer-Verlag.

Jensen, J.R. (1996) *Introductory Digital Image Processing: A Remote Sensing Perspective* (2nd edn). Upper Saddle River, NJ: Prentice Hall.

Jensen, J.R. (2000) *Remote Sensing of the Environment: An Earth Resource Perspective.* Upper Saddle River, NJ: Prentice Hall.

Koponen, S., Pulliainen, J., Kallio, K. and Hallikainen, M. (2002) 'Lake water quality classification with airborne hyperspectral spectrometer and simulated MERIS data', *Remote Sensing of Environment*, 79: 51–9.

Lanari, R., Fornaro, G., Riccio, D., Migliaccio, M., Papathanassiou, K.P., Moreira, J.R., Schwabisch, M., Dutra, L., Puglisi, G. and Franceschetti, G. (1996) 'Generation of digital elevation models by using SIR-C/X-SAR multifrequency two-pass interferometry: the Etna case study', *IEEE Transactions on Geoscience and Remote Sensing*, 34: 1097–114.

Lillesand, T.M. and Kiefer, R.W. (2000) *Remote Sensing and Image Interpretation* (4th edn). New York: Wiley.

Liu, J.G. (2000) 'Smoothing filter-based intensity modulation: a spectral preserve image fusion technique for improving spatial details', *International Journal of Remote Sensing*, 21: 3461–72.

Mao, Z., Pan, D., Huang, H. and Huang, W. (2001) 'Automatic registration of SeaWiFS and AVHRR imagery', *International Journal of Remote Sensing*, 22: 1725–36.

Mather, P.M. (1999) *Computer Processing of Remotely-Sensed Images* (2nd edn). Chichester: Wiley.

Narayan, L.R.A. (1999) *Remote Sensing and its Applications*. Hyderabad: Universities Press.

Neville, P., Coward, R.I., Watson, R.P., Inglis, M. and Morain, S. (2000) 'The application of TM imagery and GIS data in the assessment of arid lands water and land resources in west Texas', *Photogrammetric Engineering and Remote Sensing*, 66: 1373–9.

Oindo, B.O. and Skidmore, A.K. (2002) 'Interannual variability of NDVI and species richness in Kenya', *International Journal of Remote Sensing*, 23: 285–98.

Philipson, W.R. (ed.) (1997) *Manual of Photographic Interpretation*. Bethseda, MD: American Society for Photogrammetry and Remote Sensing.

Rees, G. (1999) *The Remote Sensing Data Book*. Cambridge: Cambridge University Press.

Sabins, F. (1997) *Remote Sensing: Principles and Interpretation* (3rd edn). New York: W.H. Freeman.

Schaber, G.G. (1999) 'SAR studies in the Yuma desert, Arizona: sand penetration, geology, and the detection of military ordnance debris', *Remote Sensing of Environment*, 67: 320–47.

Sunar, F. (1998) 'An analysis of changes in a multi-date data set: a case study in the Ikitelli area, Istanbul, Turkey', *International Journal of Remote Sensing*, 19: 225–35.

Tso, B.C.K. and Mather, P.M. (2001) *Classification Methods for Remotely Sensed Data*. New York: Taylor & Francis.

Wood, G.A., Welsh, J.P. and Knight, S. (2000) 'Exploring management strategies for precision farming of cereals assisted by remote sensing', *Aspects of Applied Biology*, 60: 53–60.

Yao, Z., Li, W., Zhu, Y., Zhao, B. and Chen, Y. (2001) 'Remote sensing of precipitation on the Tibetan plateau using the TRMM microwave imager', *Journal of Applied Meteorology*, 40: 1381–92.

19   The Handling and Presentation of Geographical Data

Richard Field

Synopsis

The ability to work with and present numerical data is crucial if understanding of the complex 'real world' is to be achieved and communicated. The enormous range of techniques available to geographers gives great scope for flexibility and power, but also for misunderstanding and deception. Honest graphical and numerical presentation of findings should be an integral part of quantitative study. This chapter focuses on the manipulation and presentation of data so as to aid geographical interpretation. It first considers the process of abstraction among data collection, interpretation and presentation and then considers various means of presenting data (tables and graphs). The principles of exploratory data analysis are then reviewed, followed by a discussion of data types and distributions. Finally, the chapter examines the explanation and interpretation of bivariate and multivariate graphical plots.

The chapter is organized into the following sections:

- Introduction
- Abstraction
- Types of data
- Presentation of data
- Tables
- Graphical display
- Examining how one variable affects another
- Explanation and interpretation
- Conclusion.

INTRODUCTION

Ever since people started thinking about the world they have needed ways of coping with its inherent complexity. Early scientists relied primarily on logic: simplifying the world into a series of principles. Leonardo Da Vinci (1452–1519), along with other prominent figures of 'the Enlightenment', was instrumental in shifting emphasis towards data collection and experiment. A period followed in which many leading figures, including some of the great 'natural philosophers' such as Lyell and Darwin, meticulously collected enormous amounts of information from observation and experiment. However, during much of this period there was no generally accepted, objective way of drawing inferences from data. Instead, the subjective judgement of experts was relied upon, and this could vary considerably from expert to expert.

Tables of data were used quite routinely for much of this time, but it seems that graphs did not occur to the Greeks or the Romans nor to the likes of Newton and Leibnitz. Graphs were not invented until the great work of Descartes, who set up the Cartesian system in his book *La Géométrie*, published in 1637 (Spence and Lewandowsky, 1990). After that, the use of graphical means to present data did not really advance until the seminal work by Playfair (1786; 1801), who invented many of the popular statistical graphs still in use today, including the histogram, the line graph and the pie chart (Spence and Lewandowsky, 1990). These methods became increasingly popular during the nineteenth century and were linked to the rise of statistical thinking at the same time (Porter, 1986). Much important work was done in medicine. For instance, Dr John Snow, a London anaesthetist and Queen Victoria's obstetrician, used inductive methods to work out the cause of cholera. His classic piece of work, 'On the mode of communication of cholera', was enlarged from a pamphlet published in 1849 to be published as a book in 1854. Florence Nightingale, too, systematically gathered data on illness and fatalities and presented them graphically in order to make some sense of them. Nightingale was meticulous in her attention to detail and careful to ensure that her premises were backed up statistically – that is, to grasp the essential, emergent trends from a mass of individual cases (Kopf, 1916).

Probably the greatest period of development of statistical methods was towards the end of the nineteenth century and early in the twentieth century. Probability theory (developed to a considerable extent because gambling-loving French nobles had employed great mathematicians like Pascal to give them an edge) was taken and applied to data. The work of Jevons, Pearson and Fisher was instrumental during this period. During the twentieth century, the field of statistics was becoming increasingly theoretical and inaccessible to non-statisticians until the considerable input of the great American statistician, John Tukey. Tukey argued strongly that data should be explored, visualized and played around with as much as possible, as well as being subjected to formal statistical analyses. He pointed out the 'fuzzy' nature of real-world data, which do not usually comply with theoretical ideals. He also passionately believed that statistical methods should be more accessible to ordinary people, and many of the techniques he developed go a long way towards achieving this goal. More recently, other major developments have taken place in statistical theory, such as the rise of Bayesian statistics, which takes a fundamentally different approach to data analysis.

ABSTRACTION

In trying to make sense of 'real world' complexity, we go through several processes of abstraction (Figure 19.1). The first is the dataset itself. In most cases in geography, data are collected via a sampling process (see Chapter 15) and are therefore not the entire population of

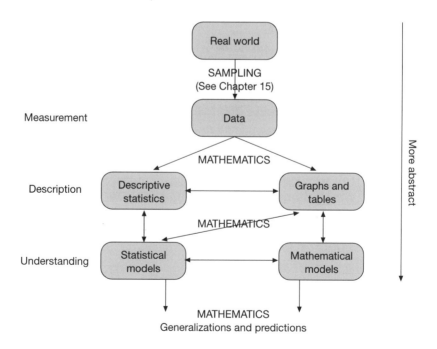

FIGURE 19.1   Abstraction of 'real world' complexity in a quantitative study. The diagram is idealized and simplified but illustrates the point that, as you move down through the levels of abstraction, complexity is reduced. Detail (information) is lost but understanding is enhanced – or, at least, that is the aim. Note that mathematics is the language that translates between all the different levels of abstraction

relevant units. These data consist of measurements that are subject to error and that cannot retain all possible information about the units sampled, no matter how careful and comprehensive the data collection. Thus the immense complexity of the 'real world' is reduced to a carefully chosen sample for which certain attributes are measured.

In order to describe the information obtained in a more meaningful way, the raw data can be plotted on graphs and summarized using descriptive statistics (e.g. averages and measures of variability). These methods represent further levels of abstraction because they summarize, but also lose, information. Graphs are powerful tools in that they can simultaneously allow both an instant impression of the data and the display of detail but, for all except the simplest datasets, graphs depict only a subset of the variables measured. Descriptive statistics allow an instant impression of the data but, again, do not convey all the detail present in the dataset.

The results of statistical analyses represent a higher level of abstraction still: the estimation of parameters within the 'real world' population based on the data collected, and the evaluation of hypotheses that aim to make generalizations about the 'real world'. Mathematical models occupy a similar level of abstraction as they incorporate estimated parameters in attempting to simulate real-world processes.

Graphical and numerical means of presenting data can therefore sit in between the 'real world' (which is typically too complex to understand immediately) and the highly abstracted results of mathematical models and statistical analyses (which are typically of little use unless interpreted in the light of the data). Such techniques are of fundamental importance to most geographical inquiry and it is essential they are used carefully and competently. As seen in Figure 19.1, mathematics is the 'vocabulary' and technique for translating between the levels of abstraction; thus numeracy and a basic grasp of mathematical principles are essential skills for geographers.

TYPES OF DATA

Most standard statistical textbooks have chapters explaining the differences between types of data. Various labels are attached to these types but the primary distinction is between categorical and continuous data. Continuous data are those that are measured on a continuous scale – i.e. any value, including fractions, between $-\infty$ and $+\infty$ (where ∞ means infinity). The most important thing about continuous data is the relational element: a meaningful progression as the numbers increase. For example, a length of 5 mm is twice a length of 2.5 mm and 6 mm less than 11 mm. This differs fundamentally from

TABLE 19.1 An example of a dataframe. This dataframe consists of 18 units (often called 'cases'; in this example they are individual people) and four variables (sex, age, occupation and income). 'Name' is not a true variable but is a list of case identifiers. Notice how the cases are rows and the variables are columns. This is the standard way of organizing data and is the format used by almost all statistical packages. Two of the variables are continuous and two categorical

Name	Sex	Age (years)	Occupation	Income (£)
David	M	36.0	Doctor	55 741
Justin	M	22.7	Social worker	19 569
Lindsay	F	46.0	Doctor	42 183
Vicki	F	60.3	Farmer	28 293
Madeleine	F	59.6	Doctor	49 658
Mark	M	63.0	Social worker	22 485
Shelley	F	18.7	Lawyer	48 627
Lizzie	F	37.1	Social worker	24 630
Jessica	F	58.6	Lawyer	45 268
Philip	M	24.5	Farmer	39 228
Charles	M	29.5	Farmer	44 165
Steve	M	20.1	Doctor	55 182
Katherine	F	19.5	Lawyer	40 677
Nicola	F	25.7	Farmer	40 607
Charlotte	F	28.3	Lawyer	61 191
Nicole	F	18.8	Doctor	50 598
Nicholas	M	31.4	Farmer	44 048
Daniel	M	34.1	Social worker	15 878

categorical data in which each value signifies membership of a particular group. An example should help to illustrate this point. Table 19.1 is a dataframe. Dataframes are essentially spreadsheets consisting of rows and columns and are the best way of organizing data in most cases – in fact, almost all statistical packages on computers require that the data are in this format. Each row represents a 'case' – i.e. a unit of sampling. In this example, each case is an individual person but it might also be a quadrant from a vegetation survey or a pebble on a shingle beach. Each entry in the row represents an attribute of that unit: an individual person's sex, age, occupation or income. The complete set of values for an attribute (one value for each case) makes up one column in the dataframe. This is called a 'variable' because the values vary between cases. Two of the variables in this dataframe are continuous (age and income), and the other two are categorical (sex and occupation).

Notice how there are no unique values for the categorical variables: members of the same group have the same value (e.g. 'lawyer'), by definition. In contrast, most or all the values for a continuous variable are unique (depending on the level of precision used for recording the data). It is important to realize that categorical variables such as sex or occupation *are* variables – there is a value assigned to each unit – they just differ in data type from continuous variables. The value assigned can be text (e.g. 'female') or numeric (e.g. a value of 1 used as code for 'female'). Statistical packages on computers tend to use the numeric form (internally at least), but some (such as SPSS) allow both numbers and text labels to be used simultaneously. Note that, if a categorical variable is recorded as numbers, these numbers are not relational as they are in a continuous variable: 'female' could be recorded as 1 and 'male' as 2, but it makes no sense to say that 'male' is one more than (or indeed twice) 'female'!

Sometimes, continuous data can be reduced to groups – e.g. fertilizer application recorded as 'low', 'medium' or 'high'. This loses information and so is not generally recommended. Further, it is important to realize that, while exact measurements can easily be converted to this form, the reverse is not true: data collected as categories cannot be converted back into exact measurements. What exact amount, in grams, corresponds to a fertilizer application recorded as 'high'? Many of the most powerful types of statistical analysis, such as regression, require data in continuous form and so, as a general rule, it is better to record exact measurements when collecting data. There are three main exceptions to this rule. The first is trivial: some variables can only be measured as categories. In Table 19.1, sex and occupation are examples of this. Secondly, there are times when the apparent precision of an exact measurement does not reflect the accuracy of that measurement. A common example is the recording of people's ages in questionnaire surveys. In many situations it has been shown that people have a greater propensity to lie about their age when asked to give a figure than when asked which age category they belong to. In such cases accuracy can actually be improved by the loss of apparent precision (the gain is in the reduced bias – see Chapter 15). The third major exception to the rule is when a better overall impression of the sampled population is likely to be gained by collecting a lot of categorical, rather than a few continuous, data. For example, recording percentage vegetation cover in quadrants is immensely time-consuming and difficult. One alternative is to record a rough estimate, done quickly by eye. While this gives an 'exact' value its accuracy is very dubious. Another alternative is to use a categorical scale, such as recording the vegetation as '<1%', '1–4%', '4–10%', '10–25%', etc. (Many such methods exist, two of the most well known being the Domin and the Braun–Blanquet scales.) This kind of method allows the rapid collection of data without producing values that are misleading in their apparent precision.

Other data types include discreet data (or integers), which are like continuous data but cannot include fractional numbers and categories that can be ordered (as in vegetation cover measured using the Domin scale), so that the difference between a '1' and a '2' is more meaningful than for purely categorical data. Some data are bounded: values cannot lie above a certain number and/or below another. A good example is percentage data (for a fuller discussion of types of data, see standard statistical textbooks, e.g. Ebdon, 1985).

It is a good idea, whenever you are thinking about data, to think first about what type of data they are. This helps in understanding of the study or dataset concerned and of the context of any manipulation of the data that is performed. Knowing what type of data you are dealing with is an essential prerequisite to deciding what statistical analysis is suitable. Always *think* about the numbers in a study, both when reading reports of other people's work and when conducting your own quantitative work. How were the values measured or arrived at? What do they mean? Are they what you expected? Do the results of an abstraction technique (e.g. a statistical analysis) make sense, or could there have been an error somewhere in the data-manipulation process? Such thought processes also help develop a feeling of being comfortable with numbers, which itself is extremely useful both for spotting errors and for spotting points of interest in the results.

PRESENTATION OF DATA

Most quantitative geographical studies involve a lot of data – too much to make much sense of simply by looking at the raw data. In almost all cases, it is a good idea to start with exploratory data analysis (EDA). EDA involves graphical depiction of the data in ways that should help to identify underlying structure or pattern. EDA informs statistical analysis of the data, and mathematical modelling of relationships between variables is also typically based on under-standing gained from EDA and/or statistical analysis. Once we have an idea what the data are actually telling us, we need ways of presenting the results in an effective and efficient way. This is usually done via two types of illustrative tool: tables and graphs. Notice that graphs (and, to a lesser extent, tables) are used both to explore data and to present results. Before examining these tools in more detail, it is worth illustrating the importance of the choice of presentation tech-nique. At its most basic, the way data are presented makes the difference between results being meaningful or not.

Look at Figure 19.2. What does it show? It is very unclear, for several reasons. First, the poor labelling makes it hard to work out

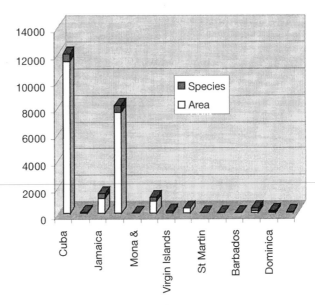

Source: Data taken from Frodin (2001)

FIGURE 19.2   An example of poorly presented data. There are many things wrong with this chart. Some of the more important ones are: (1) axis labelling – neither axis is labelled; not all the islands are named on the *x*-axis and the units are not specified for the *y*-axis; (2) it is not clear, but the point of interest here is the way in which the number of plant species increases as island area increases (a classic relationship in biogeography) for islands of the Caribbean. This type of graph is not appropriate at all for showing such a relationship; (3) the scale of the *y*-axis is such that most of the data are lost in bars too small to discern; (4) the 3-D effect adds nothing to the plot and the legend is messily placed, giving an unprofessional impression. The overall result is that we struggle to understand what message we are being given

what the point of the graph is in the first place. Secondly, many of the bars are too small to see. The most serious problem is that the chart type is completely inappropriate for the type of relationship being shown. The graph plots data on numbers of plant species and areas of islands. The relationship in question is a correlation – what we need to see is what happens to the number of plant species as island area increases. Because the numeric scale is the same for both the variables plotted and the bars are stacked, it would be very hard to discern this relationship from the chart, even if all the bars were big enough to see. What we need is a bivariate scatterplot, otherwise known as an *x–y* scatter graph. Figure 19.3 is such a plot and displays the same relationship far more clearly. The axes are properly labelled and the log–log scale allows us to see the variation in the data much more clearly. What emerges is a straight-line relationship (on these log–log axes) with a few notable exceptions – islands that deviate from the general trend. These islands are labelled (along with those at the

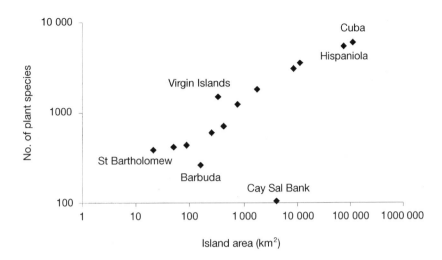

FIGURE 19.3   Better presentation of the data behind Figure 19.2. Here, the same data are shown as in Figure 19.2 but the message is much clearer. The number of species tends to increase with island area in quite a predictable manner, though there are exceptions. These exceptions, which are labelled (along with the largest and smallest islands), can be examined in more detail to see if an explanation is apparent. The type of graph is appropriate for the relationship being shown, and the log scale allows the variation within the data to be seen much more clearly than in Figure 19.2

extremes of the plot) as they are of interest. The immediate question emerging is: why are there such obvious exceptions to an otherwise strong relationship? It is beyond the scope of this chapter to explore the biogeographic reasons for this. The important point here is that both the relationship and the exceptions are immediately apparent in Figure 19.3 but are completely obscured in Figure 19.2. Some basic, guiding principles for presenting data are explored below.

TABLES

Tables are typically instruments for summarizing large amounts of information and should present information in a way which is as concise as possible, while being accurate, honest and clear. There are also circumstances (e.g. tables of raw data in the appendices of dissertations or research project reports) where very long, detailed tables are required. The guiding principle on length and presentation is to consider the needs of the reader as paramount. Think what the purpose of the table is and try to ensure that all attributes of the table help that purpose.

  Long, detailed tables such as tables of raw data are usually intended to allow the reader to access the detail, if required. This could be for the purpose of checking work, looking for anomalies in the data, or perhaps using the raw data to work on something else completely. The table should therefore be well organized and as clear and uncluttered as possible. The number of decimal places (or significant figures) displayed for values in any one column should be: (1) constant; and (2) appropriate to the accuracy and level of uncertainty of the data (see Taylor 1982: especially chap. 2) for a good explanation of uncertainty and how to report it). Rows and columns should be properly labelled and explained. When you have worked on something for some time, you become very familiar with it and it is extremely easy to take things for granted. Another reader is likely to be much less familiar with the data and so needs help in working out what everything means. Where graphs are used, all axes of graphs should be labelled (including units of measurement), and further explanation should be put into the legend of the illustration. It is better to err on the side of giving too much information than to leave the reader guessing. Generally, tables, graphs and illustrations should be understandable on their own, without reference to the text.

Summary tables are central to the presentation of data in most quantitative studies. The same principles of organization, explanation, clarity and precision apply as discussed above for detailed tables. However, in summary tables even more thought is required to select what should be included and what excluded. Also, summary tables cannot be completed until some form of data processing has been done. This might involve descriptive statistics. For example, Table 19.2 is a simple form of summary table which gives the mean and standard deviation of pollutant concentrations measured for three streams on 10 different days. Why give the standard deviations? Typically, pollutants are not a problem at very low concentrations but, above some threshold level, they may become toxic. Legislation often defines threshold levels below which pollution levels are acceptable but above which they are not. Let us assume that the important threshold pollution concentration in our example is 10 units. The mean (or average) level of pollutant is clearly important, therefore. But it is not enough simply to know the mean level. In Table 19.2, if we were only to look at the means, we would conclude that the average state of each stream is within the acceptable limit for pollution. We might be concerned, though, with stream B whose mean pollution level is only just acceptable. Streams A and C would seem to be well within acceptable limits. These would be incorrect conclusions! Look at the raw data (given in the legend of Table 19.2). While the average pollution level of stream B is quite high, the 10 measurements are remarkably consistent, and based on these data a pollution concentration of 10 on any given day seems really quite unlikely. In contrast, while the average pollution levels of streams A and C are much lower,

TABLE 19.2   An example of a simple summary table. Here, a set of sample means is given (in arbitrary units), along with their associated standard deviations and sample sizes (n), for a set of three streams, each sampled on 10 different days. A measure of variability or uncertainty should always accompany a sample statistic, as is demonstrated here. The raw data behind these statistics are as follows: stream A: 5.4, 3.4, 6.5, 8.6, 8.4, 9.5, 1.6, 5.5, 8.2, 3.8; stream B: 9.1, 8.7, 9.2, 9.2, 9.2, 9.4, 8.8, 9.1, 9.1, 9.0; stream C: 8.7, 3.1, 5.4, 3.3, 10.2, 6.0, 14.5, 4.4, 0.5, 0.0

	Stream A	Stream B	Stream C
Mean	6.1	9.1	5.6
Standard deviation	2.6	0.2	4.5
n	10	10	10

the measured values are much more variable. The highest recorded value for stream A is 9.5, but based on the data it seems quite likely that values of 10 or more might occur reasonably often. Stream C, which has the lowest mean concentration, actually has two measured values above the critical concentration of 10, suggesting that toxic concentrations of the pollutant occur very often in this stream. If we work out the actual probabilities (assuming a normal distribution of the data), it turns out that the chance of finding a concentration exceeding 10 on any given day is about 7% for stream A, about 16% for stream C, but only 0.0003% for stream B. Our conclusions obtained simply from looking at the means are therefore completely wrong.

An associated measure of either variability or uncertainty should always be given when sample statistics (i.e. parameter estimates) are presented. Whether you present a measure of variability or one of uncertainty depends on the context. In the example in Table 19.2, standard deviation measures variability, which is appropriate here because the important thing is whether or not the stream is likely to suffer toxic pollution levels. When statistical analyses are performed, the focus is usually on the likely reliability of the parameter estimate, and so the appropriate measure is one of uncertainty, such as the standard error of the mean. Further discussion of how to measure uncertainty is beyond the scope of this chapter, but it is important to note that all parameter estimates have associated measures of uncertainty. A good place to start learning about measuring and using uncertainty is Taylor (1982) and Chapter 15 of this volume.

It is very common to use summary tables to describe the results of statistical analyses, the outputs of mathematical modelling or other

TABLE 19.3 An example of a summary table reporting the results of statistical analyses. This table reports the results of a set of simple, bivariate statistical analyses. In each case, the analysis aims to account for variation in plant growth. The dataset concerned simulates an experiment investigating the effect of the mean temperature, rainfall, soil pH (all measured as continuous variables), fertilizer addition and light level (both measured as categorical variables: fertilizer added or not; low, medium and high light levels). This dataset is used for many of the graphs in the following sections, and it is given in full in the Appendix. In the table, each statistical analysis (model) is summarized with the following information: percentage of variation in the growth data accounted for by the variable (r^2); significance value for the model (p); number of data points in the analysis (n); number of degrees of freedom used by the model (df); and the parameter estimates of the model ±95% confidence interval (CI). Continuous variables were analysed using regression, which estimates the slope and the intercept of the best-fit line. Categorical variables were analysed using ANOVA, which estimates differences between means; the intercept for the fertilizer ANOVA is the mean growth of plants with fertilizer added.

Model	r^2	p	n	df	Parameter estimates Slope ± 95% CI	Intercept ± 95% CI
Temperature	0.185	0.000	60	1	5.0 ± 2.7	3.5 ± 48.0
Rainfall	0.000	0.414	60	1	—	—
pH	0.000	0.550	60	1	—	—
					Difference ± 95% CI	*Intercept ± 95% CI*
Fertilizer	0.859	0.000	60	1	−87.4 ± 9.2 g/yr	135.7 ± 6.5
Light	0.025	0.182	60	2	—	—

quite complex procedures. *A key point is that the same under-standing of statistical procedures necessary for the analyses them-selves should be used in deciding what to include in any summary table reporting their results.* When designing summary tables, always ask what information is necessary and what is not. Important information that is commonly left out includes sample sizes and/or degrees of freedom, significance values ('p values') and measures of uncertainty. When you have decided what should go in the summary table, ask what is the most efficient way of presenting it. Table 19.3 is an example of a table summarizing statistical analysis, in this case for a dataset used in the remainder of this chapter.

GRAPHICAL DISPLAY

Graphic display is an excellent way of communicating results. When done properly, it is concise, memorable, persuasive and honest. For these reasons, graphics are almost always preferable to tables in oral presentations (Ellison, 2001). However, tables are generally more useful when exact values are important and for certain types of information summary (e.g. reporting numerous statistical models simultaneously). Graphics serve three main purposes: (1) preliminary exploration of patterns within data (EDA); (2) checking the quality and assumptions of statistical and mathematical models; and (3) communicating results to the reader or audience. The first two of these require accurate, revealing depiction of the data but do not require great quality of presentation. They should be quick and easy to produce and easy to interpret. Communicating results to the target audience, on the other hand, requires high-quality graphics. Again, clarity and ease of interpretation are important, though quite a high level of complexity is often acceptable and necessary. For all these purposes, it is important to understand the principles that underlie graphical depiction of data.

Tukey (1977), Cleveland (1985) and Tufte (1983; 1990) provide comprehensive discussions of the principles of graphing. A good summary can be found in Ellison (2001: 38):

> The question or hypothesis guiding the experimental design also should guide the decision as to which graphics are appropriate for exploring or illustrating the data set. Sketching a mock graph, without data points, *before* beginning the experiment usually will clarify experimental design and alternative outcomes ... Often, the simplest graph, without frills, is the best. However, graphs do not have to be simple-minded ... and they need not be assimilated in a single glance ... Besides the aesthetic and cognitive interest they provoke, complex graphs that are information-rich can save publication costs and time in presentations.

Four principal guidelines for graphics are recognized:

1 Patterns of interest should be shown without adversely affecting the integrity of the data.
2 Graphics should be 'honest' – i.e. they should not distort, censor or exaggerate the data.
3 It should be as easy as possible for the reader to read the data (as numbers) off the graphic.
4 Figures should be efficient. That is, ink should be used only to show relevant information and not unnecessary special effects ('chartjunk') like the 3-D effect in Figure 19.2. Try to attain a high 'data-to-ink ratio'.

Tufte (1983: 51) summarizes these principles: 'Graphical excellence is that which gives the viewer the greatest number of ideas in the shortest time with the least ink in the smallest space . . . And graphical excellence requires telling the truth about the data.'

Different types of graphic are, then, suitable for different types of data and should be chosen according to the type of relationship we are trying to show. Two common examples of the importance of graphing are examining the distribution of data and exploring correlation and causation.

Examining the distribution of data

Graphs like those shown in Figure 19.4 depict the distribution of values of a single variable (usually a continuous one) within a sample. They are typically used for EDA and model checking but there are times when such graphs need to be presented formally. (The simulated dataset used for Figures 19.4–19.11 can be found in the Appendix to this chapter.)

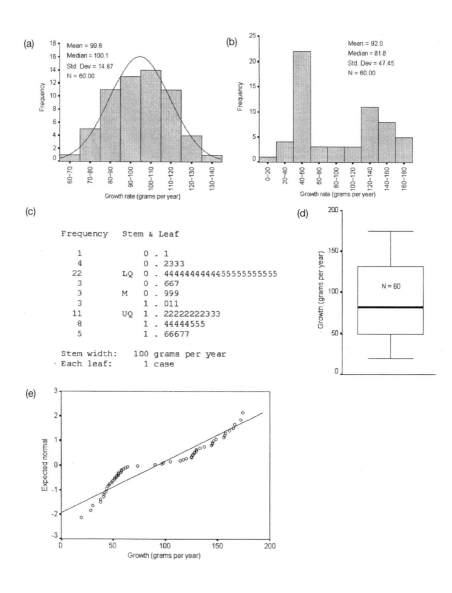

THE HISTOGRAM

The most common graph type used to examine the distribution of data is the histogram (Figure 19.4a and Figure 19.4b). Histograms are often confused with bar charts (see below) because, to the beginner at least, they look similar. However, they are used to convey fundamentally different information. A histogram consists of categories of data along the x-axis (i.e. the horizontal axis) and the number of data points in each category on the y-axis. The categories are often referred to as 'bins', using the analogy of sorting the data into bins, each of which is for a different range of values. The number of data points in each category is known as the 'frequency'. In Figure 19.4a, the data for intelligence quotient (IQ) are distributed more or less normally. More formally, we say that the difference between the distribution of the data and the theoretical normal distribution is not significant. To allow ease of comparison, the theoretical normal curve is superimposed on to the histogram. Theoretical distributions can be represented as curves like this because an exact probability can be calculated for each possible value of the variable. Such curves are called probability density functions and can be thought of as histograms with infinitesimally small bin categories. In fact, the term 'density plot' is a more general one which includes both histograms and curves representing theoretical distributions.

A key assumption of most parametric statistical techniques is that the distribution of sample data conforms to one of the known theoretical probability distributions. Because the exact probability function of each of these theoretical distributions is known, all sorts of probability statements can be made about sample data – *as long as the assumption holds*. Many of the more familiar statistical techniques

FIGURE 19.4   Examining the distribution of data: (a) a histogram showing an approximately normal distribution – a normal curve has been added to ease comparisons between the distribution of the data and the theoretical normal distribution. Histgrams suffer several disadvantages. In particular, they hide the raw data, and the division into categories ('bins') is arbitrary but can affect interpretation (Ellison, 2001). Hiding the raw data means that data processing cannot be done from a histogram – such as calculation of summary statistics (though the histogram can be annotated to include them, as here); (b) a histogram showing a distinctly non-normal distribution – in this case very bimodal; (c) a stem-and-leaf plot of the data in (b) (medium, lower and upper quartiles indicated), which is variant on the histogram. The main difference is that the raw data are included, not hidden. However, again the division into categories is arbitrary; (d) a box-and-whisker plot, of the data in (b). This is an efficient and informative way of displaying such data but can tend to hide certain features of the data, particularly bimodality – as here. See text for further explanation; (e) a probability plot of the data in (b) (in this case the plot type is called a Q–Q plot). Probability plots display the actual data against what would be expected from a given distribution (the diagonal line). See text for further explanation

(such as regression, ANOVA and t-tests) assume that the data conform to the normal distribution. Hence the value of tools such as the graphs shown in Figure 19.4 and also numerical statistical tests for normality (Sokal and Rohlf, 1995) which we can use to establish the validity of this key assumption. But herein lies a common misunderstanding.

Exactly what should we test for normality? More formally, the assumption of the relevant statistical techniques is that the residuals are normally distributed. Before expanding on this point, some technical terms need explaining. The response variable in a statistical analysis is the one under investigation – i.e. the one whose variation we are trying to explain. It is also known as the dependent variable, y variable or ordinate but the term 'response variable' is preferred. The variables we use to try to account for variation in it are called 'explanatory variables' – again, this is the preferred one of several names, others being independent variable, x variable and abscissa. The 'residuals', otherwise known as the 'errors', are the differences between the actual, observed values of the response variable and the modelled values; there is one such difference for each data point. Normal distribution of the raw data for a response variable will often mean normal distribution of residuals from a statistical model and, for this reason, many people examine the distribution of response variables prior to statistical analysis. However, results from such an exercise should be treated with caution as residuals can be normally distributed even if the response variable is not (and vice versa).

In Figure 19.4b, the distribution of the data is clearly not normal. It is bimodal (i.e. two modes or concentrations of values along the x-axis scale). When performing statistical modelling, many departures from normality of errors can be cured by transforming variables, but this may not be possible with bimodal errors. Fortunately, however, bimodality of distribution of the raw data of a response variable is usually due to underlying structure in the data and, if that can be accounted for, the bimodality disappears. For example, the distribution of people's heights is typically bimodal because men tend to be taller than women; but when men and women are analysed separately, the heights tend to be normally distributed. This simple example neatly illustrates the folly of examining only the distribution of the raw data. As we will see below, the data in Figures 19.4b–e represent a similar scenario.

STEM-AND-LEAF PLOTS

Figure 19.4c–e are different ways of depicting the distribution of the same data as Figure 19.4b. Figure 19.4c is a 'stem-and-leaf' plot. This is similar to a histogram (on its side) but has the advantage that the raw data are shown. Stem-and-leaf plots are very useful once you have got used to reading them. As with a histogram, the length of the 'bars' represents the number of data in the category. Each of these data points is not only listed but also evaluated. The 'stems' are the first column of numbers and represent the first figure(s) of the data in the row. The 'leaves' are to the right of the stems and comprise the next figure of the data value. In Figure 19.4c, the stems are in hundreds and the leaves are tens. Thus, the lowest value in the dataset is somewhere between 10 and 20 (0 hundreds and 1 ten). Its actual value is 19.2. The highest value reads off as somewhere between 170 and 180; its actual value is 174.2. As with a histogram, labels denoting the categories containing the mean, median and/or quartiles can be added to the plot to provide further information, as in Figure 19.4c.

BOX-AND-WHISKER PLOTS

Figure 19.4d is a 'box-and-whisker' plot (or just 'boxplot'). The box in the middle spans the range between the 25th and 75th percentiles (respectively, the lower and upper quartiles or 'hinges') – in other words, the box shows the interquartile range. The line in the middle of that box shows the value of the median (the 50th percentile). The whiskers extend a further 1.5 times the interquartile range, away from the median. Any points lying outside the whiskers (beyond the 'inner fence') are outliers: points that are considerably different from the rest of the data. These outliers are subdivided into two types. The less-extreme outliers lie no further from the end of the whisker than 1.5 times the interquartile range (the 'outer fence'). The more-extreme outliers are any points that are further still from the box. The two types of outlier are usually distinguished by different plotting symbols. Boxplots are extremely useful, information-rich graphics and are excellent ways of summarizing entire variables or categories within variables. However, as ways of showing distribution of data, they suffer from the disadvantages of hiding data and tending to obscure bimodality (as demonstrated by Figure 19.4d).

PROBABILITY PLOTS

Figure 19.4e is a probability plot. This is a very common way of examining how closely the distribution of a set of data conforms to a theoretical distribution. It is extremely useful for examining how well a statistical model conforms to the assumption of normality of errors. Probability plots are constructed so that the theoretical distribution plots as a straight, diagonal line, and the data are plotted as points. If the data conformed perfectly to the theoretical distribution, all the points would lie on the line. In practice, of course, this does not happen. The key thing to look out for is systematic departure from the line, especially within the range in which most of the data lie (not so much at the two ends). Some random scatter either side of the line is acceptable, but when the data clearly form a non-linear pattern (such as an S-shape, as here), it is a sure sign the distribution is not normal – and that there is something wrong with the model if the plot is being used to check residuals for normality. Often, the pattern is caused by the influence of something not accounted for by the model, such as an influential but unmeasured variable (as here, because the 'model' being tested is the null model).

EXAMINING HOW ONE VARIABLE AFFECTS ANOTHER

Bivariate analysis

Most quantitative studies in geography are concerned with trying to infer causation. The simplest case is the bivariate one: analysing the effect of one variable on one other. The pattern being examined is within the response variable, and the possible cause of that pattern is measured via the explanatory variable. If both these variables are continuous, by far the most useful type of graphic is the scatterplot (Figure 19.5). The explanatory variable goes on the x-axis and the response variable on the y-axis. Modelled relationships can be shown

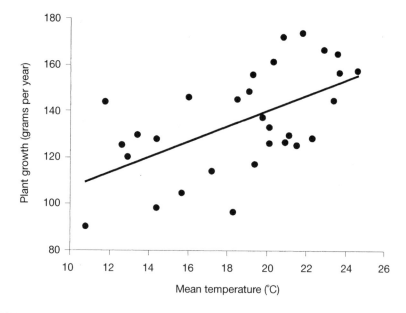

FIGURE 19.5 Scatterplot. This is one of the simplest and most useful of bivariate plots and one of the most commonly used. Values of one variable are spread along the x-axis and those of another are spread along the y-axis. Both variables are usually continuous. The convention is that the explanatory variable is plotted as the x-axis and the response variable as the y-axis (in cases where any sort of causality is inferred). A best-fit line from a simple linear regression is shown. This plot only shows data for plants to which fertilizer was added

on scatterplots as lines, one of the simplest examples being a linear best-fit line. Scatterplots can also be used where no causation is being inferred – the focus instead being simply on whether there is any correlation between the two variables.

When one variable is continuous and the other categorical, scatterplots are not particularly helpful as they obscure many of the data (Figure 19.6a). If both variables are categorical, there is usually little point in attempting to plot them against each other – devices such as tables of frequencies of each category combination tend to be more useful. Figure 19.6 shows perhaps the most common type of scenario: where the response variable is continuous and the explanatory variable categorical. Categorical explanatory variables are often called 'factors'.

There are many different ways of plotting factors. Scatterplots can be adapted to cope with factors by adding random variation along the x-axis (known as 'jitter'). This has been done in Figure 19.6b, in which

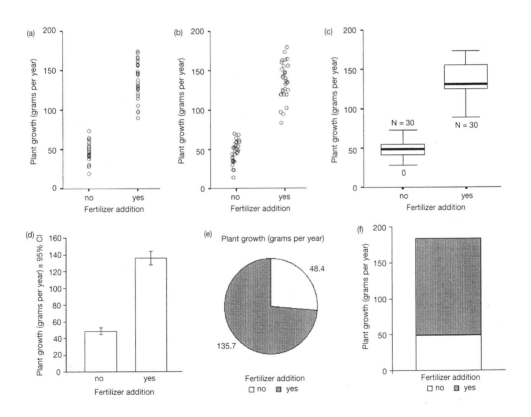

FIGURE 19.6 Ways of plotting factors. A factor is a categorical explanatory variable. All the graphs show the same relationship, which is that between fertilizer addition and plant growth. See text for further explanation: (a) scatterplot; (b) scatterplot with jitter; (c) boxplot – see above for explanation; (d) bar chart with error bars (the error bars represent the 95% confidence interval); (e) pie chart; (f) stacked bar chart

the data are still clearly categorized but the points are much less obscured than in Figure 19.6a. Another way of displaying factors, which tends to be more satisfactory, is the boxplot. In Figure 19.6c, a separate boxplot is shown for each category within the factor. Bar charts can be used to display the means, with error bars added to show the reliability of the estimates of the means; in Figure 19.6d, the 95% confidence intervals of the means are shown. Strictly speaking, the bars on the bar chart (i.e. the columns) are unnecessary but it is common practice to include them. Pie charts (Figure 19.6e) and stacked bar charts (Figure 19.6f) are more often used for percentage data but are not recommended even for that. It is not easy to read the data values from them and it is hard to display any measures of uncertainty or variability on them. Indeed, Ellison (2001: 57) goes as far as to say: 'I can think of no cases in which a pie chart should be used.'

Multivariate analyses

In most cases, more than just a single influence will cause a pattern. Analyses can be multivariate in the sense that they try to account for multiple influences on a given pattern: several explanatory variables are used to try to explain patterning in a response variable. An example could be the runoff rates of different streams in an area. The amount of rainfall, timing of rainfall, land-use types in the catchment and bedrock characteristics are just some of the things that are likely to affect runoff rates, and it would be unrealistic to expect any one of these variables to explain all patterning in them. Our example of the growth of plants is also a case in point. Sometimes the pattern itself can be multivariate: too complex to measure as one single response variable. Instead, several attributes can be measured to try to gain an overall picture. For example, we could measure a whole suite of attributes that could be construed as being related to the concept of quality of life, such as average income, equality of income, amount of free time, prevalence of disease and access to services.

Numerous statistical techniques have been developed to allow analysis of multivariate patterns, including dimensionality-reducing procedures such as principal components analysis. These are beyond the scope of this chapter but the reader is referred to statistical texts such as Scheiner and Gurevitch (2001), Crawley (2002) or the manuals of most standard statistical packages. Multiple influences on a single response variable can be analysed using extensions of the simple, bivariate techniques of ANOVA and regression. Again, it is not the purpose of this chapter to discuss these statistical methods but some of the issues related to such analyses are relevant to this section. Two issues are especially important: (1) accounting for other influences when illustrating any given effect; and (2) how to deal with inter-actions between variables.

THE SCATTERPLOT MATRIX

When a pattern results from more than one cause, a simple bivariate plot of the response versus one of the explanatory variables will not give the true picture of the effect we are trying to show. We must try to account for variation explained by other influences before we can show the 'true' modelled relationship. An extremely useful tool for exploring higher-dimensional data is the scatterplot matrix (Figure 19.7). This comprises a bivariate scatterplot of each possible combination of two variables from the list of those under consideration. Usually, each of these plots is displayed twice – one way above the diagonal and the corresponding plot with the axes transposed below the diagonal.

PARTIAL PLOTS

Scatterplot matrices are very useful for exploring data and informing statistical analysis and mathematical modelling but they do not achieve the goal of accounting for other causes when displaying the modelled effects of one influence on the response variable. To achieve

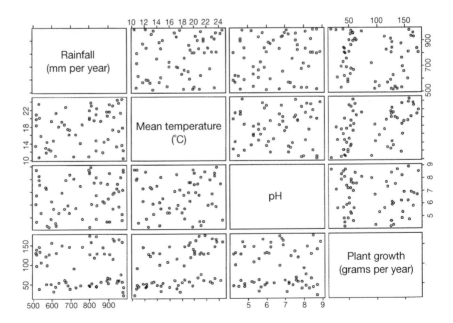

FIGURE 19.7 Scatterplot matrix. As its name suggests, this type of plot is a matrix of bivariate scatterplots. The variable names are listed in the diagonal and apply to the y-axes of the plots in the row in which the name is located, and the x-axes of the plots in the column containing the variable name

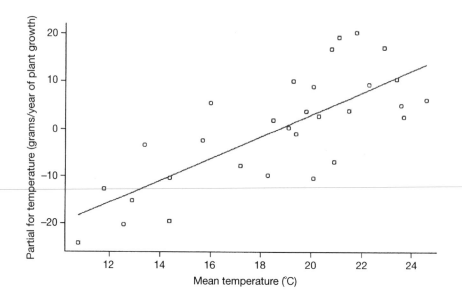

FIGURE 19.8 Partial plot. A partial plot shows the relationship between the two variables after accounting for the modelled effects of other explanatory variables on the response variable. The response variable here, as before, is plant growth. This is a partial plot from a model in which the explanatory variables are rainfall, temperature and light, and in which only the plants given fertilizer are considered. The modelled effects of rainfall and light have been corrected for, and the *y*-axis therefore displays the change in plant growth directly related to temperature. This plot is directly comparable to Figure 19.5 – note how the best-fit line now describes the data better than in Figure 19.5

this, other graphical methods are commonly used. The first is the partial plot (Figure 19.8) in which the response variable is recalculated, 'correcting for' the modelled effects of other significant explanatory variables. This is very useful for interpreting the influences of particular variables in relatively complex statistical analyses and, as such, one of its main uses is in model development. It can also be an effective way of presenting the meaning of a model to a viewer, but the scale of the response variable needs careful explanation.

3-D PLOTS AND THE SEPARATION OF CATEGORIES

A 3-D plot can also be used to good effect but only allows the plotting of one extra variable in a multiple-cause situation, so its use is restricted. Also, it is typically hard to read off the data from such plots because of the difficulty of projecting a 3-D plot on to the 2-D medium of a page or a screen. A further technique to explore multiple causation is the separation of categories in graphs. This can be achieved by plotting separate graphs for the different categories of a factor (e.g. separate graphs for men and women), or different symbols on the same graph with separate best-fit lines if appropriate (Figure 19.9). Using

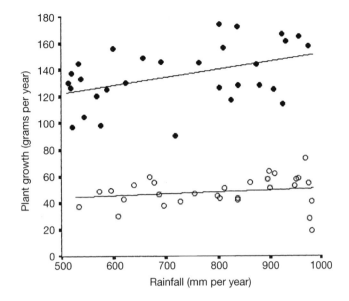

FIGURE 19.9 Using different symbols to account for the influence of a factor. Here, the relationship between rainfall and plant growth is shown separately for plants to which fertilizer has (closed circles) and has not (open circles) been given. If all the points were plotted with the same symbols and separate best-fit lines not shown, it would appear that there is no relationship between the rainfall and plant growth

different symbols on the same graph is often better for comparing between the groups.

The use of different plotting symbols on the same graph for different categories of a factor is also a good way of illustrating an interaction between a continuous and a categorical explanatory variable – a form of statistical interaction. Statistical interactions are significant when the effect of an explanatory variable on the response variable depends on the level of another explanatory variable. Figure 19.9 shows an example of this, in which the relationship between rainfall and plant growth depends on whether or not fertilizer has been added. Where fertiliser has not been added, there is no relationship between rainfall and plant growth but there is a positive relationship for plants given fertilizer. This interaction is statistically significant (p = 0.002). These results would be consistent with a situation in which nutrient availability, not water, is the primary limiting factor for plant growth.

CLUSTERED GRAPHS AND CONDITIONING PLOTS

Where a statistical interaction involves two or more categorical explanatory variables, clustered graphs are often the best way of displaying the interaction. Figure 19.10 is an example of a clustered

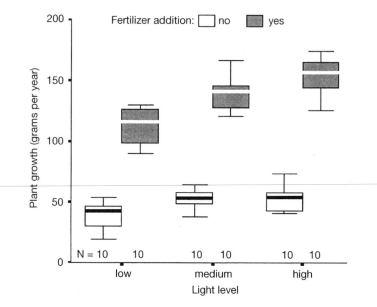

FIGURE 19.10   Clustered boxplot. This shows an interaction between the light and fertilizer treatments in their effect on plant growth. In other words, the amount of increase in plant growth with increasing light levels depends on whether or not fertilizer has been added. The corollary is that the degree of difference in plant growth between the two fertilizer treatments depends on the light level. The interaction is statistically significant ($p = 0.0004$)

boxplot, which shows an interaction between the effects of light and fertilizer on plant growth.

Finally, some statistical interactions involve two or more continuous explanatory variables. By extension of the idea of the best-fit line on a 2-D scatterplot, a two-way interaction of this sort can be displayed as a best-fit surface within a 3-D scatterplot. Again, this can suffer problems of difficulty in reading off data and only allowing for one extra dimension. In addition, such a plot tends to imply, usually unrealistically, that all parts of the surface represent the valid range of the model; in fact, some parts of such a graph are often data deficient. Instead, conditioning plots (or 'coplots') are often better ways of showing these interactions. Figure 19.11 is an example of a coplot. The range of values for one of the explanatory variables is split into segments (rather like the bins of a histogram) and the data for each segment are plotted as a separate 2-D scatterplot. This shows the way that the relationship between the x and y variables changes according to the other explanatory variable. In this case, there is no significant interaction; if there were, the fit lines would have different slopes from each other, indicating that the relationship between temperature and plant growth depends on the amount of rainfall. Advantages

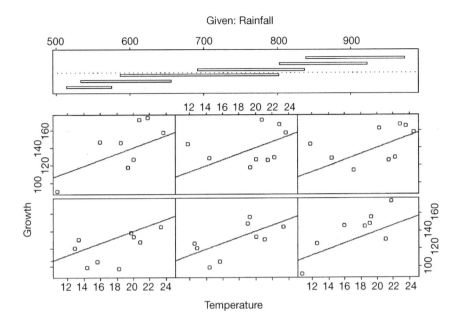

FIGURE 19.11   Conditioning plot (or coplot). This shows the relationship between temperature and plant growth for the plants with fertilizer added. Separate graphs are shown for this relationship, for different ranges of rainfall – these ranges are indicated by the top panel, which shows that the bottom left graph is for a rainfall range of about 515–570 mm/yr; the bottom middle graph is for the approximate range 530–640 mm/yr and so on. The dotted line in the top panel indicates the move from the graphs at the bottom to those at the top. There is no significant interaction between the effects of temperature and rainfall on plant growth for these data so the fit lines drawn are for the overall regression – i.e. they are the same in all the graphs

include the ability to add more explanatory variables, thereby being able to show higher-order interactions (e.g. three-way). Disadvantages include the fact that such plots focus on one of the explanatory variables rather than an equal depiction of all simultaneously.

Many other types of plot exist – triangular and time-series plots, to name but two. The interested reader should refer to more specialist texts such as Tufte (1983) and Waltham (1994) to learn more. Note also that some of the graphical techniques discussed above are not mutually exclusive. For instance, in trying to illustrate a very complex statistical model, we could draw separate graphs (to deal with one set of groups), each of which is a partial plot with different plotting symbols. All these devices are aimed at illustrating most effectively the significant effects found in the data analysis. These effects are our best guess about the cause(s) of the pattern in which we are interested. In other words, when presenting results in graphs and tables, we try to adhere to the maxim: form follows function (Ellison, 2001).

EXPLANATION AND INTERPRETATION

Figure 19.1 illustrates how our attempts to understand real world complexity by quantitative study involve various levels of abstraction, all linked together by the 'language' of mathematics. Figure 19.1 also suggests that graphs and tables lie right at the heart of quantitative treatment of data. When trying to explain patterns, present and interpret results, it makes sense to use these illustrative tools to maximum effect. Graphs, in particular, can allow us to combine raw data, descriptive statistics and the output of statistical or mathematical models within a single illustration. (Note that the output of statistical modelling has been included in some of the graphs above in the legends – e.g. Figure 19.10 where the interaction was noted as statistically significant, with a p value of 0.0004.)

Each quantitative study is unique and so a considerable amount of thought is required to interpret the results and their implications and to consider the generality of the findings. So far, our emphasis has been on the use of data in analysis and presentation of information. The flip side of this is our reaction to illustrations presented by others. An understanding of data-handling processes (including the principles of good illustrative practice) is critical if we are to be able to interpret the meaning and the context of work presented to us. We need to be able quickly to judge how well quantitative work has been done, what the value and implications of the results are and, indeed, whether we have been given enough information to make such judgements. In doing this we should consider the extent to which form follows function. Recall that illustrations are typically used to imply that the pattern we see (the form) results from the causes depicted (the function).

In Figure 19.12a, the implication is that greater area of islands tends to cause greater plant species richness. The best-fit line goes much further: it suggests that the increase of plant species richness with area is highly predictable: it follows a linear relationship on a log–log scale. What does this actually mean? Looking at the graph suggests that a tenfold change in island area is associated with much less than a tenfold change in species richness. The equation of the line quantifies this change in plant species number as 2.35-fold (which is consistent with the rule of thumb used by many conservationists that a 90% loss of habitat leads to a 50% loss of species). Figure 19.12b shows the same data on linear axes. The solid line is the same best-fit line as in Figure 19.12a and represents a power relationship. The dashed line shows an alternative model describing the data: a logarithmic function. There has been debate for decades in the ecological literature about which of these two models better represents the species–area relationship.

Further exploration of the two curves shown in Figure 19.12b gives greater insight into the relative merits of the two models on which they are based. In statistical terms, the logarithmic model provides a slightly better fit with the data, accounting for 91% of the variation in plant species numbers compared with 90% for the power model (though we must bear in mind that the sample size is small). But how do the two models compare on theoretical grounds? It is clear from the graph that the two models predict vastly different numbers of species

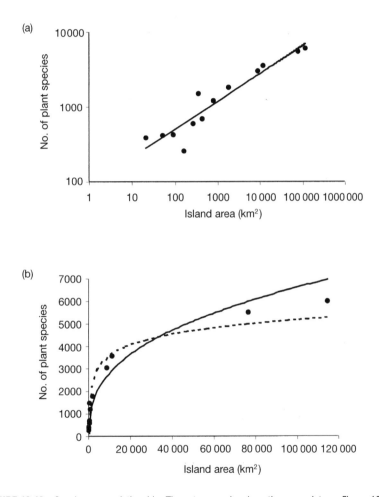

FIGURE 19.12 Species–area relationship. These two graphs show the same data as Figure 19.3 (plant species numbers plotted against island area for Caribbean islands), except that Cay Sal Bank has been removed as an outlier. Plot (a) is on log–log axes while (b) is on untransformed axes. The solid line in both graphs shows the best-fit power model. The equation for this is $y = 92.434x^{0.371}$, and the model accounts for 90% of the variation in the data ($r^2 = 0.896$). The dashed line in (b) is the best-fit logarithmic model: $y = 693.3\ln(x) - 2810.5$. This model accounts for 91% of the variation in the data ($r^2 = 0.913$)

(in absolute terms) for large islands. It is also clear that, if we extrapolate beyond the range of the data to even larger islands, this difference will increase rapidly. Such differences could be very important if we wish to use these data to predict for other islands or if we wish to predict the effect of changes in habitat area on species numbers in nature reserves. What is less clear from Figure 19.12b is what happens if we extrapolate beyond the range of the data in the other direction – to smaller islands. The equation for the logarithmic model yields negative species numbers for islands smaller than 57 km^2; for a 1 ha island, the predicted number of plant species is –6003. That is clearly non-sensical and suggests that extrapolation in this direction is highly dubious in the case of the logarithmic model. The power model, in contrast, predicts 17 plant species on an island of 1 ha, and this model reaches 0 species at an area of 0 km^2. Such predictions are at least plausible! On the basis solely of these results, we might conclude that the power model is more theoretically sound and more likely to be useful as the basis for a general model of the species–area relationship than the logarithmic model.

This example raises a number of issues. For a start, it brings to the fore the differences among interpolation, extrapolation and prediction (see also Chapter 22). Relationships that are qualitatively very different can look very similar in certain parts of their ranges. Figure 19.13 shows fundamentally different relationships, all of which look very similar in the highlighted range.

The species–area example also raises the issue of what to do with outliers. Notice the difference in data between Figure 19.3 and Figure 19.12: Cay Sal Bank has been excluded from the data in Figure 19.12. It was clearly an outlier in Figure 19.3, but this alone is not sufficient reason to exclude it when we are interested in trying to explain the real world. In fact, there are good theoretical reasons to omit the Cay Sal Bank data. This set of islands in the Bahamas is little more than coral reef and so is both qualitatively different from the other data points and has a high degree of uncertainty attached to the measurement of its area. The important point here is that the exclusion of this data point makes a big difference to the model's fit to the data.

Perhaps most importantly, the species–area example also demonstrates the important difference between a statistical fit (e.g. a best-fit line) and an underlying causal relationship. The fit of both models to the data is strong, but this does not prove there is a causal relationship. Is the geographic pattern symptomatic of one or more geographic processes? In the above analysis no mechanisms linking changes in area to changes in species numbers have been investigated, so the answer to this question is not clear. What is clear, though, is that we have to be very careful in deciding what we can, and cannot, conclude from any given study.

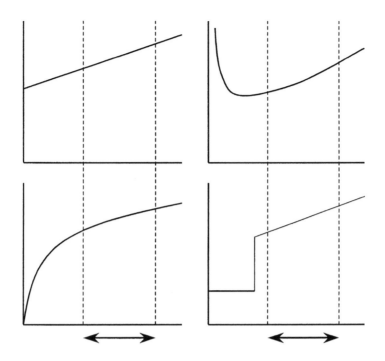

FIGURE 19.13 Qualitatively different relationships that look similar over a given data range. The four relationships shown are fundamentally different: one is linear (top left), two are curvilinear but of very different form (top right and bottom left) and the other involves a threshold (bottom right). However, all four relationships look very similar indeed over the range between the dotted lines. If the data were only collected over this range, it would be almost impossible to distinguish between these relationships (and many others not shown). In practice in this situation, we would use Occam's Razor and accept the linear relationship because this is the simplest explanation, but if the true underlying relationship were not linear we would be wrong. This highlights how dangerous it can be to extrapolate beyond the range of the data

CONCLUSION

The way that data are presented can seriously affect our inferences and conclusions. It is common knowledge that 'statistics' can be very misleading (the word 'statistics' here is usually used to mean facts and figures rather than parameter estimates). The same is true of models, graphs and tables. Whenever we consider a model, we should ask searching questions about what its purpose is – is it for prediction or merely description, for instance – and about exactly what it tells us. We should also think about what alternatives there might be. Would a different model work better – perhaps a non-linear rather than a linear relationship, or vice versa? In academic study, misleading representation of data often occurs unintentionally through incomplete or incorrect analysis of data. The risk of this can be minimized by keeping in mind the following guidelines:

- Plotting the data in a variety of ways, in conjunction with calculation of summary statistics, forms the basis of exploratory data analysis (EDA). This process is often invaluable in informing the statistical analysis.
- In turn, the statistical analysis should typically include the use of graphs to help examine the nature of the model fit and to check the validity of the underlying assumptions.
- Finally, graphical display of the results is often by far the best way of communicating the information in an elegant, concise and accurate way. This graphical display should typically incorporate elements of the statistical analysis, such as error bars (or other measures of uncertainty/variability) and indications of statistical significance on the graphs.

Whatever we do, we need to be aware of the fact that all our data-presentation tools are constructs, and all our models are wrong to at least some extent. The important question is: how far from the truth are they?

Summary

- The ability to work with data is essential, both as a prerequisite to statistical analysis and as a medium for presenting and interpreting numerical results.
- When presented with a large mass of data, a fundamental starting point is the realization that there are different levels of abstraction from the 'real world': data collection, graphical and numerical representation of data, and generalization from statistical or mathematical modelling.
- Different types of data are best presented using different numerical and graphical methods. Data (re)presentation can, therefore, affect understanding, judgement and inference and is a vital part of the examination of geographical pattern and process, and of the application of techniques such as interpolation, extrapolation and prediction.

Further reading

- John Tukey was instrumental in developing and promoting useful, accessible ways of handling data. His book (Tukey, 1977), although a bit idiosyncratic, is still well worth a read. The book, *Modern Methods of Data Analysis*, edited by Fox and Long (1990), covers the topics I have discussed in much more detail. Tufte (1983) is a clearly written book containing much wisdom about the use and presentation of data. It encourages careful, honest and imaginative display of information and is a very interesting and informative read. A more up-to-date treatment of graphics is provided by Ellison (2001). This short chapter is much more focused on research (with particular emphasis on experimental work in ecology) than Tufte (1983). Although I do not agree with everything written in Ellison (2001), I strongly recommend it along with much of the rest of the book containing it (Scheiner and Gurevitch, 2001) – even for geographers with little interest in ecology.

- Taylor (1982) provides a good introduction to error (uncertainty): how to estimate it, cope with it and report it. Even though it is written for physicists, it is a valuable read for undergraduate geographers. Taylor also has a useful section on dealing with outliers (chapter 6: 'Rejection of data').

- Although an old book, Thornes and Brunsden (1977) contains much that is still important, including a good, short section on interpreting graphs with respect to the way that physical systems operate (in chapter 7). Sayer (1992) provides a useful discussion on abstraction – from a social science viewpoint.

- Finally, it is very important to get at least some grasp of statistical methods. There are innumerable books dealing with this though not many in specifically focusing on geography. Crawley (2002) provides a comprehensive account of statistical techniques, starting from first principles. This is a general book, not just for geographers, but in my opinion it is hard to find a better statistics text than this. Rogerson (2001) has some useful material, especially on spatial statistics.

Note: Full details of the above can be found in the reference list below.

References

Cleveland, W.S. (1985) *The Elements of Graphing Data*. Monterey, CA: Wadswork Advanced Books & Software.

Crawley, M.J. (2002) *Statistical Computing: An Introduction to Data Analysis Using S-Plus*. Chichester: Wiley.

Ellison, A.M. (2001) 'Exploratory data analysis and graphic display', in S.M. Scheiner and J. Gurevitch (eds) *Design and Analysis of Ecological Experiments* (2nd edn). Oxford: Oxford University Press, pp. 37–62.

Fox, J. and Long, J.S. (eds) (1990) *Modern Methods of Data Analysis*. London: Sage.

Frodin, D.G. (2001) *Guide to Standard Floras of the World: An Annotated, Geographically Arranged Systematic Bibliography of the Principal Floras, Enumerations, Checklists, and Chonological Atlases of Different Areas* (2nd edn). Cambridge: Cambridge University Press.

Kopf, E.W. (1916) 'Florence Nightingale as a statistician', *Journal of the American Statistical Association*, 15: 388–404.

Playfair, W. (1786) *The Commercial and Political Atlas*. London: Corry.

Playfair, W. (1801) *Statistical Breviary*. London: Wallis.

Porter, T.M. (1986) *The Rise of Statistical Thinking, 1820–1900*. Princeton, NJ: Princeton University Press.

Rogerson, P.A. (2001) *Statistical Methods for Geography*. London: Sage.

Sayer, A. (1992) *Method in Social Science: A Realist Approach* (2nd edn). London: Routledge.

Scheiner, S.M. and Gurevitch, J. (eds) (2001) *Design and Analysis of Ecological Experiments* (2nd edn). Oxford: Oxford University Press.

Sokal, R.R and Rohlf, F.J. (1995) *Biometry.* New York: W.H. Freeman.

Spence, I. and Lewandowsky, S. (1990) 'Graphical perception', in J. Fox and J.S. Long (eds) *Modern Methods of Data Analysis.* London: Sage, pp. 13–57.

Taylor, J.R. (1982) *An Introduction to Error Analysis.* Sausalito, CA: University Science Books.

Thornes, J.B. and Brunsden, D. (1977) *Geomorphology and Time.* London: Methuen.

Tufte, E.R. (1983) *The Visual Display of Quantitative Information.* Cheshire, CT: Graphics Press.

Tufte, E.R. (1990) *Envisioning Information.* Cheshire, CT: Graphics Press.

Tukey, J.W. (1977) *Exploratory Data Analysis.* Reading, MA: Addison-Wesley.

Waltham, D. (1994) *Mathematics: A Simple Tool for Geologists.* London: Chapman & Hall.

APPENDIX

This table gives the simulated dataset that is used for much of this chapter. Units are mm per year for rainfall, °C for mean temperature and grams per year for plant growth. Fertilizer is recorded as applied to the plants or not, and light levels are recorded as low (shade), medium (semi-shade) or high (unshaded).

Rainfall	Temperature	pH	Fertilizer	Light	Plant growth
696	17.2	8.3	No	Low	37.9
686	17.4	5.9	No	Low	46.2
803	19.7	4.2	No	Low	43.5
609	14.6	5.6	No	Low	30.3
728	16.0	5.7	No	Low	41.1
980	10.4	8.7	No	Low	19.2
976	17.7	7.1	No	Low	28.3
755	14.0	7.4	No	Low	46.7
811	22.9	6.0	No	Low	51.4
638	18.0	5.3	No	Low	53.5
898	20.6	7.6	No	Medium	64.0
955	16.9	7.0	No	Medium	58.2
837	14.6	8.1	No	Medium	43.5
595	12.3	4.8	No	Medium	49.3
678	22.6	7.9	No	Medium	55.3
899	15.0	6.4	No	Medium	51.3
950	24.3	7.6	No	Medium	57.8
533	10.7	7.9	No	Medium	37.6
668	18.5	5.6	No	Medium	59.6

573	11.8	7.2	No	Medium	48.8
798	10.8	8.7	No	High	44.9
947	13.2	7.2	No	High	53.1
838	12.3	4.8	No	High	42.1
968	21.1	8.1	No	High	73.3
974	13.5	5.2	No	High	55.0
619	11.0	4.5	No	High	43.0
908	23.3	8.4	No	High	62.3
981	12.2	8.5	No	High	40.7
862	19.2	7.7	No	High	55.1
894	21.5	4.6	No	High	58.0
575	14.4	4.3	Yes	Low	98.3
543	15.7	5.3	Yes	Low	105.0
839	22.3	5.9	Yes	Low	128.3
906	21.5	4.6	Yes	Low	125.5
924	17.2	5.6	Yes	Low	114.4
718	10.8	8.0	Yes	Low	90.2
521	18.3	6.7	Yes	Low	96.8
624	21.1	4.3	Yes	Low	129.9
826	19.4	7.7	Yes	Low	117.4
803	20.1	8.4	Yes	Low	126.5
533	23.4	8.2	Yes	Medium	144.7
880	14.4	6.1	Yes	Medium	128.2
923	22.9	6.8	Yes	Medium	166.5
692	16.0	5.3	Yes	Medium	145.9
975	24.6	4.7	Yes	Medium	157.5
764	18.5	4.8	Yes	Medium	145.1
520	19.8	8.7	Yes	Medium	137.5
518	20.9	4.6	Yes	Medium	126.7
568	12.9	7.3	Yes	Medium	120.4
514	13.4	6.0	Yes	Medium	130.0
957	23.6	5.8	Yes	High	165.0
837	20.8	6.7	Yes	High	172.1
656	19.1	6.9	Yes	High	148.5
810	23.7	8.9	Yes	High	156.8
802	21.8	8.5	Yes	High	174.2
873	11.8	6.2	Yes	High	144.0
537	20.1	7.5	Yes	High	133.2
600	19.3	6.2	Yes	High	156.1
930	20.3	6.9	Yes	High	161.6
587	12.6	4.2	Yes	High	125.4

20   Cartography and Graphicacy

Chris Perkins

Synopsis

Maps are very powerful tools for representing our ideas and knowledge about places. As such, the skills of producing and reading maps are important within the discipline of geography. Cartography deals with the development, production, dissemination and study of maps in a wide variety of forms, whereas graphicacy is the skills of reading and constructing graphic modes of communication, such as maps, diagrams and pictures. This chapter introduces the differing roles played by mapping and the changing social significance of the map, discusses the availability of mapped information, explains how maps work as representations and offers advice about how to design maps.

The chapter is organized into the following sections:

- Introduction
- The contested terrain of mapping
- Finding the map
- How maps work
- Practical suggestions for design
- Conclusion.

INTRODUCTION

The map is a powerful medium for the representation of ideas and the communication of knowledge about places. It has been used by geographers to store spatial information, to analyse and generate ideas and to present results in a visual form. Maps are not just artifacts; mapping is a *process* reflecting a way of thinking. The quality of a printed map or map display on a screen is a reflection of the 'graphicacy' of its authors and readers. Some thirty-five years ago or so

Balchin and Coleman (1966) argued that graphicacy should be placed alongside numeracy, literacy and articulacy as educational pre-requisites – this chapter echoes that call and argues that technological, social and intellectual changes have made graphicacy increasingly important. This chapter explains the changing social significance of the map, exploring roles maps might play and discussing how these may be relevant for the geographer. It introduces sources of mapping by discussing the availability of mapped information and introducing practical ways of finding out what maps exist.

With desktop mapping packages we can increasingly create our own maps, but if we are to realize the creative power of the medium it is important to understand how maps work. Half a century of carto-graphic research has lead to some consensus but there is still debate about issues of graphical quality. Most researchers accept the continu-ing need for a holistic and artistic approach to design.

The chapter concludes with some practical suggestions of how you might design better maps that use graphicacy in a creative and useful way. It reasserts the importance of mapping for anyone studying or researching in geography.

THE CONTESTED TERRAIN OF MAPPING

The ability to construct and read maps is one of the most important means of human communication, as old as the invention of language and as significant as the discovery of mathematics (Borchert, 1987). The mapping impulse seems to be universal across cultures, time and environments (Blaut, 1991). Histories of cartography have charted changing production technologies, from the earliest surviving map artifacts dated to 3500 BC. They have described changing world views, reflected on accuracy and design and increasingly examined the social context of these images (Harley and Woodward, 1987).

Contemporary official definitions encompass a wide diversity of maps but also extend the scope of cartography well beyond earlier narrow concerns with the technology of map making. In 1995, for example, the International Cartographic Association defined a map as 'a symbolized image of geographic reality resulting from the creative efforts of cartographers and designed for use when spatial relation-ships are of special relevance'. The ICA also defined cartography as 'the discipline dealing with the conception, production, dissemination and study of maps in all forms' (International Cartographic Associ-ation, 1995: 1). This chapter also adopts a very catholic view of mapping (see Figure 20.1).

Following the influential work of Arthur Robinson, academics have sought to classify maps according to content and scale (Robinson et al., 1995). *Thematic maps* focus on one particular kind of information, such as solid geology or voting patterns. In contrast, *general-purpose* maps include diverse information ranging from large-scale planimetric coverage of a building, through official medium-scale topographic mapping, to smaller-scale maps of the world in reference atlases.

Others have sought to classify by *format* of publication. Maps were formerly only issued as printed paper publications. Mapping is now also available as digital data delivered on a wide variety of media and held locally on CD-ROM, hard drive or network server, but increasingly also distributed over the Internet (Kraak and Brown, 2000). Producers no longer control content or design; users can play a much more active role, and there is an increasing appreciation that maps might also be constructed to communicate by sound or be read by

FIGURE 20.1 The world of maps

touch. For example, tactile maps are being created for visually impaired people (Perkins, 2002).

Distinguishing maps by content becomes less significant when users can create their own maps. Emphasizing content also completely misses an essential aspect: it can be argued that *all* maps have a theme and every one represents an 'interest' (Wood, 1992). Understanding this interest may be much more useful than listing what the map appears to show. The complexity of the medium is clarified if you appreciate the different roles played by maps (see Box 20.1).

Mapping is above all else a practical form of knowledge, carried out with an end in mind. Cartography is a useful art, and the map is a tool to be used for informing, navigating, describing places, analysing spatial relationship or many other purposes. Maps work as tools by *simplifying* and by serving as guides to the much greater complexity they represent. This recognition was used by geographers in the 1970s to develop scientific approaches that focused upon how maps work as a form of communication (e.g. Board, 1984) and in the search for optimal map designs.

Maps communicate spatial information in a series of codes. It has been argued that these symbols might be analogous to natural language (Head, 1984), and that these codes not only work within the map but also allow us to use maps at a social level (Wood, 1992). By the 1990s there was an increasing concern that a narrow scientific view of the map as part of a cartographic communication system failed to reflect the diversity of roles it played. Instead, more recent research has treated the map as a representation and sign system functioning at different levels (MacEachren, 1995).

The map is also an efficient way of storing large amounts of spatial information (Tufte, 1983: 166). This 'visual inventorying' role has

---

Box 20.1   Some of the possible roles played by maps

- Practical tools.
- Models.
- Language and representation.
- Inventories and databases.
- Visualizations.
- Cultural artifacts.
- Imaginings.
- Political devices.
- Metaphors for scientific knowledge.
- Persuasive icons.
- Contested texts.
- Geography.

come to be supplanted by digital databases that no longer need to have a visual expression, but even the simplest visual display still stores and communicates information very effectively: a picture is still 'worth a thousand words'.

In the last two decades of the twentieth century, the development of the digital computer and increasingly complex software allowed graphical images to be manipulated, and more interactive map use became possible (see Figure 20.2). Mapping also increasingly moved towards the private end of the use axis. In the 1990s, scientific visualization began to reintegrate map design with science (Mac-Eachren et al., 1992).

The development and uses of mapping reflect changing social and economic contexts (Thrower, 1996). All maps are cultural, crafted works of the human mind. They are artifacts imbued with the cultural values of the society producing them. An Australian aboriginal bark painting maps out aboriginal environmental relations and might be incomprehensible to a western European (Turnbull, 1989). The tube map serves as a navigational aid for a tourist visiting London but might not be understood by a bushman. The way maps work in different cultures reflects individual imagined geographies. The power of the imagination is obvious in properly designed maps that reflect

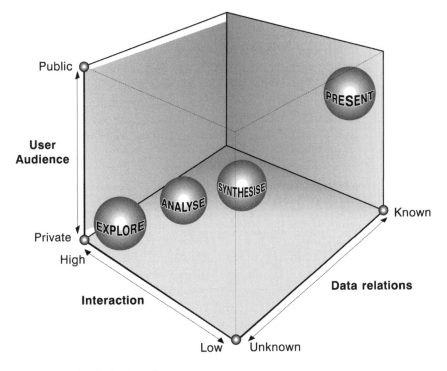

FIGURE 20.2 MacEachren's cubic map space

intuitive and individual artistic judgement (Keates, 1984): maps can decorate as well as inform. Edward Tufte (1997) has argued that this artistic aspect of design implies a quest for quality.

It has also been argued that maps are all imbued with power and that the work they carry out is always political (Harley, 1989a). The map often stands for control and acts as a synonym for order. As a form of power knowledge, the map has often represented the interests of elite groups in society and served to reinforce social norms (Black, 1997). The power of many western maps resides in their apparant objectivity (Turnbull, 1989). The map appears to show everything, to offer a neutral 'view from above' (Cosgrove, 2001) that indexes the world and allows the unknown to be known. But while they may stand for science and factual knowledge, maps may also have an inherent ability to persuade – all the evidence suggests they are more likely to be believed than words. Mapping is widely used in political propaganda, in advertising, in cartoons and in the mass media where maps are associated with news stories and reinforce the narrative of the story line (Monmonier, 1996; Perkins and Parry, 1996). Maps can also add authority to people associated with them: military leaders and politicians give their press briefings in front of maps.

Others have argued that *all* maps are inherently persuasive, that maps are best read as rhetoric and that the map is a text that needs to be deconstructed and interpreted (Harley, 1989a). The problem is how to interpret such persuasive icons (Pickles, 1992). Maps interact with other discourses, may be read in conflicting ways by different social groups and say different things to different people.

Perhaps above all else mapping is still seen as something that distinguishes geography from other disciplines. The map in the journal article or the geography dissertation sends a signal to the reader that the work is geographical (Harley, 1989b).

FINDING THE MAP

More maps are available now than at any time in human history. Mapquest.com has already delivered more maps from its web-based server than any other publisher in the whole history of cartography (Peterson, 2001). In the UK in 1996, there were 250 publishers releasing printed mapping (Perkins and Parry, 1996) and over 2500 are listed worldwide in Parry and Perkins (2000). In January 2003 over 20 000 mapping sites were indexed on the definitive list of cartography websites (Oddens, 2003).

These quantities make it difficult to find the *right* map – a problem exacerbated by the increasing diversity of types and by the infinite

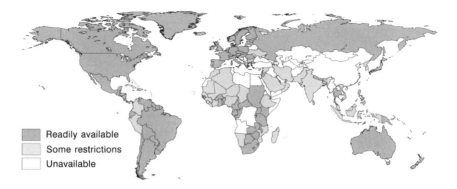

Readily available
Some restrictions
Unavailable

FIGURE 20.3 Map availability

design possibilities offered by digital mapping. How do you decide which to use and how do you find that map?

It is difficult to find out what printed fixed-format maps have been published. Publishers provide descriptive information about their mapping, and increasingly these data are available over the web. So one option is to search the home pages of mapping organizations. A few detailed guides to mapping have been published for some nations, notably for the UK (Perkins and Parry, 1996). A basic but comprehensive introduction is provided in *World Mapping Today*, including lists of URLs and publisher details, pen portraits of the state of the art in different countries, simple graphic indexes and bibliographic information (Parry and Perkins, 2000).

When searching through these sources you might be interested in such factors as the spatial and temporal coverage, resolution, currency, consistency and reliability. But a more complex appreciation also helps. For example, who produced the map for which market? What does it include or omit? Are there legal restrictions on use? How is copyright enforced? Is the map affordable? Perhaps above all else, is it available? Despite globalization the nation-state and its legal framework continue to influence public civilian availability (Barr, 2001). Figure 20.3 shows that, although most terrestrial areas have been well mapped, it is very difficult to get access to larger scales of official mapping. Maps may still be reserved for military use (e.g. in India, much of the Islamic world and even in Greece) or priced at a market cost recovery rate that maximizes revenue and makes individual access expensive (e.g. in the UK).

Having identified the best printed source, tracking down a copy involves consultation in a map library or purchase from the publisher or map seller (Parry, 1999). The quality of map libraries varies enormously according to funding, the nature of their parent organization and how they are staffed (Parry and Perkins, 2001). In the UK, many schools of geography or earth science departments hold collections of printed mapping. The currency and scope of many university collections are declining but a more comprehensive range of cartographic resources is to be found in major and national library collections, such as the six copyright libraries in the UK (British Cartographic Society, 2000). In North America most significant map collections are to be found in central libraries on university campuses, where they serve as federal depositories and where digital datasets are increasingly also being archived. National libraries such as the Library of Congress in Washington, DC, The National Library of Australia in Canberra or the British Library in London offer the most significant map holdings and may be the best sources for the most difficult items. Information about the scope of major map collections is to be found in Dubreuil (1993) and via the map collections section of the Oddens list on the web (Oddens, 2003). Descriptive information may increasingly be found in web-based online public access catalogues and may be used to discover whether the library holds a copy of the map you need. It may be necessary to visit the collection to seek advice from professional curatorial staff.

Parry (1999) suggests that buying a printed map may well be the last resort for students if resources are not available in library systems. In Great Britain and North America, mapping is relatively easy to acquire from many retailers. Wider ranges are available from specialist map shops (such as Stanfords), most of whom maintain mail-order operations and increasingly also operate e-businesses with online ordering facilities (see Box 20.2 for a list of the most significant of these dealers). For more information about the relative merits of these competing sources, see Parry and Perkins (2000).

Accessing digital map data presents different challenges. It will require GIS software, storage media, hardware and output devices. Some data are distributed with viewing and interactive mapping software, perhaps on CD-ROM, but increasingly data are being served on the Internet. Even greater disparities of availability exist between the developed and underdeveloped worlds, in part because of the economics of digital map production. Digital mapping is usually much more expensive to buy than printed maps and may only be available to use under strictly regulated licence conditions. In the UK the Joint Information Services Committee has negotiated a number of deals with data providers to release digital data to universities and colleges. Students can register to use an increasing range of digital data in teaching, learning and research. Since 2000 selected Ordnance Survey data have been available via the DIGIMAP service from EDINA. Other notable online sources of data relating to Great Britain are available through CHEST (Combined Higher Education Software Team). In the USA, federally produced digital map data are in the

Box 20.2   Map dealers

GeoPubs
Geoscience Publications Services, 4 Glebe Crescent, Minehead, Somerset
TA24 5SN, UK

Tel: +44 1643 709001
Fax: +44 1643 709002
Email: info@geopubs.co.uk
URL: http://www.geopubs.co.uk/

The Map Shop
15 High Street, Upton-on-Severn, Worcs WR8 0HJ, UK

Tel: +44 1684 593146
Fax: +44 1684 594559
Email: themapshop@btinternet.com
URL http://www.themapshop.co.uk/

Elstead Maps
PO Box 52, Elstead, Godalming, Surrey, GU8 6JJ, UK

Tel: +44 1252 703472
Fax: +44 1252 703971
Email: enquiry@elstead.co.uk
URL: http://www.elstead.co.uk/

East View Publications Inc.
3020 Harbour Lane N, Minneapolis, MN 55447, USA

Tel: +1 763 550 0961
Fax: +1 763 559 2931
Email: eastview@eastview.com
URL: http://www.eastview.com/

Four One Company Ltd
523 Hamilton Road, London, Ontario, N5Z 1S3, Canada

Tel: +1 519 433 1351
Fax: +1 519 433 5903
Email: maps@fourone.com
URL: http://www.fourone.com/

GeoCenter ILH
PO Box 80 08 30, D-70508, Stuttgart, Germany

Tel: +49 711 7 88 93 40
Fax: +49 711 7 88 93 54
Email: GeoCenterILH@t-online.de
URL: http://www.geokatalog.de/

Map Link Inc.
30S la Pactera Lane, Unit 5, Santa Barbara, CA 93117, USA

Tel: +1 805 692 6777
Fax: +1 805 692 6787
Email: custerve@maplink.com
URL: http://www.maplink.com/

Continued

Box 20.2   Map dealers   Continued

Map Swap
PO Box 1476, 9701 BL Groningen, the Netherlands

Tel: +31 50 527 85 00
Fax: +31 50 527 85 01
Email: info@mapswap.nl
URL: http://www.mapswap.nl/

OMNI Resources
1004 South Mebane Street, PO Box 2096, Burlington, NC 27216–2096, USA

Tel: +1 336 227 8300
Fax: +1 336 227 3748
Email: custserve@omnimap.com
URL: http://www.omnimap.com/

Edward Stanford Ltd
12–14 Long Acre, London, WC2E 9LP, UK

Tel: +44 020 7836 1321
Fax: +44 020 7836 0189
Email: customer.services@stanfords.co.uk
URL: http://www.stanfords.co.uk/

Treaty Oak
PO Box 50295, Austin, TX 78763–0295, USA

Tel: +1 512 326 4141
Fax: +1 512 443 0973
Email: maps@treatyoak.com
URL: http://www.treatyoak.com/

World of Maps
1235 Wellington Street, Ottawa, Ontario, K1Y 3A3, Canada

Tel: +1 613 724 6776
Fax: +1 613 724 7776
Email: bgreen@worldofmaps.com
URL: http://worldofmaps.com/

public domain and may be readily accessed over the Internet. Map libraries in North America, western Europe and Australasia increasingly offer access points to the more useful sources of digital mapping. The Internet also offers a plethora of copyright-free maps (Kraak and Brown, 2000). The majority of mapping websites still serve static images, and notably rich online libraries include the University of Texas. Increasingly the web is also a valuable source of more interactive mapping and acts as a delivery gateway to data warehouses storing digital map data that may be imported in GIS (see Chapter 18). Box 20.3 lists a number of the more important online sources of digital mapping.

Box 20.3 Online sources of mapping

Search engines and gateway sites
Oddens list
http://oddens.geog.uu.nl/index.html
Around 20 000 links to cartographic-related sites on the web. The richest and most current way in to spatial data sources.

University of Texas
http://SunSite.Informatik.RWTH-Aachen.DE/Maps/
A rich source of scanned conventionally published mapping. Huge collection of maps, mostly produced by the CIA, available as gifs, jpegs or pdf files. Also many links to other sites with maps (including historical maps, city plans and cartographic reference sources).

Atlas of Cyberspace
http://www.geog.ucl.ac.uk/casa/martin/atlas/atlas.html
A rich and multimedia grouping of virtual geographies.

Sources of online map data
Ordnance Survey
http://www.ordnancesurvey.co.uk
The national mapping agency of the UK, with a wealth of information available about mapping products.

Digimap
http://digimap.edina.ac.uk/
The EDINA Digimap site makes some OS digital datasets accessible to the higher education community in the UK.

Multimap
http://www.multimap.co.uk
A rich mapping portal, with UK maps and data at different scales, together with routing software.

Streetmap
http://www.streetmap.co.uk
UK mapping to street level searchable by postcode.

Landmark Information
http://www.old-maps.co.uk
Raster scanned versions of 85 000 historical OS maps.

Mapquest
http://www.mapquest.co.uk
The world's busiest mapping site. Road maps and routefinding software. Worldwide coverage.

Continued

Box 20.3   Continued

Topozone
http://www.topozone.com/
The 'web's first interactive topographic map of the entire United
States.' USGS mapping at 1:100 000, 1:25 000 and 1:24 000 scales.

GEOnet Names Server
http://164.214.2.59/gns/html/index.html
Access to the National Imagery and Mapping Agency's (NIMA)
database of foreign place and geographic feature names.
Approximately 20 000 of the database's 3.5 million features are
updated monthly with names information approved by the US Board
on Geographic Names (US BGN).

CASWEB
http://census.ac.uk/casweb/
The spatial interface to census mapping and data in the UK.

UK Borders
http://edina.ed.ac.uk/ukborders/subscribe.html
Boundary files for use with the census.

Data archives
UK Data Archive
http://www.data-archive.ac.uk/home/
A specialist national resource containing the largest collection of
accessible computer-readable data in the social sciences and
humanities in the UK. Through these web pages it is also possible to
search the catalogues of other national archives for computer-readable
data and to use the services of the Data Archive to acquire these data
on your behalf.

National Geophysical Data Center (NGDC)
http://www.ngdc.noaa.gov/
A wide range of science data services and information. New, well
documented databases from many sources, and value-added data
services. NGDC acquires and exchanges global data through the
World Data Center system and other international programs.

US Geological Survey
http://edcwww.cr.usgs.gov/webglis/
Provides data on climate, digital line graphs, elevation, geology,
hydrology, land cover, maps, photography, satellite imagery – mainly
for the USA.

Continued

Box 20.3   Continued

GTOPO30
http://edcdaac.usgs.gov/gtopo30/gtopo30.html
A global digital elevation model (DEM) with a horizontal grid spacing of 30 arc seconds (approximately 1 km). GTOPO30 was derived from several raster and vector sources of topographic information and was completed in late 1996.

ESRI
http://www.esri.com
The software house responsible for industry standard GIS data sources. Wide range of GIS data and digital map samples.

Air photographic data
UK Perspectives
http://www.ukperspectives.com/
Provides aerial photographs for the UK.

Cities Revealed
http://www.crworld.co.uk/
Cities Revealed supplies many different types of aerial photographs for the UK and other countries.

Millennium Map Company
http://www.getmapping.co.uk
Previews of new digital aerial coverages of the UK.

Land cover and thematic data
Soil Survey and Land Research Centre
http://www1.silsoe.cranfield.ac.uk/sslrc/
Data provided on soils, hydrology of soil type (HOST), agroclimatic data for agricultural land classification, and lowland peat survey.

British Geological Survey
http://www.bgs.ac.uk/
The national earth science agency of the UK – rich source of geoscientific data.

National Atlas of the United States
http://www.nationalatlas.gov/
A new official interactive atlas of the USA, multi-thematic layered mapping and animations.

Climate Research Unit, University of East Anglia
http://www.cru.uea.ac.uk/
Provides climatic data for the British Isles, Europe, the USA and global data too.

HOW MAPS WORK

Once you appreciate the complex roles they play, it makes sense to try to understand how maps work. This involves understanding basic spatial properties and appreciating how maps simplify and recognizing the constraints of symbolizing data.

Spatial properties

All maps are about places and represent distance, direction and location in a graphical medium. For the tool to work these spatial properties have to be mapped out in a consistent way. The surface of the earth is not flat, so a mechanism is needed to translate the relative positions of places to the flat sheet of paper in a way that minimizes distortion of scale, area, direction and shape. The science of map projection regulates this process and for many years the mathematics and technology of projection dominated cartography. More recently, the politics of projection have been emphasized – for example, in the controversy over the use of the Peters projection (Monmonier, 1996). If you need to use a small-scale map of the world then take advice on the 'best' projection to use (American Cartographic Association, 1988; 1991).

Projection maps out the position of the lines of latitude and longitude that comprise the graticule. Absolute locations of places on the earth's surface may be defined in spherical co-ordinates, but the graticule does not intersect at right angles. Grids are a rectilinear net of lines that allow eastings and northings to be defined. They allow space to be structured and were first employed by ancient Greek cartographers. Grids may be arbitrary, like the alphanumeric references used by A–Z-style town atlases, or may be mathematically related to the projection, like the British National Grid.

Consistent representation of distance implies the use of a mathematical linear scale linking the map to the world that may be expressed on the map as a representative fraction (1:50 000), as a bar scale or a verbal scale statement (one and a quarter inches to one mile). Very few maps use scale consistently for every object; most will exaggerate the size of some features so they can be read as a symbol (for example, a road). Many small-scale maps of the world will also have to distort linear scale because of the projection used.

Generalization and classification

Scale regulates how much detail the map can show. Larger-scale maps depict more detail and with greater accuracy. As scales get smaller, so features must be more generalized (see Figure 20.4). There are a number of different ways of generalizing: simplifying, enlarging, displacing, merging and selecting, etc. Classification can be another effective way to map out complexity. Alternatively, you may have to leave features off the map altogether. Matching the level of detail shown to the scale is an important aspect of design: a cluttered map will not work as well as one with the right balance between content and space. The process of selection might be seen as a technical issue but in many maps is also a political outcome, with omissions or 'silences' reflecting cultural values (Harley, 2001: 84–107).

Symbolization

The cartographer also has to symbolize the world in a regulated graphic language in which text and the visual properties of symbols are combined. This combination often takes place in quite standardized ways – common elements of a map on a screen or sheet of paper may be identified (see Figure 20.5) (Dent, 1999: 242). Objects in the

(a) **Simplification**

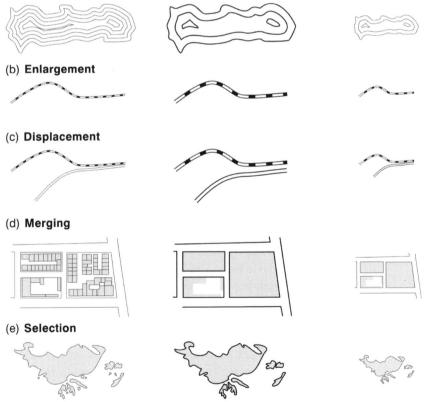

(b) **Enlargement**

(c) **Displacement**

(d) **Merging**

(e) **Selection**

FIGURE 20.4 Generalization

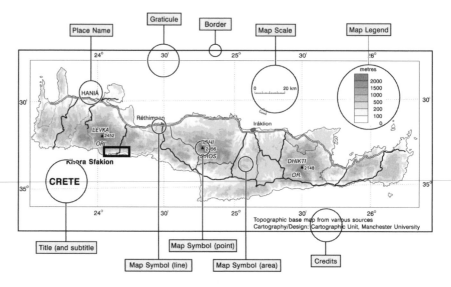

FIGURE 20.5   Map elements

map itself may be thought of as having different numerical qualities. These *measurement levels* are important for design (Monmonier, 1993). Nominal data show the presence or absence of information; ordinal data imply that a feature is larger or smaller (but do not indicate how much larger). Interval data involve ordering with known distances between observations – e.g. Fahrenheit measurement – whereas ratio measurement is an interval scale with a known starting point. Symbols and objects also have *dimensions*: points, lines, areas, volumes and duration. They may be distributed in discrete, sequential or continuous patterns.

On the map itself the geometry and measurement level of symbols have *attributes* that allow information to be communicated. For example, the road may be red and it may have a label indicating that it is the A57 Snake Pass. The effective use of lettering on maps and the rules governing how it should be used are one of the most difficult areas of cartographic design. Name placement is a complex and often intuitive process.

The map designer can use only a limited number of graphic variables. Figure 20.6 illustrates how they might be used for point symbols and suggests there are rules governing inappropriate use. For example, shape should not be used to suggest variation in quantitative data (MacEachren, 1994). Symbols constructed with these variables may be iconic, geometric or abstract (MacEachren, 1995).

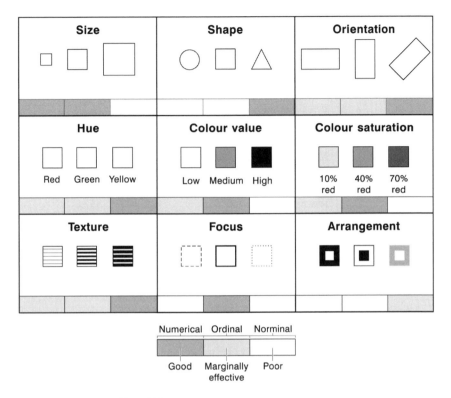

FIGURE 20.6 The graphic variables

PRACTICAL SUGGESTIONS FOR DESIGN

Having got hold of source material, how do you know whether it is a 'good' map? Putting together all the elements in Figure 20.5 does not automatically result in a map that works well: these elements need to be combined in 'a meaningful aesthetically pleasing design' (Dent, 1999: 241). Can there be universal rules defining aesthetic quality? How might these be influenced by production technologies and are they transferable to different contexts?

Universal rules?

In 1983, Edward Tufte came up with a list of qualities that might define excellence in the design of statistical graphics (see Box 20.4). A survey carried out by the British Cartographic Society Map Design Group in 1991 revealed that professional cartographers also felt that quality resided in the overall perceptual qualities of the design rather than with the individual components of the map (see Box 20.5). Maps and other graphics seem to operate as wholes, greater than the sum of their parts. Artistic qualities are important (Keates, 1984).

Box 20.4   Tufte's principles of graphical excellence

- Show the data.
- Induce the reader to think about the substance rather than the methodology, graphic design, the technology of graphic production or something else.
- Avoid distorting what the data have to say.
- Present many numbers in a small space.
- Make large datasets coherent.
- Encourage the eye to compare different pieces of data.
- Reveal the data at several levels of detail, from broad overview to the fine structure.
- Serve a reasonably clear purpose: description, exploration, tabulation or decoration.
- Be closely integrated with the statistical and verbal descriptions of a dataset.

Source: Tufte (1983: 13)

There are two key areas here. Our perceptual systems are programmed to respond to the *visual organization of images* (Dent, 1999). A good map should be balanced, with spatial layout allocated according to the Golden section. (The Golden section refers to rectangles with sides at a ratio of 1:1:6 that offer the most pleasing appearance to the eye.) It should be organized so that the eye's area of maximum attention (just above the geometric centre) corresponds to the central focus of the map. The individual elements in the design ought to work

---

Box 20.5   Maps as communication graphics

- Contrast between symbol and background and between symbols is vital.
- The symbols themselves should be clearly legible and unambiguous.
- The amount and nature of the data depicted should be appropriate to the main purpose of the map.
- The overall appearance should be clear, simple and uncluttered.
- The metrical attributes of the map should be both appropriate and clear.
- The ordering of the data should be made clear by the hierarchical organization of the map image into recognizable visual levels.

Source: British Cartographic Society (1991)

together as an integrated unit. The second key aim should be to maintain a *clear hierarchy between different visual levels* in the map (Dent, 1999). The 'figure' needs to stand out from the 'ground'. The most important objects should contrast most with their surroundings.

Production technology

By 2002 most maps produced by students in the UK were created using computer-based technologies. Choosing an appropriate type of software, and using it to best effect, is now probably the single most important impact on design. Be realistic and aware of some of the factors listed in Box 20.6. Five examples illustrate this process:

1 Serving mapping on the web requires a number of different packages, in addition to the vehicle you use to design the map. The medium delivers maps to many users independent of platform, and maps are updatable. Production depends on the configuration of the site (whether processing is client or server based), the site format (whether maps are delivered in single pages, multiple pages or a frames environment), the nature of the web interface (which

---

Box 20.6   What software should I use?

- How much time do you have?
- How ICT literate are you?
- What packages are you already aware of?
- How much support would you get for learning a new package?
- How does the software link to other packages?
- What file formats does it support, both for importing material and for pushing out completed graphics?
- What output devices does the software talk to?
- What kind of hardware environment do you intend to use: Mac or PC?
- Do you intend to use someone else's base material and edit it up or design from scratch?
- Do you have access to a fast scanner?
- What role do you want the map to play?
- What kind of use is the map intended for: presentation, analysis or exploration?
- Is the map static or dynamic?
- How complex is the information you want to show?
- In which medium is the map going to be published: printed or electronic, black and white or colour?

version of Explorer or Netscape is being used, and what plug ins),
the data type and the content interface (Cammack, 1999). Seek
advice from standard texts such as in the appendices in Kraak and
Brown (2000: 177–209).

2 GIS software, such as MapInfo, ArcView or Idrisi, is designed for
analysis but also offers a wide range of visualization options for
map design. It is complex to learn but if you know the package,
have the data and want to display results of your analysis then it
offers a realistic choice. If you need to produce many statistical
maps for a case study, designed to a common template and from a
wide range of variables, this automated kind of production will
save time. Advice and practical examples about this kind of
software are included in Madej (2001).

3 Some mapping software will also allow you to create maps automatically but will not include the range of analytical tools you
would expect in a full GIS. Mapviewer is a good example of this
kind of program, offers a compromise between analytical and
design capability, and is relatively straightforward to use. It, too,
supports a number of predefined thematic map types that automatically create maps from data held in spreadsheets. The package also
automatically creates associated support information such as legends. The designer simply has to work out which map type is
appropriate for the data (see Figure 20.7).

4 If you want real design quality, have complex ideas to map and the
time to spend learning a sophisticated piece of software then a
professional drawing package such as CorelDRAW is the best
option.

5 The majority of static maps displayed in undergraduate dissertations do not need this sophistication. A simple drawing package
will suffice. It needs to have drawing tools and to allow you to
import and edit graphics created elsewhere by adding symbols, text
and marginal information.

The context

Even if design rules are followed and the appropriate technology used,
maps still need to be matched to the medium in which they are to be
published (MacEachren, 1994). Most maps produced by students are
still designed for presentation in printed reports, but increasingly they
may be used in PowerPoint displays, slides or overhead transpar-
encies, on posters, displayed on computer monitors, delivered via the
web or as animated, dynamic visualizations. The medium may support colour, or perhaps designs will be limited to black and white. The
mapping may be static or animated. Practical decisions that need to be
taken are listed in Box 20.7.

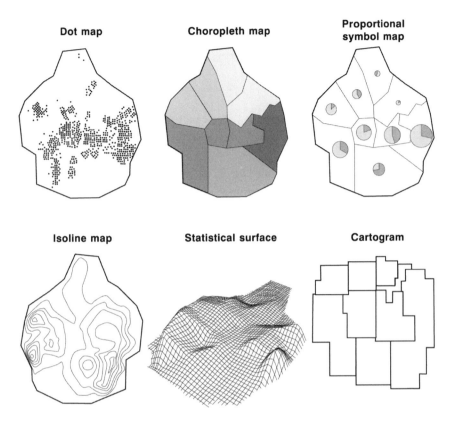

FIGURE 20.7 Different thematic map displays of the same dataset

With a coloured map resolution may be inferior, but you can use two additional visual variables to improve visual structure. It makes no sense, though, to design in colour and print on a black and white laser printer. Relating the map to other associated textual elements is also important. The more complex the medium, the more important

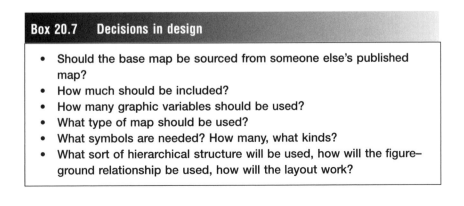

Box 20.7   Decisions in design

- Should the base map be sourced from someone else's published map?
- How much should be included?
- How many graphic variables should be used?
- What type of map should be used?
- What symbols are needed? How many, what kinds?
- What sort of hierarchical structure will be used, how will the figure–ground relationship be used, how will the layout work?

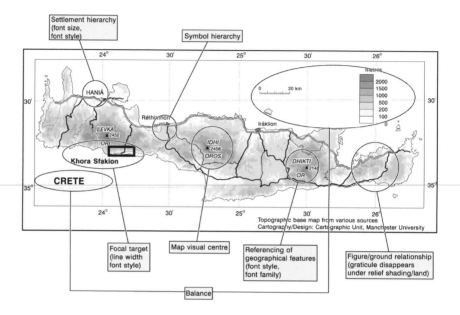

FIGURE 20.8 A location map placing a field course to Crete

the links. So include mapping close to the words that relate to the graphic rather than hidden in appendices.

Figure 20.8 shows a typical location map designed for incorporation into a report or dissertation and 'placing' the research by mapping a simple spatial context in which it is situated. The annotations draw attention to key aspects of the design. This book does not use colour, so the map has been designed in CorelDRAW and saved as a .eps file for incorporation into book design software employed by the publishers. Most maps created by students can be exported as .wmf files that preserve the visual qualities of the image, such as line weight, area tones and lettering, and can be embedded as pictures in wordprocessed documents.

The same map in a PowerPoint presentation would have to be much more generalized and use larger lettering. Maps designed for display on paper also often translate badly to screen-based displays. On-screen colours are lit from behind and usually displayed using a red–green–blue colour model, in contrast to maps printed on paper (Gooding and Forrest, 1990). Screen resolution is still inferior to process colour-printed mapping, and the viewing area is much smaller than in printed versions of the same map. More complex multimedia versions of the same map will also have to reflect the design of the graphical user interface and incorporate marginalia concerned with screen navigation and manipulation. Display is no longer the only goal for design: functionality and navigation also become important (Miller, 1999).

Above all else when designing maps, seek out examples of graphical quality and consult the standard works by Borden Dent, Mark Monmonier, Alan MacEachren and Edward Tufte. Use their advice but be very aware of the context for which you are creating the map and what you want the map to say.

CONCLUSION

The following statement concluded a text about British mapping published just as the Internet started to have a significant effect on the distribution of mapping, but well after digital map technologies had changed the world of visualization: 'Maps are important. Technology and society are rapidly changing the ways in which they are constructed, used and regarded, but as visual metaphors they will continue to provide statements which both reflect and shape our perception of the world' (Perkins and Parry, 1996: 380). The nature of the representation and its role may have altered but graphicacy is still central for geographers. We all have a responsibility to make better maps and use them in a more critical way.

Summary

- Mapping works as a social process.
- Maps are available from publishers/map sellers, libraries and online.
- Accessing digital maps requires GIS software, storage media, hardware and output devices.
- Maps work as representations, highlight spatial properties and provide a visual means of creating generalizations, classifications and symbolization.
- Good maps should be organized so that the eye's area of maximum attention corresponds to the central focus of the map, and there should be a hierarchy between different visual levels of the map.
- There is a need for more critical use and design of mapping.

Further reading

- The book by Dorling and Fairbairn (1997) is probably the single most accessible overview aimed at the undergraduate student as it introduces how maps work as images. A much more challenging series of chapters is brought together in Harley

(2001). These explore the many roles played by the medium and establish the changing social context of the history of cartography.

- Parry and Perkins (2000). By far the best source book about published map availability, providing vital publication details and contacts.

- A combination of four books and a website provides the most useful introduction to the practical issues in map design and production: Tufte (1983), Monmonier (1993), MacEachren (1994) and Dent (1999) (Tufte, 1983, is a manifesto call for design quality and includes a rich diversity of inspiring graphics). The Cartographic Communication section of the *Geographers Craft* home page (http://www.Colorado.EDU/geography/ gcraft/notes/cartocom/cartocom_ftoc.html) also offers useful practical advice on a wide range of map design issues.

Note: Full details of the above can be found in the references list below.

References

American Cartographic Association (1988) *Choosing a World Map*. Bethesda, MD: American Congress of Surveying and Mapping.

American Cartographic Association (1991) *Matching the Map Projection to the Need*. Falls Church, VA: American Congress of Surveying and Mapping.

Balchin, W.G.V. and Coleman, A.M. (1966) 'Graphicacy should be the fourth ace in the pack', *The Cartographer*, 3: 23–8.

Barr, R. (2001) 'Spatial data and intellectual property rights', in R.B. Parry and C.R. Perkins (eds) *The Map Library in the New Millennium*. London: Library Association, pp. 176–87.

Black, J. (1997) *Maps and Politics*. London: Reaktion.

Blaut, J.M. (1991) 'Natural mapping', *Transactions, Institute of British Geographers*, 16: 55–74.

Board, C. (1984) 'New insights in cartographic communication', *Cartographica*, 21: 1–138.

Borchert, J.R. (1987) 'Maps, geography and geographers', *The Professional Geographer*, 39: 387–9.

British Cartographic Society (2000) *Directory of UK Map Collections* (4th edn). London: British Cartographic Society (available at http://www.cartography.org.uk/Pages/Publicat/Ukdir/UKDirect.html).

Cammack, R.G. (1999) 'New map design challenges: interactive map products for the WWW', in W. Cartwright et al. (eds) *Multimedia Cartography*. Berlin: Springer-Verlag, pp. 154–72.

Cosgrove, D. (2001) *Apollo's Eye*. Baltimore, MD: Johns Hopkins University Press.

Dent, B.D. (1999) *Cartography: Thematic Map Design* (5th edn). Little Rock, AR: W.C. Brown.

Dorling, D. and Fairbairn, D. (1997) *Mapping: Ways of Representing the World*. Harlow: Longman.

Dubreuil, L. (ed.) (1993) *World Directory of Map Collections* (3rd edn). München: Saur.

Gooding, K. and Forrest, D. (1990) 'An examination of the difference between the interpretation of screen-based and printed maps', *The Cartographic Journal*, 27: 15–19.

Harley, J.B. (1989a) 'Deconstructing the map', *Cartographica*, 26: 1–20.

Harley, J.B. (1989b) 'Historical geography and the cartographic illusion', *Journal of Historical Geography*, 15: 80–91.

Harley, J.B. (2001) *The New Nature of Mapping*. Baltimore, MD: Johns Hopkins University Press.

Harley, J.B. and Woodward, D. (1987) *The History of Cartography*. Chicago, IL: University of Chicago Press.

Head, C.G. (1984) 'The map as natural language: a paradigm for understanding', *Cartographica*, 21: 1–33.

International Cartographic Association (1995) *Achievements of the ICA 1991–1995*. Paris: Institut Géographique National.

Keates, J. (1984) 'The cartographic art', *Cartographica*, 21: 37–43.

Kraak, M.-J. and Brown, A. (2000) *Web Cartography*. London: Taylor & Francis.

MacEachren, A.M. (1994) *Some Truth with Maps*. Washington, DC: Association of American Geographers.

MacEachren, A.M. (1995) *How Maps Work*. New York: Guilford Press.

MacEachren, A.M., Buttenfield, B.P., Campbell, J., Diabiase, D.W. and Monmonier, M. (1992) 'Visualization', in E. Abler et al. (eds) *Geography's Inner Worlds: Pervasive Themes in Contemporary American Geography*. New Brunswick, NJ: Rutgers University Press, pp. 99–137.

Madej, E. (2001) *Cartographic Design Using ArcView GIS*. London: OnWord Press.

Miller, S. (1999) 'Design of multimedia mapping products', in W. Cartwright et al. (eds) *Multimedia Cartography*. Berlin: Springer-Verlag, pp. 51–63.

Monmonier, M.S. (1993) *Expository Cartography*. Chicago, IL: University of Chicago Press.

Monmonier, M.S. (1996) *How to Lie with Maps* (2nd edn). Chicago, IL: University of Chicago Press.

Oddens, R. (2003) *Oddens Bookmarks* (available at http://oddens.geog.uu.nl/index.html).

Parry, R.B. (1999) 'Finding out about maps', *Journal of Geography in Higher Education*, 23: 265–72.

Parry, R.B. and Perkins, C.R. (2000) *World Mapping Today* (2nd edn). London: Bowker Saur.

Parry, R.B. and Perkins, C.R. (2001) *The Map Library in the New Millennium*. London: Library Association.

Perkins, C.R. (2002) 'Progress in tactile mapping', *Progress in Human Geography*, 26(4): 521–30.

Perkins, C.R. and Parry, R.B. (1996) *Mapping the UK*. London: Bowker Saur.

Peterson, M.P. (2001) 'The development of map distribution through the Internet', *Proceedings of the 20th International Cartographic Conference*, 4: 2306–12.

Pickles, J. (1992) 'Texts, hermeneutics and propaganda maps', in T.J. Barnes and J.T. Duncan (eds) *Writing Worlds: Discourse Text and Metaphor in the Representation of Landscape*. London: Routledge, pp. 193–230.

Robinson A., Morrison, J.L., Muehrzke, P.C., Kimerling, A.J. and Guptill, S.C. (1995) *Elements of Cartography* (6th edn). Chichester: Wiley.

Thrower, N. (1996) *Maps and Civilization: Cartography in Culture and Society* (2nd edn). Chicago, IL: University of Chicago Press.

Tufte, E.R. (1983) *The Visual Display of Quantitative Information*. Cheshire, CT: Graphics Press.

Tufte, E.R. (1997) *Visual Explanations*. Cheshire, CT: Graphics Press.
Turnbull, D. (1989) *Maps are Territories, Science is an Atlas*. Geelong: Deakin University Press.
Wood, D. (1992) *The Power of Maps*. London: Routledge.

21 Using Statistics to Describe and Explore Data

Danny Dorling

Synopsis

In this chapter I argue that simple statistics are better than complex statistics in the study of geography, that statistics are of themselves dull and that the majority of university geography students and staff are ill-equipped to use them, let alone enliven them. Five rough and ready rules are suggested for when and where to use statistics in the study of geography. One origin of the word statistics is that it implies 'facts about the nation-state'. The development of both 'state-istics' and the academic subject of statistics is briefly summarized to provide a context for the argument presented here that geographers should concentrate on simple statistics. Finally, the chapter ends with three examples of how things may not be as they seem when viewed statistically.

The chapter is organized into the following sections:

- Introduction
- Five rules
- What are statistics?
- Why use simple statistics?
- Three statistical examples
- Conclusion – possible futures.

INTRODUCTION

Statistics are duller than ditch water. In and of themselves they tend to be of interest only to people who are not very interesting. To me, it is only when statistics are set in wider context that they begin to come to life. In geography this wider context obviously varies widely. Suppose you are swamped by a series of floods and want to know how likely they are to happen again and how large they could be. Statistics give you techniques to make good guesses – guesses that don't necessarily rely on you knowing much about floods. Suppose you are interested in poverty and how poverty rates compare between countries. Poverty rates are statistics. You need to understand the statistics if you are interested in the issue of poverty. You could, of course, ignore the statistical study of floods or poverty while still being interested in either subject. Were you to do this, however, you would be missing out on a great deal that has been learnt about these things. Statistics on floods might miss out the nature, causes and meaning of floods. Similarly, statistics on poverty may dehumanize suffering. But if you are concerned about getting wet or about why higher levels of poverty persist in some places and not others (and hence one key aspect of what reduces poverty – variation in, say, policies that influence poverty), you are unlikely to get far without the numbers and methods which make up statistics.

Statistics are dull because they are so general – because so many different things can be turned into a percentage; because similar techniques can be used to study so many different processes. The generality of statistics is also their main weakness as well as being a strength. Often issues and problems in geography are shoe-horned into a question that can be addressed, perhaps answered, statistically. A researcher turns an interesting research question into a series of dull hypotheses, the answer to none of which quite gets to addressing his or her original question. It is, for instance, much easier with classical statistical tests to suggest that certain things are (probably) not true – rather than to assert what is true. Statistics can also have the disadvantage that many students have built up a resistance to them, an inherent dislike of their use and feel a chill wind travel down their spine when the word is mentioned. Ask yourself this (assuming you are a student studying for a geography degree): what proportion of your fellow students are likely to be reading this book and, of those, what proportion would skip this chapter? Are you in a minority of a minority to have got this far? Perhaps I am wrong, but hopefully if you read a little further I will try to show you that statistics are not quite as painful as they may be thought to be – particularly if you only use them when you need to.

There is at least one other probable reason why you may have got this far through this chapter. You bought or read this book a year or so ago when you were first being introduced to methods in geography. You read about the interesting ones, the novel and the new, but you skipped this chapter. Later, in a seminar or tutorial, someone asks 'what statistics are you using?' for your research project. That feeling of panic returns. Or you referred to some statistic in an essay and your marker irritatingly scribbled 'explain!' next to it. In this short chapter

I am going to show you only one statistical method (calculating and interpreting a confidence limit). No statistical test and no actual statistics – of the '95% of statistics are wrong' variety. Instead I'll follow this introduction with five simple rules for using statistics in geography and then talk a little about the origins of statistics, give some common misinterpretations of statistics and end with possibilities for the future use of statistics in geography.

FIVE RULES

Often there is little point in using statistics

All too often statistics are included in a study or student essay to try to show that the writer or student had been working hard. An essay on the geography of employment in Britain might begin: 'Thirty million people work in Britain today.' So what? Statistics in and of themselves are dull (unless dull things excite you!). 'There are thirty million people working in Britain and forty million jobs' would be a little more interesting – many people would have to have two jobs – but that is only interesting if it is new to you as a reader. 'There was a positive correlation between the number of people working in each county and the number of jobs' – hardly surprising, and what's your point? Perhaps what you were trying to say was:

> More people are working then ever before at more and more tasks in Britain and at even more in the USA. Work, rather than removing human drudgery, has added to it. As we work harder we have created more work. Huge numbers of people have to work at two or three jobs a week compared to our labours in the past. The proportion of both children and the 'retired' who work has risen. For the affluent, childcare and housework are increasingly subcontracted as work for others, as for many is cooking when we eat out. It is difficult to see a smooth end to the apparent unstoppable commodification of human time. Work turns time into money. We now have both more money and less time.

Only the very crudest of statistics were used in that argument. Things had simply gone up or down – were more or less. It would have helped the argument if I had listed some sources for this information. It is probably not all true and what is described is certainly not true for many people. However, the argument did not need numbers or tests to strengthen it. There is a time and place for statistics.

If you do use statistics make sure they can be understood

Most people, including a majority of human geography lecturers and a fair share of physical geography lecturers, are innumerate – as are most students of geography. By innumerate I do not mean innumerate in the sense of 'could not get an A grade at GCSE maths'. By innumerate I mean do not have an innate feel for numbers, mathematics and simple algebra in the same way that the minority of students who are dyslexic do not have an innate feel for words and their spelling. Because dyslexia is a minority condition it is given a name. Because innumeracy is the norm in America, Britain and most of the western world, we do not call it as such but save the word for people with almost no ability to handle numbers. If you are in the innumerate majority you may have thought you had some weakness and everyone else could understand maths and statistics and did not need to learn examples by rote as you did. You are wrong: you are normal. If you are in the numerate minority (that is, you could walk into an exam and get an A in maths without trying) you have a problem – many people (including often those reading or marking your work) might not understand you because they do not think like you. Statistics in geography are primarily a form of communication – a way of trying to convey information convincingly. You will only convey information convincingly if, to begin with, the person you are conveying that information to can understand the language you are using.

Do not overuse statistics in your work or methods

How many facts should there be to the page? Occasionally a great many simple statistics all pointing to the same things or suggesting a trends is going one way using many examples can be a powerful tool in making an argument. Generally, however, I'd try to stick to two or three facts on a page unless you are writing a reference book. You are trying to build up a picture in the mind of the reader/marker – explaining an argument through an essay. Similarly, how many numbers should tables of statistics contain? Part of the history of geography has been that multiple regression was a favoured statistical technique for several decades. Multiple regression allows dozens of variables to be entered into a model to predict an outcome. Each variable can have numerous parameters associated with it – for example, its 'effect', 'significance', 'error' and so on. When computers first began to be used in geography, a huge amount of work was required to enter the data and variables to undertake this technique. One result was that huge tables of 'results' were often printed in papers which in fact only referred to a few of the numbers in the tables of sometimes hundreds of figures. Have a look at a journal like *Environment and Planning A* in the 1970s and 1980s for these giant tables. The tables give the impression of thoroughness, but if you want people to *read* the numbers in your tables I'd aim to have only a dozen numbers there. Finally, concerning tests and techniques, there should be little need to use more than one statistical test or technique on a set of data. If you are using more than one, the chances are that only one is appropriate.

If you find a complex statistic useful then explain it clearly

Sometimes, for some problems, only a complex statistical method or complex statistics gets to the heart of the issue you are interested in – and it is the issue rather than the method that we are interested in. A typical case in geography might be that you have access to a whole series of river-water samples and know the level of a particular pollutant in each sample. You also know whether the sample was taken from water near the surface of the river or near its bed, whether from a pool in the river, a riffle, or a bend and what the slope and velocity of the river were at each sample point. You want to know whether some rivers are more polluted than others or whether they just appear to be so because of the nature of the way the samples of water were collected from each river. Your data are arranged in at least two geographical levels: that of the river and that of geographical location within each river. Each individual sample point also has a number of characteristics you want to keep into account. There are a number of techniques you could use to study these data. They range from various types of ANOVA (analysis of variance) tests to MLM (multi-level models). The one you are likely to choose will depend most on what you have been taught and remembered, or perhaps on what you have read. My recommendation is that, when using a complex technique in a complex situation like this, explain every step simply – but above all understand why you are using these techniques. The above techniques are interesting for their abilities to identify 'interactions'. Within a particular set of rivers you expect more pollution in riffles than in pools and vice versa in another set of rivers. If you are not interested in interactions, why are you using these complicated techniques? Furthermore, check first that there is not a very simple pattern in your data that simple statistics could under-cover more persuasively. For example: 'Pollution levels are on average ten times higher in the pools of rivers than when measured at any other point.'

Recognize and harness the power of statistics in geography

Statistics have a political power across much of geography. In physical geography their use implies you are (or are becoming) a competent 'scientist'. Many scientific (and, in particular, medical) journals use statisticians to referee papers before publication as well as referees who understand the substantive subject of the paper. Statistics has

374   **Key Methods in Geography**

become a language of scientific credibility – rather like Greek, and then Latin, was the language of religious credibility in the Christian church for most of its history. Languages are used both for their ability to communicate and to exclude the uninitiated. The power of statistics (in particular, statistical methods) in human geography has fallen in recent years. This fall was partly the result of researchers seeing through the way statistics have been used to exclude and also partly because fewer numerate human geographers choose to teach (or research) geography than once did. However, within human geography, statistics placed carefully can still add authority to an argument that, without them, appears little more than a considered rant. This is particularly true when you are arguing against the generally accepted case. As I argue (or rant!) below, statistics are increasingly used in general debates outside geography. It is possible both to appreciate the weaknesses of statistics and to use them with effect. But it helps first to know where they have come from.

Before considering the origins of statistics, in the pursuit of honesty and to sum up, it is probably worth explaining the situation in which I find myself most commonly introducing geography students to statistics (I don't teach statistics in lectures as I can't find a way of doing so that would keep 150 undergraduates awake or at least attentive for longer than a few minutes).

Here's the scenario: 'So, you've thought of your research questions, done your survey/questionnaire/measuring, got your data, and are ready for some "statistics" . . .' It would have helped if the students had thought about what they were going to do with their data before they collected them but, for many, even if they had done that they would not have known the possibilities: what are these statistics things and why do you need them? For most undergraduate students in geography, and much of the rest of the social or earth sciences, statistics are the things that change a 2.2 dissertation into a 2.1 dissertation – or at least the things they think might do that. They are the things 'taught' by the lecturer who didn't get to teach what he or she wanted to teach – or, even worse – by the lecturer who actually chose to teach statistics! They were the bane of your life while revising for A level. They are the things you dread in exams. You had to learn a few, learn to use a few techniques and learn some more general statistical methods. A dislike of statistics was the reason you studied geography rather than psychology, economics or biology. Students who can cope with statistics don't tend to opt for a geography degree. However, they find them difficult rather than repulsive. If you could not stand statistics you might have picked English, history, art or drama instead. This chapter is for these kinds of student. If you have read this far then it might well be for you too. Above I tried to explain why I think it is worth using some simple statistics. Below

I define statistics more fully, give some reasons for not using complex statistics too often and then I end this short chapter by giving a few simple statistical problems.

WHAT ARE STATISTICS?

Statistics are many things and mean different things to different people. My favourite definition is that they are 'facts about the state' – meaning the nation-state people inhabit and usually involving aspects of their lives. This is a very human geography definition but it fits the origins of statistics where the word began to be used in earnest two hundred years ago to describe the numerical portrayal of countries which was beginning in earnest then. (For more on these, consult William Playfair's writing and a little of what has been written on their origins – see the further reading section at the end of the chapter.) The first modern census – collecting useful statistics about the entire population of a country – was undertaken in the USA in 1790. It is perhaps telling that the counting of people in this way as a new technique began in the new world. Statistics as a word has come, in recent times, to mean more often than not numerical facts about the state. Most statistics people will encounter in a day in the country I am writing in (Britain) are about that country. The same is even more the case of the USA. Two centuries ago, as today, economists in various guises were and are responsible for producing the majority of these kinds of statistics. There use in political arguments, however, has involved a far wider set of disciplines than economics (Dorling and Simpson, 1999, provide examples of these).

Of course, statistics has a second meaning far removed from simply being about facts. Just over one hundred years ago a small group of people who were interested in chance (probabilities) created the academic discipline of statistics. They were also, incidentally, quite interested in society, and many had views about, say, 'criminals' and 'races' which would be seen by most today as abhorrent. This meaning of the word is a world away from the first definition. Most large universities have either a statistics department in them or statistics is part of their school of mathematics. Statistics in this sense has been dehumanized partly due to a wish to forget part of the history of statistics. You may have taken part of an A level in these kind of statistics. Half my first university degree was taken in a statistics department. There the working definition of statistics was very different from that used across the campus in geography. One of my first textbooks was entitled *Statistics: A Guide to the Unknown* (Tanur, 1989), the implication being that for those things we know about there is little need for statistics, but for anything we don't it has a purpose. The introductory examples I was taught as an undergraduate ranged from assessing the likelihood that Shakespeare was the author of each play attributed to him, to calculating how many whales were likely to be living in the seas around Japan in the future, given current trends (and not knowing how many whales were actually in the sea). The one area that examples were not often drawn from was the study of human societies, and so I often wondered whether this was because we know too much about people and how they organize their lives to make them a suitable subject for statistical study. Many human geographers would argue that people's lives are too intricate to summarize meaningfully statistically. But again, perhaps the lack of statisticians now studying societies is partly due to embarrassing links between the origins of the study of statistics and movements in the first few decades of the last century which sought to claim that some people were superior to others?

WHY USE SIMPLE STATISTICS

I have not found that the use of complex statistics has greatly enhanced my understanding of people and how their lives are organized over space. Nor have I generally found that reading other people's accounts of such models has helped me much either. Perhaps its just me, but I think it's worth saying.

From looking at complex models of who chooses how to vote for political parties, to why some people in some places are more likely to be ill than others – to me, the results of complex statistical tests and models depend in most cases more on how the tests and models are put together than on the data. This is a very personal and generalized view (which is why I have written it and why most of this chapter is in the first person). But it is half the explanation I have as to why I think simple statistics are usually better. When you have less control over the numbers and how you manipulate them it is harder to condition the answers you get to what you want to find.

A second reason for tending to use simple statistics is that you are more likely to have something to compare with. Take, for instance, researchers who study pollen in geography departments. These are people who are trying to construct past climate records from the historical record of pollen in a soil core. They tend to call themselves things like Quaternary scientists interested in palaeoclimatic reconstructions rather than people looking at specks of pollen in old mud. Note that how we choose to define ourselves is often as much to confuse as enlighten! What little I know of them and their use of some statistics suggests to me that they mostly tend to use a similar method of cluster analysis to study some aspects of their data. They do that because it worked pretty well for the first person who used it and who taught the second and, although it may not be perfect, it makes more sense if almost everyone uses a method he or she can understand rather than each invent his or her own. Cross-comparison and communication between studies are then possible. Statistics is a language that has different dialects in different places – even within different parts of one small academic discipline. A 'total fertility rate' and 'bank-full discharge' are two common simple geographical statistics but hardly anyone will know the meaning of both. If you do want to know the meaning of these things, key them into a search engine on the Internet – but be careful to confirm the definition you are given on more than one website. Anyone can write almost anything on the web!

Even if you use a simple statistic in one part of geography, it is often peculiar to that part of geography. You are choosing to speak to a very small audience. There are, however, even simpler statistics. On the physical side of the subject these are your basic rates, weights and measures. On the human side simple statistics are the stuff of social summaries. They appear in everyday settings most commonly delivered through newspapers, radio, television or the web. These are the statistics most worth using and which put a geographer into a very good position to question the facts. Many, if not most people, accept facts at face value, which is part of the reason why they are such powerful tools to use in argument (see Dorling and Simpson, 1999, if you are interested in how such things are questioned). In short, simple statistics are more easily understood and more convincing. Within a subject such as geography in the first decade of the twenty-first century – where so few geographers feel (or are) numerate – the use of complex statistical techniques has to be questioned: whom are they being presented to? So, if you as a student find them hard to understand I would say to you, don't worry, and ask yourself why you were being asked to understand them in the first place! If you find a good reason for that, *then* try to understand them.

THREE STATISTICAL EXAMPLES

It is hoped this chapter so far has given you some food for thought about the use of statistics in geography today. Rather than list a series of techniques and refer you on to books about their use, I've tried to give you a more personal view about one of the supposedly most impersonal of subjects taught within geography. My view may be a little odd: I use numbers but believe they are deadly dull; I like both to undermine them and use them in arguments I make. I think our legacy of complicated statistical techniques in geography is more of a historical accident of what was found to work at certain times than a particularly useful set of tools – at least where the study of the geography of society is concerned. I'd like to end this chapter, however, with three examples of how statisticians think differently – which, hopefully, show some things to be learnt from statistics.

Example 1: a simple prisoner's dilemma

You are playing a quiz game on TV. In front of you are three doors. Behind one door is the prize and behind the other two doors nothing. You pick a door. The quiz host then picks another door. If you picked the right door the host picks a wrong door; if you picked a wrong door the host picks the right door. You then have a chance to change your mind over which door you enter. Which do you go through?

The answer is relatively simple, or it is if you find this kind of thing easy: you pick the door the host picked. Your chance of winning the prize is 2 out of 3. Why? Because 2 out of 3 times you will have initially picked the wrong door. Why does this matter to a geographer? Well, go back to where we started this chapter. There's been a large flood and you are interested in the probability of another large flood happening. The fact that you are suddenly interested in the probability of the flood is not independent of the flood occurring. In fact, the chance of the flood having occurred after the event was 100%. You cannot say 'that was the 1 in 100 year flood', just that 'this is what it would be like' (or perhaps, better, might be like!). Probabilities are usually conditional. In geography we often make simple statistical mistakes as geographers learnt most of their statistics during the discipline's classical phase (when statisticians didn't worry about conditional probabilities so much).

Example 2: another conditional probability

You are worried you have a disease. One per cent of students have the disease. You take a test for the disease. For people who have the disease the test is accurate 95% of the time. The test is positive for you. What is the chance you have the disease?

As you might have guessed, the answer is not 95%. There are two possibilities: either you have the disease and tested positive or you didn't have the disease but the test gave you a false positive. Take 10 000 students. Of the 100 who had the disease, 95 would test positive. Of the 9900 who did not have the disease, 495 (5%) would test false positive. Thus the chance of you having the disease, having tested positive for it, is 95/(95 + 495) or roughly 1 in 6. You are unlikely to have the disease. Note that almost 6% of students would think they had the disease had they all taken the test. Of course, if you think there was a particular reason for you to be worried and could quantify that, more conditional probabilities would be introduced!

Now, substitute for disease and students, heavy metal pollution and soil samples. You are working in the labs and have a test for heavy metal pollution in a sample of soil that is 95% accurate. About 6 of your 100 soil samples appear to be polluted after you have tested them. Three of these are located in the same village – have you found a cluster? (Answer: no, probably not.)

Example 3: classical confidence limits

There is only one even mildly complicated statistic I now work out on a regular basis and even that has a more complicated interpretation than is usually taught to geography students. I often work out rates in areas – comparing how many things have happened in an area compared to how many you might have expected to have happened in that place given certain assumptions. Most commonly, this is how many people died in an area compared to how many you would have expected to die there given the ages and sexes of the people living there and national mortality rates. Given these things I might work out, for instance, that 120 people died in an area where you would expect 100 people to have died (this is called the indirect method of standardizing rates). Thus it looks as if 20% more people have died than could have been expected. However, it might just have been a bad year in that place. So with what confidence can I say that 20% more people died there? The equations to approximate the confidence limits for this statistic are as follows:

Given a standardized rate of $100 * O/E$, where O = the number observed and E = the number expected:

Lower confidence limit (95%) = $100 * O * (1 - 1/(9 * O) - 1.96/(3 * \sqrt{O}))^3/E$
Upper confidence limit (95%) = $100 * (O + 1) * (1 - 1/(9 * (O + 1)) + 1.96/(3 * \sqrt{(O + 1)}))^3/E$

Thus our standardized rate is $100 * 120/100 = 120$ with a lower confidence limit of approximately:

$100 * 120 * (1 - 1/(9 * 120) - 1.96/(3 * \sqrt{(120)}))^3/100 = 120 * (1 - 1/1080 - 1.96/32.86)^3 = 99.49$

And an upper confidence limit of approximately:

$100 * (121 * (1 - 1/(9 * 121)) + 1.96/(3 * \sqrt{(121)}))^3/100 = 121 * (1 - 1/1089 + 1.96/33)^3 = 143.49$

So what does this say? That we can be 95% sure that the real death rate in that area lies between 99 and 143? That we can't be 95% sure that the death rate in that area is not average (100)? What it actually says it that given, say, lots of years of data, 95% of the time the true death rate of that area will lie within these limits – so that if we were to work out confidence limits for the Standardized Mortality Ratio (SMR) of an area each year for 20 years, on average the true rate would lie within those limits for 19 of those years. This assumes there is an unchanging 'true' rate – that something about the area raises or lowers the mortality rates of the population living there.

Given this, these classical confidence rates do not tell us what we thought they told us. It is not the case that the true mortality rate of this area lies between 99 and 143 with 95% certainty. This has not, however, stopped researchers, including myself, from labelling areas on maps as 'significant' if their confidence limits exclude 100, but strictly speaking we have little idea what level of significance applies! (For more on this, see Congdon, 2001.)

CONCLUSION – POSSIBLE FUTURES

As this chapter is being written it is government policy to attempt to kick start a renaissance in quantitative methods in the social sciences. Evidence-based 'this and that' are all the rage, and research training money is supposed to be being redirected slightly towards the modellers and away from more qualitative studies. The intended effect is that more of your lecturers in the future should have more of a grounding in statistics. What effect is this likely to have? Very little, I suspect. Only a minority of us are confident with numbers by the time we (the third of children who go) turn up at university. The contortions of logic required to understand that significance tests and confidence limits are not quite what they seem to be – after all, you have only just managed to learn them – are a further great disincentive to use complex statistics. This should not, however, put you off using simple facts and quoting basic trends in sustaining the arguments you make in studying geography. Just because the study of statistics is becoming ever more complex does not mean the practice of using statistics within geography should cease altogether – just try to recognize the breadth of meaning to the word 'statistics'. Writing a short chapter on statistics within geography is rather like trying to define 'geography'. It means different things to different people. What you have just read has been my take on the subject. If you find contortions of logic of interest, please use the further reading section below. If you have been concerned about your insecurities with statistics, don't be – you are normal – just try to use a few more simple facts to strengthen your arguments and try to feel less intimated about the complex methods.

Summary

- Often there is little point in using statistics.
- If you do use statistics make sure they can be understood.
- Do not overuse statistics in your work or methods.
- If you find a complex statistic useful, explain it clearly.
- Recognize and harness the power of statistics in geography.

Further reading

For a chapter such as this the most useful set of further readings is probably a collection of the books and papers which inspired the point of view I've put forward here. Authors of chapters in this book were asked to supply five pieces of recommended further reading, so here they are. The more recent they are the easier they are to obtain.

- First, on the origins of statistics, why not go straight to the source? The links between geography and statistics which were there from the beginning are hard to miss, given Playfair's (1800) extremely descriptive title: *A Geographical, Historical and Political Description of the Empire of Germany, Holland, the Netherlands, Switzerland, Prussia, Italy, Sicily, Corsica and Sardinia: With a Gazetteer of Reference to the Principal Places in those Countries, Compiled and Translated from the German: To which are Added, Statistical Tables of all the States of Europe.* For students in the USA, a search for material on the 1790 census can begin with http://fisher.lib.virginia.edu/census/.

- Two hundred years on and a hugely less ambitious work is the collection of papers that encouraged me to think that statistics are still useful in the study of society: Dorling and Simpson (1999). The book contains examples of the use of statistics in studies across many aspects of life, mostly in Britain and mostly where political debate is currently raging.

- For statisticians, most have their basic reference book. Mine (before I stopped using complicated statistics!) was Breslow and Day (1980). Books such as Breslow and Day are very detailed and assume a great deal of knowledge, but its worth knowing where people using statistics refer to for their basic formulae.

- A much simpler, if now quite old introduction to the study of statistics is still available in many university libraries and, if you are interested in learning more about statistics in general, I'd start with Tanur (1989).

- Finally, for what may be both the most up-to-date and the most complex book on statistics written by a research professor in geography, see Congdon (2001). The first few pages help explain why 95% confidence limits are not what you thought they were!

Note: Full details of the above can be found in the references list below.

References

Breslow, N. and Day, N.E. (1980) *The Analysis of Case-control Studies.* Lyon: International Agency for Research on Cancer.

Congdon, P. (2001) *Bayesian Statistical Modelling*. Chichester: Wiley.

Dorling, D. and Simpson, S. (eds) (1999) *Statistics in Society: The Arithmetic of Politics*. London: Arnold.

Playfair, W. (1800) *A Geographical, Historical and Political Description of the Empire of Germany, Holland, the Netherlands, Switzerland, Prussia, Italy, Sicily, Corsica and Sardinia: With a Gazetteer of Reference to the Principal Places in those Countries, Compiled and Translated from the German: To which are Added, Statistical Tables of all the States of Europe*. London: J. Stockdale.

Tanur, J. (1989) *Statistics: A Guide to the Unknown* (3rd edn). Pacific Grove, CA: Wadsworth.

Acknowledgements

I am grateful to Chris Keylock for suggesting some of the examples used here.

22 An Introduction to Geostatistics

Adrian Chappell

Synopsis

Geographers and environmental scientists are interested in the variation of properties in space and the reasons for this variability. The investigation of spatial data and the interpretation of the arising trend or pattern are the aims of geostatistics. At its simplest, geostatistics is a suite of tools that provides a framework for that investigation. The framework is based on the notion that things closer together are more similar than things further apart and it provides the opportunity to account for the relative position of data in space. Geostatistics is different from 'classical' statistical approaches to handling data sampled in space and, consequently, it provides the potential to explore the geography of environmental variables further than previously investigated. In this chapter the key elements in the geostatistical approach are explained (the theory of regionalized variables; the use of the variogram to characterize spatial processes; and local estimation or prediction – often termed kriging). Examples are provided to illustrate the way in which information contained in the variogram (its structure and model parameters) may be used to interpret the processes and scale of spatial variation in geographical phenomena.

The chapter is organized into the following sections:

- Introduction: what is geostatistics?
- Regionalized variables
- Characterizing spatial processes using the variogram
- Local estimation or prediction
- Conclusion: how geostatistics can help you.

INTRODUCTION: WHAT IS GEOSTATISTICS?

Geostatistics is a suite of tools for solving problems with spatial data. The term 'geostatistics' is commonly applied to, but not limited to, the investigation of spatial variation of processes and landforms on or near the earth's surface (for example, de Roo, 1991; McBratney et al., 1991; Gallichand and Marcotte, 1993; Henebry, 1993; Odeh et al., 1994; Webster, 2000; Chappell and Agnew, 2001; Heuvelink and Webster, 2001). Although many applications appear to be in the realm of the geographer or environmental scientist, many other subject disciplines have used geostatistics to investigate spatial processes. Webster and Oliver (2001: 6–8) trace across several disciplines the history of developments in spatial variation that resulted in the theory responsible for geostatistics. Present-day practice is influenced most by Matheron's (1971) theoretical work and by Krige's applications in mining (Krige, 1966). Consistent with the highly applied nature of geostatistics, the intention here is to provide sufficient background knowledge to enable a rudimentary geostatistical analysis to be undertaken.

The key to this analysis are algorithms represented by flow charts or tables of procedures for decisions that are based on the aim, objectives, type of data available and the nature of the sampling, etc. For example, the flow chart in Figure 22.1 shows the highest-level decisions that must be made about the level of understanding for spatial processes, as discussed above. Since there is space here to outline only three components of geostatistics a considerable amount of information is omitted. An attempt to direct the reader to some of the other important aspects of geostatistics is provided in the further reading section at the end of the chapter.

The flow chart serves to remind us that geostatistics is a response to a system that is currently too complicated for us to make reasonable estimates about the nature of a property found at an unsampled location. In other words, we do not know a sufficient amount of information about the processes responsible for the formation of the property in space (Isaaks and Srivastava, 1989). Consequently, the remainder of this introduction will focus on the development of a stochastic, geostatistical model of spatial processes using three key elements: regionalized variables; characterizing spatial processes using the variogram; local estimation or prediction.

REGIONALIZED VARIABLES

The investigation of the spatial variation of a property (for example, those properties related to soil) may be regarded as a response to

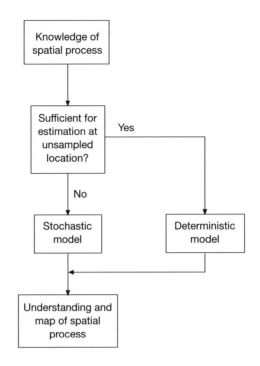

FIGURE 22.1 Flow chart of the decision-making process and consequence for model development

classification, with the implicit assumption of homogeneity within classes. In turn, homogeneity was probably assumed historically because of the apparent complexity of variation. Intuitively, soil properties and many other properties that vary in space are controlled by spatial processes. Unfortunately, our understanding of these processes is often so limited that predictions or estimates at unsampled locations are often somewhat imprecise and highly inaccurate. Geostatistics recognizes the complexity in spatial processes and utilizes a stochastic approach for analysis. This presumes that a value of a property may vary statistically at a point in space – in other words, there is uncertainty about the value. This introduction provides an outline of the stochastic approach to spatial variation (Theory of Regionalized Variables) that supports the practice of geostatistics. It overlaps the more pragmatic considerations of the following section by also providing the underlying concept and assumptions necessary for characterizing spatial variation.

With a limited understanding of the factors controlling spatial processes there is uncertainty about what happens at unsampled locations. The classical statistical approach to assessing uncertainty (see Chapter 15) is to assume that samples of a property are selected uniformly (random sampling, stratified sampling, etc.) and are independent. In this case, the statistical rules for sampling design

(Cochran, 1977; Yates, 1981) apply, such as that for the variance of the mean. The geostatistical approach to estimation at unsampled locations is based on a probabilistic model that recognizes these uncertainties (Isaaks and Srivastava, 1989). In other words, at one point in space there is not just *one* value for a property but a *whole set* of values that could be drawn:

- This is regarded as a random variable for a single location and as a single realization of an ensemble of possible values in space.
- Many randomly located samples of a property in space are regarded as a set of random variables (or realizations) that constitute a random function, a random process or a stochastic process (Webster and Oliver, 2001).
- The actual values that form the realization of the random function are known as a regionalized variable.

Values of regionalized variables tend to be related, and their variation appears to depend on the distance that separates them. It is desirable to describe the way in which they vary in space as this will better inform our estimations at unsampled locations. Regionalized Variable Theory (Matheron, 1971) is derived and explained elsewhere (cf. Webster and Oliver, 2001: 51). In practice, the assumptions of Regionalized Variable Theory are reduced to establishing that the means and the variances of differences between samples over small local neighbourhoods are very similar over the study area (called quasi-stationarity or the Intrinsic Hypothesis). In situations where this does not hold, the local mean values vary predictably or deterministically from one part of the region to another. This is evidence of trend, or drift, which violates the assumptions of geostatistics. It is common in elevation data (cf. Chappell et al., 1996) and an example of its treatment is provided in the next section.

CHARACTERIZING SPATIAL PROCESSES USING THE VARIOGRAM

A common approach to understanding the complexity of the natural environment was to simplify it and divide the variation on the basis of certain attributes. This was widespread in land resource surveys and stratification, and classification was the norm in soil survey (Webster and Oliver, 1990). This approach aimed to use characteristic information (for example, the mean and the standard deviation of the samples) for any one class to predict conditions elsewhere in space within the same class. The prediction was based on samples of the population and was dependent on a reduced within-class variance and increased between-class variance. The precision of prediction was limited by the goodness of the classification. Variation within classes (in other words, the continuous variation of a property) was ignored and local variation was not resolved. However, we know that samples of the spatial process that are closer together are more similar than those further apart – they are spatially dependent.

Geostatistics provides a different way of thinking about variation in space. It makes use of autocorrelation by using a tool (the variogram) that incorporates the separation distance and difference in magnitude between samples to characterize the variation in space. This section provides an outline of the important considerations when computing a variogram and of the importance and the procedures for fitting models to the variogram. Examples are provided to illustrate the way in which information contained in the variogram (structure and model parameters) may be used to interpret the processes and scale of spatial variation for different properties. Improvements over conventional sampling strategies are illustrated with the combined use of the variogram and nested sampling.

The semi-variogram (or commonly the variogram) is a practical tool for estimating the expected squared difference between two values and their respective separation distance. The semi-variance is commonly computed as:

$$\hat{\gamma}(h) = \frac{1}{2m(h)} \sum_{i=1}^{m(h)} \{z(\mathbf{x}_i) - z(\mathbf{x}_i + \mathbf{h})\}^2$$

where the $z(\mathbf{x})$ and $z(\mathbf{x} + \mathbf{h})$ represent actual values of a property Z at places separated by the lag vector $\mathbf{h}$ for a set of data $z(\mathbf{x}_i)$, $i = 1, 2, \ldots$ and where $m(\mathbf{h})$ is the number of pairs of data points. There are a number of important issues arising from the calculation of the variogram that are identified and revisited in the examples below:

- The variogram is the average of all pairs of semi-variances at each separation distance and consequently it is dependent on the amount of data and the consistency of those data. In other words, the variogram can be very erratic when few data are available (Webster and Oliver, 1992) and when those data include extreme values or outliers. Although Krige and Magri (1982) have shown outliers do not present a serious problem, Gilbert and Simpson (1985) suggest the conventional variogram estimator is not appropriate for this type of data.
- Note that $\mathbf{h}$ is a vector and this implies direction in addition to distance. The importance of direction in the variogram will be illustrated below, but directional variograms will not be considered here in detail. Those readers interested in anisotropic variation (spatial variation that is different in different directions) such as that found in fluvial or aeolian geomorphology are encouraged to investigate directional variograms further in Webster and Oliver

(2001). The focus here will be the model of spatial variation for a property that does not specify a direction; instead, all directions are included. This is commonly termed an omni-directional variogram.

The semi-variances ordered according to their separation distances (lag) **h** are in practice grouped according to a user-defined lag interval, averaged for each interval and plotted against the lag interval. Experimental variograms for electromagnetic (EM) surveys in soil forming a playa in Australia are shown in Figure 22.2. Measurements of EM were made approximately every 1 m over a 25 km^2 area to investigate the spatial variation of salinity in the soil. A separate complex (nested, stratified and gridded) sampling strategy of 160 locations was also used to make the same measurements. The variogram of EM from the detailed survey is shown using a very small lag interval because of the large amount of data available. The variogram of EM of a few sample locations in the same area follows a similar pattern. However, the latter is much more erratic and appears to under-represent the variation of samples spaced between 500 m and 1500 m apart (less so for lag spacings between approximately 1000 m and 1500 m). This is a consequence of the sample locations. In this case, the nested sampling strategy has not represented the spatial variation of the property as well as that of the detailed survey. However, with so few samples (about 160), the nested sampling strategy has captured a remarkably large proportion of the variation at all spatial scales. This is a consequence of tailoring the nested sampling strategy to the geomorphic information of the spatial scale of variation (for example, using the catena, topography, etc. – cf. Chappell, 1999). Conventional sampling strategies using relatively few samples are often incapable of capturing all scales of variation in a single campaign and hence would perform poorly by comparison with the detailed survey. Furthermore, where little is known about the spatial variation of a property, conventional sampling is also often inadequate and a nested sampling strategy may be the only cheap method for capturing the scales of variation (Oliver and Webster, 1986; Chappell, 1999). Improvements to the sampling strategy can later be made by using the variogram and the map of spatial variation. In this case (Figure 22.2), improvements to the nested sampling strategy could be undertaken by inserting more samples with spacings from existing samples at the under-represented lag spacings.

In order to interpret the spatial variation summarized by the variogram, it is necessary to characterize the structure and provide parameters that represent it. For example, Figure 22.2 shows that, with few samples and a large lag interval, it is difficult to know with certainty that the semi-variance decreases to zero as the lag distance

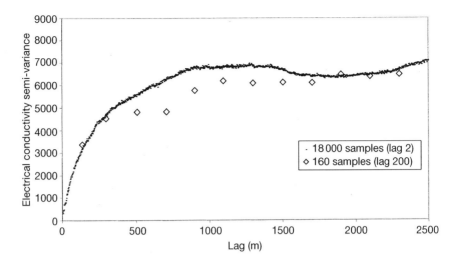

Source: Data provided by J. Leys

FIGURE 22.2   Experimental variograms of electromagnetic (EM) induction in a 25 km² area of Diamantina Lakes National Park, western Queensland, Australia, calculated for: (1) 18 000 samples approximately every 1 m; and (2) 160 samples using a nested stratified strategy

decreases and (even when many samples are available) the semi-variance does not quite reach the origin. However, the semi-variance at a very small lag spacing (tending to zero) is an important characteristic of the variogram structure. Furthermore, the experimental variograms in Figure 22.2 suggest that, as lag spacing increases (samples are further apart), the semi-variance does not continue to increase. In general, the rate of change in semi-variance decreases with increasing separation until some point at which it appears to remain relatively consistent despite lag spacing increasing.

The experimental semi-variances are generally modelled mathematically to estimate the underlying regional variogram. The characteristics of the spatial variation can then be extracted and compared with other properties, and those parameters of the variogram model can be used for kriging. Only models that meet several assumptions – i.e. are 'conditional negative semi-definite' (CNSD) – may be used for kriging (Oliver et al., 1989b). For simplicity, groups of models known to be CNSD are used and linear additions of these models may be combined for more complex structures. In general, there are two families of functions that describe the simple forms of variograms: those that are bounded and those that are not (Figure 22.3).

The steepness of the initial slope of the variogram indicates the intensity of change in a property with distance and the rate of decrease in spatial dependence. Where the extent and intensity of sampling enable the scale of spatial dependence to be determined, the variogram will reach a maximum, called the sill variance, where it flattens (Figure 22.3a). The lag at which the sill is reached is the range, or limit, of spatial dependence (Figure 22.3a). This bounded form of the variogram is generally interpreted as representing variation that consists of transition features such as different types of soil or lithology (Burgess and Webster, 1980; Oliver and Webster, 1987). It is commonly represented by the spherical model (Figure 22.3b) but the circular model (Figure 22.3c) often fits equally well. Experimental variograms with the form of the exponential model (Figure 22.3d) are expected where differences in soil type are the main contributions to soil variation and where the boundaries between types occur at random (Webster and Oliver, 2001). In an unbounded variogram, the decrease in spatial dependence may continue indefinitely (Figure 22.3). This unbound variation has been attributed to an increasing number of sources of variation as the area of interest increases (see Chappell et al., 1996; Chappell, 1998). It is typically represented by a linear model

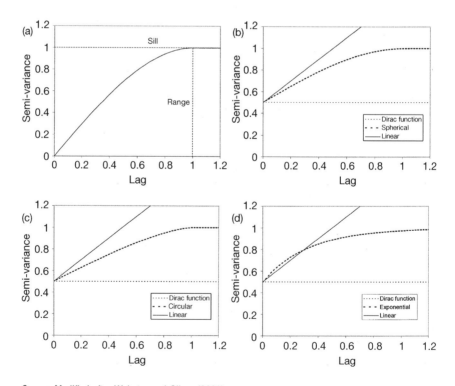

Source: Modified after Webster and Oliver (2001)

FIGURE 22.3 Theoretical models used for fitting to experimental variograms: (a) characteristic parameters of bounded models – an unbound linear model and dirac function (nugget model) – are shown for comparison with (b) spherical; (c) circular; (d) exponential models

which is a special case of the more general power model. If an experimental variogram is unbounded and concave upwards near the origin, this may indicate trend or drift which violates the assumptions of stationarity (as described in the previous section) and must be removed before the analysis can proceed. This is discussed in more detail below using elevation data as an example. The mathematical functions of these and other commonly found models in environmental applications are described in detail elsewhere (Webster and Oliver, 2001).

Experimental variograms often have a positive intercept on the ordinate called the nugget variance and this sometimes appears completely flat, exhibiting pure nugget variance (Figure 22.3b). Webster and Oliver (2001) suggest that, if the property of interest is continuous, a variogram exhibiting a pure nugget variance has almost certainly failed to detect the spatially correlated variation because the sampling interval was greater than the scale of spatial variation. The nugget variance was used successfully as an indicator of sampling suitability in rain gauge networks (Chappell and Agnew, 2001) and as a measure of noise in remote-sensing images (Curran and Dungan, 1989; Chappell et al., 2001). The other geostatistical parameters were diagnostic of the structure of spatial variation in soil geomorphology (cf. Chappell et al., 1996; Chappell and Oliver, 1997) and were essential for estimation at unsampled locations (kriging).

The presence of trend or drift violates the assumptions of stationarity in the data. This systematic trend must be removed prior to computing the experimental variogram. The trend is common in elevation data – for example, over a hillslope or along a river where the difference between the elevation of locations is not wholly dependent on location but it is also determined by the systematic change in slope. An example of trend in the experimental variogram for elevation in a river channel is shown in Figure 22.4a. The trend may be removed by fitting a low-order polynomial (for example, the polynomial $z = ax + by + c$ where z is the elevation and x and y are the location co-ordinates) to the co-ordinates of the data. The residuals of the polynomial are themselves random variables (Olea, 1975; 1977). The use of a polynomial assumes that it explains adequately the systematic variation of the slope. Some geostatisticians (cf. Chilès and Delfiner, 1999) argue that the separation between global systematic change and local variation is problematic and recommend the use of structural tools that are insensitive to drift and rely on a special theory (intrinsic random function – IRF-k – Knotters et al., 1995). However, experience in many cases suggests that the residuals of a polynomial fitted to data are adequate to identify the majority of the spatial variation (Figure 22.4a). The parameters of the model fitted to the omni-directional variogram for elevation above an arbitrary datum

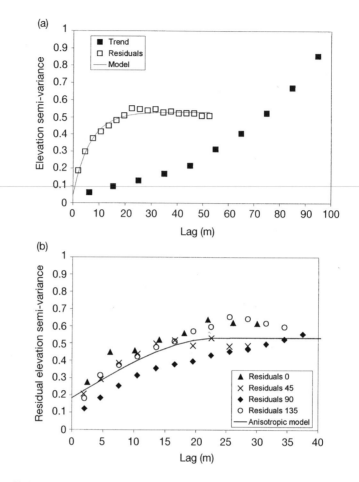

Source: Modified after Chappell et al. (in press b)

FIGURE 22.4 Experimental variograms of elevation above an arbitrary datum in a 1 km² river channel in the UK. An omni-directional experimental variogram of elevation shows evidence of trend and an omni-directional variogram of the same elevation data is shown with trend removed and an exponential model fitted (a: top). The experimental variograms calculated in four directions (0°, 45°, 90° and 135°) are fitted with an anisotropic model (b: bottom)

(Table 22.1) show that the range of spatial dependence is approximately 20 m.

This variogram is effectively the average of the variation in all directions. It hides the variation in spatial dependence in different directions (Figure 22.4b) and the range parameters of the directional variograms illustrate the difference in magnitude of the distance of spatial dependence. Most of the variation in elevation has been captured by the sampling frequency since the nugget variance for all variograms is very small (Table 22.1). This is not surprising since more

TABLE 22.1 Parameters of models (omni-directional and anisotropic) fitted to the elevation relative to an arbitrary datum in a river channel in the UK

Type	Model type	Range (m)	Variance		$C_0/(C + C_0)$ (%)
			Nugget (C_0)	Sill (C)	
Omni-directional	Exponential	19.83*	0.05	0.48	9.43
0°	Exponential	34.05*	0.22	0.45	32.84
45°	Spherical	15.68	0.14	0.36	28.00
90°	Exponential	71.96*	0.11	0.58	15.94
135°	Spherical	28.03	0.19	0.44	30.16
Anisotropic	Spherical	23.47	0.18	0.35	33.96

* The effective range is used to represent the range of spatial dependence for the exponential model.

than a thousand locations were used to sample the variation. The sill variance is also considerably larger than the nugget variance, suggesting that much of the spatially correlated variation has also been captured (see the ratio of the uncorrelated nugget variance and the correlated variance in the last column). The parameters of these models may be used for kriging. However, the anisotropic variogram fitted to all the directional experimental variograms (Figure 22.4b) is the best approximation of the variation in all directions. It may be used for kriging that takes account of different variation in different directions. A more detailed account of anisotropic modelling and the mechanism for anisotropic kriging is beyond the scope of this introduction and interested readers are directed to the core geostatistical textbooks (e.g. Webster and Oliver, 2001).

A procedure for characterizing spatial variation using the variogram is provided in Box 22.1. The aim is to model the spatial variation of a property (theoretical variogram) so that (1) interpretations of the variogram can be made to characterize variation over space; and (2) estimates may be made at previously unsampled locations. The model is estimated by the experimental variogram and a function is fitted to it. The procedure illustrates the basic decisions and considerations required when calculating, modelling and interpreting the experimental variogram and the model parameters. The calculation of the variogram is relatively straightforward and can be done with a spreadsheet or computer program unless many data are being considered. Model fitting is more complicated than the variogram computation and is best left to the bespoke geostatistical software packages (computer code and software availability is described in the further reading below).

Box 22.1   Procedure for computing an experimental variogram, fitting a model and interpreting the parameters of that model

Calculate the variogram

Omni-directional variogram   Calculate the omni-directional variogram using separation distances between samples in all directions.

Trend   Is there any systematic change in the mean of the property over space? Is there evidence in the shape of the variogram for this non-stationary behaviour (trend or drift)?

Directional variograms   Calculate the variogram separately for at least four directions (0°, 45°, 90° and 135°).

Characterize the spatial variation

Bounds   Does the variogram appear bounded – i.e. does the semi-variance reach a maximum within the distance computed or appear as though it would reach a maximum if the lag distance was extended somewhat (bounded)? Alternatively, does it look as though it would increase without limit (unbounded)?

Anisotropy   Does the variogram have approximately the same form and values in all directions? If so, then accept it as isotropic and compute an average experimental variogram over all the directions. If not, then in what way do the directions differ?

Nugget   Does an imaginary line drawn through the experimental values when projected cut the ordinate at a positive value (not 0)? If so, this intercept is known as the nugget variance.

Model the experimental variogram

Choose   Choose from simple models one or more with approximately the right shape and with sufficient detail to honour the principal trends in the expected values that you wish to represent.

Fit   Fit each model in turn by weighted least squares – i.e. by minimizing the sums of squares suitably weighted between the expected and fitted values. Calculate some statistics for goodness of it – for example, residual sum of squared (RSS) difference between expected and fitted values.

Judge   Plot fitted model on the same pair of axes as the experimental variogram and if all looks well choose from among them the smallest RSS.

Source: Modified from Webster and Oliver (2001)

The variogram is a tool to summarize spatial variation that is effective for elucidating the processes controlling landforms and a model of underlying spatial process that is used for geostatistical estimation at unsampled locations using kriging.

LOCAL ESTIMATION OR PREDICTION

When you have sampled the values of a property at many locations it should be possible to make estimates at unsampled locations. A brief history of estimation is provided by Cressie (1993). This section provides an outline of the important differences between estimation and interpolation conducted using an arbitrary mathematical function such as is found in graphical software (for example, inverse-distance squared) and of the procedures for which geostatistics was initially developed, generally called kriging (Krige, 1966).

The derivation of the kriging equations is omitted here in favour of a more pragmatic approach. Those interested in understanding the mechanisms by which the kriging conditions are conducted are referred to the further reading at the end of the chapter. Kriging estimates are often far smoother than the variability of the sample data. (Although not considered here, stochastic simulation is an alternative geostatistical mapping approach which may be very useful to some readers. It emphasizes the global attributes of a property and emulates the sample statistics and is very useful for geographical information systems (GIS) analysis – Goovaerts, 1997.)

Interpolation methods

Interpolation methods are deterministic and to that extent they accord with our understanding that the variation in the environment has physical causes – in other words, it is physically determined. However, the spatial processes are so complex that our current under-standing renders mathematical functions (inverse-distance squared, minimum curvature, etc.) far from adequate to describe any but the simplest components (Webster and Oliver, 2001). Most methods of interpolation make predictions at unsampled locations from nearby data using an arbitrarily selected mathematical function (Webster and Oliver, 1990). Burrough (1986) and Webster and Oliver (2001: 37) provide useful reviews of various methods for interpolation at unsam-pled locations. Many of these methods take an average of existing samples from a *local* neighbourhood. The way in which the nearby data are included is often based on a mathematical relationship between distance and the variation of the property in space. For example, interpolation using inverse-distance squared (IDS) takes existing data within a certain distance of the location to be predicted

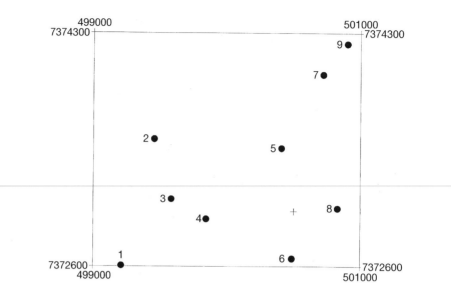

FIGURE 22.5 The location of nine samples of aeolian transport (g cm⁻¹) from Australia and the unsampled target location for estimation (+)

(Figure 22.5), calculates the influence of each sample on the prediction using the IDS function and increases the importance it will have on the average of all values (Table 22.2).

TABLE 22.2 Values of the aeolian transport data at each of the selected sample locations (Figure 22.5), the distance from the target location and the weights given to each location using homogeneity, inverse-distance squared and ordinary kriging derived from the variogram

Location	Values	Distance	Homogeneous	Inverse-distance squared*	Ordinary kriging
1	4.24	1344.34	0	0.001	0.003
2	0.03	1167.68	0	0.001	0.010
3	0.23	917.33	0	0.001	0.038
4	2.76	655.40	0	0.002	0.123
5	4.99	469.51	0	0.002	0.203
6	5.42	342.25	0	0.003	0.290
7	0.02	1021.20	0	0.001	0.036
8	208.24	330.73	0	0.003	0.291
9	0.02	1285.16	0	0.001	0.007
Estimated location	—	—	25.11	47.25	63.64

*Weights are not complete for each location since aggregation for all locations is included in the calculation.

Thus, the prediction at the unsampled location is a weighted average according to the mathematical function. The weights shown are for the distance, but the weighted values are not shown because the target value is the average of the result. An illustration of a spatial process predicted by an arbitrarily selected IDS function for aeolian transport is shown in Figure 22.5.

Geostatistical estimation: kriging

Kriging is the method of geostatistical estimation at unsampled locations. Since kriging was first formulated there have been many developments to tackle an increasing complexity of problems in a variety of disciplines. However, ordinary kriging is the most common type of kriging in practice and the discussion below is limited to this method. It is similar to methods of interpolation in its use of a weighted average of nearby samples. However, the main difference is that the weights given to nearby samples are derived from the variogram model rather than in an arbitrary way. Since the variogram model is fitted to the experimental variogram, which itself was derived from the existing samples as described above, the variogram provides an appropriate relationship between distance and the variation of the property in space.

In addition to a weighted average there are several other properties incorporated into ordinary kriging. The weights of those nearby samples required for an estimate at an unsampled location are made to sum to 1 (Table 22.2). The expectation is that there is no error of the estimate made at the unsampled location but that the estimate has variability. For example, if the estimation location happens to coincide with an existing sample the value at that location is allowed to exist and the variability at that location is reduced to zero. Unlike other methods ordinary kriging provides estimates that are an exact interpolation. The kriging procedure also has the ability to make estimates at points and over blocks of any size. The method by which the weights are computed is not straightforward and those readers interested in a mathematical explanation of the procedures for ordinary kriging are directed to the core geostatistical textbooks (Goovaerts, 1997; Webster and Oliver, 2001).

A comparison between the weights generated by an arbitrary mathematical function and ordinary kriging is provided in Table 22.2. The value at the unsampled location interpolated by inverse-distance squared and that estimated by ordinary kriging is also shown in Table 22.2. The influence of spatial variation on the prediction at the unsampled location can be seen by comparing these values with that of a simple average of the available data which implicitly assumes homogeneity over space. Webster and Oliver (2001: 155) provide further examples of the weighting procedure and also illustrate the effect that the variogram model parameters have on the weights and the estimation procedure. In general, they suggest that the nearest four or five points might contribute 80% of the total weight and the next nearest ten almost all the remainder. Goovaerts (1997: 125) provides two-dimensional examples of ordinary kriging and shows developments that include simple kriging and kriging with a trend model.

The procedure for ordinary kriging is provided in Box 22.2 and is based on that of Webster and Oliver (2001). It can be used as the basis for providing the necessary information for ordinary kriging in most computer programs. It is especially appropriate for providing some of the parameters for the GSLIB kriging code (OKB2DM or COKB3DM) (Deutsch and Journel, 1992). Examples of the parameter files are included in the GSLIB manual and information on obtaining the computer code is provided in the further reading at the end of the chapter.

For comparison with the previous inverse-distance squared interpolation, the kriging equations were used to estimate aeolian transport at points every 200 m across the 5 km × 5 km study area in Australia. Contours were threaded through the kriging estimates with the same contour frequency as the previous interpolation (Figure 22.6). The results presented here are similar to those found by many other studies (for example, Chappell, 1999). The IDS function produces concentric artifacts at unsampled locations. This is primarily because of the arbitrary distance over which nearby samples influence the interpolation. The pattern of transport produced by ordinary kriging estimates is consistent with the observed continuous spatial transport process. Kriging has been shown to be one of the most reliable two-dimensional spatial estimators (Laslett et al., 1987; Laslett and McBratney, 1990; Laslett et al., 1994) and it is expected to produce more reliable estimates of sediment transport than simple methods of interpolation. However, kriging tends to smooth the variability in the data, and other geostatistical techniques are required to maintain the variability in the original data (for example, stochastic simulation – Goovaerts, 1997). It is good practice in geostatistical analyses to validate the geostatistical model and kriging plan by cross-validation (Deutsch and Journel, 1992). Each sample location is estimated under conditions which closely resemble those used later for kriging, variogram model(s), the type of kriging and the search strategy. Actual data are removed (with replacement) one at a time and re-estimated using the remaining data. The estimation performance is commonly quantified using the root mean square error (RMSE) and mean absolute difference (MAE) between observed and estimated values.

Box 22.2	Procedure for ordinary kriging to provide the appropriate parameters for computer code such as GSLIB (OKB2DM)
Punctual or block kriging?	If the targets are points (for example, small soil cores) then punctual kriging is required. Alternatively, small blocks may be specified in block kriging. More information on block kriging and the number of points used to discretize the blocks can be found in the core geostatistical textbooks.
Number of data for each kriging estimate?	Ordinary kriging computes a weighted average of the data. The weights are determined by the configuration of the data in relation to the target in combination with the variogram model. In general, the nearest 20 points to the target should be specified. If the data points are exceptionally scattered then a local estimate can be retained by reducing the number of included points.
Search radius?	The search radius should be the same size as the range of spatial dependence. In some cases this may need to be slightly larger in order that sufficient data are included in each kriging estimate.
Transformation?	Simple linear geostatistics requires that the data conform approximately to a normal distribution. If the data are skewed and have been transformed the variogram model parameters and the kriging estimates should be of the transformed data. If you want estimates to be of the original (untransformed) data then you must transform the estimates back.
Variogram parameters?	You will need to include the number and type of each structure for your model of spatial dependence, the amount of nugget variance, the range, spatially dependent variance and any angle and ratio of anisotropy (as described in the previous section).
Kriging for mapping	Specify a finely spaced interval on a square grid that is no more than 2 mm on the final hard copy. Once the kriging estimates have been written to a computer file, pass those values to a graphics program for the final display of the results as isarithms (contours of the same level).

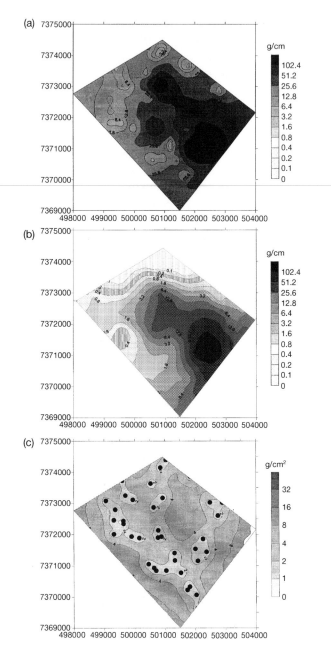

Source: Modified from Chappell et al. (in press a)

FIGURE 22.6 A universal trans-Mercator projection (in metres; north is orientated to the top) of aeolian transport (g cm^{-1}) from the south displayed using the same (doubling) isarithms for: (a) inverse-distance squared (IDS) interpolations; (b) ordinary punctual kriging estimates; and (c) sample locations and ordinary kriging estimation variance every 200 m over a 5 km × 5 km study area in Diamantina Lakes, western Queensland, Australia

It is important to note that geostatistics (kriging especially) is unsuitable where there are abrupt spatial changes in properties. Researchers have combined soil classification with kriging to utilize the advantages of both approaches (Bregt et al., 1987; Voltz and Webster, 1990; Heuvelink and Bierkens, 1992). Studies have improved kriging estimates by sampling more effectively by using regional stratifications such as soil maps (Voltz and Webster, 1990; Yost et al., 1993) or geomorphic information (McBratney et al., 1991).

An important by-product of the kriging estimation process is the estimation variance, which may be considered to be an estimation reliability map (Figure 22.6c). The kriging and estimation variance weights obtained from the variogram depend on the configuration of the sampling points and not on the observed values (Burgess et al., 1981). The estimation variances can be used to minimize the maximum kriging error and to optimize future sampling campaigns (McBratney et al., 1981).

CONCLUSION: HOW GEOSTATISTICS CAN HELP YOU

Geostatistics has a solid foundation in the statistical theory of regionalized variables. The theory is underpinned by the assumptions that the spatial process is currently too complicated to make reasonable estimates about the nature of a property found at an unsampled location; samples of the process adequately represent the model of spatial variation; and that variation is spatially dependent. There are many developments in the literature for a plethora of applications over space, time and in space-time (Heuvelink and Webster, 2001) in geography and environmental science.

Summary

*   The central tool of geostatistics is the variogram. It provides a summary for the variation of a property in space. Its structure may be used to characterize and interpret that variation in space. Repeated calculation of the variogram at different points in time enables a comparison of the spatial processes at different stages in time. More advanced geostatistics provide the opportunity to consider the variation of a property simultaneously in time and space. However, to estimate the model of spatial dependence, the mean and variance of a distribution are required to be the same everywhere. This constraint of second-order stationarity is often restrictive and may be relaxed to constraints within local areas. If this Intrinsic Hypothesis cannot be satisfied or there is evidence (in the variogram) that is not satisfied, the systematic variation over

  space must be removed so that the residuals may be used in
  investigating the model of spatial variation.

- The representation or visualization of the spatial dependence in a
  property usually requires some form of map. Kriging provides the
  ability to estimate values at unsampled locations with little
  compromise in the statistical theory, unlike other arbitrary
  mathematical interpolators. Despite this advantage, kriging tends
  to smooth the variability in the data, and other geostatistical
  techniques (for example, stochastic simulation) provide the
  opportunity to maintain the variability in the original data.

- The use of geostatistics in geography and environmental science
  is a rapidly evolving field of study.

- Software for geostatistical methods is becoming routinely
  available, and it is well worth exploring the Internet for free
  versions.

Further reading

Provided below are references to the core textbooks that have influenced my understanding
of geostatistics. Although the diversity of texts is limited to my experience of the fields
described, the annotations provide a brief outline suitable for those wishing to tackle
geostatistics in a variety of fields. As geographers or environmental scientists, we should
not limit ourselves to traditional subject boundaries and should enhance a holistic
understanding of spatial analysis by investigating geostatistical applications in many
disciplines.

- A shortage of geostatistics textbooks led several practitioners to publish in the last few
  years comprehensive and detailed information on geostatistics (cf. Cressie, 1993;
  Goovaerts, 1997; Chilès and Delfiner, 1999). Although these books are required
  reading for the more adept geostatistician, most are not suitable as introductory
  undergraduate texts because of their in-depth treatment of the subject.

- Isaaks and Srivastava (1989) provide a book which progressively develops from
  simplistic to complicated geostatistics using a common database. However, it is
  difficult to develop from it a rationale for geostatistics. Webster and Oliver (2001)'s
  book is accessible to a new student of geostatistics and gradually leads the inquisitive
  mind into greater detail and more opportunities for practice. Of particular importance
  for the novice practitioner is the section in the book on the reliability of the variogram,
  which includes discussion on sample sizes.

- The application of geostatistics to remote-sensing data and its incorporation into
  geographical information systems have made a considerable impact on land surface
  mapping (cf. Dancy et al., 1986; Atkinson et al., 1992), on the summary and
  interpretation (cf. Woodcock et al., 1988; Curran and Dungan, 1989; Oliver et al., 2000;
  Chappell et al., 2001) and classification of digital imagery (cf. Atkinson and Lewis,
  2000) and for reducing data redundancy (cf. Webster et al., 1989; Atkinson et al.,
  1990; Atkinson and Curran, 1995). There are many texts on GIS that incorporate the

use of remote sensing and geostatistics (cf. Burrough, 1986) but few which include discussion on the primary components of geostatistics outlined in the main text above that are available for use in remote sensing. Stein et al. (1999) gathered together remote-sensing researchers who utilize specialist components of geostatistics. Consequently, their book provides a wealth of information on the application of geostatistics to remote sensing. It also provides the opportunity to appreciate the difference afforded by geostatistics to the treatment of large quantities of data, which contrasts markedly with limited samples of spatial processes considered in physical geography and environmental science.

- There are many more applications of geostatistics in physical rather than human geography. This is probably because data in the latter are often not in a form that is readily analysed in this way. However, Oliver et al. (1992) illustrate the use of geostatistics in human geography in their analysis of the spatial variation of childhood leukemia. Their analysis was complicated by the form of the spatial data but, by overcoming it, they were able to provide an explanation to the factors controlling leukemia that had not previously been considered. There appear to be many more geostatistical applications in soil science than in geomorphology (Oliver and Webster, 1987). Oliver and Webster (1991) provide a brief history of geostatistics in soil science that shows that, in its earliest form, geostatistics had been considered as early as 1911. Their paper illustrates many of the areas in soil science that have improved with the use of geostatistics. Despite this shortfall in applications by comparison with soil science, there are many examples of the use of geostatistics in other fields of physical geography (cf. Oliver et al., 1989b; Leenaers et al., 1990; Chappell, 1998; Chappell et al., 1996; 1998; Chappell and Agnew, 2001).

- Sterk and Stein (1997) illustrate one of the greatest problems with the application of geostatistics to environmental problems – the shortage of data in one or more domains. This is especially the case with climate data which often derive from few stations in space that monitor climatic conditions frequently over time. This is problematic for the reliable estimation of the variogram as discussed in the main text above (Webster and Oliver, 1992). Sterk and Stein (1997) provide a method of overcoming the sampling difficulty by combining measurements made over time with variation in space. Although this may sound complicated, its application amounts to the careful design of sampling. Nested approaches to sampling have, in several conventional geostatistical studies, provided the ability to investigate and map properties that would otherwise be too expensive (cf. Chappell et al., 1996). Sterk and Stein (1997) combine samples from each (wind erosion) event to improve the modelling of the variogram and to produce reliable estimates for each event. One operational difficulty with this approach is that computer code necessary for its implementation is not readily available, but it is likely that the wide applicability of this theory will ensure its incorporation into major geostatistical software in the future.

- The proliferation of geostatics was for a long time hindered by the availability of computer code suitable for its application. Deutsch and Journel (1992) provided a suite of programs that quickly became the standard, probably because they enabled a diverse range of applications to be considered and because the programs offered the flexibility for modification for more specific solutions. The first edition of their book provides a comprehensive overview of the practical considerations for the application of geostatistical theory. The book has recently been revised to reflect a recent packaging of the programs to a more accessible operating environment. In recent years other software has been developed so that now there is a wide range of geostatistical programs. One of the best databases of available software and the most successful geostatistics email discussion list is 'Al-geostats', which can be found by searching the Internet.

Note: Full details of the above can be found in the references list below.

References

Atkinson, P.M. and Curran, P.J. (1995) 'Defining an optimal size of support for remote sensing investigations', *IEEE Transactions on Geoscience and Remote Sensing*, 33: 1–9.

Atkinson, P.M., Curran, P.J. and Webster, R. (1990) 'Sampling remotely sensed imagery for storage retrieval and reconstruction', *The Professional Geographer*, 42: 345–52.

Atkinson, P.M., Curran, P.J. and Webster, R. (1992) 'Co-kriging with ground-based radiometry', *Remote Sensing Environment*, 41: 45–60.

Atkinson, P.M. and Lewis, P. (2000) 'Geostatistical classification for remote sensing: an introduction', *Computers and Geosciences*, 26: 361–71.

Bregt, A.K., Bouma, J. and Jellinek, M. (1987) 'Comparison of thematic maps derived from a soil map and from kriging of point data', *Geoderma*, 39: 281–91.

Burgess, T.M. and Webster, R. (1980) 'Optimal interpolation and isarithmic mapping of soil properties. I. The semi-variogram and punctual kriging', *Journal of Soil Science*, 31: 315–31.

Burgess, T.M., Webster, R. and McBratney, A.B. (1981) 'Optimal interpolation and isarithmic mapping of soil properties. IV. Sampling strategy', *Journal of Soil Science*, 32: 643–59.

Burrough, P.A. (1986) *Principles of Geographical Information Systems for Land Resources Assessment*. Oxford: Clarendon Press.

Chappell, A. (1998) 'Mapping ^{137}Cs-derived net soil flux using remote sensing and geostatistics', *Journal of Arid Environments*, 39: 441–55.

Chappell, A. (1999) 'The limitations for measuring soil redistribution using ^{137}Cs in semi-arid environments', *Geomorphology*, 29: 135–52.

Chappell, A. and Agnew, C.T. (2001) 'Geostatistical analysis of west African sahel rainfall', in A. Conacher (ed.) *Land Degradation*. Dordrecht, Boston, MA, and London: Kluwer Academic, pp. i–x, 390.

Chappell, A., Heritage, G.L., Fuller, I.C., Large, A.R.J. and Milan, D.J. (in press b) 'Geostatistical analysis of ground-survey elevation data to elucidate spatial and temporal river channel change', *Earth Surface Processes and Landforms*.

Chappell, A., McTainsh, G., Strong, C. and Leys, J. (in press a) 'Temporal change in the spatial variation of aeolian sediment transport in the Queensland Channel Country, Australia', *Earth Surface Processes and Landforms*.

Chappell, A. and Oliver, M.A. (1997) 'Geostatistical analysis of soil redistribution in SW Niger, west Africa', in E.Y. Baafi and N.A. Schofield (eds) *Quantitative Geology and Geostatistics. Volume 8/2*. Dordrecht: Kluwer Academic, pp. 961–72.

Chappell, A., Oliver, M.A., Warren, A., Agnew, C.T. and Charlton, M. (1996) 'Examining the factors controlling the spatial scale of variation in soil redistribution processes from south-west Niger', in M.G. Anderson and S.M. Brooks (eds) *Advances in Hillslope Processes*. Chichester: Wiley, pp. 429–49.

Chappell, A., Seaquist, J.W. and Eklundh, L.R. (2001) 'Improving geostatistical noise estimation in NOAA AVHRR NDVI images', *International Journal of Remote Sensing*, 22: 1067–80.

Chappell, A., Warren, A., Taylor, N. and Charlton, M. (1998) 'Soil flux in southwest Niger and its agricultural impact', *Land Degradation and Development*, 9: 295–310.

Chilès, J.P. and Delfiner, P. (1999) *Geostatistics: Modelling Spatial Uncertainty*. Chichester: Wiley.

Cochran, W.G. (1977) *Sampling Techniques* (3rd edn). New York: Wiley.

Cressie, N.A.C. (1993) *Statistics for Spatial Data*. New York: Wiley.

Curran, P.J. and Dungan, J.L. (1989) 'Estimation of signal-to-noise: a new procedure applied to AVIRIS data', *IEEE Transactions on Geoscience and Remote Sensing*, 27: 620–8.

Dancy, K.J., Webster, R. and Abel, N.O.J. (1986) 'Estimating and mapping grass cover and biomass from low-level photographic sampling', *International Journal of Remote Sensing*, 7: 1679–704.

de Roo, A.P.J. (1991) 'The use of ^{137}Cs as a tracer in an erosion study in south Limburg (the Netherlands) and the influence of Chernobyl fallout', *Hydrological Processes*, 5: 215–27.

Deutsch, C.V. and Journel, A.G. (1992) *GSLIB Geostatistical Software Library and User's Guide*. Oxford: Oxford University Press.

Gallichand, J. and Marcotte, D. (1993) 'Mapping clay content for subsurface drainage in the Nuile delta', *Geoderma*, 58: 165–79.

Gilbert, R.O. and Simpson, J.C. (1985) 'Kriging for estimating spatial pattern of contaminants: potential and problems', *Environmental Monitoring and Assessment*, 5: 113–35.

Goovaerts, P. (1997) *Geostatistics for Natural Resources Evaluation*. Oxford: Oxford University Press.

Henebry, G.M. (1993) 'Detecting change in grasslands using measures of spatial dependence with Landsat TM data', *Remote Sensing Environment*, 46: 223–34.

Heuvelink, G.B.M. and Bierkens, M.F.P. (1992) 'Combining soil maps with interpolations from point observations to predict quantitaive soil properties', *Geoderma*, 55: 1–15.

Heuvelink, G.B.M. and Webster, R. (2001) 'Modelling soil variation: past, present, and future', *Geoderma*, 100: 269–301.

Isaaks, E.H. and Srivastava, M.R. (1989) *An Introduction to Applied Geostatistics*. New York: Oxford University Press.

Knotters, M., Brus, D.J. and Oude Voshaar, J.H. (1995) 'A comparison of kriging, co-kriging and kriging combined with regression for spatial interpolation of horizon depth with censored observations', *Geoderma*, 67: 227–46.

Krige, D.G. (1966) 'Two-dimensional weighted moving average trend surfaces for ore-evaluation', *Journal of the South African Institute of Mining and Metallurgy*, 66: 13–38.

Krige, D.G. and Magri, E.J. (1982) 'Studies of the effects of outliers and data transformation on variogram estimates for a base metal and a gold ore body', *Mathematical Geology*, 14.

Laslett, G.M. (1994) 'Kriging and splines: an empirical comparison of their predictive performance in some applications', *Journal of the American Statistical Association*, 89: 391–409.

Laslett, G.M. and McBratney, A.B. (1990) 'Estimation and implications of instrumental drift, random measurement error and nugget variance of soil attributes – a case study for soil pH', *Journal of Soil Science*, 41: 451–71.

Laslett, G.M., McBratney, A.B., Pahl, P.J. and Hutchinson, M.F. (1987) 'Comparison of several spatial prediction methods for soil pH', *Journal of Soil Science*, 38: 325–41.

Leenaers, H., Okx, J.P. and Burrough, P.A. (1990) 'Employing elevation data for efficient mapping of soil pollution on floodplains', *Soil Use and Management*, 6: 105–14.

Matheron, M.A. (1971) *The Theory of Regionalised Variables and its Applications. Cahiers du Centre de Morphologie Mathématique de Fountainebleau* 5. Fountainebleu: Centre de Morphologie Mathématique de Fountainebleu.

McBratney, A.B., Hart, G.A. and McGarry, D. (1991) 'The use of region partitioning to improve the representation of geostatistically mapped soil attributes', *Journal of Soil Science*, 42: 513–32.

McBratney, A.B, Webster, R. and Burgess, T.M. (1981) 'The design of optimal sampling schemes for local estimation and mapping of regionalised variables. I. Theory and method', *Computers and Geosciences*, 7: 331–4.

Odeh, I.O.A., McBratney, A.B. and Chittleborough, D.J. (1994) 'Spatial prediction of soil properties from landform attributes derived from a digital elevation model', *Geoderma*, 63: 197–214.

Olea, R.A. (1975) *Optimum Mapping Techniques using Regionalised Variable Theory. Series on Spatial Analysis* 2. Lawrence, KS: Kansas Geological Survey.

Olea, R.A. (1977) *Measuring Spatial Dependence with Semi-variograms. Series on Spatial Analysis* 3. Lawrence, KS: Kansas Geological Survey.

Oliver, M.A., Muir, K.R., Webster, R., Parkes, S.E., Cameron, A.H., Stevens, M.C.G. and Mann, J.R. (1992) 'A geostatistical approach to the analysis of pattern in rare disease', *Journal of Public Health Medicine*, 14: 280–9.

Oliver, M.A. and Webster, R. (1986) 'Combining nested and linear sampling for determining the scale and form of spatial variation of regionalised variables', *Geographical Analysis*, 18: 227–42.

Oliver, M.A. and Webster, R. (1987) 'The elucidation of soil pattern in the Wyre Forest of the West Midlands, England. II. Spatial distribution', *Journal of Soil Science*, 38: 293–307.

Oliver, M.A. and Webster, R. (1991) 'How geostatistics can help you', *Soil Use and Management*, 7: 206–17.

Oliver, M., Webster, R. and Gerrard, J. (1989a) 'Geostatistics in physical geography. Part I. Theory', *Transaction, Institute of British Geographers*, 14: 259–69.

Oliver, M., Webster, R. and Gerrard, J. (1989b) 'Geostatistics in physical geography. Part II. Applications', *Transactions, Institute of British Geographers*, 14: 270–86.

Oliver, M.A., Webster, R. and Slocum, K. (2000) 'Filtering SPOT imagery by kriging analysis', *International Journal of Remote Sensing*, 21: 735–52.

Stein, A. (1998) 'Analysis of space-time variability in agriculture with geostatistics', *Statistica Nederlandica*, 52: 18–41.

Stein, A., van der Meer, F. and Gorte, B. (eds) (1999) *Spatial Statistics for Remote Sensing*. Dordrecht: Kluwer Academic.

Sterk, G. and Stein, A. (1997) 'Mapping wind-blown mass transport modelling in space and time', *Soil Science Society of America Journal*, 61: 232–9.

Voltz, M. and Webster, R. (1990) 'A comparison of kriging, cubic splines and classification for predicting soil properties from sample information', *Journal of Soil Science*, 41: 473–90.

Webster, R. (2000) 'Is soil variation random?' *Geoderma*, 97: 149–63.

Webster, R., Curran, P.J. and Munden, J.W. (1989) 'Spatial correlation in reflected radiation from the ground and its implications for sampling and mapping by ground-based radiometry', *Remote Sensing Environment*, 29: 67–78.

Webster, R. and Oliver, M.A. (1990) *Statistical Methods in Soil and Land Resource Survey*. Oxford: Oxford University Press.

Webster, R. and Oliver M.A. (1992) 'Sample adequately to estimate variograms of soil properties', *Journal of Soil Science*, 43: 177–92.

Webster, R. and Oliver, M.A. (2001) *Geostatistics for Environmental Scientists*. Chichester: Wiley.

Woodcock, C.E., Strahler, A.H. and Jupp, D.L.B. (1988) 'The use of variograms in remote sensing. II. Real digital images', *Remote Sensing Environment*, 25: 349–79.

Yates, F. (1981) *Sampling Methods for Censuses and Surveys* (4th edn). London: Griffin.

Yost, R.S., Loague, K. and Green, R. (1993) 'Reducing variance in soil organic carbon estimates: soil classification and geostatistical approaches', *Geoderma*, 57: 247–62.

23 Using Geographical Information Systems

Michael Batty

Synopsis

Geographical information systems (GIS) are organized collections of data-processing methods which act on spatial data to enable patterns in that data to be understood and visualized. In fact, GIS is often thought of as software which takes numerical data in map form, stores it in the most efficient way in different types of computer environment and allows various techniques of analysis to process the data and then map them. However, GIS is broader than this, synonymous in some contexts with quantitative geography, and the term is increasingly being used to refer to 'geographic information science' which embraces a wide range of mathematical and statistical technique (Goodchild, 1992). This chapter first notes the origins of GIS and then defines GIS in terms of the way it represents geographical data. The various operations which enable spatial data to be analysed and visualized are referred to as 'functions', and this 'functionality' is illustrated through ways in which GIS can measure, interrogate and manipulate maps as data. The power of visualizing data through GIS is then discussed and, finally, the chapter presents issues concerning changes in the technology and its software which are driving the field. The chapter concludes by pointing to the great diversity of applications which GIS offers and the impact this is having on geography, geographers and society at large.

The chapter is organized into the following sections:

- What is GIS? Historical antecedents
- How a GIS is organized: representing data
- Functionality in GIS: geometric operations, spatial queries and map algebras
- GIS as visualization
- Technology and software
- Conclusion: applications.

WHAT IS GIS? HISTORICAL ANTECEDENTS

GIS emerged slowly from several origins. As computers became ever more powerful, computer graphics came of age and, from early and somewhat painful beginnings with large and expensive map-plotting devices, computer memories reached the point where maps could be displayed with ease on the screen. Computer cartography was complemented by the development of database theory, specifically for spatial data which involved linking geometry to geography, while quantitative models and spatial statistics gradually began to merge with the evolving GIS software (Chrisman, 1988). Remote data acquisition through various kinds of aerial and satellite sensing came to be linked with GIS, and today we stand at a threshold where the technology is being implemented in everything from hand-held devices and telephones to software for target marketing and climate modelling.

It is important to separate technology from theory. At one level, GIS is simply visualizing map data in whatever context, and in this sense, it is little different from much of the graphical computation that we see everyday on the desktop and across the web. But GIS as geographic information science is of much wider import for geography and geographical method. Currently, GIS technology has advanced to the stage where the focus is no longer on graphical representation per se but on integrating visualization with method so that both quantitative and qualitative analysis might be enriched (Longley et al., 2001). The next decade is likely to see some remarkable advances as theory and technology merge. But we are getting ahead of ourselves and before we chart the future of GIS, let us step back and illustrate how a GIS is organized and how it is used to visualize and analyse spatial data.

HOW A GIS IS ORGANIZED: REPRESENTING DATA

Two kinds of data are required. As GISs always have the capability of displaying data in map form, there must be data about the way the map is configured – as boundaries and points, for example. This is called digital data. Usually there is more than this, for the map has features or characteristics – called attribute data – and these data are associated with the map's configuration. A good example is a map of population by local authority area. The boundaries of the local authorities have to be input as digital data and the population associated with each authority is attribute data. There is a further twist to this

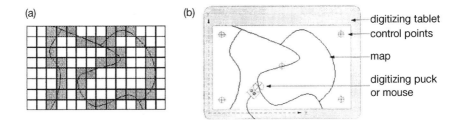

(a)   (b)

digitizing tablet
control points
map
digitizing puck
or mouse

Source: Adapted from the University of Melbourne's GIS self-learning tool
(http://www.sli.unimelb.edu.au/gisweb/)

FIGURE 23.1   Digital data types: (a) raster map with cells/pixels defining map boundaries; and
(b) vector map representation of the same on a digitizing tablet

for there are two different ways of representing the map as digital data. The simplest is to assume all areas are the same, such as if you were to represent population in grid squares. This is called a raster map. Rasters are particularly useful for data produced routinely as from satellites, where the easiest way to record it is by equal areas. However, more realistic configurations of maps are based on points and lines which are assembled into objects such as polygons. This is called a vector map such as our map of local authority boundaries. Both raster and vector maps have attributes, the raster being associated with grid squares or often pixels on a computer screen, the vector with irregular areas which are defined by assemblages of points and lines. Raster data can usually be entered into a GIS directly as numbers from the keyboard whereas vector data are usually digitized using a mouse-like device called a puck, which is centred on each of the points and lines defining the map object in question. These distinctions are shown in Figure 23.1.

Much of GIS technology is concerned with visualizing these data in map form and many low-level GIS functions are buried away in the software, no longer of any real significance to geographical analysis. In fact, there is little difference between raster and vector map data, with many systems enabling users to integrate and move easily between each. However, GIS really comes into its own when more than one set of attributes is associated with a raster or vector map. The central organizing concept is based on treating different sets of attributes as map layers. It is useful to think of geographical systems as being represented by a series of data layers which can be translated into map layers, and this leads one to consider ways in which the layers might be related. For example, in a raster map we might have layers that deal with topography, vegetation, geology, agricultural use and so on, where it is useful to relate these variables. GIS enables a first shot at such analysis by simply displaying the map layers and then by 'overlaying' them to see if there are common patterns.

Another example might be examining the relation between where people live and where they work. If these are mapped as separate layers, comparing them will reveal that people do not usually live and work in the same place. If we do this for a large city, it is likely this will reveal the classic pattern of people working at the centre and living at the edge. To illustrate the use of GIS in this chapter, the example of Greater London is presented, divided into its 33 boroughs. In Figure 23.2, different layers for the year 1991 are shown: total population, percentage of households who own their own homes and the percentage who rent from their local council. The maps showing these data have been created in the desktop software MapInfo, and the figure shows the typical desktop a user would see. Apart from the ability of this software to display such data in many different map forms, the real power of GIS comes from being able to associate, combine, analyse and model this data, and this involves us in presenting the kind of tools that are available within GIS which, in turn, form the core of geographic information science.

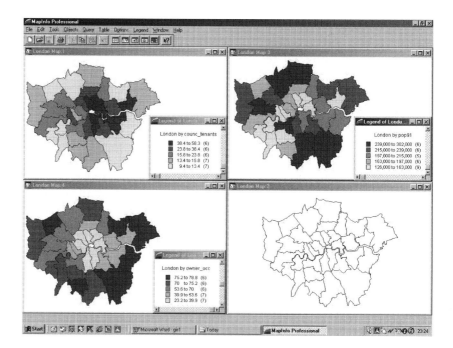

FIGURE 23.2   The digital map base and three map layers – attributes – in the GIS MapInfo

FUNCTIONALITY IN GIS: GEOMETRIC OPERATIONS, SPATIAL QUERIES AND MAP ALGEBRAS

The most basic operations of a GIS involve measuring various geometric features of the digital map data. Once such data are within a GIS, several of these measuring functions are virtually automatic. For example, most GISs have rulers which enable you to measure straight line distances, functions to compute areas of polygons and methods to count the density of points. There are many functions which derive from these, such as the ability to draw areas around points and lines – called buffers – which provide ways of computing 'nearness'. If networks are represented within a GIS, there is added functionality to find the shortest routes. Map projection into many different coordinate systems is immediate, for all such functions are routine once the map's geometric data have been represented digitally.

The other set of routine functions within a GIS pertains not to geometry but to the data themselves, and these involve various ways of interrogating it. Spatial queries of the simplest form are usually achieved by pointing at some area or point on the map and accessing its attributes directly. Much more sophisticated queries are possible, however, based on concatenating different requests. A typical one might be of the form: 'Find all areas on the map which have a population density greater than 1000 persons per square kilometre, within 5 kilometres of a main highway, and which are located on land with slopes less than 1 in 20.' Such queries can also be developed into ways of producing new data from the basic data layers and this introduces one of the most important functions of contemporary GIS software, which is called 'map algebra' (Tomlin, 1990).

The concept of data layers is useful because it is a particularly simple way of thinking about how different attributes might be combined. Such combinations are a convenient way of representing new types of derived data. For example, in the London example, we have population in 1991 – one of the maps in Figure 23.2 – and we can easily compute the area of each local authority as there is standard function in the GIS to do this. P_i is the population in each local authority i, L_i is the area of each authority and we can then calculate the density $D_i = P_i/L_i$. To do this in GIS, we can invoke a standard calculator which enables us to plot the map of density by combining the data according to this formula. We show the result in Figure 23.3 with the typical GIS dialogue boxes used to effect the calculation shown alongside. This is, in essence, map algebra, although in a sense what we are doing is ordinary arithmetic on the data and presenting it through visualizing the result as a map. This is a powerful but simple way of thinking about geographical relationships, and a further example impresses the point.

For London, we have population at two dates in time – 1981 and 1991 – which we call $P_i(t)$ and $P_i(t + 1)$. We can easily compute the growth rate for each local authority area i as $\lambda_i = P_i(t + 1)/P_i(t)$ and, using these rates, we can project the population forward one step at a time as $P_i(t + 2) = \lambda_i P_i(t + 1)$, $P_i(t + 3) = \lambda_i P_i(t + 2)$ and so on. Those familiar with ordinary algebra will see that this kind of relation is recursive and that we can project the population forward any number of time steps with the assumption of constant growth rates, of course. The formula for this would be $P_i(t + n) = \lambda_i^n P_i(t)$ which illustrates the 'compounding' or exponential nature of the growth. We can illustrate this for London in Figure 23.4 where we show the two populations, the growth rates and the population 100 years on from 1981, which

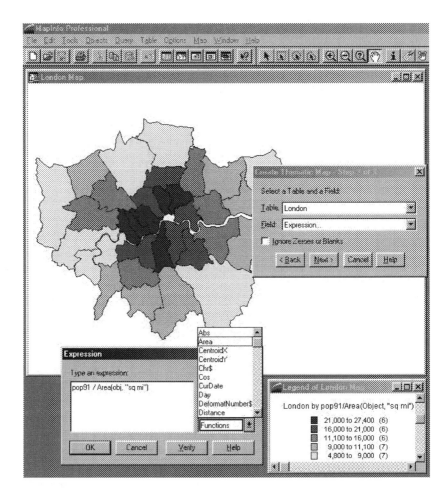

FIGURE 23.3 Computing a map of population density for London boroughs within the GIS MapInfo

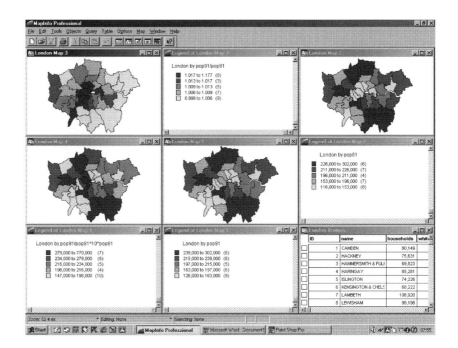

FIGURE 23.4 Projecting the population of London boroughs for 100 years from 1981. Top left: growth rates from 1981 to 1991; top right: population 1981; middle middle: population 1991; middle left: population in 2081 using 1981–91 growth rates. Legends for these maps in other windows are also shown, as well as a fragment of the data attributes table in the bottom right window

involves using the growth rate on each newly computed set of populations some 10 times. In essence, what we have done is to show how a simple population forecasting tool can be built into a GIS by thinking of this as updating a map of population.

The idea of map overlay is even more extensive. One of the original areas where GIS was developed was for landscape planning where the typical problem was to find the suitability of land for different uses (McHarg, 1969). This invariably required different factors to be considered which could be represented as maps, each providing a different and often contradictory impression of where the most suitable land was located. Methods for reducing conflict between such factors are usually based on overlaying these maps and often weighting each factor differentially to provide some sort of combined suitability surface. This is the classic example of map algebra which is illustrated schematically in Figure 23.5, where we show how maps are weighted and added. Of course, the decision to weight and add in this fashion is not something that is intrinsic to GIS for it depends on the use to which the GIS is put by those involved in such problems (Heywood et al., 1998).

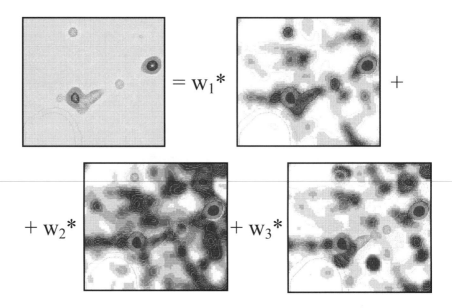

FIGURE 23.5 Map algebra: adding and weighting sustainability surfaces based on economic employment, property values and leisure potential in an area of west London

GIS AS VISUALIZATION

As we have been at pains to point out, GIS is not computer cartography although it may be quite justifiable to use the cartographic functions in a GIS to produce maps if many variants of the map are required and if the overhead of just entering map data is justified. However the visualization capabilities of most GIS are by no means restricted to presenting the two-dimensional map. There are extensive functions for presenting different types of 2D map with a distinction here between vector and raster, vector maps being associated with thematic, perhaps more abstracted mapping, and raster maps being more life-like in appearance. Thematic maps with bars, pies, density dots and so on to represent various attributes are standard, while raster maps with various types of hill shading giving the appearance of $2\frac{1}{2}$D (2D with the third dimension simply extruded), and oblique views are part of many packages. In fact, although GIS has not been traditionally used for presentation in map design, software is beginning to incorporate tools for good design.

There are usually other graphical functions for visualizing data within most GIS. For example, there are drawing capabilities which enable maps and any other graphic to be constructed in the map window. There are usually some graphing facilities. For example, MapInfo is organized around four different graphical 'views': the map itself, of course, but also a graphing routine which enables scatter graphs, pie charts and so on to be presented (of the attribute data) in non-spatial form. The other two views are of the data themselves – the numeric tables and a layout window which allows the user to place maps, graphs and tables together on a bigger canvas: a minimal form of presentational design.

The greatest advances in visualization, however, are coming with the extension of GIS to the third dimension. For a long time GISs have dealt with landscape data, and various techniques of developing 3D landscape models are included in some packages such as ArcGIS. However, the move to explicit 3D within GIS indicates the weakness of conventional forms of 3D representation. Computer aided design (CAD) models, for example, do not have anything like the data functionality that GIS has but, increasingly, those involved in 3D modelling wish to attribute other data to the 3D geometry, and then use query, overlay and spatial analysis techniques in their exploration of such models. An example of what is possible in current desktop GIS software is shown in Figure 23.6, where a 3D block model of part of inner London is illustrated, constructed from population densities at small area enumeration district level and displayed within the desktop GIS ArcView. ArcView is typical of many low-cost GISs in that it can be upgraded by purchasing plug-ins – add-on modules which extend its functionality. One of these – 3-D Analyst – enables users to extrude 2D areas into the third dimension, to then pan and zoom around such 3D scenes, and to navigate through the scene, querying the data as one goes. The extensive functionality of 2D data is not yet available for the third dimension but this will come (Batty et al., 2001).

This kind of tool is providing entirely new ways of visualizing cities and their data, and there is every prospect that, with such

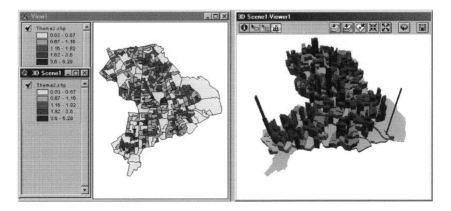

FIGURE 23.6 Representing attribute data in 3D: population density in small census areas in the London Borough of Hackney

software and their relevant data, a new urban geography of the third dimension will be constructed over the next 20 years. This is as good an example as any that the digital revolution is forcing us to think about geographies in very different ways. Finally, the extension to 3D is likely to see a much greater emphasis on realism being incorporated into GIS. It is already quite possible to embed photographs and other multimedia into such systems, but the idea of rendering 3D scenes to the highest quality possible is likely to feature strongly in future GIS. This, like much of what already exists, will be driven by the market-place as much as by the science – an illustration once again that geography and its methods are influenced as much by the wider social context as by any narrower, theoretical quest.

TECHNOLOGY AND SOFTWARE

Although the technology and software of GIS do not dominate the geographical science they support, there are remarkable changes still being worked out with respect to what this technology is able to offer. The ability to work graphically with data at immense speed, accuracy and the best visual quality is a remarkable enough phenomenon in the light of how geographers visualized and explored data a generation or more ago. But the real impact is on how software is being distributed across networks – across the Internet – and how many new users of geographic data and science are being drawn in by GIS. Extending GIS into 3D is only one cutting edge. Another is the ability to explore geographic data remotely using Internet GIS which links digital to attribute data and serves it to users in remote locations. The function-ality of such Internet map servers is rudimentary compared to desktop and workstation GIS but the ability to create maps on-the-fly is becoming an important service in many domains. Web pages are increasingly being linked to such servers as a whole new set of uses for maps as part of information in general is being devised.

An example of this is in the Internet map server constructed at University College London to deliver information about town centres to those interested in their planning. A simple overlay method based on the logic illustrated in Figure 23.5 above has been developed but, instead of being on the desktop, this is available through web pages to as many users as the network can handle at remote locations. A map overlay procedure has been designed which takes different indicators of town centre sustainability and enables users to combine these. This produces many data layers which are based on very detailed employ-ment, turnover, rents, floorspace, social composition, consumer pro-files and so on which all have different implications for the economic sustainability of different areas. Users can select, combine and differentially weight these to produce new surfaces of sustainability. This supports the idea that a very wide range of users with different viewpoints needs to discuss these issues, and this is helped by the delivery of information and the need to work actively with it in this fashion. A picture of the interface to this tool is illustrated in Figure 23.7.

Web pages are also being used to deliver geographic data, and GISs are now the main software which enables users of this data to capture, unlock and hence visualize its content. Academic users in the UK can download digital (boundary) data and census attribute data from various websites in the UK which provide user-friendly interfaces to make this painless. Such data are invariably delivered in a choice of GIS formats which means that a GIS is the essential tool in making sense of this. Education in GIS is also being delivered as distance learning through websites as in ESRI's Virtual Campus project where students/users learn by running exercises from software loaded on their local machine but with the control of the teaching and the exercises – the data, etc. – available at the host site where the educational content is stored. The logic of delivering information in this way and letting students work with data locally and at their leisure is

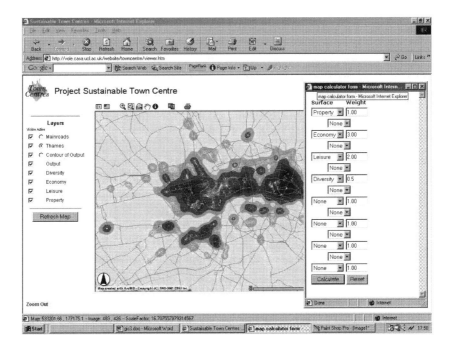

FIGURE 23.7 Combining map layers, as in Figure 23.5, to form indices of urban sustainability in Greater London using Internet GIS through a web browser

the way things will be in the near future, and GIS is in the vanguard.

The development of networked GIS is only one part of the story, however, as GIS is now being linked to other kinds of software at many different levels of sophistication. There is a facility in the spreadsheet Microsoft Excel to import maps from MapInfo. Just as it is possible to plot charts in Excel, it is now possible to plot maps. Most GISs have scripting languages that allow users to link various elements to other software, and there are various ways in which the standard GISs can be linked to statistical software at one extreme and graphics software such as CAD at the other. Many of the links are loose although, increasingly, software is merging across the desktop and the net and the days are probably numbered for the standalone all-purpose GIS. Software will increasingly come as different modules to be plugged together and this will probably mean that the current fad with the technology will lessen.

CONCLUSION: APPLICATIONS

There is virtually no geographical problem that is not touched by GIS: if not by the software or the system then by the science. Whether or not GIS is applicable will depend on many things, not least the quality and extensiveness of the data. If all that is required is computer mapping, GIS may not be worth while unless the maps are to be reworked many times. But as in all problem-solving, the problem must be well formulated for GIS to be of any use. As we have implied, the diversity of application areas is staggering and all we can do in this conclusion is to point to significant ones. Although GIS came from landscape planning and automated cartography, the biggest substantive areas currently lie in urban planning and in environmental analysis (Heywood et al., 1998). GIS began at rather coarse scales but it is gradually being applied at finer and finer scales and there is every prospect that entire new professional areas in real estate, architecture and urban design will embrace these tools during the next 10 years.

Niche areas of GIS have also emerged, one of which is 'business geographics' with a focus on marketing and retailing, utilizing new sources of data on consumers. In terms of physical systems, apart from remote sensing where GIS and remote-sensing (RS) packages are often tightly linked, there has been less development with the exception, perhaps, of geology and terrain analysis (Bonham-Carter, 1994). Much of the current functionality deals with networks, spatial data structures and, specifically, more human kinds of analysis than physical although this is changing and there is every prospect that GIS will make substantial inroads in ecology and related sciences in the near future. There are some areas that GIS has not touched and one is meteorology. This is largely because such areas have always had their own software and systems. In fact, where there are already well established models and analytical techniques, GIS has been less evident in terms of its applications. Transport is another case in point, and this illustrates an important limitation of the approach. GIS is inevitably data driven and this is often a precursor to modelling and simulation. In this sense, GIS is prior to prediction and design and, although many problems require large arrays of diverse data, those that are very focused like transportation planning or weather prediction tend to use GIS, if at all, purely as a display medium.

What has not been emphasized yet in this chapter is the difference between routine and more infrequent, more strategic uses of GIS. Routine usage is really for rather low-level queries and for scheduling, and much of the future use in hand-held systems, for example, will be of this nature. To an extent, although GIS is used in these areas, much geographic software is purpose built and often embodied directly into the control systems used in such ongoing tasks. In the strategic context, the kinds of functions illustrated for map algebra are those that are likely to be used. This emphasizes the support to decision-making and planning in the widest sense which again is an area of increasing interest and application. Finally, to anticipate the future, within a generation many of the techniques within GIS will have changed beyond recognition as GIS diversifies and fragments. What is certain, however, is that most human activities will take place in a digital environment and, in this, GIS-like functions will play an ever important part.

Summary

- GIS emerged slowly from several origins, uniting computer cartography, the development of database theory for spatial data, quantitative models, geostatistics and remote sensing.
- GIS is both technology and theory. At one level, GIS is simply visualizing map data, but GIS as geographic information science is of much wider importance for geography and geographical method.
- Currently, GIS technology has advanced to the stage where the focus is no longer on graphical representation per se, but on integrating visualization with method so that both quantitative and qualitative analysis might be enriched.
- Much of GIS technology is concerned with visualizing these data in map form. The central organizing concept is based on treating different sets of attributes as map layers.

- The most powerful new features of GIS lie in the ability to go beyond 2D map visualization into the areas of virtual reality, and to link GIS with other Internet-based applications.

Further reading

- The textbook by Longley et al. (2001) provides a good introduction, while the bigger two-volume edited reader by Longley et al. (1999) contains 72 articles on many different aspects of GIS.

- Niche areas are worth exploring – for example, Batty et al. (1999) in their chapter on 'GIS and urban design' explore how GIS might influence small-scale site design.

- A good book stressing applications in physical geography is Burrough and McDonnell (1998).

- The flagship journal with the greatest technical flavour is the *International Journal of Geographical Information Science* (*IJGIS*), while there is a host of magazines, which appear at least monthly, targeted mainly at the industry but which are useful to see how the field is developing. The main one at present in the UK is the monthly *GEO: connexion*.

- There are several really good websites where you can learn about GIS. Visit ESRI's site and the Virtual Campus (http://campus.esri.com/). The site at Edinburgh is a good resource (http://www.geo.ed.ac.uk/home/gishome.html). Our own site at UCL (http://www.casa.ucl.ac.uk/) reports a number of GIS projects. Finally, UK census data can be downloaded from Manchester (http://www.mimas.ac.uk/) and boundary data from Edinburgh (http://www.edina.ac.uk/) and read directly into various GIS packages and spreadsheets.

Note: Full details of the above can be found in the references list below.

References

Batty, M., Chapman, D., Evans, S., Haklay, M., Kueppers, S., Shiode, N., Smith, A. and Torrens, P. (2001) 'Visualizing the city: communicating urban design to planners and decision-makers', in R. Brail and R. Klosterman (eds) *Planning Support Systems: Integrating Geographic Information Systems, Models, and Visualization Tools*. Redlands, CA: ESRI Press, pp. 405–43.

Batty, M., Dodge, M., Jiang, B. and Smith, A. (1999) 'Geographical information systems and urban design', in J. Stillwell et al. (eds) *Geographical Information and Planning*. Heidelberg: Springer-Verlag, pp. 43–65.

Bonham-Carter, G.F. (1994) *Geographic Information Systems for Geoscientists*. Oxford: Pergamon Press.

Burrough, P.J. and McDonnell, R.A. (1998) *Principles of Geographical Information Systems*. Oxford: Oxford University Press.

Chrisman, N.R. (1988) 'The rise of software: a case study of the Harvard Lab', *American Cartographer*, 15: 291–300.

Goodchild, M.F. (1992) 'Geographical information science', *International Journal of Geographical Information Systems*, 6: 31–45.

Heywood, I., Cornelius, S., and Carver, S. (1998) *An Introduction to Geographical Information Systems*. Harlow: Prentice Hall.

Longley, P.A., Goodchild, M.F., Maguire, D.J. and Rhind, D.W. (2001) *Geographical Information Systems and Science*. New York: Wiley.

Longley, P.A., Goodchild, M.F., Maguire, D.J. and Rhind, D.W. (eds) (1999) *Geographical Information Systems*. New York: Wiley.

McHarg, I. (1969) *Design with Nature*. New York: Doubleday-Anchor.

Tomlin, C.D. (1990) *Geographic Information Systems and Cartographic Modeling*. Englewood Cliffs, NJ: Prentice Hall.

24 Statistical Analysis using SPSS

John H. McKendrick

Synopsis

The Statistical Package for the Social Sciences (SPSS) is a software system – a user-friendly, widely used and multifaceted means of working with numerical datasets. SPSS assists users working with quantitative data from the stage of data collection, through data access, data preparation and data management to report writing. SPSS Inc. is a private company. This chapter explores some of the common misconceptions about SPSS and explains its role in geography. It then describes the product range and provides ten steps to guide student geographers from the point of data collection to the point of data analysis.

The chapter is organized into the following sections:

- What SPSS is not
- What is, and who are, SPSS?
- SPSS, the research process and the student geographer
- What exactly does SPSS offer the student geographer?
- Beyond university . . . with SPSS.

WHAT SPSS IS NOT

If all the readers of this chapter were asked 'what are you studying at university', what proportion would be expected to answer 'essay writing skills'? Not many, one would hope! It might reasonably be assumed that those who offered this response had misinterpreted the question: this answer doesn't appear to make sense. More typically, we would have anticipated answers such as 'geography', 'human geography', 'physical geography' or one of those three descriptors, and 'geology' or another of the many cognate disciplines alongside which students pursue geographical studies. Yet, to the chagrin of many a dissertation tutor, 'SPSS' – the same type of answer – is frequently offered in response to the question 'how do you intend to analyse your [quantitative] data?' SPSS is to data analysis what essay writing skills is to doing geography: merely one possible means through which to ensure efficient and effective practice. Just as the student geographer could hone skills other than essay writing to improve geographical understanding (reading skills, observational skills, etc.), so he or she could utilize tools other than SPSS to analyse quantitative data. Just as essay-writing skills comprise many elements (paragraph construction, citation practices, etc.), so SPSS comprises many tools for working with quantitative data (frequency counts, cross-tabulations, etc.). In short, SPSS is not a method. Rather, the Statistical Package for the Social Sciences (SPSS) is a software package – a user-friendly, widely used and multifaceted means of working with numerical datasets.

Although familiar to student geographers, SPSS is not particularly popular among them and tends not to have a 'constant' presence in undergraduate geography studies. These two 'negative' statements are inextricably linked. Students tend to be 'confronted' with SPSS in limited duration, intensive, research methods modules that are set apart from their substantive geographical studies. This can render it more difficult to appreciate the wider significance of SPSS and tends to encourage learning that is narrowly focused on process rather than application (how, rather than why, to apply SPSS). Even when sequenced in a degree as a progressive programme of building competence in statistical data analysis, there is often a gap between advanced learning and preliminary studies, requiring students to spend time refamiliarizing themselves with basic statistical concepts and the SPSS software. Add to this the shift away from quantitative geography that is an inadvertent by-product of 'the cultural turn', it should come as no great surprise to find that a special place is reserved for SPSS/statistical data analysis classes at the bottom of the league table of end-of-term student module evaluations!

Sounds familiar? Why, then, should you read on? There are four reasons why student geographers should persevere with SPSS. First, many of the research questions that concern geographers are amenable to, or are best answered by, extensive comparisons between places and people, and SPSS is an appropriate tool to use to obtain answers to these questions. For example, SPSS could be used to ascertain whether there are significant differences in rates of wave erosion at different points around Britain's coastline. Similarly, SPSS could be used to calculate the proportion of families that are headed by a lone parent, and to ascertain whether there are significant differences across New York's neighbourhoods. Secondly, statutory bodies are extending the range of national survey datasets that are freely available to those involved in higher education. These datasets enable many geographical issues to be researched (see Chapter 5). Many of these datasets can be downloaded as SPSS data files, ready prepared for data analysis. Thirdly, a wide range of organizations outside academia, many of which employ geography graduates, use SPSS. These issues – utility for academic geography studies, working with data and utility beyond academia – are considered, respectively, in the next but one, penultimate and concluding sections of this chapter. Beforehand, contextual information is provided on SPSS, the company and the product. This reflects the broad aim of this chapter: to provide a general introduction to SPSS. Specialist textbooks are necessary to provide a comprehensive guide to using SPSS and some recommended key texts are introduced in the further reading section at the end of this chapter. And the fourth reason for reading the chapter? Well, it is highly probable you will encounter SPSS (again) in your degree studies: perhaps these few thousand words will leave you better motivated to make the most of this learning opportunity. Or, at least, you will be better placed to assess the relevance of SPSS . . . and perhaps even inclined to be a tad more generous in your assessment of statistical laboratory classes!

WHAT IS, AND WHO ARE, SPSS?

SPSS is both product range and company. The product is an integrated system of modules, built upon the core package, SPSS Base. SPSS Base is a comprehensive software package with an easy-to-use graphical interface that assists users working with quantitative data from the stage of data collection, through data access, data preparation and data management to report writing (Figure 24.1). While most users eventually require to supplement SPSS Base with specialist modules (e.g. SPSS Trends or SPSS Regression Models), few users would routinely use all the SPSS products. SPSS Inc., the company, describes itself as 'the industry leader in data mining technology and analytic solutions to enhance decision making'. Notwithstanding the hyperbole and lofty self-praise of company promotion, there can be no doubt that the company founded in 1975 with its headquarters in Chicago (Illinois, USA) is now a multinational corporation that services the needs of a wide range of users. With almost half its business outside the USA, over 1200 employees worldwide, operations in more than 40 countries and a portfolio that includes 95% of the Fortune 100 companies, SPSS Inc. is a significant player in statistical data-analysis markets. Oh, and if you have spare change left over from your student loan, SPSS Inc. is currently achieving an 18% return on equity!

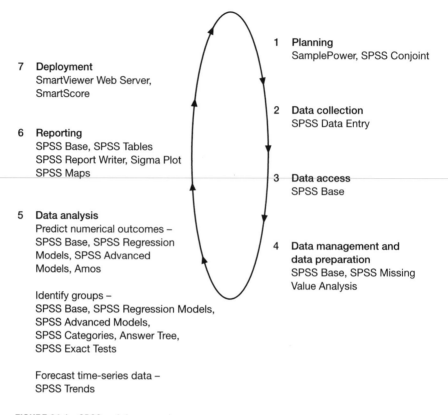

1   Planning
    SamplePower, SPSS Conjoint

7   Deployment
    SmartViewer Web Server,
    SmartScore

2   Data collection
    SPSS Data Entry

6   Reporting
    SPSS Base, SPSS Tables
    SPSS Report Writer, Sigma Plot
    SPSS Maps

3   Data access
    SPSS Base

5   Data analysis
    Predict numerical outcomes –
    SPSS Base, SPSS Regression
    Models, SPSS Advanced
    Models, Amos

4   Data management and
    data preparation
    SPSS Base, SPSS Missing
    Value Analysis

Identify groups –
SPSS Base, SPSS Regression Models,
SPSS Advanced Models,
SPSS Categories, Answer Tree,
SPSS Exact Tests

Forecast time-series data –
SPSS Trends

FIGURE 24.1   SPSS and the research process

SPSS was founded in 1968 when three graduate students from Stanford University developed the first SPSS program. It has evolved ever since with a new version being released, on average, every three years. Much has changed between the first and the current (eleventh) version and significant milestones have been achieved in passing. In 1984, SPSS/PC+ became the first mainframe-capable statistical software package that could be used on personal computers, and in 1992 SPSS launched a product for use with the Microsoft Windows operating system. Although the SP for the *Social Sciences* is perhaps reassuring to many geographers, particularly those interested in human geography, SPSS Inc. is keen to stress that times have changed and that 'the "package" has . . . evolved from its academic roots to become a leading enterprise analytical solutions provider'. This is reflected in the structure of the company and in the diversity of its markets. The SPSS Business Intelligence Division of the company serves academia alongside business, science, government and 'other organizations'. Indeed, assisting the learning of statistics is not the only way in which SPSS is applied in academia: many lecturers will use SPSS to analyse data in their academic research and university administrators may be using SPSS to count you as part of their work in monitoring changes in the characteristics and performance of student populations. Other divisions of SPSS Inc. hint at the breadth of applications beyond academia. CustomerCentric Solutions uses SPSS products to assist companies to manage their customer relations (e.g. through targeted promotional campaigns and by measuring customer satisfaction). ShowCase supports business analysts, managers and executives to monitor business practice (e.g. through performance indicator measurement). SPSS MR supports the work and entire process of market research: the company estimates that 70% of leading market research firms and corporate market research departments use SPSS products. Finally, SPSS Enabling Technologies provides custom-built solutions to meet SPSS Inc.'s needs. Yet, despite the company's concern to distance itself from its academic origins, the products continue to offer much to the academic community.

SPSS, THE RESEARCH PROCESS AND THE STUDENT GEOGRAPHER

Many human geographers would now argue against the depiction of the research process as a simple linear sequence of predefined steps from problem specification through to writing up results (see Chapter 1). In particular, those deploying multiple methods in multi-stage research (for example, using a questionnaire survey to obtain generalizable findings and following this up with key informant interviews to explore issues or processes in more detail) may find themselves writing up findings from introductory methods, while formulating research questions or supplementing their original research themes for the next stage of their research. Those working within a 'grounded theory' tradition view research as an iterative process in which themes and theoretical understanding emerge from field research rather than theory providing an overarching framework at the outset. These 'fuzzy' models perhaps better reflect the reality of undertaking research, but they contrast with those models proposed by SPSS Inc. (Figure 24.1) and by the author of this chapter for student-led research with quantitative data (Figure 24.2).

In student-led research, 'SPSS' (or any other statistical software package that could be used to analyse quantitative data) tends to be the fifth stage in a six-step process (the shaded box in Figure 24.2). Having decided upon a research theme (which might involve reviewing literature and engaging with theory and/or policy) and conceived of a research design (which involves choosing research methods and formulating specific research questions), the next task is to collect

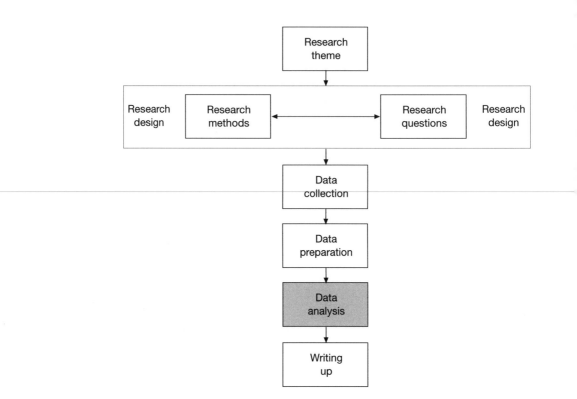

FIGURE 24.2 The research process for the student geographer using quantitative data

quantifiable data (which may, for example, be derived from systematic field measurement, content analysis of documents or survey questionnaires), prior to preparing data for analysis (which involves, for example, assigning numerical codes to categorical data). This – the data analysis stage – is the point at which student geographers tend to utilize SPSS products. Having completed the analysis, student geographers tend to move away from SPSS to write up their findings using wordprocessing software.

SPSS performs a critical role in the research process but it is far from all encompassing or all important. Effective quantitative data analysis requires clarity in the specification of the research problem (step 1), a research design that provides a meaningful sample and a dataset that is amenable to quantitative data analysis (step 2). Without these fundamental inputs – which are researcher driven – SPSS will be of little value. In short, competence in research design is as important as competence in using SPSS to analyse data. It is very easy to design a survey but very difficult to design a useful survey. Sufficient effort must be invested in what are typically pre-SPSS stages in quantitative research.

It should be noted that this 'typical' depiction of the use of SPSS by student geographers does not preclude alternative models of application. Indeed, when Figure 24.1 (the SPSS Inc. model of how SPSS products could be used from the stage of data design to writing up) is compared to Figure 24.2 (the 'normative' model for quantitative research by student geographers), it could be concluded that the full potential of SPSS is not being realized. Student geographers tend not to use SPSS products to design their research. A simpler model than Figure 24.2 is required to depict the use of SPSS in analysing existing datasets. In such secondary analysis step 2 is simplified and steps 3 and 4 are superfluous as the researcher moves straight from question formulation to data analysis (having secured data access). Finally, steps 1–4 are often by-passed in introductory statistical laboratory classes, in which the learning objective is often to 'learn SPSS'. The challenge and joy of identifying themes to research, designing research methods and formulating questions, collecting data and preparing them for analysis are far removed (and may even seem far fetched!) from the SPSS laboratory class using pre-prepared data and schedules of analysis.

WHAT EXACTLY DOES SPSS OFFER THE STUDENT GEOGRAPHER?

The answer to the question posed in this section's title is unequivocally 'far too much for one short chapter!' Another, equally valid, answer might be to note that SPSS offers the means to undertake the tasks outlined in Chapters 5 (making use of secondary data), 6 (conducting questionnaire surveys), 15 (sampling), 17 (exploring systems using mathematical models), 18 (designing numerical models), 20 (basic numeracy skills), 21 (using statistics to describe and explore data) and 22 (introductory geostatistics). Perhaps the most satisfactory answer is to direct you to the annotated guide to further reading that is presented at the end of this chapter. In the meantime, here are ten steps to guide the student geographer from the point of data collection to the point of data analysis.

File type

Two files are automatically created when you open SPSS Base (hereafter, SPSS). The Data Editor file ('untitled data') is a spreadsheet-like grid of numbered rows and labelled columns. At the outset, the row and column headings are dimmed as there are no data in the cells. It is into this file that you will enter your statistical data. The Output file ('!untitled output 1') is a blank file into which SPSS will enter the results of any analysis you undertake (e.g. if you command SPSS to undertake a frequency count). There is a third file type – a Syntax file ('untitled syntax 1') – which may be opened when you start SPSS, depending on how your department computing officer has configured SPSS. The Syntax file is a record of commands which you have either keyed in directly or have pasted in while working with SPSS's graphical user interface. Syntax files are extremely useful if you want to repeat an analysis or if you want to maintain a record of your analysis. For more information on using Syntax files, refer to one of Norusis's books (see further reading). Finally, the SPSS Application menu – the headings of File, Edit, Data, Transform, Statistics, Graphs, Utilities, Windows and Help, each of which has drop-down options – is listed at the top of the screen. The data file, output file and SPSS Applications menu are illustrated in Figure 24.3.

Structure of an SPSS data file

Each row of the Data Editor file consists of cases and each column consists of variables. Thus, each questionnaire survey (or each field site) should have all its responses (or field measurements) kept within the same row. Similarly, each question that is asked of all survey respondents (or each type of measurement that is taken for all field sites) should be kept within the same column. In effect, the SPSS data file is a structured, shorthand summary of your data in which you reduce a mountain of paperwork (questionnaire surveys) or collate several series of data (measurements for field sites) into one record.

Preparing the data file

Your first task is to translate your data into an annotated language of number, which you will always be able to translate back to its original form at any time. This is made easier if your survey (see Chapter 6) or field measurements (see Chapter 14) are well designed. For example, if you ask survey respondents to record their gender, you should be able simply to translate the responses into number for the SPSS data file: 'men' becomes '1', 'women' becomes '2' and 'rather not say' becomes '3'. You will always know what is represented by 1, 2 and 3 in the gender column of your data file. Or will you? It is surprisingly easy to forget what the numbers of your SPSS data file represent, particularly if you have a large dataset or if you return to analyse your data some time after you prepared the data file. One of the most useful functions of SPSS – and one which makes SPSS user friendlier than spreadsheets – is the function that allows the user to add descriptive labels to variables and values.

If you double-click your mouse on the dimmed header (var) at the top of any of the columns of your data file, the 'Define Variable' dialogue box will open (top of Figure 24.4). SPSS gives you the option

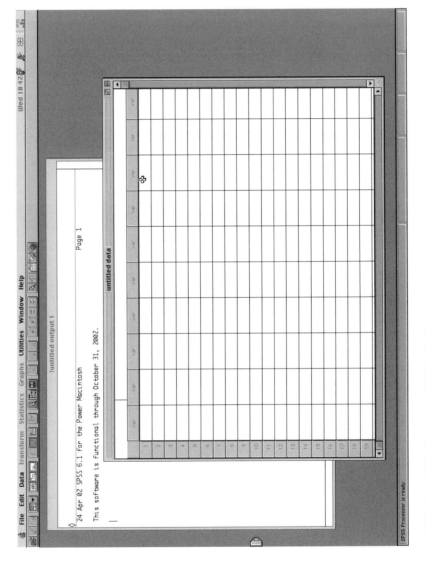

FIGURE 24.3 SPSS files and the SPSS Applications menu

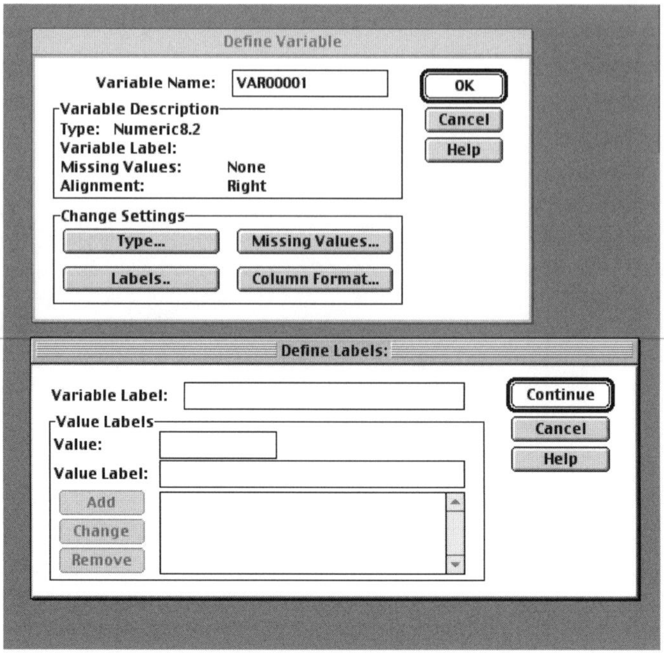

FIGURE 24.4 Defining Variables and Labels dialogue boxes

of changing the settings for variable type, missing values, labels and column format. You can also change the variable name as it appears at the top of the column on your SPSS data file. At this stage, those new to SPSS should make five changes.

First, change the variable name for the data file. You are permitted to use up to eight characters. Keep it simple and either use an obvious descriptor (e.g. 'SlopAng1' for the first slope angle that you measured) or use the ID number you used when you were collecting data for that variable (e.g. 'Q_02' for the second question of your survey). When you have accepted these changes, the dialogue box will close and your data file will now show the variable label at the top of the column. You should repeat this process for every variable in your data file.

Secondly, define the variable name and the value labels for any output files you will generate at the stage of data analysis. Click the Labels' button of the Define Variable dialogue box (top of Figure 24.4) to open the Define Labels dialogue box (bottom of Figure 24.4). You should enter a longhand description of your variable, which can be up to 256 characters in length (but keep it shorter!). You should ensure that you provide enough information to make sense of the abbreviated variable label that appears on the header row of your data file. For example, 'SlopAng1' might be described in this dialogue box as 'Slope Angle at Findhorn Creek'. For ordinal and categorical (nominal) data (refer back to Chapter 20), you should enter a descriptive label for each value. Enter the value (your numerical code) in the Value box (e.g. '1'), then in the Value Label box enter your longhand description of this value (e.g. 'men') before clicking on the Add box to register this addition. At this point, your value label and value will be added to the record box. You should repeat this process for every value of your variable (e.g. 'men', 'women' and 'rather not say' for gender) before clicking on the Continue box to confirm all the additions you have made. These changes should make no difference to your data file: it seems like a lot of effort for not much reward at this stage! You should repeat this process for every variable in your data file.

Thirdly, you should register any missing values of which you are aware – e.g. if a respondent failed to provide an answer to a survey question or if your data-recording instruments failed to provide a measure at a field site. For surveys, you may also wish to have 'don't know' or 'other' responses treated as missing values. You can specify discrete missing values or a range of missing values. Again, the utility of recording missing values is not readily apparent – no changes are instantly evident in your data file. However, this is important preparatory work for the stage of data analysis.

Fourthly, if some of your data file contains values that are alphabetic (words) rather than numeric (numbers), you will have to change the variable Type. This should be avoided if at all possible, as the value of SPSS (or any statistical data-analysis package) rests in its ability to work with numbers. However, you may find it useful to enter alphabetic responses into your data file before converting them into numerical codes (e.g. typing in job titles before translating these into social class categories). In such instances, you should change the variable type to a String variable (the label SPSS uses for alphabetic data). SPSS automatically gives you the option of increasing the number of characters (in this instance letters) from its default setting of eight. You may also find it useful to increase the width of the Column Format for the purpose of displaying your data.

Two further points are worth noting. First, you must click the OK button (in the top right-hand corner of the Variables Label dialogue box) to confirm the changes referred to above. If you don't, SPSS will not register your changes. Secondly, it may be useful to prepare your data file, as outlined above, before entering your data. This may make it easier to identify data entry errors.

Importing data

It is possible to import data into SPSS from Microsoft Excel, Lotus, SYLK, dBASE or tab-delimited files. You should ensure that the file you intend to import is structured in the same way as an SPSS data file. Imported data become your SPSS data file. Data files can be imported by using the Open drop-down subcommand from the File menu of the SPSS Applications menu (Figure 24.3). It is also possible to cut and paste data into an SPSS data file.

Entering data

You can, of course, enter your data directly into an SPSS data file. This would be the preferred option if you are working with data that have not yet been entered into a spreadsheet or database. Simply position the cursor in the first cell of the first column then enter your data (for one questionnaire survey or one field site) one variable at a time. You may find it quicker to move across columns by using the Tab button on your keyboard rather than by repositioning your cursor. As you type in your data it will appear in the bar above the row with the variable labels (Figure 24.3): it will only be added to your data file once you have pressed the Tab button or positioned the cursor in the next cell.

When you make a mistake (which you will!), simply position the cursor in the cell with the data you want to change and key in the correct value. One way to minimize these errors is to take regular breaks from data entry and to postpone data entry if you are becoming tired. Another useful aid is to switch on the Value Labels mode of data presentation. This is only helpful if you have defined your value labels prior to entering your data. When the Value Labels option is selected, your data file should comprise a list of value labels rather than a list of numerical codes (e.g. 'Women' and 'Men' instead of '2's and '1's). When a number is shown in your data file, this indicates that either you have not defined the value label for this valid code or that you have entered the wrong code. The Value Labels mode of presentation can be activated by releasing the cursor when you highlight the Value Labels option from the Utilities button of the SPSS Applications menu.

Saving a file

It is good practice to save your SPSS files at frequent intervals. SPSS has been known to freeze on networked systems and university computing networks have been known to crash from time to time! One useful rule of thumb would be to save your data file each time you have completed entering row data (that of a questionnaire survey or field site). This takes a few seconds for large data files and is a useful way of recharging the batteries before starting another round of tedious data entry!

To save an SPSS file, select either the Save Data (for an existing file) or Save Data As (for a file you have not yet saved) drop-down subcommand from the File menu of the SPSS Applications menu (Figure 24.3). It may be wise to use the SPSS filenaming suffix conventions: 'YOURNAME.*sav*' for Data files, 'YOURNAME.*spo*' for Output files and 'YOURNAME.*sps*' for Syntax files.

Identifying data errors

Having survived the marathon that is data input, it may be disconcerting to learn that you are not yet ready for data analysis. Even if you were aware of some errors you made when you entered data and managed to correct them, it is highly likely other errors in data entry will have passed unnoticed. Thankfully, there are quicker ways to check for errors than having to confirm that you have entered the correct data value by returning to your original questionnaires or field records. Although if you do not, you may have to concede that your data file will contain a small proportion of errant answers which are valid values – e.g. for gender, typing '2' (for women) instead of '1' (for men).

The most straightforward way to identify data entry errors is to undertake a frequency count of all variables and to scrutinize the results for unexpected values. This is a three-step process. First, you need to open the Frequencies dialogue box (Figure 24.5). This can be found at the top of the Summarize submenu, which can be accessed through the 'Statistics' heading on the SPSS Applications menu (Figure 24.3). Secondly, you should request a frequency count for each of your variables. This, in turn, is a four-step process. First, select the variable at the top of your list (the box in the top left-hand corner of the Frequencies dialogue box). Secondly, press the Shift key and select the variable at the bottom of your list (you may have to use the arrows to reach the bottom of this list). Now all your variables are highlighted. Thirdly, click on the arrow button to move all your variables from the untitled box in the left-hand corner to the box labelled

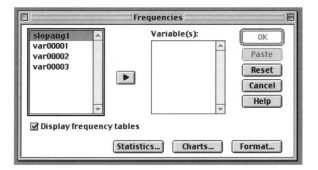

FIGURE 24.5 Frequencies dialogue box

Variable(s). Finally, click on the OK button in the top right-hand corner of this Frequencies dialogue box. At this point, your '!Untitled Output 1' file comes to the front of your screen. The third, and final, stage of the identifying data errors process is to review the results in the '!Untitled Output 1' file to identify variables for which the wrong codes have been recorded. Note the number of cases of each errant code (e.g. for instances of a code '12' for a gender variable, which should only have returned codes 1, 2 or 3).

Changing data errors

The next stage in data preparation is to find and correct the errors in the data file. To do this, you should have handy your desk notes from the 'identifying data errors' stage, the original data (questionnaire surveys or field records) and you should open your SPSS data file. Correcting data errors is a six-step process. First, select the variable (column) which contains the data error. Secondly, open the Search For Data dialogue box (Figure 24.6). This can be accessed through the Edit heading on the SPSS Applications menu (Figure 24.3). Thirdly, in the Search For Data box, enter the number of the errant value you wish to correct. Click on the Search Forward button and this will take you to the first instance of the data error. Fourthly, identify the case (the questionnaire or field site) with this data error by referring to the row number (or case ID if you added one to the data file). Fifthly, return to your original data source and identify the correct data value. Finally, correct your entry in the data file. You should repeat this process until all errors are corrected for all variables. Once you think you have amended all the errors, it would be wise to repeat the 'identifying data errors' procedure to confirm this.

Recoding existing variables

At this point you are almost ready to start data analysis. However, it may be prudent to manipulate your data to facilitate effective analysis. 'Reducing' or 'recoding' the number of values for any given variable is often essential if you are to undertake meaningful analysis.

FIGURE 24.6 Data Search dialogue box

For example, if you surveyed 100 people from two neighbourhoods to ascertain their opinion in response to the contention 'Local government should do more to encourage residential households to recycle household waste', you may find that 30 people responded 'strongly agree', 20 responded 'agree', 20 responded 'neither agree nor disagree', 10 responded 'disagree' and 20 responded 'strongly disagree'. If you wished to determine whether attitudes varied according to the area of residence, it would be helpful to reduce the number of attitudes. You could recode to two categories of agree (merging 'strongly agree' and 'agree' to give 50 cases) and disagree (merging 'strongly disagree' and 'disagree' to give 30 cases). In this instance, you would temporarily set aside the 'neither agree nor disagree' responses and code them as 'missing'. It is, of course, essential that your recoded categories maintain some substantive value.

You should use SPSS to create a new variable by recoding an existing variable – rather than amending the original variable – as you may wish to return to the original variable at a later stage (e.g. to conduct analysis on those who 'neither agree nor disagree' or to recode the original variable in some other way). To recode using SPSS is a nine-step process. First, you need to open the Recode into Different Variables dialogue box (Figure 24.7). This can be found at the bottom of the Recode submenu, which can be accessed through the Transform heading on the SPSS Applications menu (Figure 24.3). Secondly, select the variable you wish to recode (from the box in the top left-hand corner). Thirdly, click on the arrow button to move this variable from the untitled box in the left-hand corner to the box labelled Input Variable > Output Variable. Fourthly, give the new variable (the Output Variable) a name and a label by completing the box in the top right-hand corner of the dialogue box. At this point, the Change button becomes operational (it is no longer dimmed). Fifthly, click on the Change button to register the new variable name and label. At this point the new variable name is added to the Input Variable > Output

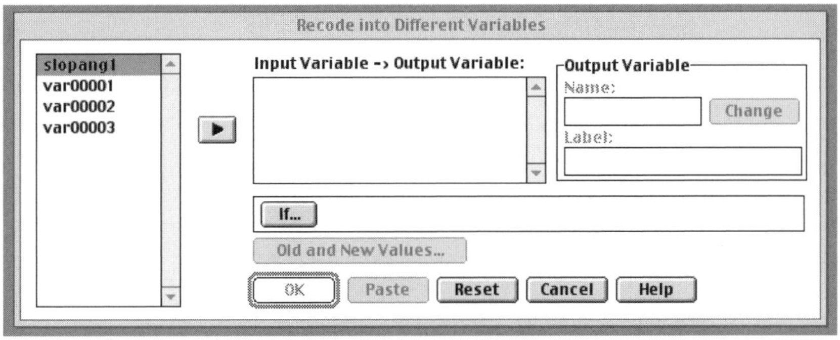

FIGURE 24.7 Recode into Different Variables, main dialogue box

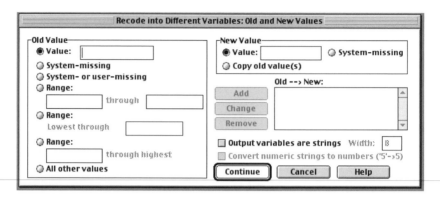

FIGURE 24.8   Recode into Different Variables: Old and New Values dialogue box

Variable box. Sixthly, click on the Old and New Values button to open its new dialogue box (Figure 24.8). Seventhly, complete this box by adding new values (for the new variable) for each of the old values (of the original variable). For example, for the attitudinal variable referred to above, old value '1' and '2' may become new value '1', old value '4' and '5' may become new value '2' and all other values should be defined as 'system missing'. Eighthly, when all transformations have been added to the Old > New window (on the right-hand side of Figure 24.8), click the Continue button to accept these changes. The Old and New Values dialogue box (Figure 24.8) closes and you are returned to the Input Variable > Output Variable dialogue box (Figure 24.7). Finally, click the OK button to create the new variable. At this point the dialogue box closes and a new variable (column) is automatically added to the right-hand side of your data file. But before you set to work on this file, you should add value labels and check for data errors.

When you have created new variables from your old variables, you are ready for data analysis!

Don't expect things to run smoothly

They won't! Trouble-shooting and problem-solving are an integral part of the SPSS experience!

BEYOND UNIVERSITY . . . WITH SPSS

This chapter has provided student geographers with a contextualized introduction to SPSS, the product and the company. In addition to outlining ten steps to guide you from data collection to data analysis, it has clarified the role of SPSS in the research process and has introduced the history of the company whose fortunes have flourished since it first introduced a statistical data-analysis package for social scientists in the late 1960s.

Just as SPSS, the company and the product have grown and developed through time, so too can the utility of SPSS for student geographers. Initial encounters with SPSS – most likely in computing laboratory classes that support introductory data-analysis modules – soon become a hazy recollection of statistical concepts, instructions, numbers, error messages, frustrations, system crashes/freezes, complicated printouts and expletives. However, it is to SPSS that many students return in the hope of making sense of questionnaire survey responses and systematic field measurements (to name but a few applications) that have been collected in the name of the honours-level dissertation. Without the backup of a laboratory class tutor and in the absence of the step-by-step guidance of a class exercise, this often is as much a test of patience as it is a test of competence in statistical data analysis. A *truly* student-friendly SPSS textbook (see the further reading section below), a dissertation tutor with much time to spare, a philanthropic postgraduate student or one who enjoys problem-solving and a shoulder to cry on/bottle to empty are pre-requisites for success.

However, familiarity with SPSS – if not, competence in its application – has value beyond university studies. In the cut-throat world of graduate recruitment, 'SPSS' on a CV serves as a marker that the graduate has grappled with statistical data, perhaps through developing skills in data classification, interpretation, analysis and reporting. The 'numerate' graduate is a sought-after commodity and it is more convincing to provide evidence of personal competence (use of SPSS) than it is to claim by proxy (through a vague recognition of having completed a degree in a 'numerate' discipline such as geography). At the very least, it might be added to a list of other computing software packages (Microsoft Word, Adobe Acrobat, etc.) to demonstrate computer literacy.

Yet, SPSS may be of even more direct value to the geography graduate's CV and long-term career prospects. In addition to serving as a marker of numeracy and computer literacy, the ability to use SPSS is, in itself, a selling point to potential employers. The SPSS home page provides a synopsis of 'success stories' – that is, reviews of how over 80 companies have applied SPSS to overcome a challenge. You might be destined for an 'SPSS future' with institutions such as the Virginia Department of Juvenile Justice (who allocate funding to crime-reduction programmes based on their effectiveness), Lloyds TSB (which has a dedicated section within the main fraud department whose sole purpose is to undertake data analysis to reduce card fraud), the San Francisco Heart Institute (who track physician performance

and patient outcomes with a view to decreasing patient length of stay)
or British Airways (who use SPSS to gauge future online sales). Then
again, you might catch the geo-research bug and your destiny may lie
in the university classroom . . . teaching SPSS to the next generation
of undergraduate geography students!

Summary

- SPSS is not a method: it is a range of products which facilitate the
  analysis of statistical data. A private company, SPSS Inc., holds
  the copyright to the product.
- SPSS assists geography students to undertake extensive
  comparisons between people and places.
- The skills acquired in using SPSS are transferable and are being
  used by geography graduates in a variety of ways in their careers.

Further reading

- Norusis (2002) is the SPSS student's best friend! This text provides chapter exercises, comprehensive examples of basic statistical techniques, solutions to selected exercises, step-by-step procedures and data files specific to chapter examples and exercises. Earlier editions remain an excellent introduction to the software, although it is preferable to consult the new publication which is released to accompany the latest edition of SPSS. The *SPSS 11.0 for Windows Brief Guide* provides a set of tutorials designed to acquaint the beginning user with the various components of the SPSS system. It is a supplement to the online tutorial that is included with the SPSS 11.0 software. The guide provides introductions to using the help system and data editor, importing your data into SPSS, working with statistics and output, creating charts with SPSS 11.0's interactive graphics, modifying data values, working with syntax and data files, calculating new data values, sorting and selecting data, and brief introductions to more advanced topics. This is an essential reference text you should access through your university library or computing laboratory class.

- Kitchin and Tate (2000) (Chapters 3–5). Although this student textbook does not offer any advice on using SPSS, it does offer hand-worked examples with clear step-by-step instructions of how to undertake most of the introductory statistical techniques for which geography students are likely to use SPSS. At the very least, these hand-worked examples will enable you to appreciate what SPSS is helping you to miss! More importantly, the hand-worked examples illustrate what SPSS is doing to your data to reach the answers you need.

- Rogerson (2001) is an introductory student-level textbook which is clearly written with examples to illustrate calculation procedures and how to interpret the results. At the end of each chapter there are instructions on how to apply statistical techniques using SPSS for Windows. It includes self-assessment exercises and downloadable datasets. Most examples are drawn from human geography, although climatological applications are also used.

- Gray and Kinnear (1998). As the title suggests, this is a student textbook which aims to take the (Macintosh) user from scratch (e.g. from 'booting up') through to advanced-level applications (regression, log linear, discriminant and factor analysis) in 15 chapters. Everything is clearly explained and amply illustrated. Notably, the book uses many screenshots from SPSS which increases the reader's familiarity with SPSS, hence making it easier to use. One significant drawback is that this particular volume has been written for psychology students and the examples are not issues with which student geographers may be familiar or comfortable.

- Fielding and Gilbert (2000) provide an extremely well written and accessible introduction to social statistics and data analysis for social science students. This book is much more of an introduction to elementary statistical applications than, for example, Rogerson (2001). It refers to SPSS throughout and uses screenshots (although not to the same extent as Gray and Kinnear, 1998) to illustrate how to use SPSS. Students with an inclination towards physical geography may find it more difficult to relate to the examples and may find the range of applications with interval level data too limiting.

- You should visit the home page of SPSS (http://www.spss.com). This supplies comprehensive information on the product range and the company. It is an enlightening read for the student geographer who, in all probability, will bemoan the utility of SPSS at some point during his or her undergraduate studies. Stato-phobes may fail to perceive any value in SPSS, while stato-purists may perceive SPSS to be too limited. The SPSS home page provides a comprehensive list of company 'success stories' which may encourage all to think again – i.e. it details many examples of private and public sector organizations that have used SPSS to reach a 'solution' to an 'information problem'. Optimistic stato-phobes may find it reassuring that their undergraduate studies are not being wasted in statistical laboratory classes; pessimistic stato-phobes may find the prospect of post-university life more daunting than ever!

Note: Full details of the above can be found in the references list below.

References

Fielding, J. and Gilbert, N. (2000) *Understanding Social Statistics*. London: Sage.

Gray, C.D. and Kinnear, P.R. (1998) *SPSS for Macintosh Made Simple*. Hove: Psychology Press.

Kitchin, R. and Tate, N.J. (2000) *Conducting Research in Human Geography: Theory, Methodology and Practice*. Harlow: Prentice Hall.

Norusis, M. J. (2002) *SPSS Base 11.0 User's Guide Package*. Chicago, IL: SPSS.

Norusis, M. J. (2002) *SPSS 11.0 Guide to Data Analysis*. Chicago, IL: Prentice Hall.

Rogerson, P.A. (2001) *Statistical Methods for Geographers*. London: Sage.

25 Coding Transcripts and Diaries

Meghan Cope

Synopsis

Coding is the assigning of interpretive tags to text (or other material) based on categories or themes that are relevant to the research. This chapter reviews strategies surrounding the coding of qualitative text, including how to evaluate your sources, identify topics and refine research questions, construct and fine tune your coding structure, build themes and maintain an accounting of the process of coding. Coding itself is broken down to include steps such as identifying patterns and forming categories as well as the procedures of defining first-level descriptive codes, developing second-level analytical codes and coding along a particular theme or concept. Different types of materials are discussed, including historical sources such as diaries and other 'pre-existing' documents, as well as 'self-generated' materials such as transcripts of interviews, focus groups and life histories.

The chapter is organized into the following sections:

- Introduction
- What is coding? A brief description and example
- Preparing to code
- Coding
- Conclusion.

INTRODUCTION

Text-based materials (such as diaries, letters, oral histories, transcripts of interviews or focus group sessions) and other similar sources are rich in information (see Chapters 7–9) but present the researcher with unique challenges for interpretation and representation. In this chapter, I discuss a series of strategies for organizing data, developing a code structure, identifying trends emerging from the text materials and building themes that connect your empirical findings with a broader literature. Although the process of interpreting and coding data does not follow a linear path, the strategies discussed here are presented in a roughly sequential order for clarity.

It is often the case that text materials can fill gaps in information that otherwise could not be known, and in that role they are extremely valuable. For example, less is known about people who have typically been left off the historical record or are located outside the gaze of the mainstream of research (e.g. women, ethnic minorities, poor people, children), and we also tend to know less about historical 'everyday' events and life conditions than about events that have large-scale or long-term significance. Many researchers in geography and other disciplines have started working in these areas of non-elites and, at more local, day-to-day levels, and are finding primary text materials to be very valuable in filling those gaps in knowledge. The question then arises of how to go about incorporating diaries, interview transcripts or other materials into a research project and how to make the most of the sources while avoiding some potential pitfalls.

One of the key debates about using these data sources surrounds the fact that, because these types of materials often represent only a small group of people's views and ideas, the data are highly specific to those individuals. Even when the n is large (i.e. you have more than 30 respondents or text documents), the open-ended nature of these data forms raises issues for interpretation and representation. Often, the traditional standards of 'good research', such as objectivity (being value-neutral) and the ability to generalize to larger populations, are not applicable for such materials. However, qualitative researchers have also developed standards for what is 'rigorous' research because the standards for quantitative inquiry do not necessarily transfer well (see, for example, Baxter and Eyles, 1997; Bailey et al., 1999). For example, as Silverman (1993) points out, the meaning of 'validity' in quantitative research is not appropriate for ethnographic work, but there are other strategies such as 'triangulation' (confirming results by consulting multiple and varied sources) that are viable and should be part of the qualitative researcher's tool kit to insure *reliability* of results. As part of making sense of these subjective and often problematic forms of data, while sticking to principles of rigorous inquiry, many researchers use strategies of coding and theme building. There are many ways to go about this, but some common principles and practices are reviewed in this chapter as a way to introduce the process.

WHAT IS CODING? A BRIEF DESCRIPTION AND AN EXAMPLE

Coding is basically a way of evaluating and organizing data in an effort to understand meanings in the text. First, coding helps the researcher identify *categories* and *patterns*. For example, in studying interviews with low-income mothers, a researcher might see a pattern of daily challenges poor mothers face (she misses the bus, her child is sick, her caregiver shows up late) and code these as 'daily challenges', keeping a memo on what has so far been included in that category, and perhaps constructing subcategories (those related to transportation, those related to childcare, etc.). By identifying categories and patterns, we can begin to make more sense of the data and start to ask new questions. Following on the example of low-income mothers, we might find several subcategories of daily challenges (housing, transportation, childcare) but then also several bigger-picture categories of challenges they face (children's future, discrimination, violence, personal and financial security, self-esteem). Coding for these can help identify how these challenges intersect and may produce new codes, new research questions and new understandings of meaning in the data.

However, as Strauss (1987) points out, the identification of patterns and the formation of categories are only the most basic level of coding. He suggests that beginning analysts consider four types of themes in the data: *conditions, interaction among the actors, strategies and tactics* and *consequences*. Many of these are indicated to the researcher directly by the subject. 'Conditions' can be indicated by such phrases as 'because' or 'on account of' (Strauss, 1987: 28) or passages like 'when I was in XYZ situation . . .' For example, if a low-income mother referred to a time when she lost her housing and had to live with her sister, we would code it as a particular type of housing 'condition'. Similarly, 'interaction among the actors' means looking for how the informants engage with others, what they think of others, what others do to them. So we might be interested in how low-income mothers interact with their children, their partners, their case workers, their employers and so on. 'Strategies and tactics' refers to what people do in certain situations or how they handle particular events. For example, poor women with children often exchange favours with friends and neighbours as a strategy to survive on low wages and/or state benefits (Gilbert, 1998). Again, there will usually be subcategories of strategies that become relevant, and these are very likely to be tied together with both 'conditions' and 'interaction' with others. Finally, 'consequences' are often easy to identify because the informant makes the connection for us: a woman says 'My child was sick so I missed work too many times and lost my job. Then I couldn't pay the rent so we got evicted and had to move in with my sister.'

These are the *consequences* of certain *conditions*, but also indicate *interactions* and *strategies*. The idea, then, is to use these special types of categories to start analysing the data and pulling out new themes.

Of course, the idea is not merely to code something as a 'strategy' but simultaneously to name that particular strategy or set of tactics that was used. In the above example, the strategy of 'depending on family' enabled the woman and her children to find a temporary place to live after being evicted. This strategy is a code that then becomes connected to other codes, such as the conditions codes for housing and the interactions codes for family members. Codes do not stand alone but are part of a web of interconnected themes and categories.

Coding enables the researcher to make new connections. For example, the connection between a child getting sick and the family losing housing may not be immediately apparent, but when it is considered in the context of the fragile balance low-income women must achieve the relationship becomes more clear. The researcher is then sensitized to such connections and seeks to identify other similar relationships within the data, and may look for related connections from other informants.

PREPARING TO CODE

Evaluating your sources

Researchers are commonly confronted with two scenarios when working with text materials – either they are working with pre-existing documents that were created without specific reference to their project (such as diaries, historical documents, secondary sources such as newspapers, oral histories and transcripts generated from others' research), or they have constructed a research project and carried out interviews or focus groups which they will then code. In my own experience I have used many different kinds of text documents which fall into one or the other of these two categories. Pre-existing sources I have studied include diaries from women living in the late nineteenth century, transcripts of oral histories that were done by a local historian and deposited in the archives of a city I was researching, and various other historical materials such as personal letters, social club newsletters, factory records and memoirs. In the self-generated category, I have used transcripts of interviews and focus groups I personally conducted, as well as transcripts of interviews I developed but which were actually conducted by others.

Clearly, these two categories of documents (pre-existing and self-generated) require somewhat different approaches, particularly in the ways they connect to your research questions. Using pre-existing material tends to be an even more inductive process than using self-generated text – that is, the researcher's initial approach must be one of broad evaluation to see what trends come out of the material. In this approach the research questions must be flexible and open to change, depending on what is contained in the documents. Imagine, for example, that you wanted to learn about migration from diaries kept by white women moving to the western USA in the 1860s – it makes sense to keep an open mind while reading through these before deciding on your specific research questions.

On the other hand, having the opportunity to generate original text documents from interviews or focus groups you yourself are conducting means you can begin by linking questions for your respondents directly to your research interests. As an example of tailoring your data collection, imagine you were interested in physically disabled people's daily negotiations of space as they pass through work, home and public space. You might ask them to keep diaries of their daily activities and mobility patterns and then interview them about what challenges they face, how they experience mobility and sources of friction in their negotiations of public space. In both these scenarios, the development of strong research questions is essential. Ideally, the research questions reflect some element of what we already know (from related literature and theory) *and* incorporate initial findings or hypotheses of the empirical component of the research.

Whether using pre-existing sources or self-generated texts, researchers doing qualitative work spend a lot of time reading and thinking about their material. By approaching the data with an open mind, researchers allow the data to 'speak' to them. This is important because even in research with self-generated materials it may require several readings for the full diversity of topics and meanings to begin to reveal itself. When allowing data to 'speak' researchers need to consider their own listening (reading) biases and decide whether to include an emerging theme or not. For example, if a diary goes into great detail on family relationships but the researcher is primarily interested in travel, a decision needs to be made on whether to stick to the original topic (travel) or shift the theme to family relationships, or perhaps combine elements of both themes and, say, view travel through the lens of family relationships.

Identifying topics and forming research questions

Reading and rereading transcripts and diary material allows the researcher to start identifying topics that are recurrent or demonstrate important insights. These may come out of a query that is generated by theoretical literature (e.g. 'Does this theory hold true in that geographical/historical context?'), or they may emerge from the data themselves (e.g. 'Many women diarists of the American West mentioned feelings of depression when confronted with the bleak prairie landscape'), or the researcher may identify a theme with both theory and data in mind (e.g. 'Do current theories of migration apply to marginalized people's experiences of relocation as domestic workers?').

Some initial familiarity with both the relevant literature and the empirical data to be used is desirable for the formation of a few key research questions (see Chapter 1). The successful creation of original academic work often begins by bringing existing literature and theory together with an empirical context or dataset in a way that has not been done before. From that point new insights start to flow as you become more familiar with the data, are engaged in existing scholarly literature and become increasingly convinced that your project has something to offer. (If things do not progress in this manner, it may be time to re-evaluate!)

Here is an example from my own experience of the development of a key topic and set of research questions. My theoretical readings in contemporary economic geography demonstrated that so-called 'contingent' labour – that which is short-term, part-time, contract or seasonal work – was perceived to be on the rise in the 1990s as the USA and other countries moved from a primarily Fordist industrial period of capitalism to a more flexible, postindustrial economy. Empirically, I was interested in the economic conditions of women's everyday lives in Lawrence, Massachusetts (a 'textile town') during the Great Depression. In reading transcripts from oral histories that were completed in the 1980s by a local historian, involving men and women who had been working in Lawrence textile mills in the 1920s and 1930s, I had seen that the women's jobs were often paid by piecework, while men's jobs carried straight salaries. This sounded like 'contingent' work to me and, indeed, on further digging I found that women were often employed by the mills on a seasonal, short-term or part-time basis, while men had more stability in their mill jobs. So my research questions became: how were women's and men's jobs constructed differently through the gender division of labour in the mills to produce both fixed and flexible components to the labour force? What impacts did this have on people's everyday lives and their household strategies? And, theoretically, what were the implications for our economic theories of industrial capitalism if there were strong elements of contingent labour within what has been perceived as standardized labour relations of the 1930s? This is an example of linking the theoretical and the empirical in one's research questions, and also of remaining open-minded in both data mining and in literature reviews.

CODING

Once you are familiar with the relevant literature, have read through your pre-existing or self-generated text materials several times and have built some initial themes that both interest you and appear in the data, you are ready to code. There are many approaches to this process but I will outline some general steps here to provide some basic guidelines.

Building and refining the coding structure

The process of developing the coding structure for your project is one that is inevitably circular, sporadic and, frankly, messy. Some scholars have tried to formalize the coding process (see, for example, Strauss, 1987; Strauss and Corbin, 1990) with some degree of success, but even they acknowledge that it is not a clear, linear process for which you can follow step-by-step instructions and at some point say that you are 'done'. Rather, coding involves reading and rereading, thinking and rethinking, developing codes that are tentative and temporary along the way, and requires a lot of enthusiasm and patience. However, coding is also really rewarding in that it enables the researcher to know his or her data intimately and see patterns and themes emerging in a way that would not be possible otherwise. For a lively account of some of the challenges and rewards of coding, see Crang (2001).

The most common way to construct the first set of codes is to start by reading through your first text document, marking important sections, phrases or individual words and assigning those a code. After reading through all your materials with a critical eye, you should have a list of codes you think are important, along with your notes about them (keeping notes or 'memos' on your coding process is very valuable). If you are using a qualitative software package, the codes are typically kept in one location and their notes are linked to them automatically (see Chapter 26). Recall that the initial set of codes will be changing: there will be some things you find you don't use and others that should be added. Strauss calls this process 'open coding'. He says open coding 'is unrestricted coding of the data. This open coding is done by scrutinizing the fieldnote, interview, or other document very closely: line by line or even word by word. The aim is to produce concepts that seem to fit the data' (1987: 28). The purpose of this stage is to 'open up' the data, fracturing them along the way if necessary, and breaking the data down so that conceptual implications can emerge in the later steps.

Strauss (1987) also recommends two other kinds of coding. One is called 'axial coding' because it proceeds along an *axis* or key category. Axial coding can be part of the open coding process but it allows the researcher to follow a particular category for a while as a way of testing its relevance. To return to the above example of interviews with low-income mothers, imagine that as the open coding progresses the axis of 'dependence on friends/family' catches our interest. We may follow that particular category in a bout of axial coding by focusing on different ways the respondents depended on their friends and family, soon returning to the freer open coding. Strauss's third type of coding is called 'selective coding'. This is a more systematic approach to coding that is done when a central or 'core' category is identified and followed. For example, after some open coding and axial coding of the interviews from low-income mothers, we might decide the core theme emerging is the struggle for a better future for their children. We say it is 'core' because we have found that most of what the mothers talk about is related in some way to this theme. From that point on other themes become secondary and the main lens through which the data are viewed is based on the core category (i.e. we are being 'selective'). While it is certainly possible to code successfully without necessarily following Strauss's method, his characterization of different types of codes and approaches to coding can be helpful in organizing what is a somewhat chaotic process.

Another way to think of different stages of coding is to consider the first level as 'descriptive' and the next level as 'analytic'. Descriptive codes contain mainly what we call *in vivo* codes – that is, they appear in the text and we use respondents' own words as codes. Analytic codes emerge from a second level of coding that comes after much reflection on descriptive codes and a return to the theoretical literature.

Here are two examples from research I have completed. The first is drawn from the historical work I mentioned above and the coding procedure is described step by step. The second example is contemporary and appears in Box 25.1. It shows how an interview question was coded in a project that looked at the impacts of welfare reform on social service organizations in Buffalo, New York, in the late 1990s.

In the first example, the following passage is from an oral history interview recorded and transcribed in the 1980s (not by me) from a woman who was recalling her working conditions in the 1930s in the textile mills of Lawrence, Massachusetts. She said:

> I was on piecework and I found my work very tedious and hard and you had to keep running and working all the time. See, the difference between my husband's salary and mine was that he was a day worker, like he got paid by the day and I didn't. I got paid piecework for the amount of work I could produce in a day. So, at the end of the week I was never sure if my pay would be larger or smaller according to the amount of work I did (Lucienne Adams, *Immigrant City Archives*).

Box 25.1	Coding an interview	
Text	***In vivo* codes/ description**	**Analytic codes**
Q: What do you expect to see happen as people hit [welfare] time limits or are cut off from programs?		
A: Yeah, the safety net thing. I see them coming into that. I suspect, and the big picture here if you look at the [welfare] population approximately a third of the individuals can probably get work on their own or with a little bit of help, another third need some fairly intensive services, basically all the services that the employment and training and other human services community can provide them, and then I believe – and this is my own opinion – that there is probably another third or so that really are going to have a very very hard time of things after the five year limit. They may be in a situation where they're really not going to be able to work. There are some people who have serious problems if you want to call them problems, if you want to label them as such, that are going to prevent them from working.	safety net who can work who needs services who is unable to work	issue of 'creaming' – early success of those ready to work role of job training and support services 'hard to reach' population respondent is sensitive to 'labeling'
Source: Interview conducted in July 1997 by Meghan Cope with a staff member of a social service organization in Buffalo, NY		

In the first-level (descriptive) coding, this passage was marked for the following codes (mostly *in vivo*): work, salary, piecework, uncertainty of income and difficulty of work. I found these themes coming up again and again in reading through over 50 oral histories and innumerable other documents and, because I was reading about similar themes in the scholarly geography journals regarding women, work, household relations and company production strategies, I felt they were codes that would lead to some interesting insights. So when I went back to the passage later in the analysis and did a second-level coding, I added analytic codes for this passage including 'gender divisions of labour', 'women's contingent work' and 'household relations'. These later codes had more connection with and significance to the broader literature and my research questions than the simple descriptive codes I had initially assigned and, indeed, they were indicative of some developing themes in the research project.

Theme building

The coding process is fluid and dynamic but it is not the end-product of analysis. As codes become more complex and more connected to the project's theoretical framework, they start building into themes that can then serve as the main topics for the final product (paper, report, dissertation). Connections between simple codes – for example, my finding 'uncertainty of income' co-occurring with 'women's work' repeatedly in multiple oral histories and other sources – can generate new paths of exploration, both into the data and back into the framing literature. Ultimately, this co-occurrence of codes turned into a major section of results from this project: women's work was purposely constructed as contingent (part time, seasonal, uncertain) within the mills in part to keep costs down (contingent workers typically earn less) and in part to take advantage of the fact that women moved in and out of the waged workforce as they reared children or attended to other family and household obligations.

The process of theme building is central to qualitative, interpretive work because it allows for the organization of information into trends, categories and common elements that are theoretically important. Themes may be based on similarities within the data or, conversely, on differences that appear and are interesting for some reason. The important thing to realize about theme building is that it is an ongoing process throughout the qualitative research project: themes may be identified before, during or after the data collection and analysis stages. Indeed, many of the best research projects are quite fluid in that they shift focus in response to the data and findings that emerge along the way.

One of the best approaches to theme building is to read *across* the materials being used rather than *solely* within them (Jackson, 2001). That is, after having spent time coding and interpreting individual documents, sit down with several or all your materials, including your notes and memos on the coding process, and work with a particular topic or code while drawing from multiple texts. This process aids in seeing trends that manifest themselves in many different ways. For example, poor mothers may use different vocabularies to talk about their struggles to maintain stable environments for their children but the theme is apparent when reading across several transcripts at a time.

Looping back to your research questions

Once you have built some themes based on interesting coding insights, revisit your research questions to evaluate and perhaps refine them. It may be that what you had hoped to find really is not apparent in the data sources you are using but that another unexpected (and – one hopes – equally compelling) theme has emerged and you will adjust the direction of your project accordingly. This can happen both when using pre-existing documents and in cases in which the researcher structures the questions for the respondents. In the former, you may discover, for example, that a theme you saw emerging in the first document or two you looked at was not evident in any other materials – somewhat of a dead end. In the latter case, you might begin a set of interviews with one idea of what is happening but find that an initial idea is not borne out by respondents but a new (and even better?) theme emerges unexpectedly.

For example, I once started a project looking at the social networks poor people use to obtain jobs and began by asking directors of social service organizations about these networks. It soon became apparent that many poor people couldn't use social networks to obtain employment because they didn't know anyone who was working (or working happily), so I had to abandon that line of inquiry. However, through discussions with the social service directors a new theme emerged: their organizations' job-training programmes were often reinforcing the existing racial and gender divisions of labour in the local workforce, and this became a new focus for the project.

Issues to consider

In the coding process there are always additional issues and challenges to consider. First, it may be helpful (or even necessary, as in multi-investigator projects) to have multiple coders. If you are working alone it may be constructive to give your code book and coded materials to a colleague to see if you are missing important themes and to test the strength of your interpretations. In group projects, the code book is ideally developed with everyone who will be involved in the coding; if that isn't possible, at least have good notes attached to each code and discuss what different codes represent so there is minimal ambiguity and unevenness.

Secondly, you may need to eliminate some codes as the project progresses. It is very easy to get caught up in designing codes for finer and finer resolutions of a theme, but having an unwieldy code book just means some codes won't be put to use very often and it may make sense to combine or consolidate them. This is another instance in which having a colleague or your tutor to serve as a sounding board is valuable.

Thirdly, if you are working on a project using self-constructed data materials (that is, your respondents are alive and available), you may want to review your interpretations with some or all of your respondents as a way to check that your findings reflect what they intended. This practice, while quite democratic and potentially enriching, is also fraught with challenges and can be very time-consuming. Some questions to think about include: how do you deal with respondents who contradict their earlier statements? How much of your project do you want to reveal to respondents (especially that which may be critical or unflattering)? At what point will you stop involving your subjects in the research process? (Some people even co-author their results with their subjects; see, for example, Pratt and the Philippine Women Centre, 1999.) What is your philosophy on the power relations set up between observer/author and 'subjects'? How much time are you willing/able to devote to collaboration with your respondents?

For an excellent review of some other issues to address in the interpretation of qualitative data, see Jackson (2001) (especially his 'checklist' in the article's conclusion). Jackson points out, for example, the need to consider the silences, hesitations, uses of humour or irony and other non-verbal cues that may not be conveyed in text transcriptions of interviews or focus groups but that are evident in the tape recordings. This is one reason it is helpful to have notes from the interviewers and/or focus group facilitators that are attached to each transcript (something that is particularly easy when using a computer analysis package). Silverman's (1993) chapter on transcripts is very useful in demonstrating how notes on gestures, facial expressions and other cues can be incorporated into a text transcript in order at least partially to convey these important non-verbal elements of interviews and focus groups.

Additionally, Jackson (2001) suggests that, at some point in the interpretation, researchers should consider what is *absent* from their respondents' accounts and think about how these might be important. In his work on masculinity, for example, he found that certain themes were rarely or never addressed by the men in his focus groups, including fatherhood, race, friendship and all things domestic.

The way respondents frame the *position* of the researcher is also important. Participants in Jackson's projects' focus groups displayed several different types of attitude towards the university researchers in his project. He also found evidence of respondents moderating what they said in his presence and in that of his co-investigators (particularly his female colleague). Indeed, feminist researchers have explored the issue of positionality (see also Chapters 8 and 12) and the relations between researcher and researched, but these issues are also a matter for all researchers and should be addressed seriously (see Cope, 2002; England, 2002; McKay, 2002; Valentine, 2002).

All these issues are important to consider, depending on your types of sources, the scope of the project, the type of data collection that was used (semi-structured interviews vs. life histories vs. daily activity diaries), who is involved in the research and who the informants are, and whether your materials were pre-existing or self-generated. The final suggestion, then, is to keep a detailed account of which techniques you have used, what problems you have encountered and how you have dealt with issues such as those discussed in this section. Qualitative research has often been critiqued for its 'hidden' methodologies: researchers need to be much more open about their procedures of data collection, coding and interpretation and should always include an explicit discussion of their methods in any presentation, whether written or oral. Coding is one way of ensuring a more systematic methodology, but the coding process also needs to be disclosed fully to readers.

CONCLUSION

Coding should be seen as an active, thoughtful process that generates themes and elicits meanings, thereby enabling the researcher to produce representations of the data that are lively, valid and suggestive of some broader connections to the scholarly literature. By approaching coding as both systematic and flexible, it becomes an enlightening, fruitful and revealing process that allows for final products that are rich with meaning and true to the initial respondents or 'subjects'. The methods presented in this chapter attempt to balance the need for a systematic approach with the advantages of remaining flexible and open to emerging themes and multiple interpretations. Coding should not be seen as tedious and boring but rather as a type of detective work – we are trying to solve mysteries using varied clues and we are open to surprises in the data that generate those 'aha!' moments of investigation and inquiry.

Summary

- Coding enables qualitative researchers to make sense of subjective data in a rigorous way.
- It is a way of evaluating and organizing data in order to understand meanings in the text.
- Coding helps researchers identify categories, patterns, themes and connections in the data.
- When preparing to code, evaluate your sources, read and reread your data.
- Building and refining code structures is a messy and dynamic process.
- There are different levels of coding: descriptive and analytic.
- When coding revisit your research questions, ask others to look at your codes, show your interpretations to informants and consider what is absent.
- Researchers need to be more open about their procedures for coding.

Further reading

There are a lot of social science books and articles about how to code qualitative data, but I have found the following particularly useful:

- Jackson (2001) is a quick and easily comprehended chapter that raises many important issues for qualitative research, specifically the interpretation of results. Jackson's conclusion 'checklist' is especially valuable.

- Silverman (1993) is a truly comprehensive treatment of how to go about coding and analysing all kinds of verbal and text-based data, with exhaustive step-by-step instructions and plenty of real-life research examples – this is the book!

- Flowerdew and Martin (1997). This is the definitive guide for geography students with a wide range of coverage and suggestions for related sources.

Note: Full details of the above can be found in the references list below.

References

Bailey, C., White, C. and Pain, R. (1999) 'Evaluating qualitative research: dealing with the tension between "science" and "creativity"', *Area*, 31: 169–78.

Baxter, J. and Eyles, J. (1997) 'Evaluating qualitative research in social geography: establishing rigour in interview analysis', *Transactions, Institute of British Geographers*, 22: 505–25.

Cope, M. (2002) 'Feminist epistemology in geography', in P. Moss (ed.) *Feminist Geography in Practice*. Oxford: Blackwell, pp. 43–56.

Crang, M. (2001) 'Filed work: making sense of group interviews', in M. Limb and C. Dwyer (eds) *Qualitative Methodologies for Geographers*. London: Oxford University Press, pp. 215–33.

England, K.V.L. (2002) 'Interviewing elites: cautionary tales about researching women managers in Canada's banking industry', in P. Moss (ed.) *Feminist Geography in Practice*. Oxford: Blackwell, pp. 200–13.

Flowerdew, R. and Martin, D. (eds) (1997) *Methods in Human Geography: A Guide for Students Doing Research Projects*. Boston, MA: Addison-Wesley.

Gilbert, M. (1998) 'Race, space and power: the survival strategies of working poor women', *Annals of the Association of American Geographers*, 88: 595–621.

Immigrant City Archives, 6 Essex Street, Lawrence, MA 01840.

Jackson, P. (2001) 'Making sense of qualitative data', in M. Limb and C. Dwyer (eds) *Qualitative Methodologies for Geographers*. London: Oxford University Press, pp. 199–214.

Limb, M. and Dwyer, C. (eds) (2001) *Qualitative Methodologies for Geographers*. London: Oxford University Press.

McKay, D. (2002) 'Negotiating positionings: exchanging life stories in research interviews', in P. Moss (ed.) *Feminist Geography in Practice*. Oxford: Blackwell, pp. 187–99.

Pratt, G. with the Philippine Women Centre (1999) 'Is this Canada? Domestic workers' experiences in Vancouver, BC', in J. Henshall Momsen (ed.) *Gender, Migration and Domestic Service*. London: Routledge, pp. 23–42.

Silverman, D. (1993) *Interpreting Qualitative Data: Methods for Analysing Talk, Text, and Interaction*. Thousand Oaks, CA: Sage.

Strauss, A. (1987) *Qualitative Analysis for Social Scientists*. Cambridge: Cambridge University Press.

Strauss, A. and Corbin, J. (1990) *Basics of Qualitative Research: Grounded Theory Procedures and Techniques*. London: Sage.

Valentine, G. (2002) 'People like us: negotiating sameness and difference in the research process', in P. Moss (ed.) *Feminist Geography in Practice*. Oxford: Blackwell, pp. 116–26.

26 Using CAQDAS in Qualitative Research

Bettina van Hoven

Synopsis

CAQDAS is the acronym for Computer Assisted Qualitative Data Analysis Software. It comprises a variety of programs with different capabilities to assist the analysis of qualitative data. Simple programs can be used for text searches, while more sophisticated ones have functions especially for theory building. This chapter examines the ways in which you might use CAQDAS.

The chapter is organized into the following sections:

* Introduction
* Computer software in qualitative research
* Key functions of CAQDAS
* Further considerations for picking a program
* An example of CAQDAS in practice: experiences with NUD.IST
* Some advantages and concerns about CAQDAS
* Conclusion.

INTRODUCTION

Qualitative research techniques, such as interviewing and participant observation (see Chapters 8 and 9), are commonly used by human geographers. They generate rich data in the form of interview transcripts and diaries. Traditionally, geographers have analysed this material by coding it manually by pen and paper (see Chapter 25) but this approach has a number of limitations. Richards and Richards (1987) note that key worries in the use and analysis of qualitative material include the volume of data generated; the complexity of data analysis; and the notion that clerical tasks required for data preparation and management may prevent thorough data analysis. As a result in recent years, computer packages have increasingly been adopted for qualitative data analysis because they are regarded as useful tools for handling and coding large amounts of written data and because they facilitate the in-depth examination of relations between and within these data.

This chapter discusses how Computer Assisted Qualitative Data Analysis Software (hereafter CAQDAS) can be used in qualitative research. Rather than dealing with the specifics of software on the market, it focuses on the fit between different types of software and the needs of research projects. The key functions of packages are outlined, such as 'text retrieval/text management', 'coding and retrieving' and 'theory building'. In addition, the specific considerations you need to take into account before choosing software are summarized. The chapter then illustrates the use of computer software by describing my own experiences of using the computer program NUD.IST in qualitative research. Finally, the chapter draws attention to concerns regarding the use of CAQDAS and its impact on the way in which qualitative research is done.

COMPUTER SOFTWARE IN QUALITATIVE RESEARCH

Most students at undergraduate level are familiar with preparing reports with a PC or Mac. Indeed, a number of activities that are part of qualitative research may have already been completed using word-processing programs such as transcribing interview tapes or making fieldnotes (using a laptop) and editing these. Equally, data display in the form of matrices or networks and graphic mapping may be familiar using 'traditional' graphics programmes. Box 26.1, compiled by Miles and Huberman (1994), gives an overview of these and other uses of computer software in qualitative research. However, for points 4, 8, 11 and 12, simple wordprocessing packages are not sufficient; rather, for speed and efficiency more specialized software is required. It is these issues this chapter addresses.

Several computer packages to aid qualitative data analysis have entered the market since the introduction of commercial programs in the 1980s. Every program has its specific focus and, of course, strengths and weaknesses. An assessment of computer packages aimed at the analysis of unstructured non-numerical data can be found in Tesch (1990), Prein et al. (1995) or Weitzman and Miles (1995). In these publications, a review of software can be found as well as their technical requirements. Keep in mind, though, that programs are constantly being improved and updated. Some have become very alike in terms of their purpose and capabilities. As a result the reviews provided by these authors (and anything else you find to read on this

---

Box 26.1   Uses of computer software in qualitative studies

1   Making notes in the field.
2   Writing up or transcribing fieldnotes.
3   Editing: correcting, extending, or revising fieldnotes.
4   Coding: attaching key words or tags to segments of text to permit later retrieval.
5   Storage: keeping text in an organized database.
6   Search and retrieval: locating relevant segments of text and making them available for inspection.
7   Data 'linking': connecting relevant data segments to each other, forming categories, clusters or networks of information.
8   Memoing: writing reflective commentaries on some aspects of the data as a basis for deeper analysis.
9   Content analysis: counting frequencies, sequence or locations of words and phrases.
10   Data display: placing selected or reduced data in condensed, organized format, such as a matrix or network, for inspection.
11   Conclusion-drawing and verification: aiding the analysis to interpret displayed data and to test or confirm findings.
12   Theory-building: developing systematic, conceptually coherent explanations of findings; testing hypotheses.
13   Graphic mapping: creating diagrams that depict findings or theories.
14   Preparing interim and final reports.

Source: Miles and Huberman (1994: 44)

---

subject) may, at least in part, be slightly out of date. Nevertheless, they remain an invaluable source and starting point when investigating the scope and functions of programs.

KEY FUNCTIONS OF CAQDAS

Three types of CAQDAS can be distinguished according to their key functions: (1) text retrievers/textbase managers; (2) coding and text retrieval; and (3) theory-builders (Fielding, 1994; Weitzman and Miles, 1995). It is useful briefly to outline what programs in these categories do as it helps to narrow down the multitude of software packages according to the analytic needs of the project. Although various programs are available for each category, I include only some. The programs chosen here have been on the market for some time and have benefited from continuous improvements through interactions between developers and users (see Richards, 1997). An extensive overview is provided in Table 26.1. The examples suggested below are

TABLE 26.1 Overview of CAQDAS

Text retrievers/textbase managers			Coding and text retrieval			Theory-builders		
PC	Mac		PC	Mac		PC	Mac	
✔		AskSam (2)	✔		Ethno (1, 4)	✔		AQUAD (1, 2, 3)
✔	✔	Metamorph (2)	✔		The Ethnograph (1, 2, 3, 4)	✔		ATLAS/ti (1, 2, 3)
✔		Orbis (2)	✔	✔	FolioVIEWS (1, 2)	✔	✔	HyperRESEARCH (1, 2, 3)
✔	✔	Sonar (1, 2)	✔		FuzzyStat (1)		✔	Hypersoft (1, 3)
✔	✔	Tabletop (2)	✔		GATOR (1)	✔	✔	NUD*IST (1, 2, 3)
✔	✔	TextCollector (2)		✔	HyperFocus (1)	✔		Qualog (1)
✔		WordCruncher (2)		✔	HyperQual (1, 2, 3, 4)	✔		QCA (2)
✔		ZynINDEX (2)	✔		KWALITAN (1, 2, 3)			
			✔		Martin (1, 2, 3)			
			✔		MAX International (1, 2, 3)			
			✔		WINMAX (1)			
			✔		QUALPRO (1, 2, 3, 4)			
			✔	✔	SQL Text Retrieval (1)			
			✔		Textbase Alpha (1, 3, 4)			

Notes: (1) Grbich (1999); (2) Weitzman and Miles (1995); (3) Kelle (1995); (4) Tesch (1990). There are several other programs on the market that have not been reviewed (see, for example, http://caqdas.soc.surrey.ac.uk/packages.htm).

also available as demo versions (distributed by Scolari) for those who wish to examine the programs first:

1   Text-retrieving programs, in brief, find words or combinations of words in so-called string searches and put these into new files. The degree of advancement of these programs varies. Some can, for example, assist in content analysis with additional functions such as counting word occurrences in the source text or creating word lists. Textbase managers are more advanced, in particular regarding their capability to organize data and their more sophisticated text search-and-retrieve functions (Fielding, 1994; Weitzman and Miles, 1995). One example of such a program is WinMAX with which data can also be coded and combined with statistical packages.

2   Code-and-retrieve packages can do what qualitative researchers do with cut-and-paste techniques using paper and scissors or coloured pencils. They are used to mark text segments for coding, cut and

sort text segments, and find, retrieve and report on data. However, as Weitzman and Miles (1995: 18) claim, this software is a 'quantum leap forward' compared with manual practice regarding its working speed as well as being more systematic and thorough. Some packages also have a memo function. In 'the Ethnograph', for example, these memos can be hyperlinked to text segments.

3   In addition to the aforementioned functions of CAQDAS, theory-builders have the capacity to test hunches, ideas and hypotheses that are pre-existing or emerge from data in the project. They allow the researcher to make connections not only between code and data but also between code and code. Relationships can often be displayed in a hierarchical index system or a graphic network. Examples of such software are NUD.IST and ATLAS/ti, respectively.

NUD.IST consists of two components: the Document System and the Index System. The first holds the source data, such as texts, video fragments and/or pictures. The second stores codes, either unstructured or arranged in a hierarchical form (tree), and text searches. It also contains memos and queries about data. A code network (i.e. the tree structure), can be displayed, and hyperlinks from nodes to source documents or memos established. In the example described below, these functions and the use of NUD.IST are illustrated in greater depth.

FURTHER CONSIDERATIONS FOR CHOOSING A PROGRAM

The choice of a particular type of software depends on such issues as the methodological orientation of the researcher (see also the further reading at the end of this chapter), the type of data, whether a project is conducted individually or in a research team, or if expectations from CAQDAS will increase over time. There are five key considerations for choosing qualitative software that are outlined below: (1) data entry/storage; (2) coding; (3) memoing/annotation; (4) data linking; and (5) search and retrieval (see also Fielding, 1995; Miles and Weitzman, 1996). Again, I refer to Weitzman and Miles (1995) for an evaluation of these functions per package in the form of a table.

Some programs, particularly older versions (which are both still useful and used often), require *data entry* in ASCII format. Text needs to be saved following formatting rules such as a limited number of characters per line, single spacing or special characters to indicate the beginning of a new text segment. Once saved, the files can be imported and stored as internal documents or left on the original PC/Mac as external documents. *Coding* can be done in several ways, of which one or more are supported by the package. Options vary from on-screen to off-screen coding or both, from coding one or several chunks of text to coding 'nested' or overlapping text. Some software permits hierarchical coding and the renaming and reorganizing of codes. Last but not least, a number of packages have hyperlinks from code to source text and are able to display relationships between codes graphically (see above). When aiming to build theory from project data, for example using a 'grounded theory' approach as described by Glaser and Strauss (1967), a function to store definitions of, and thoughts about, data and codes in *memos* is important. The capability of programs varies as the length of memos may be restricted or reference to source data is included while the actual (hyper-)link is absent. A part of theory building is the constant comparison of data (per category, case or person, for example) (see also Glaser and Strauss, 1967). In order to do this, quick *retrieval* of text segments is desirable. The time with which project data is searched can vary significantly. Furthermore, *searches* can be simple text searches, such as for specific words or phrases using Boolean functions (and, or, but not), or searches for codes or a combination of codes. Again, the level of sophistication for such searches varies. Some programs allow searches that include overlaps – 'A' in 'B' or 'A' near 'B', for instance. The results of these searches can be displayed per find, per source document or per category.

With regards to the requirements of a project, it is worth adding that some software is designed to accommodate multiple researchers, in particular those with more elaborate functions. Furthermore, software that is capable of performing all the above operations is more likely to 'grow' with increasing research demands.

AN EXAMPLE OF CAQDAS IN PRACTICE: EXPERIENCES WITH NUD.IST

Having briefly outlined the types of software available and specific consideration for choosing a particular package in a theoretical way, the remainder of this chapter explores the nuts and bolts of CAQDAS. NUD.IST is used to illustrate the discussion. It combines several functions mentioned above. It can be used for relatively simple projects in the first and second year of the course but it is also able to accommodate more demanding projects, such as an undergraduate dissertation or perhaps a thesis (it also has a very useful help network). There is a classic version (NUD.IST 4.0) and a more sophisticated 'sister version' (NVivo) for more experienced researchers. The developers of the program have an extensive website (www.qsr.au.com) with demos, downloadable handbooks and lists of literature both by the developers and users. In addition, the University of Surrey set up the CAQDAS networking project (http://caqdas.soc.surrey.ac.uk)

offering workshops, downloadable demos and literature as well as discussion forums for users. Last but not least, Kitchin and Tate (2000) give a quick overview of how data can be entered into and analysed using NUD.IST 4.0. I recommend using the latter in addition to this chapter.

In the following I will illustrate a number of the issues drawing on a project I conducted for my PhD between 1996 and 1999 (for a full account, see Hoven-Iganski, 2000). I had decided to use a modified grounded theory approach for my data analysis which had consequences for my data collection and analysis. I adhered largely to analytic steps suggested by Glaser and Strauss (1967) and Strauss and Corbin (1990) including coding, constant comparison and theory building. These steps are also supported by NUD.IST as this methodological framework is its 'bumper sticker' (Crang et al., 1997: 776).

Context of the study 'Made in the GDR'

The project 'Made in the GDR – the changing geographies of women in the post-Socialist rural society in Mecklenburg-Westpommerania' draws on data from 83 key informant interviews (e.g. politicians, civil servants at ministries or communities, chair persons of NGOs), 12 group interviews (two interviews per 'case village') and letters from 39 women throughout the wider study area. The interviews and correspondence were initiated using a topic guide and developed further based on the respondents' comments. The study had a strong exploratory character with the general aim to identify processes that help explain the formation of gendered identities and exclusionary practices in rural East Germany since unification. Although I took photographs in and around the case villages, I did not include them as data for analysis. Instead, they were used to illustrate the story or when they were pertinent to the argument. Regrettably, it was only upon my return from the last phase in the field, and when I became more familiar with the scope of NUD.IST, that I realized I could have included video recorded 'data' in my project (see Chapter 9 for discussion of video material).

Moving from paper piles to PC files

As part of my chosen methodology, I collected data in various stages, allowing time to transcribe these data before the next period of data collection. In a first stage, I corresponded with women in the study area in order to get a very general impression of the changes in East Germany and of the issues that occupied them most. I had aimed to analyse the material by hand (see Chapter 25), and started with the coding using coloured pencils and post-its. Although this helped with the development of a topic guide used for interviews as themes emerged from the data, I soon began to feel overwhelmed. I had generated so many codes in a line-by-line analysis that they had become illegible on paper and difficult to administrate. I worried that I would only be able to investigate fully some aspects while, unwillingly, ignoring others. In addition, the mere thought of the amount of data I would generate (95 × 1.5 hours tape) was almost paralysing. I had images of piles of colourful paper with post-its sticking out like a hedgehog, and of paper stacks being attacked by my cat in an idle moment. In other words, I did not feel very organized and not able to present a neatly finished project that would survive the critical scrutiny of my peers.

Preparing the data for NUD.IST

More or less at this point I became aware of NUD.IST. Learning the package at a basic level was quick. I completed the tutorial provided on the demo, did a one-day workshop through the CAQDAS networking project at the University of Surrey and was ready to go. Although NUD.IST (Version 4.0) is capable of combining the functions outlined above, it requires data entry in ASCII format. I used the text font 'Courier 10' – i.e. limiting the number of characters per line – and saved the file as 'Text with line breaks' – i.e. without any formatting. I also created a header for each document (i.e. interview or letter) and made sure to enter an asterisk (*) at the start of each speaker in a group interview, or every question or answer in individual interviews. In so doing, I produced text blocks the program recognized when searching and retrieving coded text later. Once all data were saved in this format, I imported them into the NUD.IST project.

Coding and retrieving

Two of my initial objectives were to assess the general implications of unification for women in rural areas; and to explore the dynamics behind the formulation of norms and meanings in the GDR. During my literature review, the reading of media reports and, of course, my correspondence, I identified several themes such as employment, unemployment, modernization, freedom of choice and travel. I expected these themes to emerge from 'my' data as well. Initially I had begun to 'make sense' of these, summarizing them per respondent or per theme. The 'expanded sourcebook' on qualitative data analysis by Miles and Huberman (1994) was an invaluable starting point for a research novice like me in that it illustrated various ways of reducing data. As a consequence, once I shifted to using NUD.IST, I had already generated a number of codes from the initial coding of the correspondence and my literature review. These codes were entered first and stored as 'free nodes' in the Index System, meaning that I had not ordered them in any hierarchical way. I used these nodes for subsequent line-by-line coding but generated a multitude of other nodes, 205 in total. In particular regarding the second research objective, which proved to be under-researched in the literature, the majority of codes were generated from the interviews.

At first all codes were 'open codes' (refer to Strauss and Corbin, 1990) – i.e. describing what 'my' data were about. They were relatively broad containers for the contents of one or more lines, often using the respondents' own words as code labels. For example, while reading the transcripts I noted that many respondents described experiences in the 'working collectives' – a small task-orientated unit at the workplace. Until then I had not been familiar with this theme but decided to explore this phenomenon further by including this issue in my interview topic guide. While coding a text chunk 'working collective', I also coded the same segment in many other ways, such as 'GDR', 'agricultural co-operative' (i.e. workplace), 'family', 'identity', 'democracy', 'socialist services' and others. NUD.IST stored the codes in the Index System while hyperlinking each code to the text segment, leaving my transcript uncluttered, unlike its paper counterpart before.

As a story began to evolve I saw connections between respondents and themes. I felt it would be worth exploring differences and similarities between respondents with different age groups, occupational status, gender, 'function' (i.e. key informant versus focus groups) and case villages. I was already well into my coding and had not yet coded for 'type of data'. Had I worked on paper, rereading every page of every transcript would have been required in order to assign the above labels to the relevant text chunks. In NUD.IST, however, this was less time-consuming and more thorough using a text search. For example, I entered '*130' (i.e. respondent Veronica) upon which NUD.IST retrieved everything Veronica had said. This text was stored in a separate file which I could then code '65+', 'pensioner', 'female', 'focus group' and 'Wilmersdorf'. I repeated this procedure for each respondent.

Constant comparison and theory building

Having coded 'my' data for content and 'type of data', I proceeded to compare codes. With the simple search mentioned above, coded segments can be extracted by respondent or theme.

I now wanted to combine different features to explore relationships between codes or between codes and type of data. In so doing, I would investigate hunches and ideas and test initial hypotheses. For example, I had the impression that the key informants suggested there was a difference between the experiences of men and women in the working collectives. Thinking about the narrations by women in the focus groups or accounts in my correspondence, I recalled very positive experiences of women. I therefore combined different codes to check out if this 'hunch' could be verified from my data. First, I retrieved everything women said about 'working collective'. Later, I created a file that also displayed other codes I had assigned to the same text chunk. Investigating this new file, further questions arose wherefore I needed to double check in which context something was said. As NUD.IST hyperlinks code to the source text, I could view the code in context by means of a mouse click. By so doing, my coding refined and my theory about the meaning of working collectives for gendered identities developed. This was also a part of exploring the dynamics behind the formulation of norms and meanings in the GDR, my second research objective. What I did was to 'decontextualize' data (i.e. rip bite-size chunks out of their original context) and 'recontextualize' them in a new context (see the example in Box 26.2). Tesch (1990) called this having a 'playful relationship' with data. According to the literature, I should have played more – i.e. continue this process until a state of 'saturation'. However, going back to the field collecting more data until absolutely nothing new would emerge could not be achieved due to the financial and time constraints of the project.

In Chapter 25 of this book, Meghan Cope argues that qualitative researchers need to be more transparent about how they code and analyse their material. One advantage of using NUD.IST is that it helps to create a so-called audit trail for external scrutiny (Stanjek, in Gildchrist, 1992). This is done using memos alongside my coding. Such memos can have different functions, such as keeping track of organizational issues, developing and defining coding and preserving steps in analytical thinking. I clarified why I coded a text segment in a certain way, why I related it to another code and why I felt it illustrated a theory particularly well. These thoughts were stored in easily accessible 'containers' (memos).

SOME ADVANTAGES AND CONCERNS ABOUT CAQDAS

I described several advantages of using CAQDAS in my experiential account above. NUD.IST largely worked for me but it is also important to be aware of the limitations of CAQDAS. Table 26.2 gives an overview of advantages and concerns described in the literature. The list of concerns in this table is considerable and does not yet include

Box 26.2   Decontexualizing and recontextualizing

Within a two-hour group interview during my study, respondents discussed their experiences in the GDR and those since unification. The following conversation took place sometime during the interview:

014: Well, twice a week, the butcher came to the village [in the GDR]. It wasn't like everyone had a car, like they do today. We would have to go all the way into town and especially for the pensioners that would have been too much . . . Today, the collective spirit is missing. It's missing, it used to be . . .
016: It's not there anymore.
014: We are . . . it's *missing* these days, we . . . everyone just thinks of their own job . . . one only thinks of oneself these days. We used to have a collective spirit.

The respondents were discussing their memories of the working collective. When I searched the other interview transcripts for this theme, I found that 39 documents contained the code 'collective'. I wanted to investigate more closely what insight I could gain from the women's experience of 'collective spirit' in relation to norms and meanings in the former GDR. Therefore, I displayed only those text segments that were coded 'collective'. This extraction of text segments from the interviews is called decontextualization. Other segments were, for example:

013: Women used to be able to discuss their problems and worries in the collective . . . Today they don't know where to turn and they don't dare talking about the problems.
011: It was different then, you were in a collective and somebody was asked to do this or to organize that. Like when a trip to the theatre was organized or so, you know? It was always somebody's turn and everyone participated in the organization at some point in time. Today, people find this very difficult.
120: The collective was promoted, you probably heard that already. One could get an award, a prize, and we always tried to be supportive of each other. That is what people miss so much today, the collective spirit. We were there for each other, we supported each other.

By bringing these text segments together, I recontextualized them. There were several indications that the collective was indeed an important part of the formulation of norms and meanings in the former GDR. Many women had fond memories of the collective because they experienced it as supportive and motivating. As the collective was organized according to a political plan and was also a significant part of the state's policy to form socialist identities, this finding suggested that the collective was an important means to achieve this.

TABLE 26.2   Advantages and concerns about CAQDAS

Advantages	Concerns
Managing of large quantities of data	Obsession with volume
Convenient coding and retrieving	Mechanistic data analysis/taken-for-granted mode of data handling
Comprehensive and accurate text searches	Exclusion of non-text data
Quick identification of deviant cases	Overemphasis on 'grounded theory'
More time to explore 'thick data' as clerical tasks become easier	Loss of overview
Playful relationship with data-enhanced creativity	The machine takes over – alienation from data
	Makes qualitative research look more scientific
	Limitations for connecting with geographical data such as GIS-type systems

Sources: Tesch (1990); Seidel (1991); Richards and Richards (1992); Dembrowski and Hanmer-Lloyd (1995); Lonkila (1995); Coffey et al. (1996); Fielding and Lee (1991); Kelle (1996); Crang et al. (1997); Tak et al. (1999).

the critique that the use of CAQDAS distances the researcher from the 'real world' of ethnography (Coffey and Atkinson, 1996; Coffey et al., 1996; Hinchcliffe et al., 1997).[1] The lesson drawn from the table is perhaps that a researcher will never be able to press a button and get a ready-made dissertation or project; nor is the software some kind of 'Frankenstein monster' (Lee and Fielding, 1991: 8). There is a danger that researchers may get lost or trapped in the data – particularly the logic flow of different software – and lose sight of what they are trying to write or achieve. In particular, researchers may struggle to reconcile their approach to qualitative data analysis (e.g. grounded theory) with the software (e.g. NUDIST works on a tree-branch structure). One way to avoid this is to use CAQDAS in a critical and reflective way, seeking feedback from peers and supervisors as some kind of 'rain check'. A project does not have to be analysed in a standardized way – i.e. using a grounded theory approach. The suggestions in the further reading section below describe other techniques. In sum, it is advisable to know about qualitative data analysis and its underlying methodologies before beginning to use CAQDAS, and to adopt an approach to your analysis and software package that is compatible.

CONCLUSION

In this chapter I have illustrated some uses of CAQDAS in qualitative research. Drawing on previous publications, I have outlined different types and key functions of CAQDAS and illustrated these issues using an example from my own research where I adopted a grounded theory approach to my data analysis. Overall, my opinion of the software is positive as I found the use of CAQDAS helpful, in particular for ordering and analysing data in an efficient and thorough way. However, I have also drawn attention to the dangers of seeing qualitative analysis software as a seductive technology whose benefits may be overestimated/emphasized. In particular, I have highlighted the danger of losing sight of your own methodological questions and analytical approach to the data. As such its important to retain a healthy scepticism about the value of CAQDAS and not to forget the importance of reading about approaches to qualitative data and thinking about your material (sometimes with pen and paper!). After all, the software only processes the data; it does not in itself understand or analyse the material, and it will not write your dissertation for you. At the end of the day a computer will not deliver anything that it is not commanded to!

Summary

- When using computers in qualitative research, the use of more specialized software is recommended for activities such as coding and searching text, finding relations and keeping an organized database.
- In general, CAQDAS has three key functions: text retrieval, coding and theory building. The choice of software depends on the methodological orientation of the researcher.
- More sophisticated programs usually combine several functions and accommodate different kinds of data, such as texts, statistics, pictures and video segments.
- Concerns about CAQDAS include the danger of mechanistic data analysis, the detachment from one's data and a false sense of science.

NOTE

1   It must be noted that further issues have transpired from more recent discussions about CAQDAS which should also be integrated in Table 26.2. Key concerns address the expansion from the current focus on textual data to the possibilities of hypertext and hypermedia, and the convergence towards one prevailing method of analysing these. I refer the reader to more comprehensive accounts – for example, in Lee and Fielding (1995), Coffey and Atkinson (1996: Ch. 7), Coffey et al. (1996), or Hinchcliffe et al. (1997).

Further reading

There is a growing body of literature about how to use CAQDAS. Key references include the following:

*   Weitzman and Miles (1995) should be consulted for an evaluation of what is on the market.

*   Theoretical and practical questions about how to use CAQDAS, illustrating pros and cons, are explored in Fielding and Lee (1991), Burgess (1995) and Kelle (1995).

*   Possible ways of accommodating approaches to qualitative research other than those described in this chapter – such as narrative studies, ethnography, interpretive/ hermeneutic analysis, critical theorists, collaborative or action research, or content analysis – can be found in Tesch (1991), Kelle (1996) or Miles and Weitzman (1996).

*   An invaluable Internet source is the site maintained by the CAQDAS networking project (http://caqdas.soc.surrey.ac.uk).

*   Last but not least, it is advisable to immerse yourself in the methodological literature as well. For the grounded theory approach, Strauss and Corbin (1990) provide an accessible account of 'how to . . .'

Note: Full details of the above can be found in the references list below.

References

Burgess, R.G. (1995) *Computing and Qualitative Research. Volume 5. Studies in Qualitative Sociology.* Middlesex, UK: Jai Press.

Coffey, A. and Atkinson, P. (1996) *Making Sense of Qualitative Data. Complementary Research Strategies.* Thousand Oaks, CA: Sage.

Coffey, A., Holbrook, B. and Atkinson, P. (1996) 'Qualitative data analysis: technologies and representations', *Sociological Research Online*, 1 (available at http://www.socresonline.org.uk/socresonline/1/1/4.html).

Crang, M.A., Hudson, A.C., Reimer, S.M. and Hinchcliffe, S.J. (1997) 'Software for qualitative research. 1. Prospectus and overview', *Environment and Planning A*, 29: 771–87.

Dembrowski, S. and Hanmer-Lloyd, S. (1995) 'Computer applications – a new road to qualitative data analysis?' *European Journal of Marketing*, 29: 50–62.

Fielding, N. (1994) 'Getting into computer-aided qualitative data analysis', *ESRC Data Archive Bulletin*, September (available at http://caqdas.soc.surrey.ac.uk/getting.htm).

Fielding, N. (1995) 'Choosing the right qualitative software package', *ESRC Data Archive Bulletin*, 58 (available at http://caqdas.soc.surrey.ac.uk/choose.htm).

Fielding, N.G. and Lee, R.M. (1991) *Using Computers in Qualitative Research*. London: Sage.

Gildchrist, V.J. (1992) 'Key informant interviews', in B.F. Crabtree and W.L. Miller (eds) *Doing Qualitative Research*. London: Sage, pp. 70–89.

Glaser, B. and Strauss A.L. (1967) *The Discovery of Grounded Theory*. New York: Aldine.

Grbich, C. (1999) *Qualitative Research in Health. An Introduction*. London: Sage.

Hinchcliffe, S.J., Crang, M.A., Reimer, S.M. and Hudson, A.C. (1997) 'Software for qualitative research. 2. Some thoughts on "aiding" analysis', *Environment and Planning A*, 29: 1109–24.

Hoven-Iganski, B. van (2000) *Made in the GDR. The Changing Geographies of Women in the Post-socialist Rural Society in Mecklenburg-Westpomerania*. Utrecht: KNAG/NGS.

Kelle, U. (1995) 'An overview of computer-aided methods in qualitative research', in U. Kelle (ed.) *Computer-aided Qualitative Analysis. Theory, Methods and Practice*. London: Sage, pp. 1–18.

Kelle, U. (1996) 'Computer-assisted qualitative data analysis in Germany', *Current Sociology*, 44: 225–41.

Kitchin, R. and Tate, N.J. (2000) *Conducting Research into Human Geography. Theory, Methodology and Practice*. Harlow: Pearson Education.

Lee, R.M. and Fielding, N.G. (1991) 'Computing for qualitative research: options, problems and potential', in N.G. Fielding and R.M. Lee (eds) *Using Computers in Qualitative Research*. London: Sage, 1–13.

Lee, R.M. and Fielding, N.G. (1995) 'User's experiences of qualitative data analysis software', in U. Kelle (ed.) *Computer-aided Qualitative Analysis. Theory, Methods and Practice*. London: Sage, pp. 29–40.

Lonkila, M. (1995) 'Grounded theory as an emerging paradigm for computer-assisted qualitative data analysis', in U. Kelle (ed.) *Computer-aided Qualitative Analysis. Theory, Methods and Practice*. London: Sage, pp. 41–51.

Miles, M.B. and Huberman, A.M. (1994) *Qualitative Data Analysis. A Sourcebook of New Methods*. Thousand Oaks, CA: Sage.

Miles, M.B. and Weitzman, E.A. (1996) 'The state of qualitative data analysis software: what do we need?' *Current Sociology*, 44: 206–24.

Prein, G., Kelle, U. and Bird, K. (1995) 'An overview of software', in U. Kelle (ed.) *Computer-aided Qualitative Analysis. Theory, Methods and Practice*. London: Sage, pp. 190–210.

Richards, L. (1997) 'User's mistakes as developer's challenge: designing the new NUD.IST', *Qualitative Health Research*, 7: 425–34.

Richards, L. and Richards, T. (1987) 'Qualitative data analysis: can computers do it?' *Australian and New Zealand Journal of Sociology*, 23: 23–35.

Richards, L. and Richards, T. (1992) 'Hard results from soft data? Issues in qualitative computing.' Paper presented at the British Sociological Association Annual Conference, Manchester.

Richards, T. and Richards, L. (1993) 'Qualitative computing: promises, problems, and implications for research process.' Paper presented at the British Sociological Association Annual Conference, Colchester.

Seidel, J. (1991) 'Method and madness in the application of computer technology to qualitative data analysis', in N.G. Fielding and M.R. Lee (eds) *Using Computers in Qualitative Research*. London: Sage, pp. 107–16.

Strauss, A.L. and Corbin, J. (1990) *Basics of Qualitative Research: Grounded Theory Procedures and Techniques*. Newbury Park, CA: Sage.

Tak, S.H., Nield, M. and Becker, H. (1999) 'Using a computer software program for qualitative analyses. Part 2. Advantages and disadvantages', *Western Journal of Nursing Research*, 21: 436–9.

Tesch, R. (ed.) (1990) *Qualitative Research. Analysis Types and Software Tools*. Basingstoke: Falmer Press.

Tesch, R. (1991) 'Software for qualitative researchers: Analysis needs and program capabilities', in N.G. Fielding and R.M. Lee (eds) *Using Computers in Qualitative Research*. London: Sage, pp. 14–15.

Weitzman, E.A. and Miles, M.B. (1995) *Computer Programs for Qualitative Data Analysis. A Software Sourcebook*. London: Sage.

27 Analysing Historical and Archive Sources

Iain S. Black

Synopsis

Students using historical and archive sources should clearly identify whether their research project is 'problem orientated' or 'source orientated'. Source-orientated research relies primarily upon the detailed examination of a single source, such as a diary. Problem-orientated research involves defining a research question through conceptual and theoretical reasoning before the initial engagement with archives and historical sources. This chapter considers three groups of key historical sources in detail: official and private documentary sources, visual evidence and literary sources. Each group is reviewed to assess their potential and problems in historical geographical research.

The chapter is oganized into the following sections:

* Introduction
* The nature of historical evidence
* Documentary sources
* Visual texts
* Literary sources
* Conclusion.

INTRODUCTION

Geographers have long been interested in the nature of past societies, economies, cultures and environments. This chapter introduces you to the nature of source material available to reconstruct past geographies and reviews some of the key ways in which historical research is undertaken. It begins by considering the nature of historical evidence, indicating the need to critically evaluate the creation of particular historical sources as part of any well grounded research project.

Following this, three groups of key sources are identified for more detailed scrutiny: official and private documentary sources, visual evidence and literary sources. Each group is reviewed to assess their potential and problems in historical geographical research.

Students using historical and archive sources should clearly identify whether their research project is 'problem orientated' or 'source orientated' (see Baker, 1997). Source-orientated research relies primarily upon the detailed examination of a single source, such as a diary, a set of business records or a census enumerators' book. The strength of this approach is that it allows the researcher to build up a detailed picture, or 'thick description', of a particular aspect of the past. The weakness is that such research is often led by the source material, decontextualized from wider questions, and can present problems of establishing a broader interpretive framework. Problem-orientated research involves defining a research question through conceptual and theoretical reasoning before the initial engagement with archives and historical sources – in short, establishing what questions to ask of the data. Once research questions are established it is often possible to identify a series of pertinent sources, which can be linked for consistency and greater understanding. Before commencing the research process, however, a clear understanding of the nature of the source material is necessary.

THE NATURE OF HISTORICAL EVIDENCE

Historical geographers engage with a wide range of evidence, both verbal and non-verbal, in their research. Harley (1982: 261) remarks that evidence giving a window on the past includes 'words written on paper or parchment; words or symbols carved on stone or wood; maps, paintings and photographs; monuments and landscape evidences; artefacts dug up from the ground; and the living documents of oral history'. But it is important to realize, along with Harley, that evidence is not a pre-given entity waiting to be discovered and marshalled into service to answer questions of historical research. The past can never be recovered as a solid whole, as history is always fragmented. Historical research can only deal with residues or, to use a geological metaphor, sediments left over as a result of past action. White, in *Machina Ex Deo* (1968: 4), claims that 'history does not exist; all that exists is debris – scattered, mutilated, very fragmentary – left by vanished ages'. This 'debris' – the sources – is the stuff of history.

We can never simply approach sources directly in the course of historical research, for two main reasons: first, the sources themselves have already been constructed by those who recorded the information and, subsequently, those who have selected what to preserve and what to discard; and, secondly, the construction of history depends on a process of retrieval, selection, contextualization and ascription of meaning. This is always more than simply putting 'the facts' into a chronological sequence. It involves establishing relationships of cause and effect. As Harley (1982: 272, emphasis in original) notes, 'as scholars *we* communicate with the sources and simultaneously try to recreate the way in which *they* formed part of a communication system involving people in the past'. Historical facts are, therefore, constructed by historical geographers through questioning based on an a priori understanding of what is significant. Le Goff (1992: 113) quotes the historian Lucien Febvre thus: 'to elaborate a fact is to construct it. It is, so to speak, to answer a question. And if there is no question, there is nothing at all.' Thus, without questions or hypotheses, 'facts' cannot be generated or placed in context. According to Max Weber: 'any attempt to understand [historical] reality without subjective hypotheses will end in nothing but a jumble of existential judgements on countless isolated events' (quoted in Le Goff, 1992: 113–14).

Historical sources, then, must be critically evaluated as part of any research project in historical geography. If we accept that sources are made and not simply given, a series of key questions about the nature of any particular source can be raised. First, there is a need to establish the authenticity of a source. Is it genuine? In most cases it is important to consult material in its original form rather than any subsequent transformations of the data, though evidence edited by scholars as part of a critical evaluation of a source can be very useful. Secondly, any source must be assessed for accuracy. How close is the source to the events or phenomena it is recording? How accurately was the information recorded? This highlights the need to cross-check with other sources to establish whether the information is consistent with established facts. Thirdly, it is important to understand what was the original purpose for gathering the information. How might this have influenced what was collected? Historical data are always filtered so it is necessary to establish for whom and by whom information was recorded and for what purpose. Further, this highlights the need to consider the 'silences' in the historical record whereby the actions of the powerful and the unusual have often taken precedence over the everyday life of the mass of the population. Fourthly, how has the process of archiving the information imposed a classification and order upon historical events? Archiving can be thought of as a further filter between the 'raw' information and the historical researcher. It should be clear from this that sources themselves convey residues of power structures, and this reinforces the importance of scepticism in historical research. According to Thompson (1978: 220–1), 'historical evidence is there, in its primary form, not to disclose its own meaning, but to be interrogated by minds trained in a discipline of attentive disbelief'.

A common distinction made by historical researchers is that between *primary* and *secondary* data. Primary data are generally taken to mean raw data in an unprinted or unpublished form, usually located in a record office or archive. Secondary data are generally understood to have been transformed from their raw state, often collated, classified and/or tabulated for publication. The question of superiority between these types of data has often been raised (see Dennis, 1987; 1988; Harvey, 1988) but, as Butlin (1993: 73) notes: 'there is nothing particularly sacrosanct about the nature of "primary" as opposed to "secondary" data or information, for both need careful scrutiny and verification.' In what follows, three examples of different groups of sources are presented to indicate the range of possible approaches to understanding the geographies of the past.

DOCUMENTARY SOURCES

Documentary evidence – in essence, the written word – can be usefully subdivided into official and private sources. Official data sources comprise those records produced by and for the national and local state. Such bodies had a constant need to generate information in the course of government, relating in particular to assessing population, taxation, land ownership and monitoring the state of agriculture, industry, the courts and colonial development. In Britain the legacy (contained in British parliamentary papers, reports of royal commissions and committees of the House of Commons and Lords) is vast. Examples of historical work based on such data include analysing agricultural enclosure in Britain (Turner, 1984), mapping agricultural returns (Coppock, 1984) and reconstructing the transition from domestic to factory production in the West Yorkshire woollen industry (Gregory, 1982).

Private archives are a source of growing importance for historical geographers too (Hall, 1982). Traditionally, a very high proportion of data used in historical geographical research was drawn from the official record. However, the development of new areas of research, including the construction of biographies of explorers and travellers or the analysis of private institutions, has demanded sources of greater specificity than those which official archives typically provide. Indeed, official archives, concerned mainly with aggregate information, often have relatively little to say about individuals. Even when they do appear (such as witnesses in select committee reports, for example) they usually comprise only the prominent and powerful.

Private archive sources, such as diaries or letters, can help to overcome these problems of aggregation and the recording of 'official' rather than personal views. But private archives also need care in their use. Precisely because they comprise records created by an individual or an individual institution, they possess a unique quality requiring comparison and evaluation with data from other sources. The following examples illustrate the use of documentary sources drawn from both private and public collections.

Geographies of people in the past

A concern to identify and characterize the nature of people in the past has a long tradition in historical geography. Such analyses range from the reconstruction of individual biographies, through the reconstitution of family and household units, to more general explorations of urban or rural socioeconomic structures based on large-scale official survey data contained in national censuses. The researching and writing of historical biographies are typically based upon records of population, family papers, educational and business records, and contemporary newspapers or other accounts. Clearly, the more well known the individual in question, the greater likelihood that substantial bodies of evidence on his or her life will be extant. But historical geography is not only concerned with the rich and powerful; it is also possible to reconstruct something of the lifeworlds of a whole range of people from all groups in past societies.

The start and end points for reconstructing a biography of a particular individual are conventionally the dates of birth and death, though often it is useful to locate these within a broader family history. In England and Wales systematic, though by no means comprehensive, information on the lives of individuals begins in 1538 when the recording of christenings, weddings and burials by parish began. Used as proxies for births, marriages and deaths, these parish registers provide a valuable source for the identification of the key life events of individuals in the past, despite clear problems in their use, including 'variation in accuracy through time, the exclusion of data for the non-established churches, clandestine marriages, and birth-baptism shortfalls through very early infant mortality' (Butlin, 1993: 77). From 1837 the researching of individuals becomes more straightforward as registration of births, marriages and deaths became a legal requirement.

The following example of Onesiphorus Tyndall, an eighteenth-century banker and merchant in Bristol, is based on a biographical essay commissioned for the forthcoming *New Dictionary of National Biography* (Black, forthcoming a). It indicates the range of sources available that provide details of people's past lives and how these can be interpreted. It is not possible to establish Tyndall's date of birth, but it is known he was baptized on 28 May 1689. This datum was located by a search through the parish registers of the City of Bristol. No family trees or other papers can be found, but something of his family background was traced in a nineteenth-century history of banking in Bristol. Details on his early life are opaque but it is possible to infer from the banking history that Tyndall was apprenticed to his father in the trades of grocery and dry salting. On 6 November 1717 Tyndall married Elizabeth Cowles, also of Bristol, and together they had two sons and two daughters. The date of marriage was traced from marriage licence records, while details on Tyndall's own family were reconstructed from the relevant parish registers. Tyndall was notable for his role in the formation of the Old Bank, Bristol's first bona fide banking house, which commenced business on 1 August 1750. Details on his banking and business career were found in contemporary publications on Bristol's commercial life. In addition to banking, Tyndall had a long and varied association with the activities of the port of Bristol, including the West Indies trade in sugar and tobacco. He was still listed as trading to Africa in 1755, two years before his death, in a manuscript collection on West Indian traders in Bristol City Library. On 4 June 1757 *Felix Farley's Bristol Journal* carried an obituary of Tyndall giving the details of the date of his death (30 May 1757) and burial (3 June 1757). The latter date was cross-checked in the burial registers of Christchurch Parish Church. For historians there is also life after death, a principal source for which is an individual's will. Tyndall's will, proved in the Prerogative Court of Canterbury, enabled a detailed estimate of his wealth at death and identification of his beneficiaries. Combining a range of such detailed biographies, together with wider documentary evidence and published sources, could enable a 'thick description' of a city's civic and business elite and the multiple ways they negotiated their power (see Billinge, 1982).

However, to increase the level of aggregation significantly, to look at the changing composition of the family unit for example or the demographic response of local communities to the external shocks of food shortage or disease, it is necessary to move beyond the individual. Before the nineteenth century, data contained in the parish registers can be aggregated by calculating totals of births, marriages and deaths for a specified geographical area over a given time period. By tabulating the frequency of such events, either monthly or annually, patterns of population growth or decline can be identified. A particular concern has been the specification of years of 'crisis mortality', where an unusually large number of deaths have occurred. To try to explain such events, attempts are made to correlate population dynamics with wider socioeconomic fluctuations, such as the state of harvests or the incidence of major diseases such as plague.

Family reconstitution, by contrast, involves gathering data on vital demographic events for individual families, in which the surname is the link. As Morgan (1979: 74) notes, this involves:

> recording the names of all those baptized and tracing them among the marriages 15 to 50 years later and among the burials up to 105 years later . . . [building] a figure of the actual number of people alive at any one time, the size and extent of the families, at what age people got married, had children and died.

The most ambitious and comprehensive use of data in parish registers, involving the calculation of population size and age structure and the use of backward projections to derive population trends over time where data are lacking, is Wrigley and Schofield's (1981) study of English population between 1541 and 1871. This work is a major starting point for any study of the historical geography of population prior to the census.

The nature and use of the census is given comprehensive treatment in Chapter 5. It is sufficient to note here two key forms of census data of especial significance to historical geographers. First are the census enumerators' books, which are available for consultation at ten-year intervals between 1841 and 1901 and to which a hundred-year confidentiality rule applies. The books provide a comprehensive set of data, in a standardized format, for each individual in the relevant enumeration district, including the following variables: relation to head of household, marital condition, age, sex, occupation and birthplace. This basic set of data can then be used to study the demographic, social, economic and migration structure of places ranging in scale from the individual household, through streets and districts, to large urban areas. A useful review and bibliography of work based on this source has been produced by Mills and Pearce (1989). Secondly are the census abstracts, which consolidate information derived from the original enumerators' books and are available from 1801 in the form of published parliamentary papers. They contain tables of information on population, age, sex, marital status, birthplace, occupation and housing for a range of geographical areas. From 1841 it is possible to distinguish four spatial scales of information: county, registration district, subdistrict and parish. A principal use of the abstracts has been the historical analysis of patterns of migration and the changing industrial structure of cities and regions (Lawton, 1982).

Historical geographies of money

The records of business institutions have considerable value in addressing important research questions in historical geography. Black (1995; 1996) has used data contained in financial records, for example, to research spatial patterns and processes of financial integration in early industrial England, developing a critical perspective upon regional dynamics by reconstructing flows of money, short-term credit and commercial information never recorded in official surveys. Where sufficient data exist to undertake quantitative and cartographic analysis, research based on such private archives can provide an important corrective to some of the aggregate conclusions drawn from official data sources. The following examples, drawn from work in bank archives, indicate the potential of these sources for reconstructing past monetary geographies.

Consider Figure 27.1, which shows the spatial structure of correspondence between the private bank of Peacock & Co. of Newark, Nottinghamshire, and its business and private customers. The data are drawn from letter books surviving for the period between 1809 and 1813. Though the data are relatively simple, the patterns revealed through mapping the distribution of correspondence are full of interest in the context of debates over the nature and extent of regional economic integration in the early industrial economy of England and Wales. The map shows a focused intraregional circuit of commercial information, centred on the bank's sister branch at Sleaford, and the proximate commercial centres of Nottingham and Lincoln. But more striking is the overwhelming dominance of London in the bank's business correspondence in the early nineteenth century. Given there was no formal institutional link to the capital, this pattern demands explanation. A careful reading of the content of the letters mapped in Figure 27.1 indicates that the bank had an agency link to a London private bank, which handled its business affairs on an interregional scale. Already the consideration of this small provincial firm begins to provide suggestive evidence of widespread and interlocking patterns in the circulation of commercial information on an intraregional and provincial-metropolitan level.

Moving from letter books to historical documents that recorded financial transactions, such as ledgers, bill books or remittance books, it is possible to build on the initial mapping of correspondence to reconstruct geographies of past capital flows. Figure 27.2 shows the value of bills of exchange received for discount at the Newark bank between 1807 and 1809. The bill was the key financial instrument used in the provision and circulation of short-term paper credit in the early industrial English economy (Black, 1996). The Newark bank's bill book, on which Figure 27.2 is based, includes valuable data on the

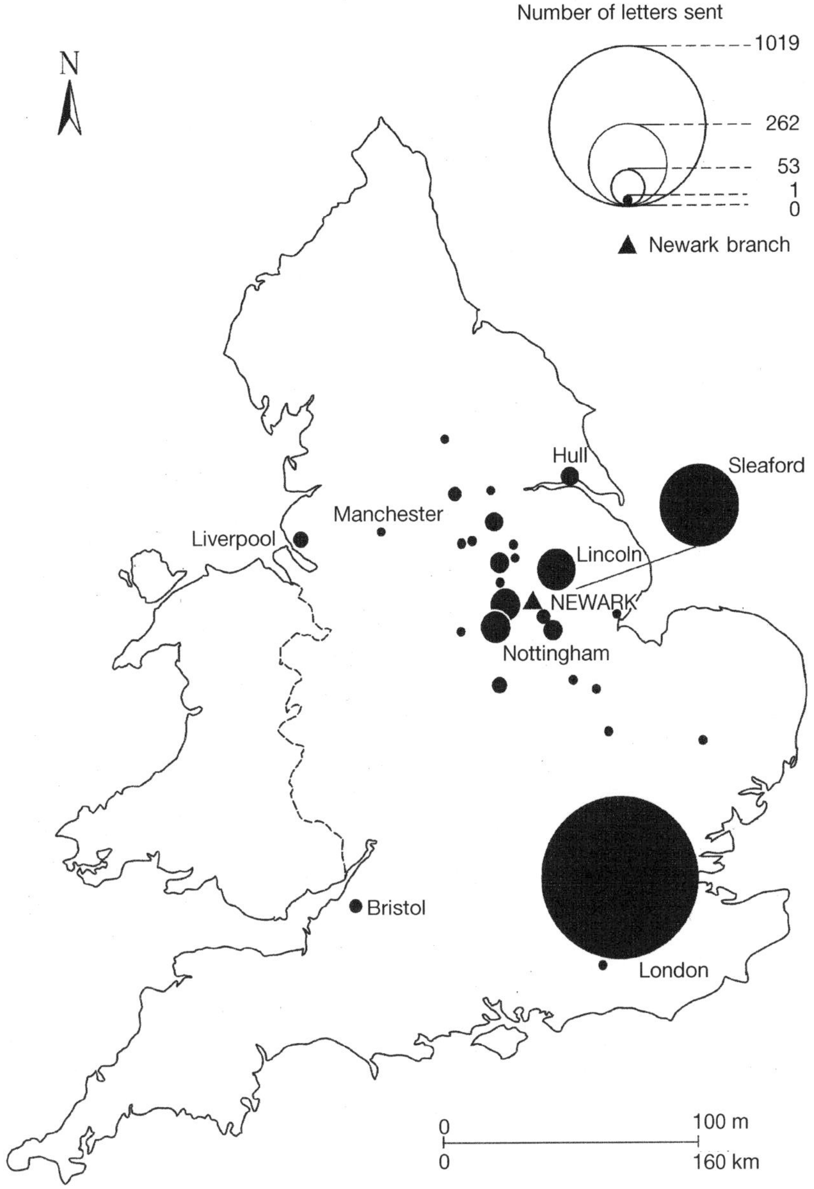

Source: Lloyds Bank Archives

FIGURE 27.1   Peacock & Co., Newark, Nottinghamshire. Correspondence with banks and private customers, 1809–13

places from which bills were received and their value. Again the patterns revealed are suggestive. Two features are immediately apparent: first, the high value of bills drawn and cashed locally; and, secondly, the flow of bills from key centres of early industrialization.

Source: Lloyds Bank Archives

FIGURE 27.2 Peacock & Co., Newark, Nottinghamshire. Value of bills received for discount,
1807–9

The first pattern is expected as the bank provided ready cash to local
businessmen by buying their bills drawn in trade. The second pattern
is more surprising and demands explanation.

What is revealed by a close analysis of the data is that bills drawn
in trade in the industrial districts of Manchester, Leeds, Sheffield and
Birmingham found their way to this small country bank in Newark

for discount (sale). The Newark district was predominantly agricultural in nature at this time and had surplus capital for investment. The industrial districts of the north and midlands were capital hungry. What we have evidence for, then, is a distinctive geography of capital flow as bills circulated from the industrializing regions of the country to regions of capital surplus in the east and the south. But a final question arises: how did manufacturers in the north and midlands know that any particular bank in the east or south had surplus funds for investment? The answer cannot be found in the data themselves, but only by linking the primary evidence shown in Figure 27.2 to a broader understanding of the role of the London money market in the early nineteenth century, where the London agents and bill brokers specialized in matching up bills seeking discount with provincial banks seeking outlets for their surplus capital. Thus, evidence drawn from an intensive study of financial documents can lead to wider conclusions about the role of London in a period of English economic history often exclusively seen as dominated by the industrialization of the provinces.

VISUAL TEXTS

Although the term 'archives' commonly conjures up an image of reams of documents, geographers also draw upon a wide range of visual evidence in their reconstructions of the past, notably in seeking to analyse and interpret the production and representation of landscape (Cosgrove and Daniels, 1988; Cosgrove, 1993; Daniels, 1993; Domosh, 2001). Paintings, sketches, engravings and architectural drawings are all valuable in the essentially archaeological practice of recovering the symbolic geography of past landscapes, be they rural or urban. Indeed, in cases where the landscape in question has been substantially modified or even erased by subsequent human agency, such visual evidence may provide the only representations extant for such a reconstruction to take place.

Though a wide range of approaches have been adopted in the historical study of landscape, two methodologies have gained prominence: one relying upon textual metaphor and the other drawing upon the techniques of iconography. The metaphor of seeing the landscape as a text, which draws upon the influential work of the cultural anthropologist, Clifford Geertz, suggests landscapes can be read as a social document, using the methods and techniques of literary theory (Barnes and Duncan, 1992). With reference to specific landscape forms, textual metaphors can be pursued to illuminate the crucial relationships between landscape and ideology, helping to identify how landscapes can transform ideologies into a concrete, visible form.

Therefore the role of the critical reader of landscapes should be, in part, to penetrate what Barthes called their 'layers' of ideological sediment (Duncan and Duncan, 1988). The second approach, that of iconography, is drawn explicitly from studies in art history. Cosgrove and Daniels (1988: 1) define iconography as, broadly, 'the theoretical and historical study of symbolic imagery'. Although the use of a formal iconographical approach to interpreting images reaches back to the Renaissance, in its modern form the concept was deployed by the discipline of art history to:

> probe meaning in a work of art by setting it in its historical context and, in particular, to analyse the ideas implicated in its imagery . . . consciously [seeking] to conceptualise pictures as encoded texts to be deciphered by those cognisant of the culture as a whole in which they were produced. (Cosgrove and Daniels, 1988: 2)

This broad conceptualization of iconography was refined and codified into a systematic approach to the study of art by Erwin Panofsky, whereby iconography in its narrower sense 'was the identification of conventional, consciously inscribed symbols', while a more inter-pretative conception of iconography (termed iconology) sought a deeper level of meaning in art by reconstructing the cultural, political and ideological context(s) surrounding its production and which the art communicated consciously or unconsciously (Cosgrove and Daniels, 1988: 2). These principles of iconographical study have become widely used in a variety of attempts by geographers to interpret symbolic landscapes in a range of different settings.

An important methodological statement on the interpretation of the built environment was set out by Domosh in 1989. In that paper she sought to develop a multilayered understanding of the design and construction of the New York World Building by linking this impor-tant early skyscraper to 'its socio-economic and aesthetic contexts, and the actors who directly produced and/or created that artifact' (Domosh, 1989: 347). Beginning with a single building, the argument is constructed to counterpose a series of inter-related 'layers' of explanation, encompassing the functional, symbolic and ideological qualities of this monumental corporate headquarters. After setting the new building within the social, economic and spatial context of later nineteenth-century New York, Domosh proceeds to link the aesthetic form and social meaning of the building to the individual aspirations of its owner (the newspaper magnate Pulitzer) and his place within the highly competitive elite of the city. Paralleling this discourse of architectural aesthetics was the changing nature of technology involved in skyscraper construction, which allowed height to be used as a competitive strategy in the buildings of new corporate capital. Domosh clearly demonstrates throughout how the 'text' surrounding the creation of this particular landscape artifact is informed by a continuous dialogue with the 'context' of the history of the city in which it was built and the social and economic conditions of late nineteenth-century American capitalism. In later work, Domosh goes on to apply this methodological framework, broadly conceived, in more detailed studies of New York's skyscrapers and in a more fully developed historical and comparative analysis of the landscapes of New York and Boston in the nineteenth century (Domosh, 1992; 1996).

Dennis, in a perceptive essay on the methodologies of historical geography and their application to the modern urban landscape, refers to Domosh's study of the New York World Building as a 'multi-layered approach to landscape interpretation' (Dennis, 2001: 24). However, he suggests her approach could have been strengthened by making it multi-*method* too. Rather than relying principally on secondary sources, chiefly newspapers, magazines and 'existing accounts of . . . land values and skyscraper technology that were not specific to the building' (2001: 24), Dennis argues that a wider range of sources and methods would have deepened her analysis. In particular, he refers to the important work of Holdsworth who, in seeking to interpret past built environments, draws upon 'corporate archives, fire insurance plans, assessment records, property transfer records, mortgage records, and the manuscript census to try to make visible, to bring out of concealment, what is not visible in today's landscape' (Dennis, 2001: 24). Indeed, Holdsworth's essay is an important reminder of the need to combine analyses of visual evidence with a wider set of documentary sources to develop a fuller understanding of the creation and functioning of built environments in the past (Holdsworth, 1997). The following example considers a way of 'reading' a monumental commercial building within a past landscape that has all but disappeared from view: the central financial district of the City of London in the early Victorian period (see Black, 2000).

Money, power and landscape

Figure 27.3 shows the headquarters built by the celebrated early Victorian architect, C.R. Cockerell, for the London & Westminster Bank in 1838. The site, in Lothbury, was at the heart of the City's central financial district, just opposite the northeast corner of the Bank of England. Careful observation of this section of streetscape suggests a number of key questions which can be used to structure an analysis of its meaning. Compared with its immediate neighbours the new headquarters was clearly dominant in both size and style, dwarfing the domestic townscape of the surrounding Georgian urban fabric. Even the business premises of the neighbouring private bank of Jones,

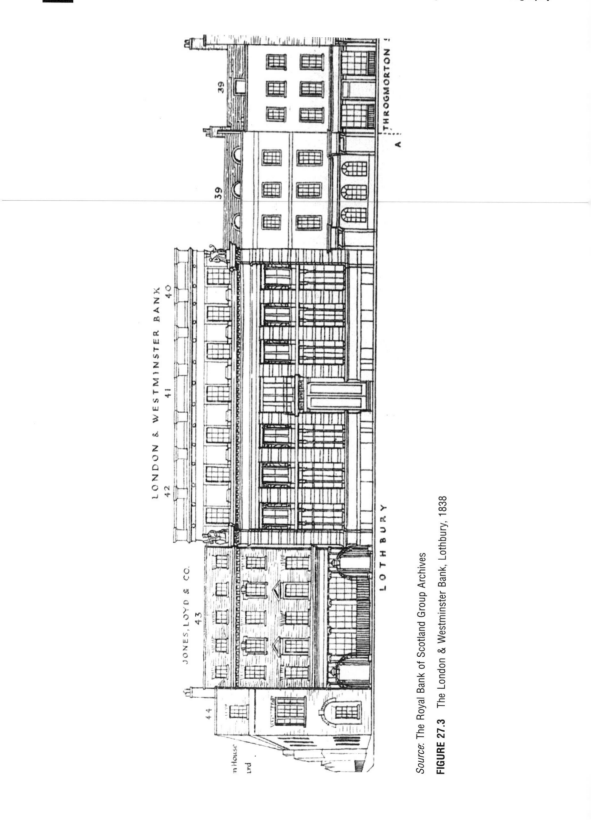

Source: The Royal Bank of Scotland Group Archives

FIGURE 27.3   The London & Westminster Bank, Lothbury, 1838

Loyd & Co., modelled on a Georgian town mansion, seem to belong to another age. A closer look at the London & Westminster's façade reveals additional features requiring interpretation: how was this break from domesticity underlined by a more formal, public quality of architecture? Why was there much greater attention paid to the visibility of the entrance doorway? What messages can be found in the deliberate use of ornamentation and statuary? To move beyond architectural description, however, requires a consideration of the place and purpose of the London & Westminster Bank in the financial world of the metropolis in the 1830s.

Prior to 1833 joint-stock banking was prohibited in the capital, with the sole exception of the Bank of England. The London private bankers were legally restricted to private partnerships of a maximum of six persons. All this was to change in 1833 when, following decades of financial instability, the state finally moved to remove the Bank of England's monopoly of joint-stock banking. Joint-stock organization was a form of corporate structure similar to today's public companies, with capital owned by shareholders and the day-to-day business of the company entrusted to a board of directors. The London & Westminster Bank was the first new joint-stock bank to be established in London, bringing with it a greatly enhanced capacity to expand its capital base through share subscription. Already, therefore, we can begin to see an *economic* rationale for the design of its new headquarters. But the story cannot be reduced to one of simple economic power. In 1830 the City was still a closed and private world where the wealth of banks was materially and symbolically embodied in the private bankers themselves. They had nothing but contempt of the *arriviste* status of the new joint-stock banking companies. The Bank of England too was hostile, jealous of the erosion of its long-standing monopoly privileges. A closer evaluation of the London & Westminster's new headquarters, in terms of style and location, therefore requires cognizance of the deep-seated differences in business culture associated with the private and joint-stock forms of banking capital.

Figure 27.4 shows an engraving of the new headquarters in 1847. The solidity, purpose and strength of the design were noted by contemporaries as befitting the office of a capital-rich banking institution providing a home for the savings of the growing metropolitan middle class. Decoration and ornament were kept to a minimum, though the first-floor windows possessed side panels between the rusticated piers available for embellishment. Here caducei and fasces were sculptured, in alternate disposition. The caduceus was a staff entwined with two serpents and bearing a pair of wings at the top which, according to Greek mythology, was carried by Hermes, messenger of the gods and representative of the divinity of commerce.

Source: Guildhall Library Print Collection

FIGURE 27.4 Engraving of the London & Westminster Bank, Lothbury, 1847

Fasces, a Roman symbol, displayed a bundle of rods with a projecting axe blade. Traditionally carried before a Roman magistrate, fasces were a symbol of authority. At the extremities of the front were bold piers, each displaying a statue of a seated female figure. These figures were emblematic of the commercial interests of the bank, with the statue at the eastern end carrying a shield decorated with the arms of the City of London, while that at the western end displayed the arms of the City of Westminster. In alluding to the two cities that lay at the heart of the expanding metropolis, they also represented the two distinct banking traditions which the new institution was seeking to incorporate. The 'City' symbolized mercantile interests and the importance of trade finance, while 'Westminster' referred to the long tradition of private deposit banking that the new institution was opening up to a wider middle-class market. The scale and boldness of the design were unprecedented in the previously closed world of metropolitan banking, save for the Bank of England. The new and more public form of money represented by the London & Westminster relied upon such distinctive architectural references to underline its legitimacy and authority.

This breakthrough to a wider middle-class market was reinforced by the scale and design of the principal banking hall shown in Figure 27.5. Gone was the simplicity of the bankers' shop, coupled with the Georgian parlour for the discussion of private financial affairs. The new banking hall provided an air of confidence to the setting of daily formalized money exchanges, where the separation of clerks and customers shown clearly by the counter underscored the mixing of public and private space so important for reputable money transactions. It was precisely these qualities of internal space that caught the attention of contemporary critics, who compared it with the baths of Imperial Rome where rounded forms of arch and dome enclosed a space in which public association reinforced moral order. The town banking hall announced a new era in the scale and purpose of banking, adding to the already-established traditions of discreet personal banking for the aristocracy and gentry, and the specialized business of the City mercantile houses, the popularization of the banking habit among the metropolitan middle class.

This example emphasizes a number of key methodological principles: first, the importance of maintaining a close relationship between visual evidence and other archival and historical sources, taking Geertz's notion of 'thick description' seriously to outline the detailed historical and geographical contexts surrounding an interpretation of particular landscape forms; secondly, to give due weight to the importance of politics and ideology embedded within visual evidence by providing a close reading of the signs and symbols

Source: *The Pictorial Times*

FIGURE 27.5   Principal banking hall of the London & Westminster Bank, Lothbury, 1845

encoded within particular landscapes; thirdly, to ensure a rigorous study of the aesthetics of landscape by drawing upon the methods and techniques of iconographical analysis; and, fourthly, recognizing the importance of linking the 'deep descriptions' of particular building projects, or other refashioning of landscape, to wider social, cultural and economic processes, maintaining a conscious dialogue between 'text' and 'context' in the social production of specific landscape forms (see Black, forthcoming b, and Chapter 28).

LITERARY SOURCES

Literature produced in past periods and places constitutes the final group of sources to be considered. Such work, especially novels and poems, has long provided important evidence for the reconstruction and interpretation of past geographies, though its potential as a source demands critical examination (for examples, see Pocock, 1981; Lucas, 1988; Daniels and Rycroft, 1993). Moretti (1998: 3), in a general survey of the geography of the European novel, suggests that space 'is not an inert container, is not a box where cultural history "happens", but an active force, that pervades the literary field and shapes it in depth'. Further, he sees the geography of the novel as comprising two distinct possibilities:

> it may indicate the study *of space in literature*; or else, *of literature in space*. In the first case, the dominant is a fictional one: Balzac's *version* of Paris, the Africa of colonial romances, Austen's redrawing of Britain. In the second case, it is real historical space: provincial libraries of Victorian Britain, or the European diffusion of *Don Quixote*. (Moretti, 1998: 3, emphases in original)

In historical geography the traditional approach to literature as a source has been rather more prosaic. A classic example is Darby's (1948) paper on the regional geography of Thomas Hardy's Wessex. Darby was concerned to correlate the 'fictional' places in Hardy's work with 'real' places in late nineteenth-century Dorset, to reconstruct something of its topography and landscape character. For him, novels were considered as a reliable source for analysis akin to many other forms of geographical data. While such topographical details can indeed be usefully combined with other types of evidence, such as maps, drawings or directories, for example, this is an unduly limited conception of literature as a historical source. One of literature's key characteristics that drew humanistic geographers to its analysis in the 1970s was its power to render palpable the essence of place in a way that many scholarly works of geography singularly failed to do (Tuan, 1978). It was argued that descriptions of people and places in the past, captured in literary form, could provide a valuable way of approaching the *experience* of the past; an experience otherwise impossible to capture by conventional ethnographic means, of course.

Although there are no simple methodological rules for the use of literature as a historical source, it is possible to set out a series of critical points to be borne in mind when approaching novels and poems written in the past. In a recent essay, Sharp (2000) outlines some key dimensions of literature that require critical scrutiny. First, literature cannot be simply appropriated and used directly as a source, decontextualized from the conditions of its own production. She notes (2000: 328) how many geographers 'display a naïveté about the form of literary writing: it is seen as unproblematic and self evident in its immediate beauty'. This cautions against the use of literature as simply a storehouse of descriptions which can be culled as exemplifications of other empirical knowledges validated in traditional ways from more orthodox sources. Literary sources themselves require a scrupulous examination of the 'positionality' of their authors, their readers and the historical context in which they were produced. Secondly, literature is too often presumed to possess a universal quality, which masks differences between individuals and social groups. The meanings and values ascribed to a piece of literature as historical evidence cannot be presumed to be valid for all people in all times and places. Such a qualification seems obvious, but the fact remains that much literature is seen to deal with the general essence of the human condition. Thirdly, geographers have largely ignored the status of literature as a form of writing, seeing it merely as another form of 'data'. Thus, according to Sharp (2000: 329), 'geographers are neglecting the most vital part of literature, which is the challenging potential of this form of expression compared to orthodox modes of geographical writing'.

This latter point has important methodological implications for the use of, say, novels as historical sources. Crang (1998) indicates, through his discussion of a range of authors writing on 'modern' cities, how their textual strategies help to shape their interpretation of modern urban life, beyond any simple attempt to capture the empirical detail of, for example, consumer cultures in nineteenth-century Paris. Thus, 'far from treating literary works as things that simply portray or describe the city, a source of data, we must look at how they construct the city in different ways' (Crang, 1998: 55). In this sense, the use of literary sources in historical geography goes beyond a search for 'evidence' to confirm or deny existing knowledges and moves towards considering how the production of the texts themselves may tell us something significant about the experience of places in the past.

CONCLUSION

This chapter has sketched out some broad parameters for the conduct of historical analysis in geography. It emphasizes the importance of defining research questions and their relationship to sources well in advance of actual engagement in the archive, as well as making a close assessment of the nature, potential and problems of any particular source material consulted. That said, archives are full of surprises and it would be a poor historical geographer who sought to apply a rigid theoretical framework or set of research questions to a particular set of sources. The relationship among theories, concepts and evidence in historical research should be open and recursive, and the researcher should remain flexible in his or her pursuit of answers to questions about the past. Following trails in archives can and often does lead to the discovery of unexpected evidence which questions or contradicts what might have been expected. The range of potential sources for the study of past geographies is undoubtedly vast and it has only been possible here to indicate three key groups of evidence for more detailed scrutiny. However, the separation of these sources into 'types' – visual, literary and documentary – is also potentially misleading. A hallmark of good historical scholarship is the ability to identify and link a series of sources, not only to deepen the analysis of interesting research questions but also to cross-check and validate different forms of historical knowledge. In short, it is in the *combination* of sources and methods that much of the most exciting historical geography is done. Finally, a word about presenting the results of an analysis of historical and archive sources. As Baker (1997: 238) has noted, many of the problems faced here are generic to all geographers, where writing involves constructing 'a series of balances between fact and interpretation, between the particular and the general, between the empirical and the theoretical, and between the objective and the subjective'. An additional problem for the historical geographer is, of course, that the past has already happened. What remains of it for interpretation is necessarily partial. Therefore, a key quality of successful writing in historical geography must be its ability to persuade others not only that it is soundly based upon reliable sources and methods but also that it accords with accepted facts and, above all, deals convincingly with the gaps in both evidence and interpretation that inevitably remain.

Summary

- Students using historical sources need to identify whether their research is problem orientated or source orientated.
- It is important to understand how and why particular sources were created and to establish their authenticity, reliability and partiality.
- Historical researchers make a distinction between primary and secondary data.
- Documentary sources include official/private archives, biographical and population data and financial records.
- As well as documentary evidence, visual evidence and literary sources can be used by historical geographers to reconstruct the past.

Further reading

- Baker (1997) provides a clear and concise introduction to the process of researching historical data in geography.

- Harley (1982) explains the need critically to evaluate sources and discusses questions to ask on the reliability of evidence in historical geography.

- Thompson (1978) is still one of the best general accounts of the need to think critically about the relationships between data and theory in historical work.

- Butlin (1993) discusses the range of documentary sources that have been used in historical geography.

- On visual evidence as a geographical source, see Burke (2001); on the built environment, see Domosh (1989) and Black (forthcoming b); and on literature, see Sharp (2000).

Note: Full details of the above can be found in the references list below.

References

Baker, A.R.H. (1997) 'The dead don't answer questionnaires: researching and writing historical geography', *Journal of Geography in Higher Education*, 21: 231–43.

Barnes, T. and Duncan, J. (eds) (1992) *Writing Worlds: Discourse, Text and Metaphor in the Representation of Landscape*. London: Routledge.

Billinge, M.D. (1982) 'Reconstructing societies in the past: the collective biography of local communities', in A.R.H. Baker and M. Billinge (eds) *Period and Place: Research Methods in Historical Geography*. Cambridge: Cambridge University Press, pp. 19–32.

Black, I.S. (1995) 'Money, information and space: banking in early-nineteenth-century England and Wales', *Journal of Historical Geography*, 21: 398–412.

Black, I.S. (1996) 'The London agency system in English banking, 1780–1825', *London Journal*, 21: 112–30.

Black, I.S. (2000) 'Spaces of capital: bank office building in the City of London, 1830–1870', *Journal of Historical Geography*, 26: 351–75.

Black, I.S. (forthcoming a) 'Onesiphorus Tyndall of Bristol, banker and merchant', in B. Harrison (ed.) *New Dictionary of National Biography*. Oxford: Oxford University Press.

Black, I.S. (forthcoming b) '(Re)reading architectural landscapes', in P. Richards and I. Robertson (eds) *Studying Cultural Landscapes*. London: Arnold.

Burke, P. (2001) *Eyewitnessing: The Uses of Images as Historical Evidence*. London: Reaktion Books.

Butlin, R.A. (1993) *Historical Geography: Through the Gates of Space and Time*. London: Arnold.

Coppock, J.T. (1984) 'Mapping the agricultural returns: a neglected tool of historical geography', in M. Reed (ed.) *Discovering Past Landscapes*. London: Croom Helm, pp. 8–55.

Cosgrove, D.E. (1993) *The Palladian Landscape: Geographical Change and its Cultural Representations in Sixteenth-Century Italy.* Leicester: Leicester University Press.

Cosgrove, D.E. and Daniels, S. (eds) (1988) *The Iconography of Landscape: Essays on the Symbolic Representation, Design and Use of Past Environments.* Cambridge: Cambridge University Press.

Crang, M. (1998) *Cultural Geography.* London: Routledge.

Daniels, S. (1993) *Fields of Vision: Landscape Imagery and National Identity in England and the United States.* Cambridge: Polity Press.

Daniels, S. and Rycroft, S. (1993) 'Mapping the modern city: Alan Sillitoe's Nottingham novels', *Transactions, Institute of British Geographers*, 18: 460–80.

Darby, H.C. (1948) 'The regional geography of Thomas Hardy's Wessex', *Geographical Review*, 38: 426–43.

Dennis, R. (1987) 'Faith in the city', *Journal of Historical Geography*, 13: 210–16.

Dennis, R. (1988) 'Comment in reply: by the waters of Babylon', *Journal of Historical Geography*, 14: 307–8.

Dennis, R. (2001) 'Reconciling geographies, representing modernities', in I.S. Black and R.A. Butlin (eds) *Place, Culture and Identity: Essays in Historical Geography in Honour of Alan R.H. Baker.* Quebec: Les Presses de l'Université Laval, pp. 17–43.

Domosh, M. (1989) 'A method for interpreting landscape: a case study of the New York World Building', *Area*, 21: 347–55.

Domosh, M. (1992) 'Corporate cultures and the modern landscape of New York City', in K. Anderson and F. Gale (eds) *Inventing Places: Studies in Cultural Geography.* Melbourne: Longman Cheshire, pp. 72–86.

Domosh, M. (1996) *Invented Cities: The Creation of Landscape in Nineteenth-Century New York and Boston.* New Haven, CT, and London: Yale University Press.

Domosh, M. (2001) 'Visual texts in historical geography', *Historical Geography*, 29: 68.

Duncan, J. and Duncan, N. (1988) '(Re)reading the landscape', *Environment and Planning D: Society and Space*, 6: 117–26.

Gregory, D.J. (1982) *Regional Transformation and Industrial Revolution: A Geography of the West Yorkshire Woollen Industry.* London: Macmillan.

Hall, C. (1982) 'Private archives as sources for historical geography', in A.R.H. Baker and M. Billinge (eds) *Period and Place: Research Methods in Historical Geography.* Cambridge: Cambridge University Press, pp. 274–280.

Harley, J.B. (1982) 'Historical geography and its evidence: reflections on modelling sources', in A.R.H. Baker and M. Billinge (eds) *Period and Place: Research Methods in Historical Geography.* Cambridge: Cambridge University Press, pp. 261–73.

Harvey, D.W. (1988) 'The production of value in historical geography', *Journal of Historical Geography*, 14: 305–8.

Holdsworth, D.W. (1997) 'Landscape and archives as texts', in P. Groth and T. Bressi (eds) *Understanding Ordinary Landscapes.* New Haven, CT, and London: Yale University Press, pp. 44–55.

Lawton, R. (1982) 'Questions of scale in the study of population in nineteenth-century Britain', in A.R.H. Baker and M. Billinge (eds) *Period and Place:*

Research Methods in Historical Geography. Cambridge: Cambridge University Press, pp. 99–113.

Le Goff, J. (1992) *History and Memory*. New York: Columbia University Press.

Lucas, J. (1988) 'Places and dwellings: Wordsworth, Clare and the anti-picturesque', in D.E. Cosgrove and S. Daniels (eds) *The Iconography of Landscape: Essays on the Symbolic Representation, Design and Use of Past Environments*. Cambridge: Cambridge University Press, pp. 83–97.

Mills, D. and Pearce, C. (1989) *People and Places in the Victorian Census. A Review and Bibliography of Publications Based Substantially on the Manuscript Census Enumerators' Books*. Cheltenham: Historical Geography Research Group.

Moretti, F. (1998) *Atlas of the European Novel 1800–1900*. London: Verso.

Morgan, M. (1979) *Historical Sources in Geography*. London: Butterworth.

Pocock, D. (ed.) (1981) *Humanistic Geography and Literature*. London: Croom Helm.

Sharp, J.P. (2000) 'Towards a critical analysis of fictive geographies', *Area*, 32: 327–34.

Thompson, E.P. (1978) *The Poverty of Theory and Other Essays*. London: Merlin Press.

Tuan, Y.-F. (1978) 'Literature and geography: implications for geographical research', in D. Ley and M. Samuels (eds) *Humanistic Geography: Prospects and Problems*. London: Croom Helm, pp. 194–206.

Turner, M. (1984) *Enclosures in Britain 1750–1830*. London: Macmillan.

White, L. (1968) *Machina Ex Deo: Essays in the Dynamism of Western Culture*. Cambridge, MA: MIT Press.

Wrigley, E.A. and Schofield, R.S. (1981) *The Population History of England, 1541–1871: A Reconstruction*. London: Edward Arnold.

28 Analysing Cultural Texts

Marcus A. Doel

Synopsis

Culture can be defined as any way of life that can be differentiated from other ways of life and texts as anything that signifies something for someone or other. So, far from being highly prized artistic and literary artifacts, most cultural texts appear mundane and are scattered among the debris of everyday life (e.g. graffiti, statutes, advertisements, photographs). As such this chapter sets out to challenge the widely held view that 'culture' is a narrow concern of literary and artistic elites but rather shows that 'culture' belongs as much to the 'low life' as it does to the 'high life,' and to 'us' as much as to 'them'. At the same time it also challenges the bookish conception of 'texts' by showing not only how everything can come to signify something for someone or other but also how texts give expression to social antagonisms.

The chapter is organized into the following sections:

- Hello culture!
- Culture is not what you think!
- Cultural texts are not what you think!
- Forget reading!
- Columbus' egg: 'mind that child!'

HELLO CULTURE!

Over the past couple of decades, more and more of human geography has taken a cultural turn. Or so the cliché goes (see, for example, Cook et al., 2000). Virtually every part of the discipline – from economic geography to medical geography, and from political geography to rural geography – has become increasingly enculturated so that it is now not uncommon for geographers to study things like subcultures, cultures of nature and cultural capital (those cultural traits within any particular social setting that confer prestige on those individuals, groups and classes who possess them and shame on those who do not – Bourdieu, 1984). Meanwhile, cultural geography itself has changed from a relatively self-contained subdiscipline into one that embraces concerns that span the sciences, social sciences and the arts and humanities. As well as cultural geographies of nations, landscapes and built environments, there are also cultural geographies of money, domestic appliances and exotic fruit (Cloke et al., 1999). Basically, there are cultures of everything just as there are histories and geographies of everything. Consequently, learning how to analyse cultural texts is fast becoming a vital skill for every human geographer. So in this chapter I want to help you set out in the right direction. I will do this by focusing on four key issues. First, I want to alert you to the fact that 'culture' and 'cultural texts' are almost certainly not what you think. Secondly, I want to encourage you to read – slowly. Very slowly. As Dr Seuss famously put it: 'Go slowly. This book is dangerous.' Thirdly, I want to demonstrate that reading cultural texts is nevertheless an easy trick to pull off – once you know the trick. Finally, I want to offer you an initial checklist for the effective analysis of cultural texts.

Now, while it would be easy to make this introduction difficult – by filling it with technical jargon and edifying quotations – I want to keep things as simple as possible. I just want to provide you with the *flavour* of textual analysis and trust that as you progress through your studies you will acquire the taste for more. My message is simple. The worst that can happen is not that you read badly. The worst would be for you to prefer not to read at all.

CULTURE IS NOT WHAT YOU THINK!

I suspect that the phrase 'cultural texts' will leave you cold. The word 'culture' still tends to conjure up images of refinement, elitism and exoticism. Little wonder, then, that culture is often assumed to pale into insignificance when compared to more pressing concerns like economic crises, regional conflicts and global warming. For those who are lucky enough to escape the numerous horrors that haunt the modern world, culture can be a sign of distinction, a weapon of class war and a way of distancing oneself from the trials and tribulations of everyday life. Whatever form it takes, culture is usually associated with the learned, the leisured and the tourist. Culture is what you experience in galleries, museums, theatres and heritage sites, not what you find in alleyways, squats, prisons and shopping centres. So, when you are offered the prospect of 'analysing cultural texts', I would not be at all surprised if you understood this to be an invitation to immerse yourself in myths, legends, scripture, poetry and literature in order to distil 'the customs, civilization, and achievements of a particular time or people', as the *Oxford English Dictionary* puts it. If you are anything like me – as I sit here on a cheap sofa, watching an inane TV sitcom while guzzling cheap supermarket wine and listening to Gatecrasher Red – culture is probably not something for the likes of you. At a stretch, you might feel vaguely at home in so-called 'popular' culture, but even this is usually experienced as something beyond the reach of everyday life. It is something for the weekend, when one is released from the banality of everyday life: working, studying, shopping, sleeping, etc. Whether it is 'high' or 'low,' 'popular' or 'unpopular,' culture seems to exist in a parallel universe. Needless to say, many applied geographers are dismayed by the apparent irrelevance of cultural geography, and especially by the 'navel-gazing' analysis of cultural texts.

Yet culture is indeed to be found in alleyways, squats, prisons and shopping centres. One may not find much 'high' culture among the dustbins, tatty furniture, slops and carrier bags, but one will certainly encounter evidence of the customs, civilization and achievements of a particular time or people (see, for example, Buckley, 1996; Bell and Valentine, 1997; Gregson and Crewe, 1997). Graffiti, for example, make for interesting reading (Cresswell, 1996; @149 St). For some, graffiti are merely signs of mindless vandalism: annoying, disrespectful and perhaps even threatening. For others, graffiti demonstrate an attempt by marginalized and alienated people to reclaim the streets: kids, gangs and wanna-be artists. For still others, graffiti are anonymous messages hurled into the world: 'MAD 4 AEP', 'Eat yourself fitter', 'Do you want to play toilet tennis?' So, when one is invited to analyse cultural texts, one should read graffiti, dustbins and carrier bags with the same rigour as one would read literature, legislation and gene sequences. Indeed, I suspect you are already well versed in reading the signs of everyday life: words, pictures, gestures, facial expressions, ambiences, street furniture, footprints, etc. You are already analysts of cultural texts.

As a point of departure, then, do not begin your analysis with a prejudice about what a geographer like you *should* and *should not* set out to read. The types of text you will need to take up will depend upon the particular culture you are interested in: British culture, urban culture, political culture, drug culture, neoliberal culture, pop culture, youth culture, etc. I could go on, but I suspect the message is crystal clear. Be guided by your interests rather than by presuppositions about what counts – and does not count – as culture. I am interested in consumer culture, and my favourite cultural text is a two-page colour chart produced by Dulux for its new range of paints. On the left-hand page there are 42 colours – with names like 'sheer amethyst,' 'blue topaz' and 'rose lacquer' – arranged into three seemingly arbitrary groups. For over an hour my students failed dismally to work out what – if anything – these colours meant. When we could no longer stand the silence, I revealed the six-word solution that appears on the right-hand page of the colour chart: the first group of colours are 'Urban Discovery,' the second 'African Discovery' and the third 'Oriental Discovery'. For the rest of the session they were able to draw out a seemingly endless series of interpretations about the geographical imagination of western consumers that ranged from ideologies of home to the colonial foundations of the discourse of 'discovery'. For example, the Urban colour scheme is predominantly made up of airy and watery blue-greys: cold, muted and vapid. It is the perfect colour scheme for a superficial, artificial and post-industrial world of anonymity, alienation and interchangeability. When one notes the names of these blue-greys – such as madison mauve^{TM}, city limit^{TM}, manhattan view^{TM}, plaza^{TM}, dot com, loft^{TM}, café latte^{TM}, platinum and brushed steel^{TM} – one can discern a highly gendered and class-specific lifestyle: single professional men immersed in the fast and furious world of e-commerce relish their cosmopolitan comforts.

Meanwhile, the African colour scheme is dominated by earthy and fiery colours: warm, grounded and vibrant. While the Urban is alien, Africa is other: exotic, untimely and tied to the earth. It is rendered through white heat^{TM}, turmeric^{TM}, fired ochre, bazaar^{TM}, ancient earth, raw umber and beaten bronze^{TM}. Finally, the pastels of the Oriental colour scheme are mainly creams and greens. Yet rather than the *coldness* of the Urban (modernity) or the *heat* of Africa (tradition), they evoke a *calming* influence (nature): silken trail^{TM}, sea grass^{TM}, palm wood^{TM}, bamboo screen^{TM}, chi^{TM}, lotus blossom^{TM}, gold veil^{TM} and eastern gold^{TM}. So, while the urban north west and the African south pit the elements of air and water against those of fire and earth, the calming spirit of nature wafts in from the Oriental east. Such is the world according to Dulux (cf. Cook and Crang, 1996). I hope this brief analysis has given you a flavour of how a few *divisions* – such as Urban/African/Oriental, North/South/East/West, modernity/tradition/nature, worldly/spiritual, cold/hot/calm – can be arranged to organize a whole range of disparate materials into a functioning whole. So, try not to get bogged down in the details (specific content). Instead, seek out the divisions that structure those details (general form). While these divisions are sometimes hard to pin down and extract, more often than not they are in plain view – and therefore all too easily overlooked.

CULTURAL TEXTS ARE NOT WHAT YOU THINK!

Three things should now be apparent. First, culture is everywhere. Secondly, we all live among innumerable cultures that may or may not have clear boundaries, stable identities and coherent practices. Thirdly, texts are not simply collections of words fixed on to paper. From now on, think of a text in terms of its original Latin root: *texere* – weave. A text is a tissue of signs. It is anything with a signifying structure. It is anything that leads one into decoding, exegesis, interpretation and translation. Quite simply, a text is anything that *refers* meaning – and thereby the reader – elsewhere: to other texts, languages, codes, situations, contexts, expectations, habits, etc. Needless to say, this process of referral can be prolonged infinitely and extended over innumerable domains, although in practice we tend to arrest this structure of dissemination in order to get things done (see Exercise 28.1).

Exercise 28.1   Signifying nothing – a flavour of moral geography

Do you believe me when I claim: (1) that a text necessarily refers you elsewhere; (2) that this process of referral is interminable; (3) that it will lead you all over everywhere; and (4) that sense, meaning and action are an interruption rather than a culmination of reading a text? No? Then consider this piece of graffito: ⊗ Once you have read it, read it again, slowly – very slowly. Now, try to answer the following two questions. What does ⊗ refer to? What *else* does ⊗ refer to? If you need some help, here is my starter for ten. First and foremost, ⊗ refers to 'no smoking'. Secondly, it refers to a *place* and a *time* for 'no smoking' (although the extent and duration of which remain unspecified). Thirdly, it refers itself to you and me: as *readers* of signs, *users* of space, *smokers, non-smokers, smoking non-smokers* (i.e. passive smokers and free riders) and *non-smoking smokers* (i.e. 'good' smokers who keep their cigarettes unlit). Fourthly, it refers above all to *actions*: to what people do. Specifically, they do not smoke. But ⊗ is much more than a simple statement of fact: there is no smoking here. It prescribes a required practice: one *must not* smoke. Fifthly, it thereby refers to *sets* of expectations, rules and sanctions that are meant to govern the relationship between people and place; to those *people* who devise, impose and enforce them; and to the *sources* of trust, expertise and authority that legitimate them – such as law, politics, commerce, culture, the media and health. Sixthly, if the sign ⊗ can manage to elicit practices of 'no smoking', it actually helps to *create* a non-smoking space. Seventhly, it is not entirely clear what 'no smoking' actually entails. Does ⊗ object to smoke, smoking or both? One can hold a lit cigarette without smoking it – as I often do when walking through 'no smoking' spaces en route from one smoking place to another – and one can smoke things other than cigarettes. Eighthly, does the prohibition against smoke and smoking occur at the sign, around the sign or from where one reads the sign? Ninthly, does ⊗ still function as a sign of prohibition when there is no one there to refrain from smoking, when it appears upside down, when it is obscured by foliage or when it is used as an example of dissemination in a textbook? Finally, although ⊗ seems to refer to 'no smoking' in general, it actually refers to particular cases. In short, we will never be finished with the interpretation of enigmatic signs. Nevertheless, we can gain a great insight into how they are socially and spatially structured along the way. So let me restate my original question. Can you think of anything that does not require interpretation and that does not lead into a spatial analysis?

Food, gadgets and clothes signify no less than written words, spoken language and The Human Genome (Barthes, 1972). Consequently, geographers can take virtually any artifact and attempt to draw out the meanings, values, dispositions, desires, knowledge, power relations and practices that are encoded into it. Consequently, what matters is not whether the artifact has a prominent place and obvious significance for the culture in question – such as Acts of Parliament, monumental architecture and famous works of art – but whether the artifact will enable you to access how that culture exists, experiences and acts in the world. Think of *the* international language of commerce: is it English, money or spreadsheet? Or think of *the* technology that holds our world together: is it the transportation system, synchronized clocks or the humble screw? To put it another way, what would have the greatest impact on our world: the disappearance of Acts of Parliament, monumental architecture and famous works of art or the absence of money, spreadsheets, clocks and screws?

Since everything can be decoded, everything is a text. Everything opens out on to a social world. But a text is not simply something that purposefully carries a message or a meaning. Rather, a text is anything that *responds* to the call of an interpretative gesture. It is anything that leads someone to suppose – either implicitly or explicitly – that it might mean something or other. For example, take a landscape. On the one hand, this landscape can be read for the material traces of the various ways of life that have been encoded into it. So a landscape can be treated as the accumulated expression of innumerable modifications over time by a host of human and non-human agents: florae, faunae, technologies, political regimes, earth-surface processes, social formations, etc. Many people call this the 'Real'. No doubt you have gazed out from hilltops, studied maps and processed data on computers in order to make sense of a landscape and extract the truth about its history and geography. On the other hand, a landscape may also function as an enigmatic blank on to which meaning and significance can be projected by different groups of people with very different kinds of interest. As with astrology, numerology and conspiracy theories, values can be read *into* things. In this way, a landscape can come to signify all kinds of things for all manner of reasons: nature, perfection, order, beauty, desolation, belonging, the sacred, wealth, alienation, eternity, woman, the future, etc. Many people call this the 'Imaginary' or the 'Symbolic'. Needless to say, while these kinds of representation usually achieve a certain consistency through habitual associations – often to the point of seeming natural, commonsensical and self-evident to the groups whose interests they serve – they are nevertheless socially constructed and inevitably contested. For what appears to be natural and self-evident invariably turns out to be the product of a long and drawn-out social struggle (Schivelbusch, 1993; Allen and Massey, 1995). Like 'heterosexuality', 'lifelong learning' and 'working for a living', things are invariably natural*ized* rather than natural. In other words, what counts as natural self-evidence is not so much a quality inherent in the world as an outcome of specific social settings. It is a social construct. The city, for example, can be represented as anything whatsoever: monumental, alien, desolate, erotic, chaotic, unruly, sacred, fun, a desert, a jungle, sociable, fleeting, eternal, fearful, artificial, second nature, a wilderness, an ocean . . . But these are not just representations since each elicits a certain set of *practices* that impact upon the city, transforming both its built environment and its social life (Clarke, 1997; Pile and Thrift, 2000). By systematically studying these material traces and representations – by reading the real *and* imagined landscape – you will be able to gain an insight into both the practices and the values that have shaped the relationship between people and place (de Certeau, 1984; Lefebvre, 1991). That will provide you with a basis for thinking about the social and spatial struggles that are articulated through a clash of representational practices: through a clash of different ways of enabling the world to make sense (e.g. painting versus photography, Creationism as opposed to evolutionism or humanism contra structuralism) (see Exercise 28.2).

Exercise 28.2   Signifying self-evidence in social space

Look around the room or space you are in. Note the people, the creatures, the plants, the objects, the ambience, the textures, the sensations and the dimensions. What do they tell you about who you are, your relationship to others and your place in society? What do they say about your social, cultural, political, economic, racial and sexual status? Take any object whatsoever. Why is it there? Where did it come from? What connections does it establish between you and the outside world? Now think of the room you are in as an ensemble. What kinds of activities does it enable? And what kinds of activities does it restrict or even preclude? What is the focal point of the room – and what does that tell you about the way of life you have become subject to? Finally, think of three activities and three objects that would be completely out of place in this room. Now imagine the way of life – the 'spatial practices' – that *would* make them as self-evidently *in place* as that focal point around which your entire existence is presently arrayed.

Most people use the word 'discourse' when they want to consider both the representations *and* practices of a particular social group. A discourse is a specific constellation of knowledge and practice through which a way of life is given material expression. It engenders a discourse-specific (i.e. partial and relative) incarnation of the world that tends to become both naturalized and taken for granted. When writers draw attention to these material and immaterial constellations of knowledge and practice, they usually do so in terms of the social and spatial power struggle between 'dominant discourses' on the one hand and 'discourses of resistance' on the other. Discourse analysis discloses how this constellation of knowledge and power is structured, and situates it within its appropriate social, cultural and geo-historical context. For instance, think of the discursive conflict between adults and children, the rich and the poor, the colonizers and the colonized, and environmentalists and capitalists. Each draws upon a very different geographical imagination, and each deploys a very different repertoire of spatial practices that jostle for supremacy. A wonderful example of such a conflict is provided by Allen and Pryke's (1994) consideration of how foreign-exchange dealers, security guards, caterers and cleaners inhabit 'the floors of finance'. Although they all more or less occupy the same physical place, they each live and work in very different spaces (cf. Daniels and Rycroft, 1993). For example, while the dealers are hooked into the hyperactive 'global space of flows' through their computer screens, telephones and social networks, the cleaners are expected to disappear without trace after attending to the materials that make up the floors of finance: glass, plastic, marble, paper, etc.

In summary, a geographical analysis of cultural texts and competing discourses needs to follow as rigorously as possible the spatial, temporal and social traces of both real and imagined signifying structures: representations and practices. The expertise you will need to draw upon will inevitably spin out, not only across the whole of human geography but also across many other disciplines. While this profusion is undoubtedly daunting, you should take comfort from the fact that the ability to make sense of the world is more likely to come from lifelong learning than divine inspiration. Perhaps the best advice is to move slowly, to remain alert for possible connections and to resist the trap of uncritically accepting common sense. So, you should try to enter into a photograph, artifact or document in the same way you would enter into a film or a novel: they lead to worlds within worlds; social spaces within social spaces. But rather than passively submit to the line that has been laid out for you to follow, you should attempt to explore and survey systematically this world within a world for yourself. Take a landscape painting, an Act of Parliament or an advertisement for new housing. Or else a bridge, a nature reserve or a display of lucky thimbles:

- How, why and for whom has it been constructed?
- What are the materials, practices and power relationships that are assumed by it and sustained through it?
- What codes, values, dispositions, habits, stereotypes and associations does it draw upon?
- What kind of personal and group identities does it promote? And how do they relate to other identities?
- What does it mean? What are its main structuring devices: oppositions, divisions, metaphors, illustrations, exemplars, etc.? And how do they overdetermine and constrain the choice and arrangement of content?
- More importantly, what kind of work does it do? And who benefits?
- What has been included, excluded, empowered and repressed?
- How might it be modified, transformed or deconstructed? How could this social space be inhabited differently?
- Finally, since nothing *ever* comes alone, what wider assemblages does it fit into and resonate with? Are these assemblages synergetic or contradictory? How does it relate to the assemblages that interest urban geographers, historical geographers, medical geographers . . .
- And since the work of contextualization and recontextualization can be carried out infinitely, there is never anything 'final' about following the manifold traces of textual analysis.

A good place to start is with a map – not least because, like a photograph or a measurement, most people tend to assume that a well prepared map tells the truth. However, as with everything else, maps are made to serve particular purposes in specific social, cultural, economic and political settings. They edit, transform and remake the world in a way that suits the interests to hand (Wood, 1993; Pinder, 1996).

FORGET READING!

So far I have tried to clarify how you should *approach* cultural texts: slowly, attentively, broadly, openly and with an eye towards the struggles between competing discourses for representational and practical supremacy over social space. However, I have been asked by the editors of this book – Nick Clifford and Gill Valentine – to do something strange: perhaps even pointless. Since this is a 'how to . . .' book I have been asked to *instruct* you in – of all things – *reading*. But how could you have got to Chapter 28 without already knowing how to read? What concerns readers like you is not learning *how* to read but *whether* to read, how *much* to read and *what* to read. And you don't need me to tell you the answers to these dilemmas because you have heard them repeated ad nauseam by teachers and tutors: 'You must read . . . as much as possible . . . of what is on your reading lists.' And the difficulties you will face will probably have less to do with reading per se and more to do with the nature of the writing. Academic texts are notoriously dry, long-winded, boring, jargon-ridden, humourless, irrelevant, pompous, obtuse and turgid.

Now, while I get endless queries from students about how things should be *written* and *presented* – structure, flow, balance, objectivity, referencing, quoting, exemplification, contextualization and especially length – I cannot recall anyone *ever* asking me how to read. This is odd: not only because many students experience an entirely understandable difficulty in reading 'academic' texts but also because reading in general is actually a demanding and skilled activity that requires training. Indeed, you should study techniques of reading in the same way as you would study any other qualitative or quantitative research technique. And believe it or not, there are countless ways in which to read – few of which boil down to reading word after word for page after page! I will *mention* just a few: hermeneutics, semiotics, psychoanalysis, structuralism, deconstruction, discourse analysis, postcolonialism, Marxism, feminism and reader response. To see them in action, see journals such as *Environment and Planning D* (a.k.a. *Society and Space*), *Gender, Place and Culture* and *Space and Culture*. For a more formal encounter, select a few introductory texts from your library's 'cultural studies' and 'literary theory' sections. For the flavour of things, try some of Icon's comic-book series designed 'for beginners'. I especially enjoyed *Introducing Baudrillard, Introducing Cultural Studies, Introducing Derrida, Introducing Postmodernism* and *Introducing Semiotics*. For now, however, I just want to give you a feel for the power of innovative reading strategies (see Exercise 28.3).

Exercise 28.3   **A speedy literature review**

My second-year tutees have been struggling to perform an onerous but nevertheless important task: a so-called 'literature review'. Basically, rather than write an essay on something or other out there in the world, they need to write about how geographers have written about something or other. Fifteen journal articles, several books and four weeks later the reviews are in. Although they have read a vast amount they still find it hard to get a grip on what academics think about things. Imagine their dismay when I suggested this short-cut after the event. Take any textbook – perhaps this one – and go and get a very old textbook on the same subject from the library (say something from the 1980s). Open their indexes and make three lists of words: (1) those that appear in both; (2) those that only appear in the old book; and (3) those that only appear in the new book. What are the dominant themes within *each* of the lists? Now, compare what has been *preserved* within the subject (list 1) what has been *purged* from the subject (list 2) and what has been *added* to the subject (list 3). What does this tell you about the changing interests of geography and geographers? Do these changes amount to *progress*? If you are really pressed for time, try the same procedure with the contents pages instead. These will give you a sense of how the subject is structured and of the relative importance of various issues and debates.

Note: When faced with the inconvenience of long indexes, feel free to restrict your textual analysis to a workable range of letters: say M–R. For those with a *quantitative* bent, try drawing *inferences* about the *population* of the entire index from this sample, write down how confident you are about your inferences and then compare your expected results with what is actually the case. This should alert you to the fact that quantitative analysis is a particular way of *thinking* rather than a fixation on numbers per se.

COLUMBUS' EGG: 'MIND THAT CHILD!'

Let me bring this chapter to a close with an anecdote and a checklist. Apart from a few advertising slogans, the only phrase that I can remember from my childhood is: 'Mind that child!' It was written on the back of the ice-cream vans that toured the streets where I lived. Although there was nothing difficult about the words, I always felt unnerved by them. On every occasion I encountered the phrase, the anticipated pleasure of eating ice cream was tainted by anxiety over these enigmatic and sinister words. Now, believe it or not, it was only a couple of months ago that this childhood association between ice-cream vans and foreboding dissipated. I was about to overtake an ice-cream van when I noticed those words: 'Mind that child!' Suddenly, I realized that the words were addressed to *adults* rather than children. They simply warned drivers not to run over children distracted by the pleasures of ice cream rather than to warn *children* about the dangers of 'that!' – something so terrible that adults dare not even write its name! I had assumed that the statement 'Child: mind *that!*' was meant to warn me of evil or distracted car drivers who routinely ran people over (this was the moral panic par excellence of my infant years: soon to be followed by the horrors of sexually transmitted diseases, heroin abuse and glue-sniffing), but it always struck me as odd that one actually needed to be *in* the road to read the warning – as if the whole thing were a terrible ruse to expedite the killing of children. Years later I would be reminded of these ice-cream vans when I read about the Nazis' highly effective discourse of dissimulation: 'Work makes free' (death camp), 'Showers' (gas chambers) and 'vehicles' (mobile gas-vans) (Lanzmann, 1985). The comeuppance is clear. When you start to analyse a cultural text, do not take its message at face value or assume that it was meant for people like you. It almost certainly wasn't. Once you have worked out for whom the text was produced you will have gone a long way to reading it effectively. So here is a final checklist to get you going. Investigate *who* produced the text, *why* they produced it, *how* they produced it and for *whom* they produced it (cf. du Gay et al., 1997). Sometimes all of this will have been consciously intended. On other occasions, however, it will have been unconscious and unintended. Investigate the *form*, *content* and *assumptions* of the texts in question, paying just as much attention to what is *absent* from the text as to what is present. Set the texts in appropriate *contexts* and compare and contrast them with other relevant material: texts, lifestyles, belief systems, practices, artifacts, etc. Investigate how they have been used and abused as mechanisms for articulating *power* and *resistance* by a range of people and groups. But above all else, investigate what *work* these texts do: how do they impact upon and affect society and space? Once you have done that you can begin to approach the analysis of cultural texts according to a wide variety of criteria that may interest you: power, knowledge, desire, truth, fidelity, meaning, motivation, intention, exclusion, class, race, gender, sexuality, etc. And if anyone mocks your scholarly and geographical interest in *reading* cultural texts, remember the story of Columbus' egg. In reply to a suggestion that *anyone* could have discovered America, Columbus challenged the guests at a banquet held in his honour to make an egg stand on end. When all had failed, Columbus did so by flattening one end with a sharp tap on the table, thus demonstrating that while others might follow, he had discovered the way. It is true that reading, like egg-standing, is an easy task to pull off – but only when you have learnt the trick.

Summary

- Cultural texts are not what you think: everything has a signifiying structure, everything can be decoded, everything is a text.
- 'Culture' and 'textuality' have become key terms in geography.
- There are many different kinds of textual analysis (e.g. discourse analysis).
- It is important to distinguish between the producers and consumers of cultural texts and the form and content of cultural texts.
- Power relations are encrypted into cultural texts.
- Cultural texts matter because they shape and inform social practices and social spaces.
- Effective reading is invariably more difficult than one would think.
- Take time to read – and to learn how to read.

Further reading

This guide includes an eclectic selection of examples of different forms of cultural analysis:

- *@149 St. New York City Cyber Bench* (http://www.at149st.com) (accessed January 2003). This website archives the history of the graffiti art form developed on New York City's subways. It includes a wide range of artists, crews, images, texts and links.

- Clarke (1997) is a collection of essays that explores how cities have been depicted in film and how cities themselves have been affected by the cinematic form.

- Cook et al. (2000) addresses the nature, significance and impact of the so-called 'cultural turn' in contemporary human geography which leads into a full-scale exploration of the relationship between culture and space.

- Cresswell (1996). From graffito to peace protests, this is a wonderful analysis of how power, transgression and resistance get played out in the geography of everyday life.

- Du Gay et al. (1997). Everything you ever wanted to know about the way in which the 'turn to culture' impinges on our lives using the Sony Walkman as an example.

- Rybczynski (2000), in attempting to identify the most significant invention of the past millennium, uncovers the mind-boggling secret history of the screwdriver and screw. This is a wonderful example of why one should never take anything for granted.

- Schivelbusch (1993) presents an astonishing world of social conflict and cultural struggle that surrounded the introduction of pepper, coffee, chocolate, tobacco and opiates into Europe. It is a remarkable study of the modernization and industrialization of culture.

Note: Full details of the above can be found in the references listed below.

References

@149 St. New York City Cyber Bench (http://www.at149st.com) (accessed January 2003).

Allen, J. and Massey, D. (eds) (1995) *Geographical Worlds*. Oxford: Oxford University Press.

Allen, J. and Pryke, M. (1994) 'The production of service space', *Environment and Planning D: Society and Space*, 12: 453–76.

Barthes, R. (1972) *Mythologies*. London: Jonathan Cape.

Bell, D. and Valentine, G. (1997) *Consuming Geographies: We are Where we Eat*. London: Routledge.

Bourdieu, P. (1984) *Distinction: A Social Critique of the Judgement of Taste*. London: Routledge.

Buckley, S. (1996) 'A guided tour of the kitchen: seven Japanese domestic tales', *Environment and Planning D: Society and Space*, 14: 441–61.

Clarke, D. (ed.) (1997) *The Cinematic City*. London: Routledge.

Cloke, P., Crang, P. and Goodwin, M. (eds) (1999) *Introducing Human Geographies*. London: Arnold.

Cook, I. and Crang, P. (1996) 'The world on a plate: culinary culture, displacement and geographical knowledges', *Journal of Material Culture*, 1: 131–53.

Cook, I., Crouch, D., Naylor, S. and Ryan, J. (eds) (2000) *Cultural Turns/Geographical Turns: Perspectives on Cultural Geography*. Harlow: Prentice Hall.

Cresswell, T. (1996) *In Place/Out of Place*. Minneapolis, MN: University of Minnesota Press.

Daniels, S. and Rycroft, S. (1993) 'Mapping the modern city: Alan Sillitoe's Nottingham novels', *Transactions, Institute of British Geographers*, 18: 460–80.

de Certeau, M. (1984) *The Practice of Everyday Life*. Berkeley, CA: University of California Press.

du Gay, P., Hall, S., Jones, L., Mackay, H. and Negus, H. (1997) *Doing Cultural Studies: The Story of the Sony Walkman*. Milton Keynes: Open University Press.

Gregson, N. and Crewe, L. (1997) 'The bargain, the knowledge and the spectacle: making sense of consumption in the space of the car-boot sale', *Environment and Planning D: Society and Space*, 15: 87–112.

Lanzmann, C. (1985) *Shoah: An Oral History of the Holocaust*. New York: Pantheon.

Lefebvre, H. (1991) *The Production of Space*. Oxford: Blackwell.

Pile, S. and Thrift, N. (2000) *City A–Z*. London: Routledge.

Pinder, D. (1996) 'Subverting cartography: the situationists and maps of the city', *Environment and Planning A*, 28: 405–427.

Rybczynski, W. (2000) *One Good Turn: A Natural History of the Screwdriver and the Screw*. New York: Simon & Schuster.

Schivelbusch, W. (1993) *Tastes of Paradise: A Social History of Spices, Stimulants, and Intoxicants*. New York: Vintage.

Wood, D. (1993) *The Power of Maps*. London: Routledge.

29 Writing Essays, Reports and Dissertations

Michael Bradford

Synopsis

The essay, report and dissertation are three different forms of presenting (and assessing) an argument with evidence. The essay is discursive but short compared to the dissertation, which is a major piece of work usually based on original empirical research. The report is concisely and precisely written. Unlike the other ways of writing, it is often used outside academia. The chapter is organized into the sections listed below. The sections are each written in the style of the form of writing they describe:

- Introducing the three forms of writing and the common skills required
- Guidelines on the essay, written as an essay
- Guidelines on the report carried out by a team, written in report style
- Guidelines on a dissertation: a model with various deviations.

INTRODUCTION

This chapter discusses three major forms of presenting written work when studying geography at university: the essay, report and dissertation. Students are expected to learn some common and some different things from completing these three types of assessment or, in the current educational jargon, particular 'learning outcomes' are associated with them. The learning outcomes may also vary for each of the three, depending on the exact requirements. Some of these learning outcomes are discussed in this introduction together with the variations in the forms before the three are separately explored. The discussions on the essay and the report are presented as an essay and a report, respectively.

Essays vary in length. Some are completed to time within an unseen exam and fast thinking is necessary. Some comprise about 1500 words and are submitted as part of tutorial work, which may or may not count towards the overall assessment. Others are extended essays of about 3000 words, which are part of the coursework requirement of the degree programme. These last two types allow more considered thought. Essays are discursive while reports are written in a precise and concise style, with paragraphs numbered for easy reference. They have an executive summary that comprises the main findings of the report, which can vary in length. The dissertation or project is the longest of these forms varying from perhaps 6000 to 15 000 words at undergraduate level. It is divided into chapters. It is usually research orientated, based either on both primary and secondary evidence or on secondary evidence alone. In some universities it can be entirely based on a review of literature and not involve the collection of primary data or the analysis of primary and secondary evidence.

Essays and dissertations are very much academic modes of writing, while reports are used widely outside academia – for example, in government and business. Gaining experience in writing reports is therefore very valuable for later life. The skills associated with all three, however, are very valuable both during and after university. All involve skills of organizing material and presenting it in a structured way. All involve developing and sustaining an argument and providing evidence to support that argument. All involve considerable preparation in terms of thinking and reading. The dissertation particularly requires preplanning. It is a large piece of work and occupies a long period of time, so organizational and time management skills are very important. Networking skills can also be very useful when collecting evidence, whether through interviews, data gathering or searches of archives. Using these skills helps find the right people to answer questions or to give access to information. So many types of skills are developed and assessed through these pieces of work.

All three will mostly be carried out as individual tasks but the extended essay, the report and the project can be completed in a team. Here the guidance on writing a report is given for a team. The process of working as a team is very significant with organizational skills and time management becoming even more important to a successful outcome. In most situations, however, only the end-product will be assessed, and the process by which it was achieved will be assessed indirectly through the end-product. Sometimes the process is directly assessed – for example, the important stage of choosing a topic for a dissertation is often separately assessed by a written proposal. In this chapter only the presentation of the end-products is considered.

Whichever of the three is being completed, the first thing to do is to study what is expected. There are separate guidelines available in most departments on dissertations and often on essays. Good guidelines will include the marking schemes involved and the grade-related criteria that indicate what is expected to obtain certain levels of marks (Clark and Wareham, 2003). The structure of the argument used is likely to be one criterion on which all three forms are judged. It will be reflected in the grade-related criteria through a gradation of preceding adjectives – for instance, from basic to excellent. Certain aspects of the grade-related criteria may appear at only the top levels – for example, originality and evidence of synthesis. It is important to study any guidelines and to relate work to the relevant criteria. Within these guidelines will be tips on time management that are crucial to all three forms. They are particularly relevant to team reports and the dissertation. Many advise students to work backwards from deadline dates in establishing a timetable of work (Kneale, 1999). Once having studied the guidelines and the intended learning outcomes, students are ready to approach the tasks.

THE ESSAY

Discuss the major challenges of writing an essay

University Challenge has been the British television audience's main view of universities for decades. The questions posed by Bamber Gascoigne and Jeremy Paxman may have changed but the format remains the same. Over an even longer period of time, the essay has formed one of the challenges for students at university. Here, too, the questions may have changed, but the format of the essay has remained essentially the same. It is still the main mode of communication and assessment at university for geography students, even though its relative significance has declined. Since the importance of the essay has declined much more within schools, writing an essay, in itself, has become a major challenge to many geography students at university.

Going to university presents many challenges, from the more general ones of managing finance and time, to developing specific study skills such as writing essays. The extent of the challenge depends on students' previous experience and their ability to transfer it into a new context, and the standard to which they aspire. Experience varies in that, for example, some may have been trained at school or college to introduce and conclude an essay. Others may have been encouraged to be more instrumental, given the shortage of time within exams, and to score points without the overview provided by introductions and conclusions. Here they would be dispensing with the two ends of a traditional view of the structure of an essay:

- Say what you are going to say (introduction).
- Say it.
- Say what you have said (conclusion).

Many, however, would argue that an introduction catches the attention and sets the essay in perspective as well as indicating the structure of its argument, while a good conclusion is more than the sum of the preceding paragraphs, synthesizing points, developing implications and perhaps ending with related questions. At university such introductions and conclusions are expected. Indeed, their quality may contribute very significantly to the difference between marks awarded. Understanding what is expected by the general and specific task is then fundamental to producing a high-quality essay. Preparing and gathering evidence for the argument to be used are crucial early stages that inform the writing and editing of the essay. The final stage is learning from the process and the assessment so that other essay-based tasks might be completed more effectively. These various stages will be discussed below in order to help students achieve the highest possible marks and, more generally, to help them learn from the overall process which underlies many assessment tasks.

The first challenge is understanding the task set, whether the essay is unseen within an exam, a tutorial essay or an extended essay submitted as a piece of coursework required for a course unit. Many essays will consist of one or more 'command' words such as 'discuss' or 'evaluate' or 'describe and explain', together with the subject such as 'the challenge of essay writing' or (evaluate) 'the success of regeneration schemes in British cities'. Understanding the 'command' word can be quite challenging because academics who set the essays may have somewhat different ideas about what they expect by the word. 'Discuss', for example – a very common command word – means to some 'presenting a balanced argument' while to others it requires 'presenting an argued case, preferring some viewpoints to others'. Expectations may also vary with the level of the task. 'Evaluate' is quite a high-level task because it requires deciding the criteria against which something (e.g. regeneration schemes) will be judged, as well as critically reviewing evidence that will permit judgements to be made. Matching students' expectations to those of staff is more generally one of the major challenges of university. Some degree programmes will articulate expectations very transparently and have guides to command words within handbooks.

What expectations are associated with the subject of the essay title? Discussing the title 'the success of regeneration schemes in British cities' will illustrate the required thought processes. 'British cities' sets the geographical context, but the timing of the schemes has been left open for the writer unless, of course, the course unit is about very recent change. If recent change is concerned then the regeneration schemes might include urban development corporations, City Challenge, the Single Regeneration Budget, and the New Deal for Community. If a wider time perspective is required, or could be argued, postwar comprehensive redevelopment might also be included. 'Success' is the key word because it begs the questions 'for whom?' and 'on what terms?' Success could be evaluated narrowly against the aims of the schemes – did they achieve their objectives? It could be viewed more widely against more general criteria, such as 'did they reduce social exclusion?' So schemes might have been successful for the eventual residents but not for the displaced original residents. The challenge of the question lies not just in knowing the details of the schemes but also in understanding the more general economic, social, political and environmental contexts on which regeneration may be judged. A good answer might argue that some schemes have been successful in regenerating the built environment, but not in improving the economic opportunities and social inclusion of the original residents.

The task is not always fully set by staff. For an extended essay, students may be asked to decide the topic and/or the exact title. Even some unseen exam essays include 'write an essay on (global warming)'. To produce a good answer to this very open task the student has to decide and execute the task – e.g. 'evaluate the evidence for global warming' or 'discuss the degree to which global warming has been produced by recent human activities'. The setting of the task, as well as the degree to which it is achieved, becomes part of the assessment.

So understanding and making the task explicit are the crucial first step – defining the terms (e.g. regeneration schemes, global warming), the spatial and temporal contexts or scales that set the limits to the discussion and interpreting key words (e.g. the meaning of 'success').

Once the task has been understood, the next step is to construct a possible argument or approach so that students know which 'literature' to consult or upon which revised areas to draw. The process is made easier by having already clarified the task. 'Evaluating success' in the regeneration essay has suggested that there are many possible criteria to be considered, and these may be viewed from different perspectives – for example, have the schemes helped the city to position itself economically in a globally competitive way? Have they aided social inclusion? Have old buildings been refurbished for new functions in a way sensitive to the needs of conservation (Imrie and Thomas, 1998)? Just reading about or recalling the details of the regeneration schemes will not help to develop an argument for such an essay. Once the key areas of literature have been identified, there is often a further challenge when completing non-exam work: that of getting access to the appropriate journals and books and selecting appropriate web-based material. Another very significant, associated challenge is approaching this material in a critical way so that the argument developed is soundly based.

The next challenge is central to the task: writing the essay. Writing an essay in an exam requires a linear development of the argument from introduction to analysis and conclusions, having established a structure through an initial short plan. Some regard the analytical part as the meat within the sandwich of the introduction and conclusion. When a wordprocessor may be used the development does not have to be linear. Some would write the meat of the analysis first and only then complete the introduction. The roles of the introduction are to catch the attention, set the topic in perspective and outline the points of the argument that need to be considered and thus the structure of the essay. It is not necessary to outline the argument here, though some people advocate this. An outline at this stage does not have any evidence to support it.

The analytical meat addresses each point of the argument, ensuring that each paragraph contributes to answering the question and that the points occur in a logical order. Perhaps the two most often seen negative comments in the margin of essays are 'irrelevant' and 'order'. Editing on the wordprocessor allows the paragraphs to be reordered and the argument within paragraphs to be sharpened once drafted. In an extended essay, subheadings may be used to help the reader (and often the writer) to follow the structure.

When making the major points of an argument, it is important to present supporting evidence and to use examples to illustrate the points being made. The challenge is to demonstrate the breadth and depth of understanding possessed through the examples without allowing the argument to be dominated and obscured by the examples. If the example is dominating the essay and the reader is being left to infer the argument, it is time for a major rewrite. Developing and sustaining the argument are the major goal. Examples are used to help attain that goal.

The conclusion not only draws the points of the argument together but it also develops their implications and may raise further questions. For example, it might summarize the arguments on the degree of success of regeneration schemes against their objectives and against wider societal aims, and consider the degree to which there have been connections made between other schemes within the city, particularly those that are spatially contiguous. The possibility of schemes remaining 'islands of regeneration' raises the question whether spatially targeted regeneration alone can regenerate cities that have experienced major deindustrialization.

After proofreading and ensuring that all the references are included at the end of the essay, the final stage of the essay task is for the student to stand back and consider its evaluation. Have the aims been achieved? What are the positive points? What could have been improved? The submission of a brief evaluation helps the reader in his or her feedback but more importantly it helps students to be more generally reflective and to learn from their experience. This final challenge is only completed when feedback from the essay is received. Some tutors will give comments without a mark and expect the student to read the comments, compare them with grade-related criteria and arrive at his or her own assessment of the mark before learning and discussing the mark given by the tutor.

The overall challenge, of course, is learning from the process of completing the essay: learning about the subject matter, for example, points that might have been missed on urban regeneration; learning about interpreting command words and key words; learning about ways of introducing an essay, structuring an argument, using evidence and examples, and drawing together conclusions; writing references in the bibliography in a consistent and accepted fashion; and learning how to draw a line under the completion of the task. It is too easy with modern technology for a perfectionist to spend too long attempting to sharpen the argument or to achieve the best order. Too much time spent on one essay or, more generally, on one assessment task may limit the performance achieved on other tasks. It may be galling to a geographer to conclude that the management of time, rather than space, is one of the most important skills necessary to succeed at *University Challenge*.

Reference

Imrie, R. and Thomas, H. (1998) *British Urban Policy*. London: Sage.

Self-evaluation of this essay

I think I have achieved my aim of writing an essay about writing an essay in a way that has conveyed the essential parts of the process. Perhaps I have dwelled too much on command words relative to key words. I would have liked to have given more examples of the process, especially one from physical geography, but I have concentrated on one, urban regeneration, to improve clarity, given the limitations of space. I have attempted to place the essay in the wider perspective of the challenge of university and linked it to *University Challenge* to catch attention. Perhaps, though, the eye-catcher is too stretched.

WRITING A REPORT AS A TEAM

1   Introduction

1.1   This report aims to outline the purpose and structure of reports and to present guidance on the ways they are produced by a team.

1.2   A report is the conventional method of presenting precise information. It may be used to convey an assessment of any situation or the results of an investigation.

1.3   A report is concise and succinct. It has clearly stated aims. It is tightly focused on the subject of investigation. A really effective report will also be compelling and stimulating to read.

1.4   This report is structured to give guidance on working in a team, to prepare, plan and write a report.

2   The importance of the team

2.1   This particular report is a team responsibility. As a team, you should be supportive of each other and be able to give constructive criticism on the project and each other's work.

2.2   You need to recognize and make best use of the talents of the team and to work in an organized way. The successful outcome of good teamwork will be more than the sum of the parts of the individuals' efforts.

3   Producing the report

Three stages in the process of producing a report have been identified in order to help you tackle key issues and understand the task set. These three stages are:

(a)   preparation;
(b)   planning; and
(c)   writing and editing.

4   Preparation

4.1   Careful preparation is vital. It allows you to make the best use of the time available to the team. During this period you should collectively decide the aims (why), the process (how) and the contents (what) of your writing.

4.2   Establish the broad aims of the project and a draft plan of the report first. This enables you to identify the type of reading that each member of the team needs to undertake. Your aims and argument may subsequently be more precisely set.

4.3   When you have completed your individual reading you must meet as a team to discuss the key issues of your report.

4.4   You must always remember that report writing is a continuous process of decision-making. Your initial decisions will include:

(a)   the identification of the specific topic;
(b)   the aims of your report;
(c)   the precise message your team wishes to convey;
(d)   the appropriate structure and format; and
(e)   the correct vocabulary, style and tone.

4.5   When you have completed your discussion note the decisions you have agreed as guidelines for your work.

5   Planning

5.1   Planning saves time and promotes clarity in collecting the information you require, organizing the material and writing the report.

5.2   Writing a report encompasses a number of autonomous activities. You will find it easier if you break the whole process down into a number of tasks:

(a)   collection and analysis of evidence can be organized according to the source or the subsection of the report; and

(b)   the writing process can be organized into writing subsections and presentation of evidence into graphs, tables, maps and quotations.

5.3   Planning and deciding on the method of work will eliminate much reproduction of effort. It will save time and it will also help you create a logical structure for the report. Good organization is the key to success.

5.4   The following sequence may help you to plan your report:

(a)   decide upon a working conclusion and the order of presenting your findings to create a preliminary structure;

(b)   identify the sources of evidence (data and/or literature);

(c)   decide what is the most appropriate and relevant evidence to collect and how to analyse and present it;

(d)   decide on the tasks to be completed and their order of priority;

(e)   draw up a realistic timetable for the completion of each task, including writing the draft of the report; and

(f)   allocate the tasks among the team members.

6   **The writing process**

6.1   There are three main factors to consider at this stage to give your report a sound framework, clear style and an attractive appearance:

(a)   structure;
(b)   style and language; and
(c)   presentation.

6.2   Structure

6.2.1 A clear structure with headings and subheadings helps the reader to digest the report. It also helps you write and organize your material logically.

6.2.2 A sensible structure is:

(a)   title page
(b)   contents
(c)   executive summary
(d)   introduction
(e)   sources and methods of analysis
(f)   analysis and interpretation
(g)   conclusion
(h)   appendices
(i)   reference list.

6.2.3 The following order for writing is suggested:

(a)   *Analysis and interpretation.* To help you present your findings have your material sorted, selected and arranged in note form. This section includes:

(i)   the results of your analysis; and
(ii)   your interpretation of them.

You should help the reader by ending each part with its own conclusion.

(b)   *Methods.* In this section you should discuss:

(i)   the sources of evidence;
(ii)   the collection and analysis of evidence; and
(iii) the limitations of the sources and methods of collection and analysis.

(c)   *Conclusions.* This section is a summary of all the major findings made throughout the report. No new evidence should appear here. The conclusion considers the evidence presented in the main body of the text, draws out the implications and brings it to one overall conclusion or an ordered set of final conclusions or recommendations.

(d)   *Introduction.* After having written your findings and conclusions you now know clearly what you want to introduce. The introduction is where you indicate the purpose and structure of your report.

(e) *Appendices*. These are set aside for supplementary evidence not essential to the main findings, but which provide useful backup support for your main arguments.

(f) *Contents*. All the sections of the report should be listed in sequence with page references.

(g) *References*. This section covers the books and articles that have been used in your research. It must include every reference mentioned in the text and be presented correctly.

(h) *Title page*. The title should indicate the central theme of the report. The page must also include the authors' names and the date of completion of the report.

(i) *Executive summary*. This is a very important part of the report and should be the last thing you write. Refer to the guide on executive summaries in section 7 of this report.

6.3 Style and language

6.3.1 Even though written by a team, the report should read as though one person has written it. So careful editing is needed.

6.3.2 Focus on the specific purpose of the report. Every part of the report should relate to it and this will help to keep the report concise and coherent.

6.3.3 Clarity and accuracy are vitally important so always be precise. You must know precisely what you want to say in each paragraph and sentence.

6.3.4 Each paragraph should contain one significant point so that, if the paragraph is referred to, its subject matter would be unambiguous.

6.3.5 Your sentences must be grammatically correct and well punctuated. Words must be spelt accurately.

6.3.6 Other important things to remember:

(a) Keep sentences short and simple. Long complex sentences slow the reader down, confuse and impede understanding. The same applies to paragraphs; and

(b) spell check and proofread the final document so that the reader is not deflected by poor spelling.

6.4 Presentation

6.4.1 The presentation of the report requires the same level of care that went into composing the text.

6.4.2 Adequate headings and numbering make it easier for the reader to comprehend what you are saying.

6.4.3 Numbering allows the writer and reader to refer to specific points. Using more than a three-number sequence can be unwieldy (e.g. 6.4.1.1) so use a letter or bullet point (e.g. 6.4.1. a).

6.4.4 The presentation of statistics is often more informative and eye-catching if they are shown visually – for example, by using graphs, pie charts or histograms. Remember that you should always discuss any tables or diagrams you use in the text.

6.4.5 Layout is important. This is the relation between print and space on the page. A crowded page with dense blocks of print and little space looks unattractive and is off-putting. Always ensure that there are:

(a)   adequate margins;
(b)   either double or 1.5 spaced lines; and
(c)   headings that stand out clearly from the page.

7   Writing the executive summary

7.1   An executive summary is defined as 'the main points of a report'. The purpose is to provide the briefest possible statement of the contents of the report, any significant findings, conclusions and recommendations.

7.2   It must cover all the essential points. It must be fully comprehensible when read independently of the full document. It is not a list of extracts, highlights or notes on the original.

7.3   The executive summary must:

(a)   introduce the subject of the full report, its aims, methods, findings and/or recommendations;
(b)   help the reader to determine whether the report is of any interest; and
(c)   save time for busy executives/officials in that it may be used by the reader to make initial decisions, so ensure that it fulfils its purpose.

7.4   The authors can use the executive summary as a rigorous check on the success of the full report – i.e. that the full report is a clear, concise statement that meets its aims.

7.5   The method of constructing an executive summary is to:

(a)   read the whole document;
(b)   isolate and summarize its central theme;
(c)   read each section and eliminate all repetition, lists, examples and detailed description;
(d)   identify and summarize the main statement of each section;
(e)   combine (b) and (d) into a set of major points. Write these as a continuous narrative because your aim is to convey

the overall impression of the full document in as clear and brief a way as possible; and

(f) read through your summary to check that it gives a fair impression of the original while ensuring that it will make sense to the reader as a separate document from the full report.

(I would like to acknowledge the original work of Peter Shirlow and Laura Smethurst in writing this report. This is an amended version.)

THE DISSERTATION

There are many ways in which to present a dissertation but a model will be discussed with some deviations from it so that students can select the most appropriate form for their particular topic. In this model it is assumed that some original gathering of evidence is expected although, as noted in the introduction to this chapter, sometimes only secondary data are needed or, in the extreme case, only literature is reviewed. An outline of the structure of the dissertation is discussed first, then the deviations and, finally, the order in which it is often written.

Most dissertations will begin with a preface in which acknowledgements are made to those who have helped, with perhaps some comments on the process, particularly if the topic has changed in a major way.

Introduction

The first chapter introduces the general topic or problem and sets it in perspective within geography. It might include the reasons for the author's interest and the overall aims of the dissertation. It should outline the structure of the dissertation as a guide for the reader as to how those aims will be achieved.

Review of the literature

The second chapter reviews the literature that establishes a context for the topic. It might identify a gap in the literature or a debate that needs to be addressed. It comprises a critical review that argues towards the specific research questions considered in the dissertation, which may well be stated at the end of this chapter. They may consist of questions or hypotheses, the latter being used more often in physical geography. It is through the analysis of these questions that the aims of the dissertation are achieved. Their identification is, therefore, crucial to the success of the dissertation.

Methodology

The third chapter discusses the methodology used to address the specific research questions posed. It explains the research design and justifies the methods used to collect and analyse evidence. It can also include a justification of the choice of area in which the study has been carried out. The precise format will depend upon the approach to research used – whether, for example, it is extensive (seeking associations within representative samples) or intensive (seeking 'causes' and insights from a few in-depth investigations) (Sayer, 1992). If it is the former then the type of sampling used and the response rate should be discussed. If it is the latter then, for example, the original cues that were used in the first interview might be discussed, along with the way interviewees were chosen and the interviews changed.

Analytical chapters

The next few chapters form the main analytical sections of the dissertations where the results are presented and discussed. Each chapter, for example, may address a particular research question posed. The chapters are closely argued with evidence presented in maps, graphs, diagrams, tables, photographs and/or quotations to support the argument. The discussion will also feed back to the literature where it is appropriate to do so.

Conclusion

The conclusions to these chapters are brought together in the final chapter. It should include the limitations of the work, whether based on the methods used to collect or analyse the evidence. The overall conclusion is not just a summary of the individual conclusions to the research questions. It attempts to synthesize the findings (make more of them than just the sum of the parts) and to present their implications. It relates them back to the literature reviewed in chapter two. Finally, it poses further questions that arise from the research.

The references to all the work mentioned in the text follow. These are usually presented according to the Harvard convention, in alphabetical order, as in this book.

The appendices complete the dissertation. They may include such details as a questionnaire, an example of a transcribed interview or the description of a standard analytical method or laboratory technique used. It is too easy where there are word limits to put lots of extra work in the appendices. This should be avoided. If it is important to the argument, the material should appear in the body of the dissertation. Material that is irrelevant to the research questions should not appear anywhere.

In many physical geography dissertations the structure of the analytical chapters is further divided. It is customary to present the results of the research separately from their interpretation (Parsons and Knight, 1995). The results are considered to be objective and therefore are presented in the form of maps, tables, graphs and diagrams. The interpretations of the results follow in a separate chapter. They relate back to the literature and are considered to be those of the author.

In most research in human geography (Flowerdew and Martin, 1997), the 'results' and their interpretation are presented together because they are not seen to be so clearly separate. An extensive approach to human geography research will identify the research questions, decide the methods and choose appropriate analytical techniques to address the research questions before the collection of evidence is begun. The three sets of decisions are decided together so that, for example, questions in a questionnaire are asked in ways that allow the appropriate form of analysis to answer the research questions. It is not appropriate, for example, to collect answers in yes/no form if the research questions require degrees of agreement or disagreement. Although the decisions are reached in an iterative way, the research is carried out and appears in a consecutive manner – i.e. research questions, methods and data collection, and then analysis. Here a standard data-collection technique will be used such as a questionnaire. The same one will be given to the whole sample and the analysis carried out after all the questionnaires are completed. In an intensive approach to research, on the other hand, the analysis is often carried out at the same time as the evidence is collected. An in-depth, semi-structured or unstructured interview may raise a number of questions or lines of inquiry that were not anticipated but are pertinent to the research. Further interviews will be amended to include them. Further research questions may also emerge. It is still possible to write up the dissertation in a similar order, methods and collection of evidence and then analysis, but some would prefer to discuss the methods alongside the analysis, reflecting the rather different approach to research.

These are the major deviations from the model presented here and they involve different approaches to research. There are some other variations that depend upon the topic. In some dissertations the topic and research questions are decided and then a study area is chosen in which the evidence will be collected. In this case, the study area is like a laboratory. Its choice may be due to accessibility as well as appropriateness. In other dissertations, the study area is part and parcel of the topic. It is one of the reasons for studying the topic. In this case the study area may well be discussed as part of the literature review rather than within a chapter on the methodology.

It is worth discussing the literature review in a little more detail to give some idea of its nature. As with all the chapters it is helpful to emphasize the structure of the review using subheadings. Sub-subheadings may also be used, but it is advisable not to divide the text any further because it becomes too complicated to read. In a dissertation on 'Selling Birmingham: consensus or conflict?' a student began his review by setting the intellectual context. He discussed the re-emergence of 'civic boosterism' as the economy moved from being characterized by Fordism to flexible accumulation. He discussed 'the rise of the entrepreneurial city' and the 'selling of the city'. Later he moved on to discuss the ways cities compete through 'flagship developments' and 'spectacle', as events such as major games and carnivals. He ended by discussing 'whose city?' it was and some of the 'false promises' that have been made in regenerating and selling cities. The phrases in parentheses were subheadings within the review. Each was reviewed in a critical way.

Two excerpts from the dissertation demonstrate how quotations may be used within a review and within the dissertation, how the Harvard system of referencing works and how an argument may be sustained:

> Indeed Harvey (1989: 3) suggests that as a result of these economic changes, there has been an important shift in the way that cities in advanced capitalist societies are governed:
>
>> In recent years, urban governance has become increasingly preoccupied with the exploration of new ways in which to foster and encourage local development and employment growth. Such an *entrepreneurial* stance contrasts with the *managerial* practices of earlier decades, which primarily focused on the local provision of services, facilities and benefits to urban populations.
>
> Indeed, Colenutt (1993) asserts that 'place marketing' is now firmly embedded in the vocabulary of urban regeneration in Britain, used as a technique by city leaders to create or illustrate 'the niche in the world of inter-urban competition which their city occupies' (p. 187).
>
> (From Mole, J. (1998) 'Selling Birmingham: consensus or conflict?'
> University of Manchester.)

Note how the first large quotation (from Harvey) is indented, as would be the case with a quotation that was introduced as evidence from an interview in an analytical chapter, whereas the shorter quotation (from Colenutt) is within the text. In both cases the page number of the quotation is given. Figures and tables should be integrated into the text in an analogous way to quotations.

Finally it is worth considering the order in which a dissertation may be written. Many will write it inside out, as in the report above, completing the methods and analytical chapters first, followed by the final version of the review of the literature, the conclusions and, lastly, the introduction, when what is to be introduced is known. The final touches ensure that the chapters lead on from one another with links forward in the conclusions or links back in the introductions. Each introduction to a chapter sets out the structure of the chapter, while each conclusion brings together the main points of the chapter. Just as in the essay, each paragraph within a chapter should relate to the general argument. The final act will probably be the completion of an abstract, often no more than 200 words, which summarizes the aims and findings of the dissertation.

With all three forms of presentation the process of writing will in itself yield or clarify ideas. It is a creative process that should excite the author as well as the reader.

Summary

- Understanding the assessment task is crucial.
- Planning the organization of the tasks to be tackled when writing essays, reports and dissertations is one key to successful completion.
- Developing a clear, structured argument that is supported with appropriate evidence is essential for all three.
- It is important to recognize the different formats, styles and audiences for the three forms.
- For all three, catch the attention in an introduction and make the conclusion more than a summary.

Further reading

There are very few articles or books written about how to write as a geographer or about different styles of writing for different forms of assessment/audiences. However, the following books are useful places to start:

- Kneale (1999) provides a practical guide to a whole range of study skills.
- Parsons and Knight (1995) focus specifically on how to do a dissertation in geography, from the design to the writing-up stages.

Note: Full details of the above can be found in the references list below.

References

Clark, G. and Wareham, T. (2003) *Geography @ University*. London: Sage.
Colenutt, B., (1993) 'After the Urban Development Corporations: development elites or people-based regeneration?' in R. Imrie and H. Thomas (eds) *British*

Urban Policy and the Urban Development Corporations. London: Paul Chapman, pp. 175–85.

Flowerdew, R. and Martin, D. (1997) *Methods in Human Geography: A Guide for Students Doing a Research Project.* London: Addison-Wesley-Longman.

Harvey, D. (1989) 'From managerialism to entrepreneurialism: the transformation in urban governance in late capitalism', *Geografiska Annaler*, 71: 3–17.

Kneale, P. (1999) *Study Skills for Geography Students: A Practical Guide.* London: Arnold.

Parsons, T. and Knight, P.G. (1995) *How to Do your Dissertation in Geography and Related Disciplines.* London: Chapman & Hall.

Sayer, A. (1992) *Method in Social Science: A Realist Approach.* London: Routledge.

30   Understanding Assessment Criteria

Robin A. Kearns

Synopsis

Assessment criteria are measures used by educators to judge the quality of students' work. There are two main approaches: norm-based and criterion-referenced assessment. In the former, work is assessed relative to that submitted by others working at the same level. In the latter, work is evaluated relative to standards (rather than a set of students). This chapter explains what academics are looking for when they assess your work.

The chapter is organized into the following sections:

- Introduction
- Why is work assessed?
- How is work assessed?
- Student and university perspectives on assessment
- Conclusion.

INTRODUCTION

Earlier chapters in this book have surveyed issues associated with preparing for, and undertaking, geographical research, then analysing the information yielded by these activities. With the previous chapter's emphasis on presenting your findings, the emphasis shifted to the *product* of these labours. It is the purpose of this, the final chapter, to consider a frequently overlooked stage in the research process: the assessment of your written work.

To provide an understanding of the criteria used by university teachers, this chapter discusses three sets of ideas: why work is assessed; how work is assessed (i.e. what criteria are commonly applied in assessment and why they are used); and how assessment is

perceived and received. The central argument of the chapter is that
assessment involves more than having your work judged. Rather, it
can be a process through which your learning as a geographer can be
enhanced. In the chapter I pay particular attention to two different
approaches to grading students' work: norm-based and criterion-
referenced assessment. In the first of these approaches, teachers assess
your work in relation to that submitted by others in your class. This
can be contrasted with the second approach, criterion-based assessment, which involves evaluating how you perform according to sets of
measures (rather than a set of students). I conclude that most assessment in universities uses a mix of these approaches. Further, if each
party in the academic teaching relationship (i.e. students and lecturers) can come to appreciate the other's motivations and perspectives better, the assessment process will contribute to a more fruitful
educational experience for all involved.

WHY IS WORK ASSESSED?

The radical educator and philosopher, Ivan Illich (1971), believed all
students should receive a university degree simply for participating in
courses. While a few lecturers may perhaps be sympathetic to this
idea, very few are likely publicly to proclaim this view. This is
because universities maintain their prestige through the common
requirement that certain courses of study must be completed to a
satisfactory standard and within an established time frame for a degree
to be granted. Written work is assessed in order to determine whether
students meet expected standards of scholarship. Therefore while you
may choose not to adopt one or more of the research methods outlined
in earlier chapters of this book, there is no avoiding the experience of
assessment.

Beyond the university, the criteria applied in assessing written
work vary according to context. A poet, for instance, might assess his
or her work as successful because it gets published and people attend
his or her public readings. Alternatively, an editor might apply an
economic criterion and assess his or her newspaper as successful if it
sells well. In universities, the research work produced by students is
assessed for a range of purposes. Gibbs (1999: 153) lists six purposes:

1   Capturing student attention and effort.
2   Generating appropriate learning activity.
3   Providing feedback to the student.
4   Developing in students the ability to monitor their own learning
    standards.
5   Allocating marks (to distinguish between students or degree
    classification).

6 Ensuring accountability (to demonstrate to outsiders that standards are satisfactory).

You might note that purposes 1–4 relate to student learning goals, whereas purposes 5 and 6 relate more to the functioning of the university as an institution. The question of why work is assessed leads us towards the closely related question of *how* assessment occurs. For, as Gibbs (1999) points out, larger classes and increasing marking loads have shifted the emphasis in assessment back towards tests and computer-based assessment (i.e. the fifth purpose above), whereas it is purposes 1–4 that actually work best to encourage and support student learning.

We can think more broadly than the Gibbs (1999) categorization, however, and follow Neil et al. (1999) in recognizing that there are crucially different perspectives on assessment according to whether your vantage point is that of a student, teacher, institution or member of the public. To help you gain a rounded view of assessment at the outset, Box 30.1 lists some features of each of these perspectives. As the contents of Box 30.1 portray, there is clearly a range of 'stakeholders' who each serve to benefit from the existence of assessment. To develop an understanding of assessment criteria we must move from the 'why' question to examine the issue of 'how' more closely.

HOW IS WORK ASSESSED?

Within the university environment, standards are maintained through agreed-upon methods of assessment being employed to judge the quality of students' work. At the crudest level, an essay or research report is regarded as either passing or failing to meet the accepted standards of the university. While students are concerned in the first instance with simply 'passing', anyone with an ounce of ambition is likely to be concerned with *how well* he or she passes. This degree of passing a minimum set of standards is usually conveyed either by a numeric grade (expressed as a percentage of full marks) or, more commonly, as a letter grade. In some countries passing grades range from A to C with a D or E being a fail. When part-letter grades are employed in a grading scheme, brackets of numerical percentage marks can be summarized with, for instance, 90–100% being otherwise expressed as an A+. In Britain, passing grades tend to be expressed as 'first', 'upper second (2i)', 'lower second (2ii)' and third class.

The next question is: what do these grades represent? While officially they symbolize an assessment made by the university itself, in reality it is the individual judgements of lecturers or tutors that eventually contribute to an official and overall course grade issued in the name of the university. The question of 'judgement' deserves

---

Box 30.1   Purposes of assessment – from student, teacher, institution and community perspectives

Student's needs

- To know how you are doing in general ('Am I on track?').
- To know whether you are reaching a required standard ('Will I pass/graduate?').
- To have something to show to others ('Please employ me/give me a scholarship').

Teachers' needs

- To know whether students are achieving the intended learning outcomes – particularly at the subject level ('Am I getting through? Are they doing the work?').
- To prove to others that they are effective teachers ('Please give me tenure/promote me').
- To certify that students can proceed to subjects for which theirs is a prerequisite ('A student who passes my subject should be able to cope with yours').

Institution's needs

- To know whether students are achieving the intended learning outcomes ('Are our courses effective?' 'Are our staff effective?').
- To prove to others that graduates have achieved what the institution claims they have ('Please continue to fund/support us').
- To certify students can proceed to employment, professional practice or further study ('Our graduates meet your requirements').
- To know whether to accept applicants into programmes of study ('Do you have the required prerequisite knowledge?' 'What is your level of achievement compared to other applicants?').

Community needs

- To know whether the institutions and the teachers are effective ('We will continue to fund/support you').
- To know whether individual graduates are employable, capable of practising, etc. ('You may work here, teach here, etc.').

Source: Nightingale et al. (1996: 8)

---

comment for, while a mathematical answer might be easily deemed to be objectively right or wrong, a geographical research report is not so easily categorized. Rather, what underlies the lecturer's assessment is

Box 30.2   A letter-based grade distribution

A+   Outstanding – fulfils all the general assessment criteria to an unusually high standard.

A   Excellent – fulfils all the general assessment criteria to a very high standard consistently.

A–   On the verge of excellence – fulfils all the general assessment criteria to a high standard intermittently.

B+   Very good – fulfils most of the general assessment criteria to a very good standard.

B   Good – fulfils most of the general assessment criteria to a good standard.

B–   Very reasonable – fulfils some of the general assessment criteria to a good standard intermittently.

C+   Reasonable – fulfils some of the general assessment criteria to a competent standard.

C   Fair – fulfils some of the general assessment criteria to an adequate standard.

C–   Marginally passable – fulfils some of the general assessment criteria to an adequate standard.

D+   Inadequate – fails to fulfil enough of the general assessment criteria to a competent standard.

Grades below D+ indicate greater degrees of inadequacy.

Source: Department of Geography, University of Otago

usually the result of a *combination* of the application of identifiable assessment criteria and the harder to define issue of professional judgement. A measure of each of these processes (application of criteria and intuitive judgement) is commonly at work in grading systems. An example of a commonly used differentiation of grades is listed in Box 3.2.

The verbal descriptions of difference between the letter-based grades contained in the example in Box 30.2 are helpful in guiding teachers in applying grades and students in interpreting their meaning. However, it is also immediately apparent that there is subjective meaning attached to words used in the grade distribution such as 'good' and 'reasonable'. What do these words *really* mean? Educators who have been assessing student work for some time will often speak of being able to 'know an A-grade report' after only reading the first few pages of student work. This aspect of assessment may, on face value, seem a case of jumping to conclusions. However, the insight that comes with experience means a lecturer can be applying criteria subconsciously as he or she reads a research report. But regardless of

---

Box 30.3   Advantages of the criterion-referenced approach to assessment

For students

- You can compete with your own previous performance rather than with your peers.
- Your grades are based on your performance, rather than that of a cohort of your peers.
- You have a clear understanding of the standards required for a given outcome.
- Your outcomes are based on demonstrated competence rather than arbitrary standards.
- You can exercise greater judgement and choice regarding the outcome you target.
- There is greater transparency in the assessment process.

For lecturers

- It encourages the setting of clear goals and objectives.
- It encourages planning coherent courses that will achieve such goals and objectives.
- It encourages greater transparency in the assessment process.

Sources: Adapted from and Abbiss and Hay (1992); Neil et al. (1999)

---

whether these criteria are being painstakingly weighed up one by one, or being intuitively considered as a whole set, what are the actual criteria applied by markers?

To answer the foregoing question, it will be useful first to distinguish between two distinctive ways of assessing student work: *norm-based* and *criterion-based* assessment. Explaining these might be helpful to you in recognizing the ways in which you are being, or will be, assessed by your lecturers or tutors.

Norm-based assessment '. . . uses the achievement of a group of students to set the standards for specific grades or simply passing or failing. The best X per cent of students get the best result and the worst X per cent fail' (Nightingale et al., 1996: 9). Another term for this type of assessment is 'grading on the curve'. In other words, regardless of the individual achievements of students, a set proportion end up attaining a particular allocation of grades (e.g. 20% get As in any one year). This approach assesses your work by comparison with the work of others in your class.

By contrast, *criterion-based assessment* establishes criteria for specific grades or for passing and failing. As a result of these standards being applied, '. . . a student who meets the criteria gets the specified result' (Nightingale, 1996: 9). This direct link between fulfilment of

---

Box 30.4   Disadvantages of criterion-based assessment

For students

- You are encouraged to conform, perhaps discouraging innovative styles of communication.
- You can work on satisfying individual criteria of a project but lose sight of the overall objectives of the assignment.

For lecturers

- Little room for professional judgement.
- Inflexibility (it is hard to take account of the circumstances of individual students).
- Establishing valid and weighted marking systems is a very challenging task.
- Even with criteria, different markers can assess criteria in different ways, but still agree on a final mark.

For both lecturers and students

- This approach can lead to excessive regulation of learning that can lower morale.

Sources: Adapted from Hay (1995); Neil et al. (1999)

---

criteria and attainment of a result signals some of this approach's immediate appeal: it appears more ethically sound to give a student a grade based on his or her performance according to a standard rather than according to how an entire class has performed. Whereas norm-based assessments judge student work by comparison with the work of others, criterion-referenced assessment is made with pre-established standards. What, commonly, are such standards? We will come to this, but for now let us consider advantages of criterion-based assessment (see Box 30.3). Box 30.3 clearly signals advantages of criteria-based assessment for both lecturers and students. However, notwithstanding these advantages, there are also drawbacks in using assessment criteria. Some of these disadvantages are listed in Box 30.4.

In summary, conformity, inflexibility and regulation describe key disadvantages of this approach, as noted by commentators. As Hay (1995) suggests, the criterion-based approach must be treated with caution because there are no true standards, in the strictly psycho-metric sense, to separate out those who have mastered some skill from those who have not. In other words, we are brought back to the issue of professional judgement and the ability of an experienced

> ### Box 30.5   Descriptions of an 'A' report in geography
>
> 'I think there is a real quality to an "A" piece of work that is hard to decompose into different parts, but which involves a combination of insight, energy, inherent authority and rationality of argument and focus . . . It is the voice of an articulate, autonomous, independent mind engaged in deep thought . . . To borrow from an Incredible String Band song title, an A answer belongs to someone who knows all the notes but also understands the song.' (P. Forer, pers. comm., 2001).
>
> '. . . a student must show three things: a very good understanding of the issues, significant effort (in terms of research, structure and presentation for example), and some spark of originality or critical insight' (R. Law, pers. comm. 2001).
>
> '. . . an A answer demonstrates flair, originality, insight, extension of ideas, is well structured and completely written, statements are substantiated, examples used to illustrate points, well-referenced, and goes beyond the requirements of the set question' (J. Mansvelt, pers. comm., 2001).

marker to spot an 'A' or, alternatively, a fail essay. So exactly what, to the experienced eye, constitutes an 'A' report in geography? Box 30.5 offers three examples drawn from communications with experienced university teachers. As the quotes in Box 30.5 imply, any university teacher might be expected to provide clearly stated and understood criteria for the assessment of your work. However, *how* that assessment is reached may have much to do with how familiar the lecturer is with the academic context in which the assignment is set.

Few universities use entirely norm-based or criterion-based assessment practices. Indeed, most use both. As a very structured and transparent approach, criterion-based assessment is increasingly favoured for the extent that it both spells out expectations to you, the student, and requires us, as lecturers, to be more accountable in our teaching (and specifically, grading) practices. Given the likelihood you will encounter this form of assessment in your courses, there is merit in briefly reviewing some common components of assessment criteria. Scott et al. (1978) outline four key dimensions, or 'competencies', they see as central to the assessment of all university-level work (see Box 30.6).

To focus on the first competency in Box 30.6, a key element of the *process* of completing a research report is demonstrating communication skills. There are six fundamental criteria for demonstrating communication competence that are applicable across all subjects and disciplines (Neil et al., 1999). These criteria can be divided into aspects

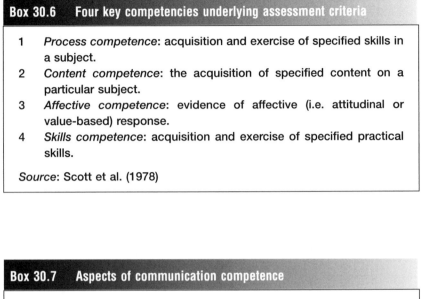

Box 30.6   Four key competencies underlying assessment criteria

1   *Process competence*: acquisition and exercise of specified skills in a subject.
2   *Content competence*: the acquisition of specified content on a particular subject.
3   *Affective competence*: evidence of affective (i.e. attitudinal or value-based) response.
4   *Skills competence*: acquisition and exercise of specified practical skills.

Source: Scott et al. (1978)

Box 30.7   Aspects of communication competence

Textual features

1   Spelling, paragraph and sentence construction, punctuation and grammar.
2   Word choice.

Contextual features

3   Purpose.
4   Language.
5   Information sequencing.
6   Ideas.

Source: Neil et al. (1999), adapted from Anon (1998)

of *text* (the material you are communicating) and *context* (the setting in which you are communicating) (see Box 30.7).

With respect to assessing the text, communication can be greatly improved by minimizing mistakes in the writing. Yet it is only rarely that university teachers have the time to correct every page of students' written work meticulously. It is therefore a good idea to ask a friend, family member or fellow student to 'proof' or check through your final draft before submitting it. Alternatively, it can be a good idea to seek out the services offered by your university's 'Writing Centre' or 'Student Learning Centre', as they are sometimes called. You should have no sense of reticence or shame in enlisting the help offered by such centres, for writing is as much an acquired skill as driving a car. Other sources of advice on writing include books such as *The Elements of Style* (Strunk and White, 1979). Ultimately, the way you write will serve as the gateway to your argument and ideas. Poor writing will turn a reader off and distract an assessor from your ideas through encountering grammatical and other errors. Compelling writing, conversely, can entice a reader into greater curiosity even about a topic he or she is not completely interested in. A good 'rule of thumb' is that you should be aiming for work that you would be proud to show a potential employer as typical of the quality and standards you are capable of attaining and maintaining (Neil et al., 1999).

Based on the criteria for communication competencies listed in Box 30.6, a piece of work meeting a high and 'proficient' standard would typically contain the following features:

- Rare spelling or grammatical errors.
- A lively and effective choice of words.
- A clearly stated purpose.
- An appropriate choice of language for the audience and type of communication.
- Clearly sequenced information.
- A thoughtful progression of ideas.

For each of the key competencies identified in Box 30.6 (process, content, affective and skills) there are criteria that are equally applicable whether you are writing up research in human or physical geography (see Box 30.8). To elaborate, the first criterion in Box 30.8 deals with the *content* of your report. You might ask yourself: how appropriate are the materials I have drawn on and what is their scholarly quality? A good piece of work situates its argument or findings within a context built on the work of others. Usually, such work has been published in scholarly journals. Remember that there is a hierarchy of status among published sources. Referring to articles in *National Geographic*, for instance, tends not to be as highly regarded and considered as appropriate as work published in journals such as the *Transactions of the Institute of British Geographers*.

The second criterion in Box 30.8 relates to *process*: how you use the material you have assembled. The potter's craft is in working with clay to form an object of use and elegance. So, too, the writer's task is to work with the materials gathered and to construct a relevant and appealing report. Here the challenge is to explain findings clearly and reflect on them in light of the wider literature (Hay, 1999). As Holloway and Valentine (2001) point out, findings alone cannot carry a research project. Instead you need to create an argument. This involves integrating your empirical findings with theoretical material to show the reader their importance, demonstrating how your findings

Box 30.8   Typical assessment criteria for written work in geography

1 **Content** (i.e. the nature and scope of the materials used):

• Quality, relevance and depth of information and references.

2 **Process** (i.e. your comprehension and how you have made use of your materials):

• Definition of the topic and/or problem in the broader context.
• Analysis of key issues.
• Logical sequence of argument.
• Constructive discussion and consideration of contrary arguments.
• Adequacy of supporting argument for recommendations and conclusions.

3 **Affect** (i.e. the originality of your insights and your handling of value positions):

• Demonstration of original and independent thinking.
• Presentation and examination of personal views and conclusions (supported by literature and logical argument).

4 **Skills** (i.e. your proficiency in the requirements for clear presentation):

• Layout of the paper/essay.
• Proficiency in language and communication.
• Graphic and cartographic proficiency.
• Accuracy, completeness and consistency of the way references and sources are cited.

Source: Adapted from Neil et al. (1999)

relate to previous work and indicating what new insights your work adds to your area of scholarship.

The third area, *affect*, relates literally to how effectively you are able to affect the reader. All research is value laden and it is up to you to find ways to signal 'where you are coming from'. Thinking critically and knowing where you stand on issues will help you signal this perspective in your writing. This, in turn, will assist your readers to appreciate what you see and why. While there are generally agreed-upon structures for research reports you should work within (discussed in the previous chapter), you should strive to be creative within the acceptable format.

The fourth area of assessment criteria deals with the practical skills demonstrated in the course of presenting your work. Questions such as the neatness and accuracy with which work is presented are considered important skills. The layout and organization of your work are also important, and a good essay or report will always have clearly demarcated beginnings, 'bodies' and conclusions (see Fitzgerald, 1994). Precision in referencing is one skill considered to be crucial (Mills, 1994).

As the quotations included in Box 30.5 imply, there are some well accepted qualities that distinguish a top from an average answer or piece of work. We can summarize by saying that key criteria for a first-class written response include: clear reference to relevant literature and examples; evidence of independent thought; a demonstrated ability to distinguish between ideas and arguments; and a high standard of writing, making it interesting to read. The difference between these criteria and those applicable to the next highest level of assessment ('Upper second' in the British system) is subtle yet clear. The University of Sheffield criteria, for instance, refer to this level of achievement as signalled by '. . . reference to *particular* authors . . . a *reasonable* breadth of knowledge, but lacks real original thought . . . has an ability to write *good* English' (emphasis added). Note that the highlighted terms signal a qualified level of excellence (e.g. the use of 'good English' rather than 'well written and interesting to read').

One point worth making is that different sets of expectations tend to apply to assessment situations according to the level of constraint imposed on the exercise. For instance, students often ask whether they are required to provide citation details in a written examination. Most lecturers will accept a name and perhaps a date of publication as sufficient and certainly not expect the attention to detail required in a formal essay. In summary, therefore, some assessment criteria (e.g. writing and referencing standards) are often relaxed when applied by markers of invigilated examinations. In other words, evidence of scholarship is always sought, but the time and resource constraints of some assessment settings mean that less attention can reasonably be devoted to issues of presentation and detail.

STUDENT AND UNIVERSITY PERSPECTIVES ON ASSESSMENT

Despite both students and university lecturers being part of the same overall enterprise of tertiary education in geography, and despite all lecturers having once been students themselves, the two groups inhabit distinct and differently experienced 'worlds' or 'subcultures' within the university. How is assessment seen from a student perspective? You, the reader, may be in the best position to answer this question because I, the writer, have not been a student in many years.

However, in talking with students I have increased my appreciation for your perspective.

The experience of being assessed is a cumulative learning process. As one student told me, it is wise to '. . . remember the comments made by a marker and take these into account recognizing that assignments are linked in your course of study'. In other words, you are likely to have multiple assignments from any one lecturer in any course or set of courses, so it is important to get to know their particular expectations, as well as the general expectations of your department or university. Research shows that students often fail to transfer the feedback on one essay or report to the demands of the next one very well. Too often there is an assumption that a disappointing result has been caused by not getting the answer right rather than a more general failure to understand how to write up a piece of research (Nightingale, 1996).

Students also speak of the 'glass ceiling' – a metaphor signifying that at some universities there can be a poorly defined upper limit to what grade you can achieve. In some places, this upper limit means that 'A+' grades are simply not given or only very rarely awarded. While it is generally recognized that it is harder to get 'As' in the humanities and social sciences than in the natural or physical sciences, there can also be an institutional reluctance to award grades signalling excellence. In the words of one graduate student working as a teaching assistant in Canada, 'professors say you can't give grades above 90%'. This 'no-go zone' within a norm-based grading scale is one way that some universities, or schools within universities, maintain their prestige.

In the past, university-level assessment has involved ranking students according to the knowledge they have gained in a programme of learning. The methods of assessing this learning were seen to allow students to demonstrate their knowledge in ways that could easily be measured so as to ease comparison between students. Student achievements were almost always measured in quantitative terms raising questions such as 'how much do they know'. While this approach persists in universities in many forms, there are pressures for change. According to Nightingale et al. (1996: 14), these pressures are:

a) *'a desire to broaden university education and develop (and hence assess) a broader range of student abilities'.*

Research into student learning has shown that assessment can be powerful in determining what students perceive to be the 'real' curriculum in a subject area, as opposed to the 'espoused' curriculum. By this they mean that how a course is assessed will, inevitably, affect what you as students consider are the important components of it. If a number of lectures are delivered in a course, but there is no assessment of what you have learnt from them, these topics tend to be disregarded as priority areas. Creative ways to involve students in demonstrating what they have learnt across the curriculum of a course are therefore an important challenge for instructors. This recognition has, in turn, led to the development and use of a much broader range of assessment methods:

b) *'A belief that university education should lead to students having the capacity to apply independent judgement and an ability to evaluate their own performance.'*

In universities, lecturers are encouraged to monitor their own performance as well as that of their peers (e.g. through the refereeing of research papers). This recognition has led some university-based teachers to advocate approaches to assessment that help students develop their own capacity for self-evaluation. The purpose in this is to develop in students the recognition that they have the capacity and can be trusted to evaluate their own learning so that they will graduate as professionals capable of this useful career skill. Despite the competitive atmosphere that often prevails at universities, one path towards evaluating your own learning is to participate in the informal evaluation of others' work. For example, you might consider trying out the idea of student writing groups in which members give each other feedback on drafts prior to submission (Hay and Delaney, 1994).

c) *'A desire to use assessment in ways that might support learning processes themselves.'*

Some assessment methods can too easily lead students into what they call 'surface learning', a '. . . rote learning of isolated facts and formulae – quickly acquired to meet exam pressures and just as easily forgotten' (Nightingale et al., 1996: 6). Recent developments in assessment attempt to go beyond this 'cramming/mental regurgitation' sequence that too easily typifies how students respond to traditional examinations. A more favourable situation is surely when you find a range of assessment tasks (including writing research reports) to be significant learning experiences in themselves that reinforce positive attitudes to learning.

In summary, it is worth bearing in mind that, while there have been huge strides towards taking teaching more seriously in universities, the emphasis given to teaching compared to research in the training of lecturers is still minimal (Gibbs, 1999). You may well, therefore, encounter teachers experimenting with assessment methods and who may be as keen to learn from the experience as you are. Your teachers, however, have no better learning resource than you, their students. You should therefore offer your lecturers feedback on their assessment practices through routine avenues such as evaluation questionnaires. You would also be wise to make use of information, guidelines and criteria provided to you in written and verbal form at the beginning of any course or assignment. You should also acquaint yourself with any relevant university policies on matters such as late submission and plagiarism. The latter term warrants some explanation. Plagiarism is the use of another person or organization's ideas or words without giving the source appropriate recognition (i.e. through standard referencing). It is, put simply, adopting, adapting or copying material without due attribution. Plagiarism is widely regarded as one of the more serious breaches of academic conduct and pleading ignorance is an unacceptable defence when policies have been spelt out in programme handbooks or course outlines.

Students can reasonably expect that they will be given both criteria to guide the writing of a research report and feedback after work is assessed. In the words of one student, it is simply unfair to be told that, after a research project is returned, '. . . to get an "A" you should have done this'. You are entitled to request clarification if the guidance you are provided seems insufficient. For instance, it is only reasonable to expect a course outline that supplies details of the goals, the assessment requirements and the criteria that will be used for the grading of your work. Just as I found it helpful to talk with students about assessment to gain the foregoing insights, so too it may be helpful to talk to your tutor or lecturers about assessment. Ask them how they apply assessment criteria and, if not immediately evident, what criteria they apply. Communication about this and other matters is surely the key to geographical education being a learning experience for all.

CONCLUSION

This chapter has attempted to 'unmask' some of what underlies the assessment of research in human and physical geography. It has discussed why work is assessed and how it is assessed, arguing that most assessment involves a mix of norm-based and criterion-based approaches. In conclusion, we can revisit the responsibilities of both parties in the learning dynamic that you are part of as university geography students. For us as lecturers, it is essential to improving student learning that we establish assessment criteria, communicate these clearly to students and take them into close account in marking student work. For you, as students, it is important that you identify and familiarize yourselves with what assessment criteria have been set, and reflect back on them when considering the feedback you receive after submitting work.

Summary

- Within universities, standards are maintained though agreed-upon methods of assessment used to judge the quality of students' work.
- There is a range of 'stakeholders' who each serve to benefit from the existence of assessment.
- It is useful to distinguish between two distinctive ways of assessing student work: norm-based and criterion-based assessment.
- Most universities use both approaches to assessment.
- Competence in communication is central to assessment across all subjects.
- Criteria for communication can be divided into aspects of text (the material you are communicating) and context (the setting in which you are communicating).
- Common criteria used in assessment of geographical work relate to skills, content, process and affect.
- The experience of being assessed is a cumulative process.
- The way criteria are applied varies according to the situation in which work is produced (e.g. an examination essay vs. a non-invigilated essay).
- Communication with teachers about how they apply assessment criteria and what criteria they apply is important.

Further reading

While there is not a great deal written about assessment criteria per se, what is to be recommended is that you read some of the geographical education literature that provides guidance on aspects of constructing written reports. Four useful articles are as follows:

- Neil et al. (1999) discuss assessment criteria and present sets of criteria you might wish to take into consideration not only for your writing but also for the maps and graphics you include in research reports.
- Mills (1994). Correctly referencing the material that provides the context for your work is an important skill geography students are assessed on. This article is a helpful guide for citing sources in your academic writing.
- Hay (1999) provides helpful guidance on report-writing from a geographer who has contributed a great deal to thinking through the process of assessment.

- Holloway and Valentine (2001). Having an argument is a key criterion in the assessment of geographic research reports. This paper provides a very readable guide to the nature of an academic argument and how to create one in the course of writing up your research.

Note: Full details of the above can be found in the references list below.

References

Abbiss, J. and Hay, I. (1992) 'Criterion-based assessment and the teaching of geography in New Zealand universities', *New Zealand Journal of Geography*, 94: 2–5.

Bean, J.C. (1996) *Engaging Ideas: The Professor's Guide to Integrating Writing, Critical Thinking, and Active Learning in the Classroom*. San Francisco, CA: Jossey-Bass.

Fitzgerald, M. (1994) 'Why write essays?' *Journal of Geography in Higher Education*, 18: 379–84.

Gibbs, G. (1999) 'Improving teaching, learning and assessment', *Journal of Geography in Higher Education*, 23: 147–55.

Hay, I. (1995) 'Communicating geographies: development and application of a communication instruction manual in the geography discipline of an Australian university', *Journal of Geography in Higher Education*, 19: 159–75.

Hay, I. (1999) 'Writing research reports in geography and the environmental sciences', *Journal of Geography in Higher Education*, 23: 125–35.

Hay, I. and Delaney, E. (1994) 'Who teaches, learns. Writing groups in geographical education', *Journal of Geography in Higher Education*, 18: 217–34.

Holloway, S.L. and Valentine, G. (2001) 'Making an argument: writing up human geography projects', *Journal of Geography in Higher Education*, 25: 127–32.

Illich, I. (1971) *Deschooling Society*. New York: Harper & Row.

Mills, C. (1994) 'Acknowledging sources in written assignments', *Journal of Geography in Higher Education*, 18: 263–8.

Neil, D.T., Wadley, D.A. and Phinn, S.R. (1999) 'A generic framework for criterion-referenced assessment of undergraduate essays', *Journal of Geography in Higher Education*, 23: 303–25.

Nightingale, P. (1996) 'Communicating', in P. Nightingale et al. (eds) *Assessing Learning in Universities*. Sydney: Professional Development Centre, University of New South Wales, pp. 204–20.

Nightingale, P. and O'Neil, M. (1994) *Achieving Quality Learning in Higher Education*. London: Kogan Page.

Nightingale, P., Te Wiata, I., Toohey, S., Ryan, G., Hughes, C. and Magin, D. (eds) (1996) *Assessing Learning in Universities*. Sydney: Professional Development Centre, University of New South Wales.

Scott, E., Berkeley, G.F., Howell, M.A., Schunter, L.T., Walker, R.F. and Winkel, L. (1978) 'A review of school-based assessment in Queensland schools'. Brisbane: Brisbane Board of Secondary School Studies.

Strunk, W. and White, E.B. (1979) *The Elements of Style* (3rd edn). New York: Macmillan.

Glossary

Abstraction The effect of moving from the 'real world' through collection and manipulation of data to generalizations and predictions.

Active sensors Emit electromagnetic radiation and record the amount of radiation scattered back from the Earth's surface.

ANOVA Analysis of variance is a method whereby variation within the data being studied is partitioned into components that correspond to different potential sources that can explain that variation.

ASCII American Standard Code for Information Exchange. The 'translation' of characters into numbers the computer will 'understand'. When an ASCII format is required, this normally means a plain text without formatting (i.e. no bold or underlined texts, no bullets or borders).

Attribute data Information about a geometric feature on a map that identifies what the feature represents.

Axial coding Coding along an 'axis' or theme; researchers may suspend 'open coding' in order to pursue a particular theme.

Bias The extent to which sample values deviate systematically from the population value precision (the size of the deviations between repeated estimates of a given statistic). Good sampling therefore aims to minimize bias and maximize precision.

Bivariate graph A graph which illustrates a single cause–effect relationship (e.g. scatterplot) between the response variable (data being examined) and the explanatory variable (possible cause).

CAQDAS The acronym for Computer Assisted Qualitative Data Analysis Software. It comprises a variety of programs with different capabilities to assist the analysis of qualitative data. Simple programs can be used for text searches while more sophisticated ones have functions especially for theory building.

Closure The requirement to simplify a problem to obtain a solution by removing uncontrolled influences 'outside' the system under consideration.

Conditioning plot (coplot) A plot in which the range of values for the explanatory variable is split into segments and then data for each segment are plotted on a separate scatterplot.

Confidence interval The probability that the sample mean lies within a given standard error of the true population mean.

Confidence limits Limits which describe an interval that contains plausible values for a statistic and within which that statistic may lie for a certain proportion of the time, given repeated sampling.

Cross-cultural research Researching cultures other than one's own, which may be spatially distant or closer to home. This requires sensitivity to cultural similarities and differences, unequal power relations, fieldwork ethics, the practicalities and politics of language use, the position of the researcher and care in writing up the research.

Dataframe Spreadsheets of data organized into rows and columns. Rows represent an attribute of the data 'case'; columns represent values of the attribute between all cases.

Data types Continuous data are measured on a continuous scale and bear a relational element; categorical data values signify membership of a group.

Dendroclimatology Analysis of annual variations in tree-ring widths.

Diaspora The dispersal or scattering of a population. It can also refer as a noun to dispersed or scattered populations, such as the black diaspora or Jewish diaspora. In more theoretical terms, the idea of diaspora, or diasporic communities (*see also* transnational community), challenges notions of fixed connections among cultures, identity and place.

Digital data Data about the way the map is configured, such as boundaries and points.

Discretization In field studies, the ability to isolate a particular field site from 'outside' influences which may have an uncontrolled effect on observations and measurement.

Ecological fallacy Inferring the characteristics of individuals from information about the aggregate populations of which they are part. A common problem in the use of secondary data.

EDA Exploratory data analysis – graphical representation of data to help identify structure and pattern, often as a precursor to statistical inference or modelling.

Electromagnetic radiation A system of oscillating electric and magnetic fields that carry energy from one point to another and used to measure the properties of features in an image.

Equifinality (or convergence) When a similar end-point in a system (e.g. a landform) arises from the operation of similar processes but from different initial conditions.

Essentialism/essentialist/essentialize The assumption that particular characteristics, such as culture, gender or 'race', are determined by a single unchanging 'essence', often assumed to be a biological or 'natural' difference. The term is often used to criticize the stereotypes or oversimplifications such views produce.

Ethnocentric/ethnocentrism Prioritizing one's own worldview, experiences and culture as the 'norm' against which others are measured for their strangeness, lack of development, difference or exoticism. Usually implies taking western experiences as the norm.

Factors A categorical explanatory variable (e.g. gender, location).

Feedback The positive or negative effects of process and process–form linkages in physical systems.

Field test-bed Field site chosen to collect data with the purpose of answering a given research hypothesis or in choosing between competing hypotheses.

Generalization A set of processes to reduce complexity on smaller-scale maps.

Geodemographics The analysis of data on the social, economic and population characteristics of areas as part of an exercise leading to the classification of areas in different categories. Often used as a basic step towards targeted marketing.

Geographical information systems (GIS) Organized collections of data-processing methods which act on spatial data to enable patterns in those data to be understood and visualized.

Geostatistics   A suite of tools for solving problems with spatial data. The term is commonly applied (but not limited) to the investigation of spatial variation of processes and landforms on or near the earth's surface.

Globalization   A highly contested term which expresses the ways economic, technical, political, social and cultural processes connect people and places at a global scale, albeit in highly uneven and unequal ways. While these processes have been occurring for several hundred years, it is generally argued that current processes have intensified the scale and extent of these globalized connections.

Graphicacy   Skills in reading and constructing graphic modes of communication, such as maps, diagrams and pictures.

Graphic variables   The visual attributes of a symbol that may be changed by a map designer.

Human Development Index   An index designed for use in the international comparison of standards of life, based on a limited set of variables (usually life expectancy at birth, adult literacy and gross national product per capita).

Hybrid/hybridity   Hybrids result from the mixing of two or more distinct things. In relation to cultures, the terms emphasize the positive and productive results of mixing different cultures, producing new and distinctive cultural forms, often out of the experiences of diaspora populations or transnational communities. Used to reject the negative ideas of cultural assimilation or pollution.

Hyperlink   An invisible link that connects you to text segments in the same or other pictures or HTML pages. Through (double) clicking on the link, you can access the text, picture or web page.

Hypothesis   A testable research statement or question for investigation, often linking observation to an assumed cause or set of causes.

Image analysis   Either manual or digital analysis including filtering, enhancement, combination of images and classification of data.

Indices of Dissimilarity and Segregation   Statistical measures of differences and similarities in population distributions. The Index of Dissimilarity measures the difference between two distributions of population subgroups. The Index of Segregation measures the difference between the distribution of a subgroup and the total population of which it is part. The scale for both indices runs from 0 (no difference) to 100 (total difference).

Inductive Generating understandings or theories from the data themselves ('from the ground up'), as opposed to 'deductive' research in which a theory is already adopted and a hypothesis is tested.

Inferential statistics Procedures applied to sample data in order to make generalizations, validated by probability statements, about the entire population from which the sample was drawn.

Instrumental record Direct measurements of environmental variables, such as temperature and precipitation.

Interval data A measurement level indicating order and precise distances between observations.

In vivo codes Codes that use words or phrases that appear in the text.

Kriging A method of geostatistical estimation at unsampled locations, in contrast to arbitrary mathematical functions used for interpolation.

Layers (overlays) These refer either to data or to maps where one set of attributes is compared with another to examine strength of associations.

Map projection A mechanism for the systematic rendering of geographic co-ordinates of features on the earth's surface on to a flat medium.

Microdata Data on individuals as opposed to 'traditional' census data which relate to the aggregate populations of areas.

MLM (multi-level modelling) A method from studying variation within data where that data are known to be organized within a particular hierarchical or clustered structure.

Model A simplification of reality used for conceptual, analytical or statistic purposes. Models of systems and system operation are used to obtain closure and, hence, predict future outcomes in otherwise intractable situations.

Multiple regression A statistical technique to analyse the possible relationships between several predictor variables and one possible response variable.

Multispectral images Images with relatively high information content (i.e. using many wavebands) which can be represented in numerous colour combinations.

Multivariate analysis Analysis designed to examine multiple influences on a given dataset or pattern.

Natural kinds An ancient concept relating to the 'real' essence of things, allowing causal mechanisms, powers or processes to be identified as well as natural categories (e.g. species, landforms) that are independent of the context of inquiry.

Nominal data A measurement level indicating the presence or absence of information.

Non-stationarity The presence in a series of observations of trend or cyclicity which makes the identification of representative conditions (e.g. mean/variance) time dependent.

NUD.IST Stands for Non-numerical Unstructured Data Indexing Searching and Theorizing. A rather sophisticated yet user-friendly computer package for qualitative data analysis. Its tools include functions for coding texts, exploring documents and ideas about them and developing theories.

Nugget variance Positive intercept on the ordinate of the semi-variogram indicating the unexplained variance.

Numerical models Models which use a conceptual model to define links among fundamental physical, chemical and occasionally biological principles, which are then represented mathematically in computer code.

Open coding (also called 'free coding') Refers to the first stage of coding when the text is read through thoroughly and anything of interest is coded as the researcher goes along; used to 'open up' the text.

Ordinal data A measurement level indicating the relative significance of a feature, but not its precise dimensions.

Orientalism/orientalist Drawing on the work of Edward Said, 'orientalism' describes and critiques the set of ideas common among Europeans in their depictions of 'the Orient' during the era of imperialism. These emphasized both the apparent attractions and the exotic nature of such people, places and cultures, and their supposed danger and lack of civilization, helping to justify European colonial endeavours. The terms have subsequently been applied more widely to all colonial situations and also to contemporary representations of 'other' cultures, peoples and places.

Other Often used in cultural analysis with inverted commas, and sometimes an initial capital letter (e.g. the 'Other' or 'other' cultures), to imply that such cultures, social groups or societies may not be as different as is implied by cultural and social norms.

Palaeoclimatic reconstruction The reconstruction of environmental conditions over timescales from hundreds to millions of years based upon direct and proxy evidence.

Palaeoclimatic signal and noise These arise in interpreting proxy environmental records where the effects of climate (the signal) have to be separated from the effects of all other non-climatic influences (the noise).

Parameterization Use of a numerical quantity in a model to represent the effect of a more complex process or property, or set of process interactions.

Partial plot A graphical plot between two variables after accounting for the modelled effects of other possible explanatory variables.

Participatory research Involves working in a collaborative fashion to develop alternative ways to generate data based on joint agenda setting, analysis and control of outcomes.

Passive sensors These receive electromagnetic radiation from an external source (primarily reflected sunlight).

Philosophy of science Studies of science traditionally focusing on instructing practising scientists on the manner in which scientific research is appropriately conducted but more recently concerned with how practising scientists pursue their research.

Positionality Recognizing and trying to understand the implications of the social position of the researcher with respect to the subjects, particularly with regard to power relations or cultural differences that may influence the process of the research and its interpretation. For example, how we are positioned in relation to various contexts of power (including gender, class, 'race', sexuality, job status, etc.) affects the way we understand the world. Likewise, the information given by informants to a researcher may depend on how the researcher is viewed in that particular context (threatening, insignificant, powerful).

Precision   The size of the deviations between repeated estimates of a given statistic.

Process colour printing   Output derived by overprinting selected values of additive primary colours.

Proxy data sources   Historical records or measurable properties of biological, chemical or physical systems that provide quantitative information about past environmental conditions. Biotic proxies are based on the composition of plant and animal groups and/or measures of their growth rates; geological proxies quantify changes (physical or chemical) in the earth's materials that have accumulated through time, most commonly sediments in oceans or lakes or ice in polar or alpine glaciers; historical proxies are records of events, such as droughts, floods or harvest yields.

Range   The limit of spatial dependence in the semi-variogram – that is, the lag at which the sill is reached.

Raster map   Particularly useful for data produced routinely from satellites where values are recorded for equal areas associated with grid squares or pixels on a computer screen. Raster data can usually be entered into a GIS directly as numbers from the keyboard.

Ratio data   A measurement level indicating precise distances between observations and with a non-arbitrary starting point.

Reflexivity   Critical and conscious introspection and analytical scrutiny of oneself as a researcher. Reflexivity is not simply 'navel gazing'; it is examining our own practice in order to gain new insights into research.

Regionalized variable   The outcome of many randomly located samples of a property in space, assumed to be the statistical realizations of a set of random variables or a stochastic process.

Remote sensing   The collection of images of parts of the earth's surface using specialized instruments, commonly aerial cameras and satellite sensors.

Sampling   The acquisition of information about a relatively small part of a larger group (population), usually with the aim of making inferential generalizations about the larger group.

Scale-linkage A phenomenon whereby things of one size and/or timespan are composed of objects and time periods that are smaller and, in turn, are themselves formative components of larger assemblages and/or time periods.

Selective coding Selecting the 'core category' or primary theme of the research and relating other categories/themes to it, then coding along that core category.

Semi-variogram A graphical tool to summarize spatial variation and also a model of underlying spatial process used for geostatistical estimation at unsampled locations.

Significance levels The probability (set by the researcher as part of the testing procedure) of incorrectly rejecting a true hypothesis.

Sill variance The maximum value in a semi-variogram determining the scale of spatial dependence. The steepness of the initial slope of the semi-variogram indicates the intensity of change in a property with distance and the rate of decrease in spatial dependence.

Snowballing A technique used by researchers whereby one contact or participant is used to help recruit another, who in turn puts the researcher in touch with another. The number of participants soon increases rapidly or 'snowballs'.

Spatial resolution The size of features discernible from remotely sensed data.

SPSS A widely used and very comprehensive statistical package in the social sciences. It stands for Statistical Product and Service Solutions. It contains a Data Editor to enter and handle data as well as extensive graphic possibilities for the display of data in tables and graphics.

Standard error The standard deviation of a sampling distribution.

Structured approach A sequential process to site selection and the subsequent measurement programme designed to yield robust, reproducible and general conclusions from field study.

System A set of objects, together with the relationships between the objects and between their attributes.

Townsend Index A composite index attempting to measure affluence and deprivation. The index was devised for use in the UK in health studies but its principles have been adapted for use elsewhere and on other topics.

Transition   May refer to any number of processes of societal change (from dictatorship to democracy, for example) but most often used to refer to the contested and problematic transformation of postcommunist societies and economies.

Transnational community   Social and cultural relations that transcend and escape the bounded spaces of the nation-state, possibly as a result of migration and diaspora or because the social group does not fit existing national boundaries (such as the Kurdish population).

Triangulation   Using multiple data sources and/or research methods to strengthen your results. For example, interviewing different populations around the same issue (multiple sources) or combining focus groups, surveys and published data such as the census (multiple sources and methods).

Uneven and unequal social relations/power relations   The idea that groups of people and places are differently positioned in social, economic, political or cultural terms. Social relations are uneven (across space and time) and they both reflect and shape unequal social and power relations between people and places (for example, along lines of class and gender or between developed and less developed countries).

Validation   The process by which a model is compared with reality.

Vector map   A more realistic configuration of maps based on points and lines which are assembled into objects such as polygons. Vector data are usually digitized using a mouse-like device called a puck, centred on each of the points and lines defining the map object in question.

Verification   The process by which a numerical model is checked to make sure it is solving the governing equations correctly.

Index